SOLVING POLYNOMIAL SYSTEMS USING CONTINUATION FOR ENGINEERING AND SCIENTIFIC PROBLEMS

Alexander Morgan

Senior Staff Research Scientist
Mathematics Department
General Motors Research Laboratories
Warren, Michigan

PRENTICE-HALL, INC.

Englewood Cliffs, New Jersey 07632

Library of Congress Cataloging-in-Publication Data

MORGAN, ALEXANDER, (date)
    Solving polynomial systems using continuation
for engineering and scientific problems.

    Bibliography: p.
    Includes index.
    1. Equations—Numerical solutions.   2. Poly-
nomials.   3. Fractions, Continued.   I. Title.
QA214.M67 1987   512.9′42   86-30685
ISBN 0-13-822313-0

Editorial/production supervision and
    interior design: Richard Woods
Cover design: Edsal Enterprises
Manufacturing buyer: Gordon Osbourne

Printed in the United States of America

10   9   8   7   6   5   4   3   2   1

ISBN   0-13-822313-0   025

PRENTICE-HALL INTERNATIONAL (UK) LIMITED, *London*
PRENTICE-HALL OF AUSTRALIA PTY. LIMITED, *Sydney*
PRENTICE-HALL CANADA INC., *Toronto*
PRENTICE-HALL HISPANOAMERICANA, S.A., *Mexico*
PRENTICE-HALL OF INDIA PRIVATE LIMITED, *New Delhi*
PRENTICE-HALL OF JAPAN, INC., *Tokyo*
PRENTICE-HALL OF SOUTHEAST ASIA PTE. LTD., *Singapore*
PRENTICE-HALL DO BRASIL, LTDA., *Rio de Janeiro*

To Roz
with love, respect, and appreciation

Contents

Part I   The Method

Part II   Applying the Method

Appendices

Bibliographies and References

Index

Preface

A simple, reliable numerical method for solving small polynomial systems has recently been discovered. This book explains how to implement the method in practical settings. Featured are complete computer codes in FORTRAN for solving up to 10 equations in 10 unknowns.

The problem of solving small systems of polynomial equations is faced daily by scientists and engineers working in diverse applications areas: physicists studying laser optics; computer scientists developing systems for computer-aided design; chemists concerned with coal gasification, forensics, and pharmacokinetic reactions; and engineers involved in the design of electrical circuits, nonlinear least-squares data analysis, and mechanical design, among others. In most cases these scientists and engineers want to settle the mathematical problem without having to focus attention on it. This book will provide such persons with the basic knowledge needed to do this.

The new method differs from more classical techniques in that it is *global* and *exhaustive*. It does not require a choice of initial solution estimates, and it finds all solutions. This is in contrast with *local* methods, such as Newton's method.† Local methods are reliable only if provided good initial solution estimates. The freedom from the need to find initial solution estimates is a principal feature of the new approach. The new method is one of a class of continuation methods (also called embedding or homotopy methods). These methods have undergone revolutionary development by a number of researchers over the last 10 years, thanks in part to breakthroughs in the theoretical mathematical area of differential topology and in part to the stimulus provided by the increasing availability of computers.

† See Appendix 1.

I have adopted a conversational tone in this book on the grounds that an informal approach makes it easier to get into a new subject area. By considering examples and experimenting with methods implemented in computer codes, you can obtain a good sense of what the area is about. I believe that the loss in formal rigor is more than made up for by the excitement of discovery. (On the subject of "experimental mathematics," see the articles by [Grenander and Shisha, 1985] and by [Hazewinkel, 1985].)

The minimal mathematical prerequisites for this book are a working knowledge of multivariable calculus, linear algebra, and computer programming. Applications-oriented training in complex variables, differential equations, and numerical analysis would be helpful, but it is not necessary.

The book is in two parts. Part I, which consists of Chapters 1 through 6, deals with the method itself, while Part II, Chapters 7 through 10, is devoted to applying the method. The first two chapters introduce the basic ideas with a minimum of mathematical fuss, leading to the more general material in Chapters 3 through 6. Chapter 1 deals with one equation in one unknown and Chapter 2 with two equations in two unknowns. Many fundamental issues of both theory and implementation are introduced, with seven short computer codes used to illustrate various ideas and give a feeling for the numerical characteristics of the approach. If you work through these two chapters slowly, setting up the codes and experimenting with them as suggested in the text, you will develop a good sense of what the method is all about. (This book is more a lab manual for a course in experimental mathematics than a standard textbook. The proposed experiments and projects far outnumber the theorems and propositions.) Chapter 3 describes the method for any number of equations, and Chapter 4 discusses implementation. The focus here is on one good approach, rather than a survey of all approaches. Chapter 5 treats "scaling," which is important for many problems. In Chapter 6, some alternative continuation methods are discussed which may be useful in certain cases, such as when the equations are not polynomial (i.e., contain sin, cos, exp, or log) or when the system is polynomial but too large to allow for finding all solutions in practical computing times. The theory is less well developed in these cases, and the implementation and use are less routine than for small polynomial systems.

Chapter 7 deals with practical considerations of system reduction. Chapters 8 through 10 are devoted to case studies from real applications. These chapters make clearer how the mathematics looks in context and what sorts of model reduction should be done before using the method.

For readers more interested in scientific or engineering models than in solution techniques, Chapters 1 through 3, followed by Part II, will give a qualitative overview of polynomial systems and the associated modeling issues.

Aside from providing immediately usable material for working scientists and engineers, this book can be used for advanced undergraduate and graduate courses in mathematical methods for engineering applications.

An industrial mathematician does not work in isolation. This book documents my continuing interaction with computer scientists, mechanical engineers, control theorists, and others. I especially acknowledge those with whom I have pursued joint projects at General Motors Research Laboratories: Keith Meintjes, Ramon Sarraga, Vadim Shapiro, Lung-Wen Tsai, and Mason Yu. The backing of the Mathematics Department management—Gary McDonald, Jim Cavendish, and the late Allen Butterworth—has been wholehearted and significant. My group leader and friend, Rob Goor, has been a constant source of enthusiasm and encouragement. Professor Eugene Isaacson read my first draft and made a number of useful suggestions, which I appreciate. I also appreciate the valuable conversations I have had with Professor Andrew Sommese on topics in algebraic geometry. My friends Jeannie and John have been very supportive, and I want to acknowledge the importance to me of their support.

Introduction: Beyond Newton's Method

The other day an electrochemist friend came by my office with a problem. He was trying to work out part of a battery-plate manufacturing process. He had set up a math model to determine the amounts of various metal compounds that would be present in the plating bath at various times. He ended up with a system of 10 polynomial equations in 10 unknowns. His problem was that Newton's method kept converging to nonphysical solutions.

The system was presented in a form that was not very good for numerical solution. This is typical of chemical equilibrium problems (see Chapter 9). I reworked his equations, reducing them to a system of two sixth-degree equations in two unknowns. Then I applied the method described in this book. It worked the first time, without parameter adjustments or fooling around. I gave my friend a copy of the code so that he could run it on his own. He left, able to turn his mind back to what he was really interested in: chemistry instead of mathematics.

This incident has been repeated in various guises many times. The most commonly used numerical method for finding solutions to systems of polynomial equations is Newton's method. A brief description is given in Appendix 1. There are many variations on this method (see [Rheinboldt, 1974] and [Ortega and Rheinboldt, 1970]). They are all local methods. This means that sometimes they work well, sometimes they work with a lot of effort, and sometimes they don't work at all.

Now there is a new method for solving polynomial equations. I call the new method CONSOL, since it uses CONtinuation to SOLve systems of equations. The good news is that CONSOL is often more reliable than

Newton's method. (In theory, it never fails.) Also, it is not difficult to implement in computer programs. Several such programs are included in this book. Furthermore, it finds all solutions, not just one solution. The bad news is that it takes more time, especially if the system is "large." In theory, there is no limit on the size of the system that can be solved, but in practice I would hesitate to use the method on more than about 10 equations in 10 unknowns. You will learn more about practical limits on the size of the system by conducting the experiments given in the text.

The power of the theory behind the method is that we can set up an algorithm that requires no prior knowledge of the solutions of the system, no "start points" or "initial guesses." This is in contrast with Newton's method, which succeeds or fails depending on the initial guess. Small polynomial systems come up in many applications, and this book is for people who want a reliable way to solve them.

PART I

The Method

CHAPTER 1

One Equation in One Unknown

1-1 A SIMPLE CONTINUATION FOR SOLVING QUADRATIC EQUATIONS

Let's start by considering quadratic equations. A quadratic equation looks like

$$2x^2 + 7x + 5 = 0.$$

We already know how to solve it; use the quadratic formula. (Numerical considerations for implementing the formula are discussed in [Forsythe et al., 1977, pp. 20–23].) However, seeing how CONSOL works for quadratics will illustrate the method and lead directly to the general case.

Quadratic equations often come up in applications. Here is an example from computer graphics and computer-aided design: finding the intersection of a line and a cylinder. (Finding the intersections of lines and surfaces is fundamental to "ray casting"; see [Roth , 1982].) The equation for a cylinder is

$$x^2 + y^2 = r^2, \tag{1-1}$$

assuming that it has the z-axis as its centerline (see Fig. 1-1). In other words, a cylinder in (x, y, z)-space is a collection of points, and a particular point (x, y, z) is on the cylinder exactly when x and y satisfy equation (1-1). The geometrical fact (being on the cylinder) is equivalent to the algebraic fact [satisfying equation (1-1)].

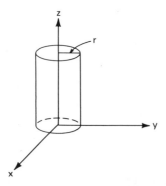

Figure 1-1 Cylinder with radius r, as given by equation (1-1).

We describe a line with parametric formulas

$$x = u_1 s + v_1$$

$$y = u_2 s + v_2 \qquad\qquad (1\text{-}2)$$

$$z = u_3 s + v_3,$$

where s is the free parameter. [Each choice of s produces a point (x, y, z) on the line. For (1-2) to really represent a line, one of u_1, u_2, or u_3 must be nonzero.] Thus (1-2) describes a line through $v = (v_1, v_2, v_3)$ parallel to the vector from the origin to $u = (u_1, u_2, u_3)$ (see Fig. 1-2). Now, the line described by (1-2) hits the cylinder described by (1-1) exactly when an (x, y, z) point generated by a choice of s in (1-2) satisfies (1-1). In other words, the geometric intersection of the cylinder and the line corresponds to the algebraic simultaneous solution of (1-1) and (1-2).

Substituting (1-2) into (1-1) gives the equation

$$(u_1 s + v_1)^2 + (u_2 s + v_2)^2 - r^2 = 0,$$

which, when multiplied out and simplified, becomes

$$s^2 + as + b = 0, \qquad\qquad (1\text{-}3)$$

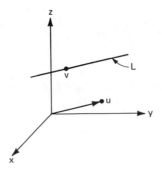

Figure 1-2 Line L is parallel to vector u and contains the point v.

where

$$a = \frac{2(u_1 v_1 + u_2 v_2)}{u_1^2 + u_2^2}$$

$$b = \frac{v_1^2 + v_2^2 - r^2}{u_1^2 + u_2^2},$$

and I am assuming that $u_1^2 + u_2^2 \neq 0$. (If $u_1^2 + u_2^2 = 0$, the line is parallel to the z-axis.)

When we solve (1-3) for s, we can substitute into (1-2) to find the (x, y, z) coordinates of points where the line hits the cylinder. Figure 1-3 should convince you that there are four possibilities. The line might be completely contained in the cylinder, although I have ruled that out by the assumption $u_1^2 + u_2^2 \neq 0$. This geometric fact corresponds to the algebraic fact that a quadratic equation can have all coefficients zero, in which case any number is a solution. Assuming that $u_1^2 + u_2^2 \neq 0$, the line can miss the cylinder, hit just once, or hit twice. That is, equation (1-3) can have two complex solutions, a repeated real solution, or two different real solutions. Here we are confronted for the first time by one of the prime nuisances and profundities of solving polynomial systems: It is difficult to get very far without running into the complex numbers. (Notice that "complex solutions" corresponds to "missing the cylinder.")

We have just seen an example of how a quadratic equation can come up. It is not difficult to find other examples. We know that in practice we use the quadratic formula to solve quadratic equations, but let's see how CONSOL does it.

We start with an explicit example. Consider

$$x^2 + 3x - 4 = 0, \tag{1-4}$$

and consider the simplification of (1-4) that results from deleting the middle term:

$$x^2 - 4 = 0. \tag{1-5}$$

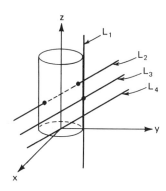

Figure 1-3 Line L_1 (parallel to the z-axis) is contained in the cylinder. Line L_2 hits the cylinder twice. Line L_3 hits the cylinder once. Line L_4 misses the cylinder.

We can solve this "simplified" version of (1-4) simply by taking square roots. We get $x = 2$ and $x = -2$. Now we begin to think along the following lines. We *want* to solve (1-4) and we *can* solve (1-5). Is there a way to gradually turn (1-5) into (1-4) with the solutions of (1-5) gradually becoming solutions to (1-4)? We can accomplish this by setting up a new equation.

Define a "continuation equation" by the formula

$$h(x, t) = x^2 + 3tx - 4 = 0. \tag{1-6}$$

Note (1-6) looks like (1-4), but it has a new variable, t, as well as x. Note further that $h(x, 0) = 0$ is equation (1-5) and $h(x, 1) = 0$ is equation (1-4). Finally, note that as t is changed from 0 to 1, $h(x, t) = 0$ changes from equation (1-5) into equation (1-4). Further, we can change the solutions to (1-5), 2 and -2, into solutions to (1-4) as follows. Choose a small increment, Δt. To be explicit, let's take $\Delta t = 0.01$. With $x = 2$ as a start point, solve $h(x, \Delta t) = 0$ using Newton's method (see Appendix 1†). That is, $h(x, \Delta t) = 0$ is an equation in the one variable, x:

$$x^2 + 0.03x - 4 = 0.$$

Since Δt is small, $h(x, \Delta t) = 0$ is not too different from $h(x, 0) = 0$. Therefore, the solutions to $h(x, 0) = 0$ can reasonably be supposed to be close to those of $h(x, \Delta t) = 0$.‡ Since 2 is close to a solution to $h(x, \Delta t) = 0$, Newton's method converges quickly, finding a solution x_1 to $h(x, \Delta t) = 0$. Next, we solve $h(x, 2\Delta t) = 0$, using x_1 as a start point. In other words, x_1 is a solution to $h(x, 0.01) = 0$, and we use it as a start for Newton's method to solve $h(x, 0.02) = 0$, where

$$h(x, 0.02) = x^2 + 0.06x - 4 = 0,$$

obtaining a solution, x_2. Then we solve $h(x, 3\Delta t) = 0$, using the x_2 as a start point, and continue in this way. Eventually, we solve $h(x, 1) = 0$, which is equation (1-4), the system we are interested in. Then we go through the same process again, beginning at $x = -2$. We end up with the two solutions to equation (1-4). Table 1-1 summarizes what we have done. It lists the two "continuation paths," one beginning at 2 and the other beginning at -2.

Appendix 6 contains a computer code, CONSOL1, to implement the continuation method as we have outlined it, for any choice of constants, a and b. In other words, it solves the continuation given by

$$h(x, t) = x^2 + atx + b = 0, \tag{1-7}$$

---

† The Appendices appear at the end of the book.

‡ You may be concerned about how small Δt needs to be. The method can be developed to adapt Δt to the problem, so that Δt is automatically made smaller, if necessary. However, setting the initial Δt is still a significant choice. More on this in Chapter 4.

TABLE 1-1 The two continuation
paths for (1-6), as generated by
CONSOL1[a]

t	x	x
0.0000D + 00	0.2000D + 01	− 0.2000D + 01
0.5000D − 01	0.1926D + 01	− 0.2076D + 01
0.1000D + 00	0.1856D + 01	− 0.2156D + 01
0.1500D + 00	0.1788D + 01	− 0.2238D + 01
0.2000D + 00	0.1722D + 01	− 0.2322D + 01
0.2500D + 00	0.1660D + 01	− 0.2410D + 01
0.3000D + 00	0.1600D + 01	− 0.2500D + 01
0.3500D + 00	0.1543D + 01	− 0.2593D + 01
0.4000D + 00	0.1488D + 01	− 0.2688D + 01
0.4500D + 00	0.1436D + 01	− 0.2786D + 01
0.5000D + 00	0.1386D + 01	− 0.2886D + 01
0.5500D + 00	0.1338D + 01	− 0.2988D + 01
0.6000D + 00	0.1293D + 01	− 0.3093D + 01
0.6500D + 00	0.1250D + 01	− 0.3200D + 01
0.7000D + 00	0.1209D + 01	− 0.3309D + 01
0.7500D + 00	0.1170D + 01	− 0.3420D + 01
0.8000D + 00	0.1132D + 01	− 0.3532D + 01
0.8500D + 00	0.1097D + 01	− 0.3647D + 01
0.9000D + 00	0.1063D + 01	− 0.3763D + 01
0.9500D + 00	0.1031D + 01	− 0.3881D + 01
0.1000D + 01	0.1000D + 01	− 0.4000D + 01

[a] The "D" indicates powers of 10. Thus 0.2000D + 01 denotes 0.2 × 10.

yielding solutions to the polynomial equation

$$h(x, 1) = x^2 + ax + b = 0.$$

You should run this program so that you can try the method firsthand.† The only input values needed are the numbers a, b, and Δt. The bigger Δt is, the faster the method works but the more chance that it will "go astray" (see Experiment 1-4).

There is something else to note. Maybe you have already noticed it. For some values of a and b, the method will not work at all; namely, when b is positive. One reason for this is that equation (1-7) with $b > 0$ and $t = 0$ has imaginary solutions, $x = \pm \sqrt{b}\, i$, where $i^2 = -1$. Further, (1-7) may have complex (nonreal) solutions for some $t > 0$, including when $t = 1$. In other words, the original system (1-3) may have complex solutions.

Unfortunately, even when we are not really interested in complex solutions, we often want to know that there are no real solutions. For example, in the ray-casting example, we wanted to know that the line did not hit the

---

† All the codes given in this book are available in machine-readable form from the author. See the card included in this book.

cylinder. The best way to know this for sure is to find all solutions and verify that none of them are real. However, there is a more basic reason why we must face complex solutions and complex computer arithmetic. This is inherent in the theory of the method, as outlined in Chapter 3 and Appendices 3 and 4. You can see yourself what happens in CONSOL1. Without complex arithmetic, the method "gets stuck" at once. You will see, as we develop CONSOL2 and CONSOL3, that complex arithmetic is essential to keeping the method from "getting stuck" in general. For languages (or compilers) that do not support complex arithmetic, we must simulate complex operations in real arithmetic. In principle this is easy, but it can be a nuisance to do. To see what is involved, consider the codes CONSOL2 and CONSOL2R, given in Parts A and B of Appendix 6, respectively. Both of these are complex arithmetic versions of CONSOL1. CONSOL2 differs from CONSOL1 in only a few parts of the code: the complex variable declaration, the generation of start points to allow for imaginary points, and two statements in the Newton's method part of the code to prevent underflow errors. (Underflow tends to occur when pure real or pure imaginary points come up at the end of paths. In connection with this, two short utility routines, DREAL and DIMAG, are also used.) CONSOL2R is logically identical with CONSOL2. However, it represents complex numbers as two-dimensional vectors, and it uses subroutines to emulate multiplication and division of complex numbers. (See Appendix 2 for a discussion of what is involved in simulating complex arithmetic using real arithmetic.) CONSOL2R performs complex addition as two-dimensional vector adds. This essentially doubles the lines of code involving this operation. Multiplying and dividing with subroutine calls forces us to name the results of multiplication and division. Thus "XSQ" is used in CONSOL2R for "X**2". For algorithms using complex arithmetic I provide two codes in Appendix 6, one version in Part A using complex declarations and a corresponding one in Part B that does not. This is for the convenience of readers whose FORTRAN does not support double-precision complex arithmetic. I will refer to the CONSOL updates without citing the versions.

Try out CONSOL2 with $\Delta t = 0.1$, $a = 3$, $b = 4$ [Experiment 1-1, part (a)]. It works! If using complex computer arithmetic bothers you, take a look at Appendix 2. Even people with lots of numerical experience sometimes find using complex arithmetic a barrier. Unfortunately, it cannot be avoided. It is easy to use, once you accept it.

1-2 SINGULAR PATHS AND THE INDEPENDENCE PRINCIPLE

Even though CONSOL2 succeeds for problems that CONSOL1 cannot solve, it still fails sometimes. Try $\Delta t = 0.1$, $a = -4$, $b = 1$ [Experiment 1-1, part (b)]. You will see that both continuation paths find the same solution, $x = 0.2679$, even though the equation has two distinct solutions, the other being $x = 3.732$. What is happening is that the paths are crossing when $t = 0.5$. In other words, even though there are two paths for the two solutions, the paths intersect at $t = 0.5$ and CONSOL gets confused and switches paths for one solution (see Fig. 1-4 and Table 1-2). The basic problem here is that $h(x, 0.5) = 0$ is "singular." That is, the derivative of h is zero at a place where h is zero. Explicitly,

$$h(x, 0.5) = x^2 - 2x + 1 = (x - 1)^2$$

and

$$\frac{dh}{dx}(x, 0.5) = 2(x - 1),$$

so $\frac{dh}{dx}(1, 0.5) = 0$ and $h(1, 0.5) = 0$. This really messes up CONSOL. Luckily, there is a wonderfully efficient maneuver to handle the problem. We change h so that it will never go singular in the middle of a continuation path. Here is the new h:

$$h(x, t) = x^2 + atx + [tb - (1 - t)q^2], \tag{1-8}$$

where q is a complex constant, chosen as indicated below. First, let's note that this h is the same as the h we have been using, except that the constant term is now $tb - (1 - t)q^2$ instead of b. We still have $h(x, 1) = x^2 + ax + b$. Using this new h does not give us any particular trouble as far as implementation is concerned. I have updated CONSOL2 to CONSOL3, and as you can see in Appendix 6, only a few lines of code had to be changed.

But how does this new h avoid the problem of continuation paths going singular? Equation $h = 0$ can go singular only if the constants a, b, t, and q obey a specific algebraic relationship. To work out what this relationship is, we merely have to solve the equations $h = 0$ and $dh/dx = 0$ simultane-

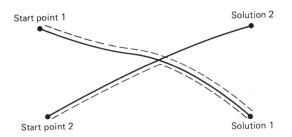

Figure 1-4 Schematic diagram to illustrate path crossing. The solid lines are the continuation paths. The dashed lines indicate what CONSOL2 does when $a = -4$ and $b = 1$.

TABLE 1–2   The two CONSOL2 paths when $a = -4$ and $b = 1^a$

	Path 1		Path 2	
t	Re (x)	Im (x)	Re (x)	Im (x)
0.0	0.0000D + 00	0.1000D + 01	0.0000D + 00	− 0.1000D + 01
0.1	0.2000D + 00	0.9798D + 00	0.2000D + 00	− 0.9798D + 00
0.2	0.4000D + 00	0.9165D + 00	0.4000D + 00	− 0.9165D + 00
0.3	0.6000D + 00	0.8000D + 00	0.6000D + 00	− 0.8000D + 00
0.4	0.8000D + 00	0.6000D + 00	0.8000D + 00	− 0.6000D + 00
0.5	0.9998D + 00	0.5859D − 03	0.9998D + 00	− 0.5859D − 03
0.6	0.5367D + 00	0.0000D + 00	0.5367D + 00	0.0000D + 00
0.7	0.4202D + 00	0.0000D + 00	0.4202D + 00	0.0000D + 00
0.8	0.3510D + 00	0.0000D + 00	0.3510D + 00	0.0000D + 00
0.9	0.3033D + 00	0.0000D + 00	0.3033D + 00	0.0000D + 00
1.0	0.2679D + 00	0.0000D + 00	0.2679D + 00	0.0000D + 00

^{a} Note "path crossing" after $t = 0.5$

ously. From equation (1-8) we get

$$\frac{dh}{dx}(x, t) = 2x + at = 0,$$

which implies that $x = -at/2$. Substituting this value into (1-8) yields

$$-a^2t^2 + 4(b + q^2)t - 4q^2 = 0. \tag{1-9}$$

Now let's consider what this means. Equation (1-9) must hold if something bad is going to happen: namely, if h is going to be singular in the middle of a continuation path. Therefore, we want to choose q so that (1-9) cannot ever hold.

There are several ways to do this. One way focuses on the fact that t is between 0 and 1. The idea is to choose a real number q so that for t between 0 and 1, the expression above is always positive or always negative, and therefore never zero. For each a and b, we would compute q, then proceed with the method. I am going to take a different tack, one that will work for higher-degree equations and for systems as well as for a quadratic. I will focus on the fact that t is a real number and try to choose q "complex enough" so that (1-9) cannot be zero because its imaginary part will be too big. The advantage of this approach is that we can choose one *fixed* q that works for all a and b. The fact is that if we choose q "independently" of a and b, then h will never go singular. For example, if I choose

$$q = 0.928\ 746\ 354 + 0.265\ 465\ 386i$$

and let the a and b come up without being based on this choice of q, then h will never go singular. I realize that this sounds a little mysterious, so let's consider the entire situation more closely.

I will sketch through a plausibility argument here. It reflects the flavor of the mathematical proof. First, I would like you to accept the

INDEPENDENCE PRINCIPLE.   Two equations in one unknown, chosen independently, do not have a common solution.

For example, consider the following system of two linear equations in one unknown:

$$Ax + B = 0$$

$$Cx + D = 0.$$

Assuming that A and C are nonzero, this system can have a simultaneous solution only if a very special relationship holds among the coefficients A, B, C, D: namely, $B/A = D/C$. So generally, two linear equations in one unknown do not have a solution. We can express this another way. Suppose that you choose A and B, and I choose C and D. Further, assume that we choose *independently*. Your choice is made without consulting me, and mine without consulting you. What are the chances of the resulting linear system having a solution? What are the chances that we would choose so that $B/A = D/C$? Realistically speaking, it would not happen. This Independence Principle is true for polynominal equations in general. If you choose the coefficients for one polynomial and I choose those for another, it would be virtually impossible for the resulting system of two polynomials to have a simultaneous solution. The key idea is *independence*. If the coefficients are chosen independently, there can be no simultaneous solution.

Now fix t and q. The system

$$h(x, t) = 0 \tag{1-10}$$

$$\frac{dh}{dx}(x, t) = 0$$

is two equations in the one unknown x. Therefore, it can have a solution only if h and dh/dx happen to be related to each other in a special way. (Note that q occurs in $h = 0$, but does not appear anywhere in $dh/dx = 0$.) The idea is that by choosing q ''independently'' of the other coefficients of h, we can prevent from occurring the special relationship necessary for (1-10) to be an exception to the conclusions of the Independence Principle. Further, we will be dealing in our method with only a finite number of t values, and the comments above remain valid over the range of t's that we would choose. We explore the Independence Principle more in later chapters.

Now we have a full-fledged solver for quadratic equations. CONSOL3 solves any equation (1-3) for any a and b (within the limitations of computer arithmetic), the only proviso being that a and b are to be chosen without reference to q. However, we are not quite done with "path crossing." Although paths cannot truly cross using the CONSOL3 continuation [equation (1-8)], paths might cross *numerically* if the step size, Δt, is too big. Unfortunately, "too big" is problem dependent, and there is no effective way to know if Δt is too big except by trial and error. If you suspect the possibility of path crossing, you should rerun with Δt smaller. Path crossing is discussed further in Chapter 4.

1-3 SOLVING HIGHER-DEGREE EQUATIONS

So far we have focused on quadratic equations. But we can immediately extend the method to higher-degree equations. Let's consider equations of the form

$$x^5 + ax + b = 0. \tag{1-11}$$

The beauty of CONSOL is that it will handle (1-11) almost as easily as it handles a quadratic. But now there is no general fifth-degree equation formula we can use as an alternative. The continuation equation is

$$h(x, t) = x^5 + atx + [tb - (1 - t)q^5], \tag{1-12}$$

which is the same as (1-7), except that x^2 and q^2 have been replaced by x^5 and q^5. We can use the same q as before. CONSOL4, a modification of CONSOL3, solves equation (1-11). Note in Appendix 6 that we have changed only a few lines of code to convert CONSOL3 into CONSOL4. The main difference is generating the solutions to $h(x, 0) = 0$. There are five such solutions, the five fifth roots of q^5:

1. $0.034\ 525\ 822 + 0.965\ 323\ 588i$
2. $-0.907\ 408\ 223 + 0.331\ 137\ 401i$
3. $-0.595\ 334\ 945 - 0.760\ 669\ 419i$
4. $0.539\ 470\ 992 - 0.801\ 256\ 956i$
5. $0.928\ 746\ 354 + 0.265\ 465\ 386i$

These are generated as q times the five fifth roots of unity.

   Try out CONSOL4 on $a = 1, b = 1$ and $a = -16, b = 17$ [Experiment 1-3, parts (a) and (b)]. With a fifth-degree equation, it is easier to run into numerical trouble than with a quadratic. For example, $a = -16, b = 17.119$ might give you trouble, depending on your computing environment [Experiment 1-3, part (c)]. We will never be free of the possibility of computer-arithmetic-generated problems. I will have more to say about this in later chapters. Right now, it is a good idea for you to test out CONSOL3 and CONSOL4 on various problems. See how small you have to make Δt, and see if either code gets stuck. In theory, you will always be able to set Δt small enough to make the method work. In fact, the limited-precision computer arithmetic restricts the truth of this to a narrow range of values for a, b, q, and Δt (see Chapter 5).

The method for general equations

$$x^d + a_d x^{d-1} + \cdots + a_2 x + a_1 = 0 \tag{1-13}$$

is exactly like CONSOL3 and CONSOL4. It is implemented in CONSOL5. The main difference is that we need a general way of generating solutions to $h(x, 0) = 0$. To solve (1-13), the user inputs the degree of the equation, d; the coefficients of the equation,

$$a_d, \ldots, a_2, a_1;$$

and Δt. That is all there is to it. For the record, here is the continuation equation:

$$h(x, t) = x^d + a_d t x^{d-1} + \cdots + a_2 t x + [t a_1 - (1 - t)q^d]. \tag{1-14}$$

The constant q is as before. The coefficients

$$a_d, \ldots, a_2, a_1$$

can be real or complex numbers, but for simplicity, I have assumed in the CONSOL codes that the coefficients are real. We have

$$h(x, 0) = x^d - q^d = 0,$$

and its solutions are given by the formula

$$x = q\left(\cos \frac{2\pi j}{d} + i \sin \frac{2\pi j}{d} \right)$$

for $j = 1, 2, \ldots, d$; that is, q times the dth roots of unity.

CONSOL5 is a good code, although it has some drawbacks as far as efficiency goes. Rather than deal with those here, let's wait until later to introduce efficiencies. Try CONSOL5 on various problems (see Experiments 1-7 and 1-8). Note how the speed to solve varies with various choices of $a_d, \ldots, a_2, a_1, d$, and Δt. And, of course, you can now cause the code to fail totally by setting d very large or by choosing very large or small values for $a_d, \ldots, a_2, a_1$. (In order to try $d > 99$, you will have to change the dimension of the array A.) Go ahead and try some of these extremes. CONSOL5 will handle many equations for d greater than 10, but numerical problems will arise in connection with taking high powers or extreme coefficient values (see Chapter 5). Good performance is much easier to obtain for $d \leq 10$, and many physical problems generate equations of low degree. This leads me to the following observation. A frequent practice is to reduce a system of, say, five second-degree polynomials in five unknowns to one polynomial of degree 32 in one unknown, and then proceed to try to solve the one polynomial. This is sometimes a bad strategy. The system of five second-degree equations can often be dealt with much more easily than the one polynomial. Note especially the numerical consequences of taking thirty-second powers rather than second powers.

1-4 THE GEOMETRY OF CONTINUATION

Let's finish the chapter with some observations on the geometry of continuation. It is convenient (and mathematically accurate) to view continuation as generating a path in space from each start point [solution to the initial equation $h(x, 0) = 0$] to each solution to the original equation. The continuation happens by changing $(x_0, 0)$ to $(x_1, \Delta t)$ to $(x_2, 2\Delta t)$ . . . to $(x^*, 1)$, so we visualize the paths being in the space $C^1 \times [0, 1]$, where C^1 denotes the complex numbers. If we identify C^1 with R^2 (pairs of real numbers), then we visualize the three-dimensional space $R^2 \times [0, 1]$ with each continuation path moving from the $R^2 \times \{0\}$ plane to the $R^2 \times \{1\}$ plane. Figure 1-5 shows this for a (hypothetical) quadratic example.

In the one-variable case, we can display three-dimensional graphs such as Fig. 1-5 for actual continuation runs. (For two or more variables, of course, we cannot.) However, it is convenient to use a shorthand or "sketch" version of Fig. 1-5, in which the $R^2 \times \{t\}$ planes are collapsed to

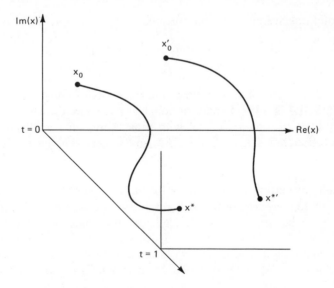

Figure 1-5   Two continuation paths going from start points x_0 and x_0' to end points x^* and $x^{*\prime}$ respectively, as t goes from 0 to 1. The axes are labeled t, Re(x), and Im(x) (for the real and imaginary parts of x).

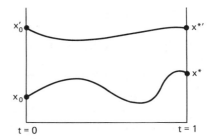

Figure 1-6 Schematic of two continuation paths corresponding to Fig. 1-5. The vertical axis represents "x" seen one-dimensionally, even though it is actually two-dimensional.

a line. Let's call these shorthand figures "schematics" (see Fig. 1-6). If $(x(t), t)$ denotes a continuation curve, you might think of the schematic as representing $(|x(t)|, t)$, although sometimes this interpretation is inconvenient. The basic behavior indicated by a schematic is:

- How many paths there are
- Whether the paths touch and for what values of t they touch
- Whether paths stay bounded or "diverge to infinity"

We will be using schematics throughout the book. To illustrate their usefulness in categorizing behavior, consider Fig. 1-7. This displays the two basic modes of behavior for CONSOL in solving a quadratic equation.

In Fig. 1-7b the two paths merge at the end. As noted above, paths cannot cross. However, they can come together when $t = 1$. Let's consider this phenomenon a little further. It will turn out to be important. Paths come together at a solution, x^*, to (1-13) exactly when it is a *singular* solution, that is, when x^* satisfies the derivative equation

$$dx^{d-1} + (d-1)a_d x^{d-2} + \cdots + a_2 = 0, \qquad (1\text{-}15)$$

as well as (1-13). Up to d paths might come together at x^*. The number of paths converging to x^* equals the *multiplicity* of x^*. By definition, the mul-

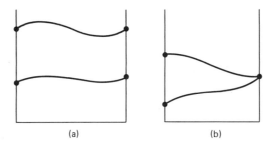

(a)   (b)

Figure 1-7 Schematics indicating possible behavior for CONSOL in solving a quadratic equation: (a) two distinct paths; (b) two paths coming together at a singular solution.

tiplicity of x^* is a positive integer, m, such that $(x - x^*)^m$ is a factor of (1-13) and $(x - x^*)^{m+1}$ is not. We sometimes call singular solutions "solutions of higher multiplicity." Proving these facts about multiplicity and continuation is difficult (see Chapter 3). If x^* is a solution, it is relatively easy to tell if it is singular [by checking (1-15)]. However, it is more difficult to find the exact multiplicity of x^*. In fact, counting the continuation paths converging to x^* is one way (see Exercises 1-2 and 1-3). Singular solutions are generally (much) more difficult to compute accurately than nonsingular solutions (see Appendix 1).

EXERCISES

1-1. Do a complete set of schematics for the cubic case (see Fig. 1-7).

1-2. Let $p(x)$ denote a polynomial.

    (a) Show that if x^* is a multiple solution of $p(x) = 0$, then x^* is a singular solution. [For this problem use the definition of "multiple solution" to be that $p(x) = (x - x^*)^m q(x)$ for some polynomial q, where $m > 1$ is the multiplicity of x^*. By "singular solution" is meant that $p(x^*) = 0$ and $\dfrac{dp}{dx}(x^*) = 0$, where $\dfrac{dp}{dx}$ denotes the derivative of p.]

    (b) Show that if $p(x)$ has only real number coefficients and x^* is a complex solution of $p(x) = 0$, the complex conjugate of x^* is a solution also. What are the implications for continuation?

1-3. Consider $p(x) = x^6 - 2x^5 + x^4$.

    (a) What are the solutions to $p(x) = 0$?

    (b) Are they singular?

    (c) What are their multiplicities?

    (d) Draw a schematic for solving $p(x) = 0$ using (1-14).

    (e) Run CONSOL5 to solve $p(x) = 0$. Observe the continuation paths "coming together" to give the multiplicities of the solutions. Also note the reduced accuracy obtained for singular solutions.

EXPERIMENTS

1-1. Run CONSOL2 with the following choices of input values:

    (a) $\Delta t = 0.1$, $a = 3$, $b = 4$

    (b) $\Delta t = 0.1$, $a = -4$, $b = 1$

    (c) $\Delta t = 0.1$, $a = -3$, $b = 2$

    (d) $\Delta t = 0.1$, $a = -2$, $b = 1$

Did you get the correct solutions? If not, what happened?

1-2. Do Experiment 1-1 using CONSOL3 instead of CONSOL2.

1-3. Run CONSOL4 with the following choices of input values:
(a) $\Delta t = 0.01$, $a = 1$, $b = 1$
(b) $\Delta t = 0.01$, $a = -16$, $b = 17$
(c) $\Delta t = 0.01$, $a = -16$, $b = 17.119$
Did you get the correct solutions? (How can you tell if your solution list is complete?) If not, what happened?

1-4. Run CONSOL1 for all 27 combinations of inputs given by
(a) $a = 3$, 300, 30000
(b) $b = -4$, -0.04, -0.0004
(c) $\Delta t = 0.1$, 0.01, 0.0001
For each run, answer the following questions.
(1) Was the run a success? That is, were two distinct solutions found? If not, why not?
(2) How many Newton iterations were used? What was the speed of the run in clock time?†
Summarize the 27 runs. What does it cost you in iterations or time to go from $\Delta t = 0.1$ to $\Delta t = 0.01$ to $\Delta t = 0.0001$? What happens as a gets big and/or b gets small? Make a table or graph of these relationships.

1-5. Do Experiment 1-4 with CONSOL3. What is the overall difference in terms of success and time (or work) spent?

1-6. Do Experiment 1-4 with CONSOL4. (Each run should yield five distinct solutions.)

1-7. Here is a collection of polynomials that arises in a computer vision application (design of Gaussian filters) as described in [Schunck, 1986]:

$$p_m(x) = \sum_{k=1}^{m} (-1)^k x^{k^2} + 0.5$$

For example,

$$p_4(x) = x^{16} - x^9 + x^4 - x^2 + x + 0.5,$$

and note that

$$p_{m+1}(x) = (-1)^{m+1} x^{m+1^2} + p_m(x).$$

Use CONSOL5 to solve $p_m(x) = 0$ for $m = 2, 3, 4, \ldots$ until you cannot find the complete solution set. What is the biggest value of m for which you *can* find the complete solution set? What goes wrong for the next m? As m grows, do you need to set Δt smaller to get all solutions? What patterns do you observe in the solution sets you can compute?

1-8. Consider the following collection of polynomials:

$$w_m(x) = (x - 1)(x - 2)(x - 3) \cdots (x - m).$$

---

† Counting how many Newton's method iterations the entire run took gives us a way of comparing the computer work needed to solve different problems. If you do not have access to an internal clock in your computer system, you can just look at your watch.

For example,

$$w_5(x) = x^5 - 15x^4 + 85x^3 - 225x^2 + 274x - 120,$$

and note that

$$w_{m+1}(x) = w_m(x) [x - (m + 1)].$$

The polynomial $w_{20}(x) = 0$ is famous as an example of extreme sensitivity of solutions to perturbations in the coefficients [Wilkinson, 1963, pp. 42–43]. Use CONSOL5 to solve $w_m(x) = 0$ for $m = 2, 3, 4, \ldots$ as in Experiment 1-7. You should at least do this for m up to 5.

NOTES

1. I have implied that path crossings are poison for continuation, but some continuation approaches can deal with them. See [Allgower, 1981]. In this book, we always insist that paths not overlap.
2. CONSOL is a method for polynomial systems, not single equations. Although a great deal can be learned about CONSOL by considering the single-polynomial case, and although CONSOL will solve single polynomials, more efficient techniques exist for this case. See, for example, the algorithm described in [Jenkins and Traub, 1970].

CHAPTER 2

Two Equations
in Two Unknowns

2-1 INTRODUCTION

In this chapter we focus on systems of equations in two unknowns. As in Chapter 1, we begin with second-degree equations. After a full development of CONSOL for the second-degree case, the extension to general-degree equations is routine. We will encounter "solutions at infinity," which are an important new source of complications for CONSOL. Otherwise, the basic issues are the same as those in the one-variable case.

2-2 SECOND-DEGREE EQUATIONS

A system of two second-degree equations in two unknowns looks like

$$ax^2 + by^2 + cxy + dx + ey + f = 0 \qquad (2\text{-}1)$$
$$a'x^2 + b'y^2 + c'xy + d'x + e'y + f' = 0,$$

where x and y are the variables and a through f, a' through f' stand for constants, called "coefficients." In other words, when we pick specific numbers to substitute for the coefficients, we have a system of two equations given by (2-1). Moreover, any *second-degree* system in two variables can be represented in this way. After we have chosen coefficients for (2-1), we want to find the list of all solutions to (2-1). Let's denote the first equation by $EQ = 0$ and the second by $EQ' = 0$. For real coefficients, the solutions to EQ $= 0$ alone (or EQ$' = 0$ alone) form either a curve in the plane, or two straight lines, or if all the coefficients are zero, the whole plane. If neither equation has all zero coefficients, the solutions of (2-1) (i.e., the simultaneous solutions of EQ $= 0$ and EQ$' = 0$) correspond geometrically to the intersection of the curves or lines associated with the two equations (see Fig. 2-1). In the exceptional case that one or both equations have all coefficients zero, the solutions to (2-1) are still given by the intersection of the solutions of each equation, which in this case will be either a curve, two lines, or all of the plane.

Let's consider an example. If we choose $a = 1$, $b = 1$, $c = 0$, $d = 0$, $e = 0$, $f = -25$, and $a' = 1$, $b' = 0$, $c' = 0$, $d' = 0$, $e' = 0$, $f' = -9$, we get

$$x^2 + y^2 - 25 = 0 \qquad\qquad (2\text{-}2)$$
$$x^2 - \qquad\quad 9 = 0.$$

Then the list of (x, y) solutions pairs is $(x, y) = (3, 4)$, $(-3, 4)$, $(3, -4)$, $(-3, -4)$. You can work out the algebra of this by substituting the second equation into the first and taking square roots. Geometrically, we are considering the intersection of a circle with two parallel lines (see Fig. 2-2).

If we leave EQ as it is but change EQ$'$ by choosing $a' = 1$, $b' = 0$, $c' = 0$, $d' = 0$, $e' = -1$, $f' = -5$, the system is

$$x^2 + y^2 - 25 = 0 \qquad\qquad (2\text{-}3)$$
$$x^2 - y - 5 = 0$$

and the solutions are $(x, y) = (3, 4), (-3, 4), (0, -5)$. The associated picture is a circle intersecting a parabola, as shown in Fig. 2-3a.

This is a good place to define singular solutions. When two curves intersect by "just touching," the resulting solution is said to be "singular." By "just touching" I mean the tangents to the two curves are equal at the solution. For system (2-3), this is not the case at $(3, 4)$ or $(-3, 4)$, but it is at $(0, -5)$ (see Fig. 2-3b).

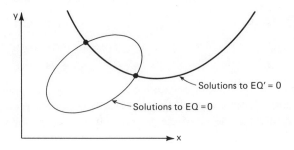

Figure 2-1 Geometry of two equations in two unknowns. Each equation's solutions form a curve. The intersection of the curves gives the simultaneous solutions.

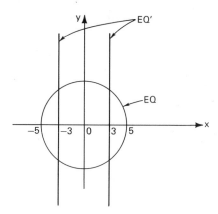

Figure 2-2 Graph for system (2-2).

We can detect singular solutions algebraically as follows. The matrix of partial derivatives of system (2-1) is called the "Jacobian" of (2-1), denoted $J(x, y)$, and defined by

$$J(x, y) = \begin{bmatrix} \dfrac{\partial\,(EQ)}{\partial x} & \dfrac{\partial\,(EQ)}{\partial y} \\[2mm] \dfrac{\partial\,(EQ')}{\partial x} & \dfrac{\partial\,(EQ')}{\partial y} \end{bmatrix}$$

$$= \begin{bmatrix} 2ax + cy + d & 2by + cx + e \\[2mm] 2a'x + c'y + d' & 2b'y + c'x + e' \end{bmatrix}.$$

If (x, y) is a solution, this solution is "singular" whenever $\det\,(J(x, y)) = 0$,

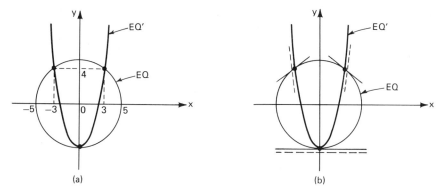

(a)   (b)

Figure 2-3   (a) Graph for system (2-3); (b) graph for system (2-3) showing tangents at solutions of EQ = 0 (solid lines) and of EQ' = 0 (dashed lines).

where "det" denotes the determinant. Thus

$$\det (J(x, y)) = (2ax + cy + d)(2b'y + c'x + e')$$
$$- (2by + cx + e)(2a'x + c'y + d').$$

For system (2-3) we have

$$J(x, y) = \begin{bmatrix} 2x & 2y \\ 2x & -1 \end{bmatrix}$$

and

$$\det (J(x, y)) = -2x - (2y)(2x) = -2x(1 + 2y),$$

so

$$\det (J(0, -5)) = 0$$
$$\det (J(3, 4)) = -6(1 + 8) = -54$$
$$\det (J(-3, 4)) = 54.$$

This confirms that $(0, -5)$ is a singular solution to (2-3), and (3, 4) and $(-3, 4)$ are not.

Here is another example, one without singular solutions. Letting $a' = 1$, $b' = 1$, $c' = 0$, $d' = 0$, $e' = -16$, $f' = 39$ yields the system

$$x^2 + y^2 - 25 = 0 \qquad\qquad (2\text{-}4)$$
$$x^2 + y^2 - 16y + 39 = 0,$$

with the two solutions $(x, y) = (3, 4), (-3, 4)$ (see Fig. 2-4).

These examples illustrate that we want to be able to handle a variety of situations. In particular, we have examples of systems with four, three, and two solutions. In fact, for all possible choices of a through f and a' through f', (2-1) will have no, one, two, three, four, or an infinite number of solutions. No other possibilities exist. Here are illustrative examples. Let EQ be unchanged; that is, let $a = 1$, $b = 1$, $c = 0$, $d = 0$, $e = 0$, and $f = -25$. Then, defining EQ' by the following choices, we get

- no solutions when $a' = 1$, $b' = 1$, $c' = 0$, $d' = 0$, $e' = 0$, $f' = -36$ [System (2-5)]
- one solution when $a' = 1$, $b' = 1$, $c' = 0$, $d' = 0$, $e' = -20$, $f' = 75$ [System (2-6)]
- an infinite number of solutions when $a' = 1$, $b' = 1$, $c' = 0$, $d' = 0$, $e' = 0$, $f' = -25$ [System (2-7)].

See Figs. 2-5, 2-6, and 2-7, respectively.

It is important to have in mind the types of solution sets we might encounter before we discuss techniques for solution. For example, if we are

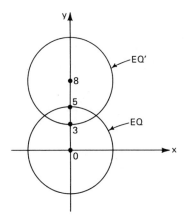

Figure 2-4 Graph for system (2-4).

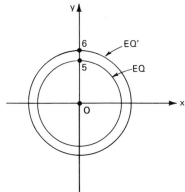

Figure 2-5 Graph for system (2-5).

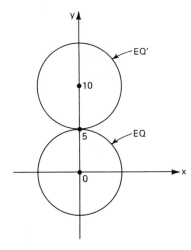

Figure 2-6 Graph for system (2-6).

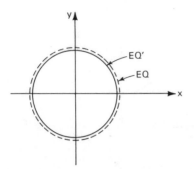

Figure 2-7 Graph for system (2-7).

setting up a code to handle every possible case, we would have to deal with the case that all coefficients are zero. This has every (x, y) pair as a solution. System (2-7) illustrates another infinite-solution case. By setting $b = 0$ (leaving $a = 1$ and $f = -25$), and $a' = 0$, $b' = 0$, $c' = 1$, $d' = -1$, $e' = -5$, $f' = 5$, we get a more interesting "infinite number of solutions" case. This system,

$$x^2 - 25 = 0 \qquad\qquad (2\text{-}8)$$
$$xy - x - 5y + 5 = 0,$$

has a straight line and a point not on the line as its solution set.

If we needed separate codes for each of the cases of no, one, two, three, four, or an infinite number of solutions, this would be unfortunate. It is generally as hard to decide how many solutions a system has as it is to solve it. Further, characterizing systems by the number of solutions is incomplete. The system

$$xy + 9x - 6y + 6 = 0 \qquad\qquad (2\text{-}9)$$
$$x^2 - y + 2 = 0$$

has exactly three solutions: $(1, 3)$, $(2, 6)$, and $(3, 11)$, but none of them is singular. Thus (2-9) is fundamentally different from (2-3), and, in fact, most solution methods behave differently on these two systems.

One of the advantages of continuation over some other approaches is that breaking into cases is not necessary. Basically, there are two cases: finite number of solutions and infinite number of solutions. The first case we can handle. The second case we can *partially* handle. Note that the only way a single equation in a single variable can have an infinite number of solutions is for every coefficient to be zero. In the two-variable case, it is sometimes less immediately obvious that there are an infinite number of solutions. For example, see (2-8). For now, we will *assume* that the systems under study have only a finite number of solutions. We will discuss later what happens if this is false.

Recall that we want to construct a computer program that will take as input the numbers a through f, a' through f', and print out the solution list to the system (2-1). We do not want to have to customize the program to each problem. We do not want to have to divide the inputs into cases. However, for now, we accept the one limitation: The choice of inputs should yield a system that does not have an infinite number of solutions.

EXERCISES

2-1. Which of the solutions of systems (2-4), (2-6), and (2-7) are singular? Figure it out from the associated figures and confirm your guesses by computing the Jacobians.

2-2. Draw a figure for system (2-8).

2-3. Draw a figure for system (2-9). (*Hint:* Do some algebra first.)

2-3 CONTINUATION EQUATIONS

In solving polynomials in one variable (i.e., that have only one variable), we used a "natural" continuation equation: namely, the one defined by multiplying terms of intermediate degree† by the continuation parameter t. When $t = 0$, the equation was solved merely by taking roots. After some modifications, to avoid singularities, we arrived at a continuation equation that always works [equation (1-14)]. It is less clear how to proceed in the two-variable case.

Systems (2-2), (2-3), . . . , (2-9) have a variety of forms. Should we customize our continuation equation to each system, or look for a way to define it in general? There is nothing wrong with conjuring up special continuations for special systems. Using the insights already obtained (and to be obtained as we continue), we can construct reasonable "customized" continuations, find out when they fail, and guess modifications that will make them succeed. In fact, this is to some extent how we must proceed for nonpolynomial nonlinear systems (see Chapter 6). But for polynomial systems we *can* find generic continuations. We can obtain algorithms which are guaranteed to work in exact arithmetic and which are robust in computer arithmetic.

How do we "discover" such generic continuations? Suppose that we try to find some generalization of "multiply the terms of intermediate degree by t" to generate continuation equations, just as we solved (1-13) by generating the continuation equation (1-14). We are already in trouble with system (2-2), since it has no terms of intermediate degree. But not only can we

---

† By "terms of intermediate degree" I mean terms in the equation that do not have the biggest or smallest degree compared to the other terms. If there is a constant term, this is the term of smallest degree [see (1-6) and (1-14)].

modify existing terms by multiplying them by t, we can add new terms multiplied by $(1 - t)$. We have already seen this in the term involving q of equation (1-14). Using this idea, we can generate a variety of continuation systems. In essence, we can start with any system and turn it into our original system as t goes from 0 to 1. Not all such continuations will work, but many do. Here is one of the simpler ones.

Regardless of the form of our given system [(2-1)], we will take the initial continuation system to be

$$p^2x^2 - q^2 = 0 \qquad\qquad (2\text{-}10)$$
$$p'^2y^2 - q'^2 = 0,$$

where p, p', q, q' are "random" complex numbers. Thus (2-10) has four solutions, $(x, y) = (\pm q/p, \pm q'/p')$. To link (2-10) to (2-1) as simply as possible, we use

$$h(x, y, t) = (1 - t)(p^2x^2 - q^2) + t\text{EQ} = 0 \qquad (2\text{-}11)$$
$$h'(x, y, t) = (1 - t)(p'^2y^2 - q'^2) + t\text{EQ}' = 0,$$

which written out is

$$h(x, y, t) = [(1 - t)p^2 + ta]x^2 + tby^2 + tcxy$$
$$+ tdx + tey + [tf - (1 - t)q^2] = 0$$
$$h'(x, y, t) = ta'x^2 + [(1 - t)p'^2 + tb']y^2 + tc'xy$$
$$+ td'x + te'y + [tf' - (1 - t)q'^2] = 0.$$

Note that (2-11) is as generic as we could want, since its form is independent of the specific form of (2-1). The use of random complex constants and the way they enter into the equations is reminiscent of the use of the constant q in Chapter 1. We discuss their utility later. The Chapter 1 considerations are still relevant, along with a new issue. It is this new issue that I would like to focus on. Let us assume for now that (2-11) yields well-defined continuation paths that do not cross and which we can follow as t goes from 0 to 1.

Here is a summary of important facts about the continuation paths (see the theorem in Section 3-4). System (2-11) always generates four paths, no matter how many solutions (2-1) has. Each solution to (2-1) will have a path converging to it. [We are assuming that (2-1) does not have an infinite number of solutions.] When (2-1) has four solutions, the four paths will converge to the four solutions. In this case, the solutions will be nonsingular. If (2-1) has a singular solution, (x^*, y^*), at least two paths will converge to it, perhaps more. The number of paths converging to (x^*, y^*) is the *multiplicity* of (x^*, y^*). [Proving that the multiplicity of (x^*, y^*) is independent of the choice of p, p', q, and q' is difficult. It is, in fact, difficult to give an obviously unambiguous definition of multiplicity, except for polynomials in one variable. I will defer further discussion of multiplicity until Chapter 3.] Sometimes paths diverge to infinity. That is, as t gets close to 1, these paths become unbounded. What we see is that as t gets close to 1, x or y gets bigger and bigger, until eventually the computer arithmetic breaks down, generating overflows or other anomalies, such as "path crossing." It is important to understand this phenomenon and to see why paths cannot become unbounded except when t approaches 1. This is the "new issue" that we need to consider. To begin, we must learn about "solutions at infinity," in the next section.

2-4 SOLUTIONS AT INFINITY

Sometimes a system has "solutions at infinity." We discuss this intuitively here and give a procedure for computing solutions at infinity. In Chapter 3 we reconsider the topic from a more general viewpoint. Suppose that (x, y) is some point in the plane, and r is a nonzero number. Consider

$$\text{EQ}(rx, ry) = ar^2x^2 + br^2y^2 + cr^2xy + drx + ery + f.$$

Then, dividing through by r^2 yields

$$\frac{\text{EQ}(rx, ry)}{r^2} = (ax^2 + by^2 + cxy) + \frac{1}{r}\left(dx + ey + \frac{1}{r}f\right).$$

Now, if r is very large compared to d, e, f, x, and y, and if $(ax^2 + by^2 + cxy) = 0$, the expression above is very small. This suggests the following definition. Let

$$\text{EQ}^0 = ax^2 + by^2 + cxy$$

and

$$\text{EQ}'^0 = a'x^2 + b'y^2 + c'xy.$$

Then the system

$$\text{EQ}^0 = 0 \tag{2-12}$$
$$\text{EQ}'^0 = 0$$

is the "homogeneous part" of (2-1), and its nonzero solutions are called "solutions to (2-1) at infinity." Using this definition we do not have to make "letting r go to infinity" be an official part of the definition. This makes the concept of "solution at infinity" more precisely defined and easier to use.

Note that if (x, y) is a solution to (2-12), then (rx, ry) is a solution to (2-12) for any number r. It is convenient to pick out solutions in such a way that no two are multiples of each other. We will use the convention of choosing solutions in either the form $(1, y)$ or $(0, 1)$. Any solution at infinity is a scalar multiple of a solution in one of these two forms.

Let us check some systems for solutions at infinity. The homogeneous part of system (2-2) is

$$x^2 + y^2 = 0 \qquad\qquad (2\text{-}13)$$
$$x^2 = 0.$$

Since $x^2 = 0$, $x = 0$. Thus $y = 0$. But $(0, 0)$ cannot be a solution at infinity (by definition, solutions at infinity are nonzero), so (2-2) has no solutions at infinity. The homogeneous part of (2-3) is (2-13) also, so (2-3) has no solutions at infinity. The homogeneous part of (2-4) is

$$x^2 + y^2 = 0 \qquad\qquad (2\text{-}14)$$
$$x^2 + y^2 = 0.$$

We must check for solutions at infinity in both of the two forms $(1, y)$ and $(0, 1)$. If $x = 1$, then (2-14) becomes $1 + y^2 = 0$ and thus $y = \pm i$. Therefore, (2-4) has the two solutions at infinity, $(1, i)$ and $(1, -i)$. These are the only solutions at infinity, since $(0, 1)$ is not a solution to (2-14).

It turns out that solutions at infinity can be singular and thus have a multiplicity greater than 1, like ordinary solutions. I will wait until Chapter 3 to clarify this. However, I cannot wait to state (for second-degree systems in two variables) a beautiful theorem.

THEOREM OF BEZOUT (VERSION 1)
1. Unless (2-1) has an infinite number of solutions, it has four or less.
2. Unless (2-1) has an infinite number of solutions or an infinite number of solutions at infinity, the number of its solutions and its solutions at infinity adds up to exactly four, counting multiplicities.

"Counting multiplicities" means that if a solution (or solution at infinity) has multiplicity "2," it is counted twice; if it has multiplicity "3," it is counted three times; and so on.

If we look back at some examples, we see the relationships predicted by Bezout's theorem:

- System (2-2): 4 solutions + 0 solutions at infinity = 4
- System (2-3): 2 nonsingular solutions + 1 singular solution counted twice = 4 (in this case, we infer the multiplicity from Bezout's theorem)
- System (2-4): 2 nonsingular solutions + 2 solutions at infinity = 4

Note that generally it is easier to find solutions at infinity than solutions because when we take the homogeneous part of our system, we drop out all but the maximum degree terms for each equation (simplifying the system).

Then we set $x = 1$ or we set $x = 0$ and $y = 1$, leaving us with no more than a system of two equations in one variable to solve.

Solutions at infinity have a surprising significance for us. For each (isolated) solution to (2-1) at infinity, there will be at least one divergent continuation path. There will be divergent continuation paths only if (2-1) has solutions at infinity. The multiplicity of a solution at infinity determines the number of associated divergent paths.

Bezout's theorem suggests why paths diverge if (2-1) has solutions at infinity. We begin at $t = 0$ with exactly four paths. If these paths do not diverge, then at $t = 1$ they will define four finite solutions (counting multiplicities). Assuming that (2-1) does not have an infinite number of solutions or an infinite number of solutions at infinity, the existence of a solution at infinity means that (2-1) cannot have four finite solutions. Hence at least one path has nowhere to go, except to infinity.

Let us now investigate why the continuation paths can become unbounded only as $t \to 1$. Suppose that a path becomes unbounded in the middle. That means that we have a sequence of solutions to (2-11), (X_n, Y_n, t_n), $n = 1, 2, 3, \ldots$ with $|(X_n, Y_n)| \to \infty$ and t_n bounded away from 1. Let's write $(X_n, Y_n) = r_n(x_n, y_n)$, where $r_n = |(X_n, Y_n)|$ and the (x_n, y_n) are unit vectors in C^2. (By "unit vectors" I mean the vectors have length 1.) Since $\{t_n\}$ is bounded away from 1, we may as well assume that $t_n \to t_0$ for some $t_0 < 1$. Since the (x_n, y_n) are unit vectors, we may as well assume that $(x_n, y_n) \to (x_0, y_0)$, for some unit vector (x_0, y_0). [The point is that since the (x_n, y_n) are bounded, we can find a converging subsequence.] Then plugging into the first equation of (2-11) yields

$$0 = h(X_n, Y_n, t_n) = [(1 - t_n)p^2 + t_n a]r_n^2 x_n^2 + t_n b r_n^2 y_n^2$$

$$+ t_n c r_n^2 x_n y_n + t_n d r_n x_n$$

$$+ t_n e r_n y_n + [t_n f - (1 - t_n)q^2].$$

Thus

$$[(1 - t_n)p^2 + t_n a]x_n^2 + t_n b y_n^2 + t_n c x_n y_n$$

$$+ \frac{t_n d x_n}{r_n} + \frac{t_n e y_n}{r_n} + \frac{t_n f - (1 - t_n)q^2}{r_n^2} = 0$$

for all n, so taking the limit as $n \to \infty$ yields

$$[(1 - t_0)p^2 + t_0 a]x_0^2 + t_0 b y_0^2 + t_0 c x_0 y_0 = 0.$$

Similarly, from the second equation of (2-11) we get

$$t_0 a' x_0^2 + [(1 - t_0)p'^2 + t_0 b']y_0^2 + t_0 c' x_0 y_0 = 0.$$

Our conclusion is that the system

$$h(x, y, t_0) = 0$$
$$h'(x, y, t_0) = 0$$

(2-15)

with t_0 fixed has a solution at infinity, namely (x_0, y_0). To summarize: If a path becomes unbounded in the middle, (2-11) has a solution at infinity, for some $t_0 < 1$.

Now let me demonstrate that (2-11) cannot have solutions at infinity with $t_0 < 1$, because of the random choice of p and p'. As noted above, we may take a solution at infinity in the form $(1, y)$ or $(0, 1)$. What if $(1, y)$ is a solution at infinity? Then we have (from the homogeneous part of 2-11)

$$[(1 - t_0)p^2 + t_0a] + t_0by^2 + t_0cy = 0$$

and

$$t_0a' + [(1 - t_0)p'^2 + t_0b']y^2 + t_0c'y = 0.$$

Thus we have two quadratic equations (in one variable) with a common solution y. I claim that the random (i.e., *independent*) choice of p and p' makes this impossible. More precisely, we have the following variant of the Independence Principle stated in Chapter 1.

INDEPENDENCE PRINCIPLE.   If $Q(y)$ and $Q'(y)$ are two quadratic polynomials and p_0 and p_0' are independently chosen constants, the system

$$Q(y) + p_0 = 0$$
$$p_0'y^2 + Q'(y) = 0$$

(2-16)

does not have a solution.

In our case we have

$$Q(y) = T_0by^2 + T_0cy + T_0a,$$

where $T_0 = t_0/(1 - t_0)$,

$$p_0 = p^2, \qquad p_0' = p'^2,$$

and

$$Q'(y) = T_0b'y^2 + T_0c'y + T_0a'.$$

Clearly, the existence of a solution at infinity of the form $(1, y)$ would contradict the Independence Principle.

You can convince yourself of the Independence Principle as follows. Fix p_0' so that

$$p_0'y^2 + Q'(y) = 0$$

(2-17)

is a nonzero quadratic equation and has only two solutions, y^* and y^{**}. If y^* also solves $Q(y) + p_0 = 0$, then $p_0 = -Q(y^*)$. Thus there is only one choice of p_0 that makes y^* a solution to (2-16). Similarly, only one choice of p_0 allows y^{**} to be a solution to (2-16). Therefore, only two choices of p_0 (after p_0' is fixed) will allow (2-16) to have a solution. Now let's play a game like the one we played in Chapter 1. You choose quadratic equations Q and Q'. I will choose constants p_0 and p_0' without reference to Q and Q', perhaps by using a random number table. Do you think the resulting quadratics as given in (2-16) will share a solution? That is, do you think that I will happen to choose one of the two choices of p_0 defined by $p_0 = -Q(y^*)$ or $p_0 = -Q(y^{**})$, where y^* and y^{**} are the solutions to (2-17)? The answer is clearly "no." It is in this sense that the independence principle is applied.

Can (0, 1) be a solution at infinity? If so, we have [from the homogeneous part of (2-11)]

$$[(1 - t_0)p'^2 + t_0 b'] = 0.$$

Thus

$$p' = \pm\sqrt{-T_0 b'},$$

and this is inconsistent with the independent choice of p'. This completes the demonstration that continuation paths cannot become unbounded in the middle. A rigorous proof is given in Appendix 4.

EXERCISES

2-4. Compute solutions at infinity for systems (2-5) through (2-9) and using Bezout's theorem, sort out the multiplicities of all the solutions and solutions at infinity.

2-5. Although this section is about second-degree systems, we have not ruled out $a = b = c = 0$ or $a' = b' = c' = 0$ in (2-1). That is, one or both equations might "really" be linear. Carefully review the definition of "solutions at infinity" and convince yourself Bezout's theorem still holds. What is the significance for CONSOL?

2-5 SCHEMATICS

Schematics were introduced in Chapter 1 as a way of visualizing the basic behavior of continuation paths. For systems of two equations in two variables, continuation paths move through the space $C^2 \times [0, 1]$, which we identify with the five-dimensional real space $R^4 \times [0, 1]$. As before, to generate schematics we condense the R^4 to R^1 while preserving (in the picture) the basic properties of "touching" and "boundedness" in the paths. Using schematics, we can classify all the possible modes of behavior for

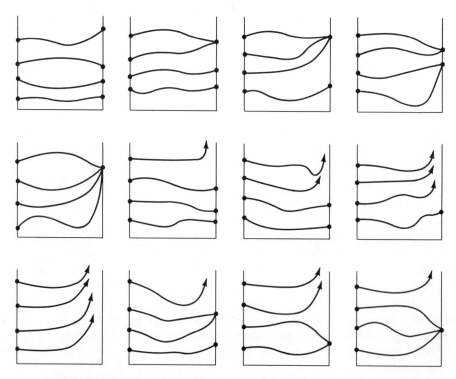

Figure 2-8 Schematics showing all modes of behavior for systems of two second-degree equations.

CONSOL on two-dimensional systems, including "divergence," which is shown by an arrow (see Fig. 2-8). The multiplicity of solutions at infinity is not shown.

EXERCISE

2-6. Identify systems that exhibit each mode of behavior given in Fig. 2-8.

2-6 CONSOL EXPERIMENTS

CONSOL6, given in Appendix 6, implements the continuation system (2-11). This code is the two-variable analog of CONSOL3. It is, like CONSOL3, a simple implementation, and thus it is less efficient and reliable than we would like. However, in Chapter 4 we will see how to develop more sophisticated methods. In this section we run some experiments to see how the method behaves for various choices of parameters.

Since CONSOL increases t in fixed increments, it will generally behave as follows when tracking a path that diverges to infinity. The path will be computed correctly until t is near 1 and then CONSOL will lose the path, perhaps crossing to another (finite-length) path, but not necessarily. The result will be either that a solution will be found twice, once by the mistracked diverging path, or the final point found will be a nonsolution point (the "h" and "Newton residual" values will not be small). CONSOL prints out the last five steps before $t = 1$ and you can sometimes spot this mistracking behavior by examining these final path points and the h and residual values. The exception is that when Δt is large (say, $\Delta t = 0.1$), the fine structure in the tracking behavior will be lost and the only clue will be that a nonsingular solution may be found twice. I would say in this situation that you try and count the solutions at infinity "by hand" so that Bezout's theorem will tell you the proper solution count. Alternatively, you can rerun with Δt smaller. In Chapter 3 we will see that the "projective transformation" generally eliminates diverging paths altogether. Also, the more sophisticated variable-step-size path-tracking methods discussed in Chapter 4 make diverging paths easier to spot and more efficient to compute.

EXPERIMENTS

2-1. Solve (2-2) through (2-9) using CONSOL6. Set $\Delta t = 0.1$, 0.01, 0.0001. Note the behavior of CONSOL in terms of:

- Clock time
- Total number of Newton iterations
- Success or failure

Call a run a "failure" if some solution is missed, a "success" if reasonable approximations to all solutions are found. Try to explain the failures. Note the behavior especially when there are an infinite number of solutions, when there are solutions at infinity, and when there are singular solutions.

2-2. I chose random values for the constants p, p', q, and q' in CONSOL6. Make a copy of CONSOL6, and change these values to other reasonable choices.†
Rerun some of Experiment 2-1. How does the behavior of CONSOL6 change? Do some runs go faster? Slower? Now choose some unreasonable values: say, all equal to "1" or some equal to "1" and some equal to "0." What happens when you rerun some of Experiment 2-1? Now pick one of the systems that you have been solving and find p, p', q, and q' specifically to make continuation fail. Can you make it fail in two distinct ways? (See Chapter 1.)

---

† Putting primes on variable names is not convenient in a FORTRAN program. Subscripting is more natural. Thus, in CONSOL6 and CONSOL7, p, p', q, and q' are represented as P(1), P(2), Q(1), and Q(2), respectively.

2-3. Solve the linear system

$$x + y + 1 = 0 \qquad (2\text{-}18)$$
$$x - y + 1 = 0$$

using CONSOL6. What happens? Why?

2-7 HIGHER-DEGREE EQUATIONS

How can we solve a system like

$$EQ = ax^5 + bx^2y + cy^4 + dx + ey + f = 0 \qquad (2\text{-}19)$$
$$EQ' = a'x^3 + b'x^3y^4 + c'xy^6 + d'x + e'y + f' = 0,$$

where a through f, a' through f' are constants?

I hope that at this point you have a good idea of how to approach the problem! In this section we consider higher-degree equations in two variables: what they look like, the structure of their solution sets, solutions at infinity, Bezout's theorem, and a general-purpose continuation.

A general equation in two variables, E, is a sum of terms:

$$E = \text{TERM}_1 + \text{TERM}_2 + \text{TERM}_3 + \cdots + \text{TERM}_l,$$

where each term looks like

$$\text{TERM}_j = a_j x^{k_j} y^{m_j}$$

and a_j is a complex constant (possibly zero), k_j and m_j are integers, perhaps zero but not negative, and x and y are the variables. Our interpretation of a zero exponent is that it removes the variable from the term: "$a_j x^2 y^0$" means "$a_j x^2$." (Another way of looking at this is the definition: $z^0 \equiv 1$ for any number z, even $z = 0$.) The "degree" of TERM_j is "$k_j + m_j$," and the degree of the equation, E, is the biggest of the degrees of the terms. Thus the first equation of (2-19), EQ, has six terms, as follows:

TERM_1: $a_1 = a$, $k_1 = 5$, $m_1 = 0$, degree $= 5$

TERM_2: $a_2 = b$, $k_2 = 2$, $m_2 = 1$, degree $= 3$

TERM_3: $a_3 = c$, $k_3 = 0$, $m_3 = 4$, degree $= 4$

TERM_4: $a_4 = d$, $k_4 = 1$, $m_4 = 0$, degree $= 1$

TERM_5: $a_5 = e$, $k_5 = 0$, $m_5 = 1$, degree $= 1$

TERM_6: $a_6 = f$, $k_6 = 0$, $m_6 = 0$, degree $= 0$.

Thus the degree of EQ is 5. The degree of EQ' is 7.

We have a last and very important definition. The "total degree" of a system is the *product* of the degrees of the two equations. The total degree of (2-19) is 35.

I must now discuss a nuisance item. Frequently in practice, we wish to set up a code to solve systems of a fixed form [e.g., (2-19)], so we can input various coefficient values (a through f, a' through f' in the example) and get the solutions as output. It is convenient if our solution method will work without branching into cases when key coefficients are zero (say, if $a = 0$ or if $b' = c' = 0$). Codes that branch into cases depending on whether or not coefficients are zero are difficult to program and debug. Thus, given any polynomial equation, it is acceptable to declare it to have a degree higher than it "really" has. That is, a given polynomial may be viewed as having a higher degree, but with the coefficients of the higher-degree terms set equal to zero (see Experiment 2-3). Thus we may declare EQ to be degree 5 even with $a = 0$. CONSOL will "work" under this convention, and this fact is very convenient.

Even though CONSOL allows us to overestimate the degree of equations, efficiency and simplicity compel us to prefer to know the actual degree of the equation and not overestimate. If $a = 0$ always, we should omit the term so that EQ can be labeled "fourth degree." Incidentally, if $a \neq 0$, there is no reasonable sense in which EQ can be said to have degree less than 5.

We will now define solutions at infinity and state Bezout's theorem. The *declared degree* of the equations affects the solutions at infinity. Overestimating the degrees of equations can negate the valuable insights we generally expect to gain from Bezout's theorem. The homogeneous part of an equation is defined to be the equation with the "lower-degree terms" dropped out. That is, the equation has a declared degree, and we get the homogeneous part by omitting terms that have degree less than that. We generally denote the homogeneous part of EQ by "EQ^0." Thus for (2-19) we get:

$$EQ^0 = ax^5 = 0$$
$$EQ'^0 = b'x^3y^4 + c'xy^6 = 0, \tag{2-20}$$

which is the homogeneous part of the system (2-19). Now, as in Section 2-4 we *define* the "solutions at infinity" of a system to be the solutions to the homogeneous part, with the understanding that we will choose only solutions of the form $(1, y)$ or $(0, 1)$. Then we get

THEOREM OF BEZOUT (VERSION 2)
1. Unless a system has an infinite number of solutions, the number of its solutions is less than or equal to its total degree.
2. Unless a system has an infinite number of solutions or an infinite number of solutions at infinity, the number of its solutions and its solutions at infinity adds up to exactly the total degree, counting multiplicities.

Thus (2-19) can have no more than 35 solutions unless it has an infinite number. It is easy to solve (2-20), but we must consider cases:

1. If $a \neq 0$, then $EQ^0 = 0$ implies that $x = 0$. Our only candidate for a solution is $(0, 1)$, which satisfies $EQ'^0 = 0$ and so is a solution at infinity for (2-19).

2. If $a = 0$, then $EQ^0 = 0$ "disappears." We have only $EQ'^0 = 0$. The pair $(0, 1)$ is a solution. Let's try $(1, y)$. Thus

$$b'y^4 + c'y^6 = 0. \qquad (2\text{-}21)$$

Clearly, $y = 0$ works, so $(1, 0)$ is another solution at infinity. To see if any nonzero y will work, let's assume that $y \neq 0$. Then we get

$$b' + c'y^2 = 0. \qquad (2\text{-}22)$$

  a. If $c' \neq 0$, (2-22) yields $y = \pm\sqrt{-b'/c'}$. Thus

$$\left(1, \sqrt{\frac{-b'}{c'}}\right) \quad \text{and} \quad \left(1, -\sqrt{\frac{-b'}{c'}}\right)$$

are additional solutions at infinity [unless $b' = 0$, in which case they equal $(1, 0)$, which we have already included].
  b. If $c' = 0$ and $b' \neq 0$, then (2-22) has no solutions.
  c. If $c' = 0$ and $b' = 0$, then (2-22) has any y as a solution. Thus $(1, y)$ is a solution for any y and (2-19) has an infinite number of solutions at infinity.

Here is a summary:

Case	Solutions at Infinity
1. $a \neq 0$	$(0, 1)$
2. $a = 0,\ c' \neq 0$	$(0, 1),\ (1, 0),\ \left(1,\ \pm\sqrt{\dfrac{-b'}{c'}}\right)$
3. $a = 0,\ c' = 0,\ b' \neq 0$	$(0, 1),\ (1, 0)$
4. $a = 0,\ c' = 0,\ b' = 0$	$(0, 1),\ (1, y)$ for all complex y

This example illustrates several points. First, in case 4, we see that part 2 of Bezout's theorem gives no information, since we have passed over into the category "infinite number of solutions at infinity." Let us assume that (2-19) does not have an infinite number of solutions. Then from Bezout's theorem in cases 1 and 3 we see that (2-19) can have *at most* 34 and 33 solutions, respectively, where we must say "at most" because we do not know the multiplicities of the solutions at infinity. If $b' \neq 0$, case 2 implies that (2-19) can have at most 31 solutions; if $b' = 0$, it can have at most 33 solutions. In case 4, the most degenerate case, the "true degree" of both equations is less than the "declared degree." If (2-19) has an infinite number of solutions, both parts 1 and 2 of Bezout's theorem tell us nothing. Notice that it is not particularly easy to specify exactly what conditions on a through f, a' through f' cause (2-19) to have an infinite number of solutions.

Now we will define the continuation equations. Let's agree that

$$EQ = 0$$
$$EQ' = 0$$

$$(2\text{-}23)$$

will denote an arbitrary polynomial system in two variables. Let d and d' be the declared degrees of EQ and EQ', respectively. Then we define

$$h(x, y, t) = (1 - t)(p^d x^d - q^d) + tEQ = 0$$
$$h'(x, y, t) = (1 - t)(p'^{d'} y^{d'} - q'^{d'}) + tEQ' = 0$$

$$(2\text{-}24)$$

where p, p', q, and q' are random complex constants. This is, of course, very similar to (2-11). The initial system

$$p^d x^d - q^d = 0$$
$$p'^{d'} y^{d'} - q'^{d'} = 0$$

$$(2\text{-}25)$$

can be solved easily. Define

$$u(s, j) \equiv \cos \frac{2\pi j}{s} + i \sin \frac{2\pi j}{s},$$

$$(2\text{-}26)$$

where $i^2 = -1$, establishing the notation "$u(s, j)$" for the sth roots of unity. (Thus $[u(s, j)]^s = 1$.) Then the solutions to (2-25) are given by

$$(x, y) = \left(u(d, j) \frac{q}{p}, u(d', k) \frac{q'}{p'} \right)$$

for $j = 1, \ldots, d$ and $k = 1, \ldots, d'$.

The system (2-25) has total degree $d \cdot d'$, and CONSOL7 [implementing (2-24)] generates $d \cdot d'$ paths. If (2-23) does not have an infinite number of solutions, CONSOL7 will generate paths converging to all the solutions, with multiple paths converging to paths of higher multiplicity, as discussed for second-degree systems. If (2-23) has solutions at infinity, paths will diverge, one for each solution at infinity (counting multiplicities), if there are not an infinite number.

Unlike the previous programs, CONSOL7 requires a user-coded subroutine, FFUN7. FFUN7 generates the equation and partial derivative values for the system that is being solved, (2-23). I have included with CONSOL7 three versions of FFUN7, called FFUN7A, FFUN7B, and FFUN7C. These are for Experiments (2-4), (2-5), and (2-6), respectively. To run one of these experiments, merely change the name of the appropriate subroutine to FFUN7. These subroutines can be used as models of FFUN7 for solving other systems.

In the next chapter a general continuation will be defined, and solutions at infinity will be developed more rigorously. The specific mathematical conditions and restrictions that justify CONSOL will be given, although some proofs are reserved for Appendices 3 and 4.

EXERCISES

2-7. Discuss† solutions at infinity for the following system:

$$ax^2y + bxy^2 + cxy + dy^3 + ey^2 + fy + g = 0 \qquad (2\text{-}27)$$
$$a'x^2y + b'x^2 + c'xy + d'x + e'y = 0.$$

2-8. Discuss solutions at infinity for the following system:

$$ax^2y^2 + bxy^2 + cxy + dy^3 + ey^2 + fy + g = 0 \qquad (2\text{-}28)$$
$$a'x^2y + b'x^2 + c'xy + d'x + e'y = 0.$$

2-9. How will CONSOL behave when the given system has an infinite number of solutions? An infinite number of solutions at infinity? (Guess.)

2-10. We have not presented an argument that paths cannot "go singular." What would you observe in a CONSOL run if a path did go singular? It turns out that we can, in fact, count on paths not going singular. From the Chapter 1 discussion, what part of the structure of the continuation system (2-24) prevents paths from going singular?

2-11. What would "solutions near infinity" mean? Can (2-27) have solutions near infinity?

2-12. Construct an argument to show that continuation paths for (2-24) cannot diverge except as $t \rightarrow 1$ (see Section 2-4).

EXPERIMENTS

2-4. Solve (2-19) with all coefficients equal to "1" using CONSOL7 and with the various choices of Δt suggested for Experiment 2-1. Note the run time and final solution set. How confident are you that you have found the complete solution set?

2-5. Solve (2-27) with the following coefficients, using CONSOL7, as in Experiment 2-4.

$$a = \quad 15.05 \qquad a' = -4.398$$

---

† "Discuss" is an invitation to explore and report what you find. In this case, find the solutions at infinity if you can, or at least count them. Since the number of solutions will depend on the values of a through f and a' through f', there will be cases just as in Section 2-7. Note especially the "Summary" above.

$$b = \quad\ 4.365 \qquad b' = \ 39.75$$

$$c = \ -25.45 \qquad c' = -1.039$$

$$d = \quad\ 2.007 \qquad d' = \ 11.21$$

$$e = \ -63.84 \qquad e' = -6.738$$

$$f = \quad 488.3 \qquad f' = \quad 0.0$$

$$g = -2096.0 \qquad g' = \quad 0.0$$

2-6. Using CONSOL7, solve (2-28) with the choice of coefficients given in Experiment 2-5.

CHAPTER **3**

General
Systems

3-1 INTRODUCTION

We will be less exploratory and experimental in this chapter. Basically, we know what we are after: a generic continuation for systems of polynomials with any degree and any number of variables. Implementation of the continuation in a computer code will be discussed in Chapter 4. Here we focus on two things:

1. A more precise development of "solutions at infinity"
2. A clear description of the general method and why it works

Beyond developing a generic continuation method, our goal is to eliminate divergent paths completely. We will see that this is possible if the system does not have an infinite number of solutions or an infinite number of solutions at infinity.†

3-2 GENERAL SYSTEMS

Let n denote the number of variables in the system. Our problem is to solve a system of n polynomial equations in n variables. We denote the variables

---

† Recent developments show that divergent paths are eliminated entirely without qualification, using the "projective transformation" described in section 3 (see [Morgan and Sommese, 1987]).

by $x_1, x_2, \ldots, x_n$, and the equations by $f_1, \ldots, f_n$, so that the system is

$$f_1(x_1, \ldots, x_n) = 0$$

$$f_2(x_1, \ldots, x_n) = 0$$

$$\vdots$$

$$f_n(x_1, \ldots, x_n) = 0$$

or in more compact notation, simply

$$f(x) = 0,$$

where $x = (x_1, \ldots, x_n)$ and "0" is allowed to stand double duty as a column of zeros.

In Chapter 2 we had $n = 2$ with $x_1 = x$, $x_2 = y$, $f_1(x_1, x_2) = EQ$, $f_2(x_1, x_2) = EQ'$, but that notation is too awkward to carry over to the general case.

The polynomial equations $f_1, \ldots, f_n$ are sums of terms. Thus

$$f_j(x_1, \ldots, x_n) = TERM_1 + TERM_2 + \cdots + TERM_{l_j},$$

where l_j is the number of terms in f_j. Each term is a product of a constant and of variables raised to powers. Thus the general form of a term is

$$TERM = ax_1^{m_1} x_2^{m_2} \cdots x_n^{m_n},$$

where a is complex constant, possibly real, possibly zero, and $m_1, \ldots, m_n$ are nonnegative integers. As in Chapter 2, if $m_k = 0$, then $x_k^{m_k} \equiv 1$, even if $x_k = 0$. The *coefficient* of the term is a. The *degree* of the term is $m_1 + \cdots + m_n$, the sum of the degrees of each variable. The *degree* of an equation is defined to be the biggest of the degree of its terms. The *total degree* of the system is the product of the degrees of the individual equations.

Sometimes, the degree of f might seem ambiguous. For example, if we are considering a class of equations of the form

$$f(x_1) = ax_1^2 - 5x_1 + 6, \tag{3-1}$$

then apparently the degree of f is 2. But what if $a = 0$? In this situation we use the convention of *declared degree*. Given the form of (3-1), we take the declared degree to be 2. For computations, it is more efficient to declare degree minimally, but the theoretical results hold even if "declared degree" is greater than "actual degree."

We denote the degree of f_j by d_j, sometimes writing $d_j = \deg(f_j)$. Then the total degree is denoted d, so that

$$d = d_1 \cdot d_2 \cdot \cdots \cdot d_n.$$

For illustration let's consider a general system of three second-degree equations in three unknowns. Thus $n = 3$, $d_1 = 2$, $d_2 = 2$, $d_3 = 2$, and the total degree is 8. Here is what the system looks like:

$$f_1(x_1, x_2, x_3) = a_1x_1^2 + a_2x_2^2 + a_3x_3^2 + a_4x_1x_2 + a_5x_1x_3$$
$$+ a_6x_2x_3 + a_7x_1 + a_8x_2 + a_9x_3 + a_{10}$$
$$f_2(x_1, x_2, x_3) = b_1x_1^2 + b_2x_2^2 + b_3x_3^2 + b_4x_1x_2 + b_5x_1x_3$$
$$+ b_6x_2x_3 + b_7x_1 + b_8x_2 + b_9x_3 + b_{10} \qquad (3\text{-}2)$$
$$f_3(x_1, x_2, x_3) = c_1x_1^2 + c_2x_2^2 + c_3x_3^2 + c_4x_1x_2 + c_5x_1x_3$$
$$+ c_6x_2x_3 + c_7x_1 + c_8x_2 + c_9x_3 + c_{10},$$

where the a_j, b_j, and c_j are the coefficients. Each equation has 10 terms. If we take

$$a_1 = a_2 = a_3 = 1$$
$$a_4 = a_5 = a_6 = a_7 = a_8 = a_9 = 0$$
$$a_{10} = -1,$$

we get

$$f_1(x_1, x_2, x_3) = x_1^2 + x_2^2 + x_3^2 - 1 = 0.$$

whose solutions give the unit sphere in 3-space. If we transpose the sphere one unit in the x_1 direction, we have the associated equation

$$(x_1 - 1)^2 + x_2^2 + x_3^2 - 1 = 0.$$

Letting this be f_2, we can put it in the form of (3-2) by multiplying out the $(x_1 - 1)^2$. This yields

$$f_2(x_1, x_2, x_3) = x_1^2 + x_2^2 + x_3^2 - 2x_1 = 0,$$

so $b_1 = b_2 = b_3 = 1$, $b_7 = -2$, and the other $b_j = 0$. Taking f_3 to be a cylinder of unit radius along the x_3-axis yields

$$f_3(x_1, x_2, x_3) = x_1^2 + x_2^2 - 1 = 0,$$

with, therefore, $c_1 = c_2 = 1$, $c_{10} = -1$, and the other $c_j = 0$. Figure 3-1 shows the two spheres and cylinder whose common intersection points correspond to the solution of our system.

I have chosen the system to make it easy to solve by hand. The intersection of the two spheres is the circle

$$x_2^2 + x_3^2 - \left(\frac{\sqrt{3}}{2}\right)^2 = 0$$

with $x_1 = \frac{1}{2}$, as shown in Fig. 3-1. You can see this as follows: $f_1 = 0$ implies

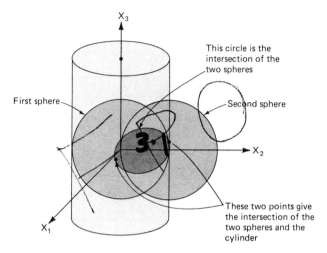

Figure 3-1. Graph for system (3-3). The intersection of the two spheres alone is a circle.

that

$$x_1^2 + x_2^2 + x_3^2 = 1.$$

Substituting this in $f_2 = 0$ yields

$$1 - 2x_1 = 0,$$

implying that $x_1 = \frac{1}{2}$. Substituting $x_1 = \frac{1}{2}$ in $f_1 = 0$ gives the equation of intersection. The solutions of the system are

$$(x_1, x_2, x_3) = \left(\frac{1}{2}, \frac{\sqrt{3}}{2}, 0\right) \quad \text{and} \quad \left(\frac{1}{2}, -\frac{\sqrt{3}}{2}, 0\right).$$

We will come back to this example later. For reference, let's display and number it:

$$x_1^2 + x_2^2 + x_3^2 - 1 = 0$$
$$x_1^2 + x_2^2 + x_3^2 - 2x_1 = 0 \qquad (3\text{-}3)$$
$$x_1^2 + x_2^2 - 1 = 0.$$

The *Jacobian* of f is an $n \times n$ matrix of partial derivatives denoted by J or df. The Jacobian is defined by

$$J_{r,s}(x) \equiv \frac{\partial f_r}{\partial x_s}(x) \equiv f_{rs}(x) \qquad (3\text{-}4)$$

or, more schematically,

$$J(x) = \begin{matrix} f_1 \\ \vdots \\ f_n \end{matrix} \begin{matrix} \rightarrow \\ \\ \rightarrow \end{matrix} \begin{bmatrix} f_{1,1}(x) & f_{1,2}(x) & \cdots & f_{1,n}(x) \\ \vdots & \vdots & & \vdots \\ f_{n,1}(x) & f_{n,2}(x) & \cdots & f_{n,n}(x) \end{bmatrix}$$

$$\begin{matrix} \uparrow & & & & & \uparrow \\ \dfrac{\partial}{\partial x_1} & \cdot & \cdot & \cdot & \cdot & \dfrac{\partial}{\partial x_n} \end{matrix}$$

with "equations along the rows and partial derivatives along the columns" (see e.g., [Kaplan, 1973] or [Guillemin and Pollack, 1974]). Let x^* denote a solution to $f(x) = 0$. Then x^* is "singular" if $\det (J(x^*)) = 0$. Otherwise, it is "nonsingular."

For (3-3) we have

$$J(x) = \begin{bmatrix} 2x_1 & 2x_2 & 2x_3 \\ 2x_1 - 2 & 2x_2 & 2x_3 \\ 2x_1 & 2x_2 & 0 \end{bmatrix} \tag{3-5}$$

and thus

$$\det (J(x)) = -8x_2 x_3.$$

Therefore, both solutions given above are singular.

Singular solutions are numerically significant because they are generally more difficult to compute than nonsingular solutions (see Appendix 1). In terms of theory, singular solutions have *higher multiplicity* and complicate such results as Bezout's theorem. The concept of *multiplicity* cannot be fully clarified without using tools from algebraic geometry which are beyond the scope of this book. However, multiplicity is too important a concept to avoid. We complete this section with a brief presentation on multiplicity.

A solution to a polynomial system is called *geometrically isolated* (or simply *isolated*) if there is a ball around the solution that contains no other solution. A solution that is not geometrically isolated is singular, but an isolated solution can be singular also. For example, (2-8) has a line of solutions. Each of the solutions on the line is not geometrically isolated. However, the solutions of (2-3) are all geometrically isolated, and the specific solution $(x, y) = (0, -5)$ is singular.

Let x^* be a geometrically isolated solution to $f(x) = 0$. Perturb f by adding arbitrarily small complex numbers to each coefficient of f (including the zero coefficients) in such a way that the perturbed system has only nonsingular solutions. [For example, by the Transversality Theorem (Appendix 4), it suffices to add small random numbers to the constant terms.] The perturbed system has a number, m, of solutions arbitrarily close to x^*.

Then m is (by definition) the multiplicity of x^*. See Experiment 3-1 for a specific illustrative example.

This definition of multiplicity is valid, but some mathematical development is required to make it precise and show that it is unambiguously defined. You can read further about multiplicity in [Fischer, 1977], [Fulton, 1984], [Schafarevich, 1977], and [Van der Waerden, 1953], but this material is difficult. We have the following final result.

PROPOSITION 3-1.   A solution x^* to $f(x) = 0$ has multiplicity greater than 1 exactly when it is singular.

I will not present a proof of this. Showing that a nonsingular solution has multiplicity 1 is essentially a result from advanced calculus: say, a consequence of the Taylor expansion. The converse requires algebraic geometry.

3-3 SOLUTIONS AT INFINITY

From the point of view of clarifying the structure of the solution set of $f(x) = 0$, this section's purpose is a more precise statement of a general Bezout's theorem. From the computational point of view, this section aims to provide an efficient means of avoiding divergent continuation paths. We begin by restating some facts from Chapter 2 in the context of general systems.

The *homogeneous part of* f, denoted f^0, is the polynomial system derived from f by setting lower-degree terms equal to zero. More precisely, each equation, f_j, in the system has a degree, d_j, and f_j^0 is the sum of terms of f_j that have degree d_j. See, for example, (2-2) and (2-13) or (2-19) and (2-20).

A *solution at infinity* of polynomial system $f(x) = 0$ is a nonzero solution of $f^0(x) = 0$ whose first nonzero entry is equal to 1. Any nonzero multiple of a solution at infinity will satisfy $f^0 = 0$. (Why?) We fix the first nonzero entry equal to 1 as a convenient convention. The significance of solutions at infinity for continuation is that paths diverge to infinity exactly when the continuation system has solutions at infinity (see the theorem in Section 3-4).

We can use $f^0 = 0$ to compute solutions at infinity, but f^0 is inadequate for two reasons. First, f^0 is deficient in its derivatives; we cannot use it to compute the singularity or multiplicity of solutions to $f = 0$. Second, using continuation to compute solutions and solutions at infinity of $f = 0$ via $f^0 = 0$ requires separate computer runs to solve $f = 0$, $f^0(1, x_2, \ldots, x_n) = 0$, $f^0(0, 1, x_3, \ldots, x_n) = 0$, and so on. If we do this, we have not eliminated divergent paths, because any of these systems might have solutions at infinity.

We want to generate a single system that itself has no solutions at infinity but from which all the solutions and solutions at infinity of $f = 0$ can be calculated. It turns out this is easy to do. But first we need to consider *projective representations* of f.

A polynomial system can be expressed in a variety of "projective co-ordinate bases," just as linear systems can be expressed in a variety of linear bases. Also in analogy with linear systems, it is sometimes helpful to choose representations for polynomial systems in special bases. Let's consider some definitions and examples.

Given $f(x) = 0$ in the n variables $x_1, \ldots, x_n$, we define $\hat{f}(y) = 0$ to be a system of n equations in the $n + 1$ variables $y_1, \ldots, y_{n+1}$, where

$$\hat{f}_j(y_1, \ldots, y_{n+1}) = y_{n+1}^{d_j} f_j\left(\frac{y_1}{y_{n+1}}, \ldots, \frac{y_n}{y_{n+1}}\right) \tag{3-6}$$

with $d_j = \deg(f_j)$. We call $\hat{f}$ the *homogenization* of f. This definition of $\hat{f}$ works out to be the same as replacing x_j by y_j for $j = 1, \ldots, n$, and making every term of f_j have degree d_j by multiplying by the appropriate power of y_{n+1}. For example, if we put (2-19) into the notation of this chapter, we have

$$f_1(x_1, x_2) = a_1 x_1^5 + a_2 x_1^2 x_2 + a_3 x_2^4 + a_4 x_1 + a_5 x_2 + a_6 = 0 \tag{3-7}$$
$$f_2(x_1, x_2) = b_1 x_1^3 + b_2 x_1^3 x_2^4 + b_3 x_1 x_2^6 + b_4 x_1 + b_5 x_2 + b_6 = 0.$$

Then $\hat{f} = 0$ is

$$\hat{f}_1(y_1, y_2, y_3) = a_1 y_1^5 + a_2 y_1^2 y_2 y_3^2 + a_3 y_2^4 y_3 + a_4 y_1 y_3^4$$
$$+ a_5 y_2 y_3^4 + a_6 y_3^5 = 0 \tag{3-8}$$
$$\hat{f}_2(y_1, y_2, y_3) = b_1 y_1^3 y_3^4 + b_2 y_1^3 y_2^4 + b_3 y_1 y_2^6 + b_4 y_1 y_3^6$$
$$+ b_5 y_2 y_3^6 + b_6 y_3^7 = 0.$$

If y is a solution to $\hat{f} = 0$, and $y_{n+1} \neq 0$, then $(y_1/y_{n+1}, \ldots, y_n/y_{n+1})$ is a solution to $f = 0$. The solutions to $\hat{f} = 0$ with $y_{n+1} = 0$ generate the solutions to $f = 0$ at infinity, because $\hat{f}(x_1, x_2, \ldots, x_n, 0) = f^0(x)$.

When f is presented in general coefficients, there might be some confusion as to what $\hat{f}$ is associated with f. For example, if f is given by (3-1), then apparently

$$\hat{f}(y_1, y_2) = a y_1^2 - 5 y_1 y_2 + 6 y_2^2, \tag{3-9}$$

but what if $a = 0$? Should we take

$$\hat{f}(y_1, y_2) = 5 y_1 + 6 y_2 \tag{3-10}$$

instead? In this situation, we use the convention of *declared degree* as noted in Section 3-2. Given the form of (3-1), we take the declared degree to be 2, in which case (3-9) is appropriate rather than (3-10).

We use $\hat{f}$ to define various representations of f in projective coordinate bases. Let $L(x) = a_1x_1 + \cdots + a_nx_n + a_{n+1}$ be a linear equation where the constants a_j are complex numbers and $a_{n+1} \neq 0$. Denote by $\hat{f}^{L,j}(x) = 0$ the polynomial system of n equations in n unknowns defined by

$$\hat{f}^{L,j}(x) = \hat{f}(x_1, \ldots, x_{j-1}, L(x), x_j, \ldots, x_n).$$

We call $\hat{f}^{L,j}$ the *projective representation* of f given by L and j. The *standard projective representations* of f are defined by taking $L(x) = 1$ with any j. We denote these representations by $\hat{f}^j$, and we refer to $\hat{f}^j$ as the jth projective representation of f.

Taking $a = 1$ in (3-1) and (3-9), we have

$$\hat{f}^1(x_1) = \hat{f}(1, x_1) = 1 - 5x_1 + 6x_1^2 \tag{3-11}$$

and

$$\hat{f}^2(x_1) = \hat{f}(x_1, 1) = f(x_1) = x_1^2 - 5x_1 + 6. \tag{3-12}$$

If $L(x) = -x_1 + 1$ and $j = 2$, then

$$\hat{f}^{L,j}(x_1) = \hat{f}(x_1, -x_1 + 1) = 12x_1^2 - 17x_1 + 6. \tag{3-13}$$

In two variables, consider

$$f(x_1, x_2) = \begin{cases} x_1x_2 - 6x_2 + 11x_1 - 6 = 0 \\ x_1^2 - x_2 = 0. \end{cases} \tag{3-14}$$

Then

$$\hat{f}(y_1, y_2, y_3) = \begin{cases} y_1y_2 - 6y_2y_3 + 11y_1y_3 - 6y_3^2 \\ y_1^2 - y_2y_3 = 0 \end{cases} \tag{3-15}$$

and

$$\hat{f}^1(x_1, x_2) = \hat{f}(1, x_1, x_2) = \begin{cases} x_1 - 6x_1x_2 + 11x_2 - 6x_2^2 = 0 \\ 1 - x_1x_2 = 0 \end{cases} \tag{3-16}$$

$$\hat{f}^2(x_1, x_2) = \hat{f}(x_1, 1, x_2) = \begin{cases} x_1 - 6x_2 + 11x_1x_2 - 6x_2^2 \\ x_1^2 - x_2 = 0 \end{cases} \tag{3-17}$$

with

$$\hat{f}^3(x_1, x_2) = \hat{f}(x_1, x_2, 1) = f(x_1, x_2).$$

Letting $L(x) = x_1 + 2x_2 - 1$ and $j = 2$, we get

$$\hat{f}^{L,j}(x_1, x_2) = \hat{f}(x_1, x_1 + 2x_2 - 1, x_2)$$
$$= \begin{cases} x_1^2 - 18x_2^2 + 7x_1x_2 - x_1 + 6x_2 = 0 \\ x_1^2 - 2x_2^2 - x_1x_2 + x_2 = 0. \end{cases} \tag{3-18}$$

Each projective representation of f is, in essence, an expression of f

in a new basis. (Note that $\hat{f}^{n+1} = f$.) Given f, there are many choices of representations, $\hat{f}^{L,j}$, that are essentially equivalent to f; that is, the continuation process behaves qualitatively the same in solving $f = 0$ and in solving $\hat{f}^{L,j} = 0$. But not always. In particular, we can find a representation $\hat{f}^{L,j} = 0$ that does not have any solutions at infinity, whether $f = 0$ has solutions at infinity or not (under certain hypotheses). This result (stated as Proposition 3-7) is what we are aiming for. But first we need to clarify some basic facts about projective representations.

We will be comparing projective representations, so we adopt the following notational conventions. Let $\hat{f}^{L,j}$ and $\hat{f}^{M,k}$ denote projective representations with

$$l(x) = \sum_{i=1}^{n} a_i x_i \quad \text{and} \quad L(x) = l(x) + a_{n+1}$$

$$m(x) = \sum_{i=1}^{n} b_i x_i \quad \text{and} \quad M(x) = m(x) + b_{n+1},$$

where a_i and b_i are complex numbers for $i = 1$ to $n + 1$, $a_{n+1} \neq 0$, $b_{n+1} \neq 0$, and j and k are integers with $1 \leq j \leq n$, $1 \leq k \leq n$. Usually, z will denote a solution to $\hat{f}^{L,j} = 0$ and w a solution to $\hat{f}^{M,k} = 0$. Then the point $y \in C^{n+1}$ is defined by

$$y = (z_1, \ldots, z_{j-1}, L(z), z_j, \ldots, z_n)$$

and $y_{[k]} \in C^n$ is defined by

$$y_{[k]} = (y_1, \ldots, y_{k-1}, y_{k+1}, \ldots, y_{n+1}).$$

Observe that $\hat{f}^{L,j} = 0$ can have solutions at infinity as well as solutions. In fact, solutions (with first nonzero entry equal to 1) of $\hat{f}^{l,j} = 0$ give the solutions at infinity of $\hat{f}^{L,j} = 0$, where $\hat{f}^{l,j}$ is defined analogously to $\hat{f}^{L,j}$, using $l(x)$ instead of $L(x)$. For example, any standard representation, $\hat{f}^{j}$, is $\hat{f}^{L,j}$ with $L(x) = 1$, and therefore $l(x) = 0$. Thus

$$\hat{f}^{l,j}(x) = \hat{f}(x_1, \ldots, x_{j-1}, 0, x_j, \ldots, x_n).$$

But this is exactly $\hat{f}^{j}$ with lower-degree terms set equal to zero, that is, $(\hat{f}^{j})^0$, the homogeneous part of $\hat{f}^{j}$. Check this for systems (3-15) through (3-18). [Note for (3-18) that $l(x) = x_1 + 2x_2$.]

Let z and w be solutions to $\hat{f}^{L,j} = 0$ and $\hat{f}^{M,k} = 0$ as above. Then we say that "z corresponds to w" if $(z_1, \ldots, z_{j-1}, L(z), z_j, \ldots, z_n)$ is a scalar multiple of $(w_1, \ldots, w_{k-1}, M(w), w_k, \ldots, w_n)$. If, on the other hand, z is a solution at infinity of $\hat{f}^{L,j} = 0$ and w is a solution of $\hat{f}^{M,k} = 0$, then "z corresponds to w" if $(z_1, \ldots, z_{j-1}, l(z), z_j, \ldots, z_n)$ is a scalar multiple of $(w_1, \ldots, w_{k-1}, M(w), w_k, \ldots, w_n)$. Correspondence between solutions at infinity is defined similarly.

For example, the solutions of (3-11) are $\frac{1}{2}$ and $\frac{1}{3}$, and the solutions of (3-12) are 2 and 3. Thus 2 corresponds to $\frac{1}{2}$ and 3 corresponds to $\frac{1}{3}$. The solutions of (3-13) are $\frac{2}{3}$ and $\frac{3}{4}$, corresponding to solutions 2 and 3 of (3-12), respectively. We see that solving $f = \hat{f}^2 = 0$ yields a solution set equivalent to those obtained by solving $\hat{f}^1 = 0$ and $\hat{f}^{L,j} = 0$ with $L(x) = -x_1 + 1$ and $j = 2$.

But the example given by systems (3-14) through (3-18) is different. System $\hat{f}^1 = 0$ has solutions $(1, 1), (2, \frac{1}{2}), (3, \frac{1}{3}), \hat{f}^2 = 0$ has solutions $(1, 1),$ $(\frac{1}{2}, \frac{1}{4}), (\frac{1}{3}, \frac{1}{9}), (0, 0)$, and $\hat{f}^3 = 0$ has solutions $(1,1), (2, 4), (3, 9)$, respectively. The solutions correspond in the order listed, except that $(0, 0)$ has no corresponding solution in the other systems. In fact, $(0, 0)$ corresponds to a solution at infinity of the other two systems. In particular, $(0, 0)$ corresponds to the solution $(0, 1)$ to $f^0 = 0$, that is, to a solution at infinity of $f = 0$. The $\hat{f}^{L,j} = 0$ given by (3-18) has solutions $(\frac{1}{2}, \frac{1}{4}), (-\frac{3}{4}, -\frac{1}{4})$, and $(0, 0)$. These correspond to solutions $(1, 1), (\frac{1}{3}, \frac{1}{9})$, and $(0, 0)$ of $\hat{f}^2 = 0$, respectively.

We ought to note that "corresponds" is an equivalence relation. That is, if z corresponds to w and w corresponds to x, then z corresponds to x (see Exercise 3-3).

We view corresponding solutions as being equivalent. A solution, z, of one projective representation of f can be converted into a corresponding solution of another if a certain condition holds. Otherwise, z corresponds to a solution at infinity. Consider the following.

PROPOSITION 3-2.   Let z be a solution or a solution at infinity of $\hat{f}^{L,j}$ $= 0$. Let $\hat{f}^{M,k}$ be given. Then $\hat{f}^{M,k} = 0$ has a solution, w, corresponding to z exactly when $y_k - m(y_{[k]}) \neq 0$, where we take

$$y = (z_1, \ldots, z_{j-1}, L(z), z_j, \ldots, z_n)$$

if z is a solution of $\hat{f}^{L,j} = 0$, and we take

$$y = (z_1, \ldots, z_{j-1}, l(z), z_j, \ldots, z_n)$$

if z is a solution at infinity of $\hat{f}^{L,j} = 0$. In fact, we have $w = cy_{[k]}$ where $c = b_{n+1}/[y_k - m(y_{[k]})]$.

On the other hand, $y_k - m(y_{[k]}) = 0$ exactly when $\hat{f}^{M,k} = 0$ has a solution at infinity, w, corresponding to z.

COROLLARY 1.   If z is a solution to a projective representation, $\hat{f}^{L,j}$ $= 0$, then z has a corresponding solution, w, to the standard representation $\hat{f}^k = 0$ exactly when

$$y = (z_1, \ldots, z_{j-1}, L(z), z_j, \ldots, z_n)$$

has its kth entry nonzero, and in this case $w = (1/y_k)y_{[k]}$.

COROLLARY 2.   If z is a solution to $\hat{f}^{L,n+1} = 0$, then $f = 0$ has a corresponding solution, w, exactly when $L(z) \neq 0$. In this case, $w = [1/L(z)]z$.

Proof of Proposition 3-2.   Suppose that z is a solution or solution at infinity of $\hat{f}^{L,j} = 0$ and $y_k - m(y_{[k]}) \neq 0$. If w corresponds to z, we have

$$(w_1, \ldots, w_{k-1}, M(w), w_k, \ldots, w_n) = cy$$

for some complex number c. Thus

$$cy_k = M(w) = m(w) + b_{n+1}$$

$$= \sum_{i=1}^{n} b_i w_i + b_{n+1}$$

$$= \sum_{i=1}^{k-1} b_i(cy_i) + \sum_{i=k}^{n} b_i(cy_{i+1}) + b_{n+1}$$

$$= c(m(y_{[k]})) + b_{n+1},$$

implying that

$$c(y_k - m(y_{[k]})) = b_{n+1},$$

and the result follows. The case when $y_k - m(y_{[k]}) = 0$ is proven by a similar argument.

PROPOSITION 3-3.   Let $\hat{f}^{L,j}$ be a projective representation of f and assume that $\hat{f}^{L,j} = 0$ has no solutions at infinity. Then each solution and each solution at infinity of $f = 0$ corresponds to a solution of $\hat{f}^{L,j} = 0$.

Proof. This follows easily from Proposition 3-2.

In Proposition 3-7 we show how to find projective representations of $f = 0$ that have no solutions at infinity.

PROPOSITION 3-4.   Let z be a solution or solution at infinity of some $f^{L,j} = 0$. Then z corresponds to a solution of some standard representation $f^k = 0$.

Proof. Whether z is a solution or solution at infinity, y is defined in the statement of Proposition 3-2. But y_k cannot equal zero for all k. The result follows from Corollary 1 to Proposition 3-2.

COROLLARY.   Let z be a solution at infinity of $f = 0$. Suppose that k is the index of the first nonzero entry of z, where we take $z_k = 1$, as usual.

Then z corresponds to the solution

$$w = (0, \ldots, 0, z_{k+1}, \ldots, z_n, 0)$$

of $\hat{f}^k = 0$.

We are now in a position to define the singularity of solutions at infinity. A solution at infinity, z, of $f = 0$ is *singular* if z corresponds to a singular solution, w, of some projective representation $\hat{f}^{L,j} = 0$. The following proposition shows that "singular solution at infinity" is well defined.

PROPOSITION 3-5.   Let z be a solution to $\hat{f}^{L,j} = 0$ and w a corresponding solution to $\hat{f}^{M,k} = 0$ for some L, M, j, k as usual. If z is a singular (nonsingular) solution of $\hat{f}^{L,j} = 0$, then w is a singular (nonsingular) solution of $\hat{f}^{M,k} = 0$, respectively.

The proof is given at the end of this section.

We also have a result on multiplicity. First, let's define isolated solutions at infinity. A solution at infinity, z, of $f = 0$ is *geometrically isolated* if z corresponds to a geometrically isolated solution, w, of some projective representation $\hat{f}^{L,j} = 0$.

The following proposition justifies the definition and shows that multiplicity of solutions at infinity can be defined in terms of projective representations.

PROPOSITION 3-6.   Let z be a geometrically isolated solution to $\hat{f}^{L,j} = 0$ and w a corresponding solution to $\hat{f}^{M,k} = 0$ for some L, M, j, k. Then w is geometrically isolated and the multiplicity of w is the same as that of z.

Sketch of Proof. It is easy to see that a sequence of solutions to $\hat{f}^{M,k} = 0$ converging to w yields a sequence of solutions to $\hat{f}^{L,j} = 0$ converging to z. Thus w is geometrically isolated. If we perturb the coefficients of $\hat{f}^{L,j} = 0$ by a small random perturbation, then a singular solution, z, resolves into a collection of nonsingular solutions near z, whose number gives the multiplicity of z. The corresponding perturbation of $\hat{f}^{M,k} = 0$ has the corresponding nonsingular solutions.

Recall that $d = d_1 \cdots d_n$, the total degree of f. Now we can state Bezout's theorem.

THEOREM OF BEZOUT (VERSION 3).   Let $f = 0$ be a polynomial system. Then:
1. The total number of geometrically isolated solutions and solutions at infinity of $f = 0$ is no more than d.

2. If $f = 0$ has neither an infinite number of solutions nor an infinite number of solutions at infinity, then it has exactly d solutions and solutions at infinity, counting multiplicities.

By "counting multiplicities" is meant that we add the multiplicities of each of the solutions and solutions at infinity instead of adding the number of solutions and solutions at infinity. If all the solutions and solutions at infinity are nonsingular, then counting multiplicities is the same as counting the solutions and solutions at infinity.

Consider system (3-3), which we will call $f = 0$. We discovered in Section 3-2 that this system has two solutions, $(\frac{1}{2}, \pm\sqrt{3}/2, 0)$, and both are singular. Let's compute the solutions at infinity. The homogeneous part of (3-3), $f^0 = 0$, is

$$x_1^2 + x_2^2 + x_3^2 = 0$$

$$x_1^2 + x_2^2 + x_3^2 = 0 \qquad\qquad (3\text{-}19)$$

$$x_1^2 + x_2^2 = 0,$$

which has solutions $(1, \pm i, 0)$. There are, then, two solutions at infinity. If these are singular solutions, each of the solutions and solutions at infinity of (3-3) has multiplicity 2, and we have a full "Bezout accounting" of the solution set. But how do we confirm that both solutions at infinity are singular? We find a projective representation with solutions corresponding to the two solutions at infinity and check for singularity of these corresponding solutions. By the corollary to Propositon 3-4, the standard representation $\hat{f}^1 = 0$ will work. This system is

$$1 + x_1^2 + x_2^2 - x_3^2 = 0$$

$$1 + x_1^2 + x_2^2 - 2x_3 = 0 \qquad\qquad (3\text{-}20)$$

$$1 + x_1^2 - x_3^2 = 0,$$

which has solutions $(\pm\sqrt{3}, 0, 2)$ and $(\pm i, 0, 0)$. The first pair of solutions corresponds to the solutions of $f = 0$, and the second pair to the solutions at infinity of $f = 0$. Let's check their singularity. The Jacobian of (3-20) is

$$J(x) = \begin{bmatrix} 2x_1 & 2x_2 & -2x_3 \\ 2x_1 & 2x_2 & -2 \\ 2x_1 & 0 & -2x_3 \end{bmatrix}$$

and we see that $\det(J(\pm i, 0, 0)) = 0$. This confirms that the solutions at infinity of $f = 0$ are singular. Note that $\hat{f}^1 = 0$ has no solutions at infinity.

Before stating Proposition 3-7, I would like to clarify the concepts of "independence" and "independent choice of parameters" used in Chapters

1 and 2. One way to make these concepts more precise is via the idea of "sets of measure zero."

Let A be a subset of V^n, where V denotes either R or C. We want a way of saying that "A is small in V^n," to clarify the idea that if you choose a point x at random from V^n, you almost surely would not discover that you had picked an x in A. Consider that if A were a finite set, it would be small in this sense. Unfortunately, we cannot restrict our "bad sets" to be finite. However, we can restrict them to being "measure zero."

A subset A of V^n has measure zero if for every $\epsilon > 0$, there exists a collection of boxes in V^n, $\{S_i : i = 1, 2, 3, \ldots\}$, such that

$$A \subset \bigcup_{i=1}^{\infty} S_i \quad \text{and} \quad \sum_{i=1}^{\infty} \text{vol}\,(S_i) < \epsilon,$$

where vol (S_i) is the volume of box S_i. Thus "A has measure 0" means something like "A has no volume." So straight lines and (smoothly embedded) curves in the plane have measure zero, but disks and rectangular areas do not. Proper hyperplanes in V^n have measure zero, that is, subsets of V^n of the form

$$A = \left\{ x \in V^n : \sum_{i=1}^{n} a_i x_i + a_{n+1} = 0 \right\}$$

as long as some $a_i \neq 0$. See [Guillemin and Pollack, 1974, p. 39] for more detail (see also Appendix 4).

In Chapters 1 and 2 "independence" is presented as a choosing game. If the coefficients of a test system of $k + 1$ equations in k unknowns are chosen at random (or "independently"), what is the chance that the system will have a solution? The answer, we have said, is "none." We can be more precise now. The set of coefficients for which such a system has a solution turns out to be a set of measure zero in the space from which coefficients are chosen ("parameter space"). This is good news, because the test system is not the system we want to solve. It is, rather, a system we need not to have solutions, so that our method will work (see Appendix 4).

If the bad parameters are all in a set of measure zero, we say that "almost all" parameters will work, where the space from which parameters are chosen must be understood or specified.

Two warnings:

1. The set of numbers represented in any computer is finite. The fact that bad parameters are contained in a set of measure zero does not automatically mean that we will not encounter them. Tests must be made. As always, theory is the beginning, not the end of what we want to know.

2. If our system of equations is generated by a physical application, we cannot choose its coefficients. They are given by nature. Measure zero or "almost all" conditions are useful only when they concern parameters that we can choose.

The following proposition is the main result of this section.

PROPOSITION 3-7. Assume that $f = 0$ has neither an infinite number of solutions nor an infinite number of solutions at infinity. Then, for almost all $a \in C^n$ and almost all $a \in R^n$, if j is an integer between 1 and n and $a_{n+1} \in C$, $a_{n+1} \neq 0$, then $\hat{f}^{L,j} = 0$ has no solutions at infinity, where $a = (a_1, \ldots, a_n)$ and

$$L(x) = \sum_{i=1}^{n} a_i x_i + a_{n+1}.$$

This proposition tells us that if we choose $a_1, \ldots, a_n$ at random and $a_{n+1} \neq 0$, we can solve the associated projective representation without encountering diverging continuation paths (unless $f = 0$ has an infinite number of solutions or solutions at infinity). Note that we can choose the a_i to be real numbers, if convenient.

We can be more precise about which parameters, a, are to be avoided. Define $A \subset C^n$ as follows. For $a \in C^n$ we have

$$l(x) = \sum_{i=1}^{n} a_i x_i,$$

as usual. If w is a solution to $f = 0$, define

$$A_w = \{a \in C^n : l(w) = 1\}.$$

If w is a solution at infinity of $f = 0$, define

$$A_w = \{a \in C^n : l(w) = 0\}.$$

Then define $A = \cup A_w$, where the union is over all solutions and solutions at infinity of $f = 0$. Now Proposition 3-7 holds without the finiteness of solutions and solutions at infinity hypothesis if we replace the "for almost all" statements by "for all $a \notin A$." With the finiteness hypothesis, A has measure zero in C^n and $A \cap R^n$ has measure zero in R^n, because each A_w is a (proper) hyperplane. However, A may not have measure zero in general.

Proof of Proposition 3-7. Let A be the set defined in the paragraph above. Let z be a solution at infinity of $\hat{f}^{L,j} = 0$. Without loss of generality, take $j = n + 1$. Then z has first nonzero entry equal to 1 and $\hat{f}(z, l(z)) = 0$. We shall see that $a \in A$, and the proof will be established.

If $l(z) = 0$, then z is a solution at infinity of $f = 0$. Then $l(z) = 0$ implies that $a \in A_z \subset A$.

If $l(z) \neq 0$, then $z/l(z)$ is a solution, w, of $f = 0$. We may as well assume that $z_1 = 1$. Then $w_1 = 1/l(z)$ and $w_i = z_i/l(z) = z_i w_1$ for $i = 2$ to n. Then, substituting $z_i = w_i/w_1$ into $l(z)w_1 = 1$ yields $l(w) = 1$. So $a \in A_w \subset A$, and we are done.

I sometimes refer to $\hat{f}^{L,j} = 0$ with an independently chosen $a \in C^n$ as "the projective transformation of $f = 0$," although (of course) it is far from uniquely defined.

Let's summarize the computational significance of the projective transformation. We can solve $\hat{f}^{L,j} = 0$ using continuation and know that no paths will diverge unless $f = 0$ has an infinite number of solutions or solutions at infinity. We can then recover the solutions of $f = 0$ from those of $\hat{f}^{L,j} = 0$. (We can also recover the solutions at infinity.) This follows from Proposition 3-3. Solving $\hat{f}^{L,j} = 0$ does not increase the dimensionality of the solution process, since $\hat{f}^{L,j} = 0$ has n variables (the same as f) and $\deg(\hat{f}_i^{L,j}) = \deg(f_i)$ for $i = 1$ to n.

I always substitute $L(x)$ into $\hat{f}$ numerically (rather than symbolically). Consider that $\hat{f}^{L,j}$ is the composition of two functions:

$$\hat{f}^{L,j} = \hat{f} \circ v^{L,j} \colon C^n \to C^n,$$

where

$$\hat{f} \colon C^{n+1} \to C^n$$

is defined by (3-6) and

$$v^{L,j} \colon C^n \to C^{n+1}$$

is defined by

$$v^{L,j}(x) = (x_1, \ldots, x_{j-1}, L(x), x_j, \ldots, x_n)$$

with $L(x)$ given by $a \in C^n$ and $a_{n+1} \neq 0$, as usual. Then $d\hat{f}^{L,j}(x)$, the Jacobian of $\hat{f}^{L,j}(x)$, can be obtained by the chain rule as

$$d\hat{f}^{L,j}(x) = d\hat{f}(y) \circ dv^{L,j}(x),$$

where $y = v^{L,j}(x)$,

$$d\hat{f}(y) = \begin{matrix} \hat{f}_1 \to \\ \vdots \\ \hat{f}_n \to \end{matrix} \begin{bmatrix} \hat{f}_{1,1}(y) & \hat{f}_{1,2}(y) & \cdots & \hat{f}_{1,n}(y) & \hat{f}_{1,n+1}(y) \\ \vdots & \vdots & & \vdots & \vdots \\ \hat{f}_{n,1}(y) & \hat{f}_{n,2}(y) & \cdots & \hat{f}_{n,n}(y) & \hat{f}_{n,n+1}(y) \end{bmatrix}$$

$$\begin{matrix} \uparrow & \uparrow & & \uparrow & \uparrow \\ \dfrac{\partial}{\partial y_1} & \dfrac{\partial}{\partial y_2} & \cdots & \dfrac{\partial}{\partial y_n} & \dfrac{\partial}{\partial y_{n+1}} \end{matrix}$$

and

$$
dv^{L,j}(x) =
\begin{matrix} v_1^{L,j} \rightarrow \\ \vdots \\ v_j^{L,j} \rightarrow \\ \vdots \\ v_{n+1}^{L,j} \rightarrow \end{matrix}
\left[
\begin{array}{ccc|c|ccc}
 & & & & & & \\
 & I_{j-1} & & 0 & & 0 & \\
 & & & & & & \\
\hline
a_1 & \cdots & \cdot & a_j & \cdots & \cdot & a_n \\
\hline
 & & & & & & \\
 & 0 & & 0 & & I_{n-j+1} & \\
 & & & & & & \\
\end{array}
\right],
$$

$$
\begin{matrix} \uparrow & & & & & \uparrow \\ \dfrac{\partial}{\partial x_1} & \cdots & \cdots & \cdots & & \dfrac{\partial}{\partial x_n} \end{matrix}
$$

where the I_{j-1} and I_{n-j+1} denote square identity matrices of the indicated sizes, and the 0's indicate blocks of zeros.

If $\hat{f}^{L,j} = \hat{f}^j$ is a standard representation, the jth row of $dv^{L,j}$ is identically zero. Thus $d\hat{f}^j(x)$ is $d\hat{f}(y)$ with the jth column omitted and with $y_j = 1$, where $(y_1, \ldots, y_{j-1}, y_{j+1}, \ldots, y_{n+1})$ is then identified with x. This fact is useful in that it allows us to "read off" the Jacobians of the standard representations from $d\hat{f}(y)$. We sometimes use the notation $\hat{f}^j$ for $d\hat{f}^j$. The subroutine FFUNT, referenced in Chapter 5 and given in Appendix 6, Part A, Section 4, evokes a projective transformation using the formulas above.

The projective transformation is the simplest and most effective way I know of to eliminate divergent paths. For an alternative approach, however, see [Wright, 1985].

You may be concerned about the case that $f = 0$ has an infinite number of solutions at infinity. This does come up in some application areas (see Chapter 10). I know of no general technique that eliminates all difficulty with this case. Sometimes, solving $\hat{f}^{L,j} = 0$ is numerically desirable, even when Proposition 3-7 does not apply, and sometimes not. This is discussed in Chapter 4. The effect of the projective transformation as a scaling transformation is a separate relevant issue (see Chapter 5).

Let's complete this section with a proof of Proposition 3-5. The argument is broken into three main parts, as Lemmas 1 through 3 below. Then the proof is given. You may, if you would like, skip to Section 3-4 at this point; we will not be using any of the following material later.

Let $L: C^n \rightarrow C$ and $v^{L,j}: C^n \rightarrow C^{n+1}$ be defined as above. Define $s_k: C_k^{n+1} \rightarrow C^n$ by

$$
s_k(y_1, \ldots, y_{n+1}) = \frac{1}{y_k}(y_1, \ldots, y_{k-1}, y_{k+1}, \ldots, y_n),
$$

where $C_k^{n+1} \equiv \{y \in C^{n+1}: y_k \neq 0\}$.

LEMMA 1.   Let $\alpha = \hat{f}^{L,j}$ and $\beta = \hat{f}^k \circ s_k \circ v^{L,j}$. Suppose that z is a solution of $\alpha = 0$, and assume that $L(z) \neq 0$ if $j = k$ or $z_k \neq 0$ if $j \neq k$. Then z is a solution of $\beta = 0$, and $d\beta(z) = \delta\, d\alpha(z)$, where δ is a nonsingular diagonal matrix.

Proof. It suffices to take $k = 1$.

Case 1.   Suppose that $j = k$. Then

$$\alpha(x) = \hat{f}^{L,1}(x) = \hat{f}(L(x), x)$$

and

$$\beta(x) = \hat{f}^{L,1} \circ s_k \circ v^{L,1}(x) = \hat{f}\left(\frac{1}{L(x)}(L(x), x)\right).$$

Thus

$$\beta_r(x) = \left[\frac{1}{L(x)}\right]^{d_r} \alpha_r(x) \tag{3-21}$$

for $r = 1$ to n, where $d_r = \deg(f_r)$. Therefore,

$$\frac{\partial \beta_r}{\partial x_s}(z) = \left[\frac{1}{L(z)}\right]^{d_r} \frac{\partial \alpha_r}{\partial x_s}(z) \tag{3-22}$$

for $r = 1$ to n. It follows from (3-21) that z is a solution to $\beta = 0$, and (3-22) implies that $d\beta(z) = \delta\, d\alpha(z)$, where $\delta_r = [1/L(z)]^{d_r}$.

Case 2.   Suppose that $j \neq k$. It suffices to take $j = 2$. Then

$$\alpha(x) = \hat{f}^{L,2}(x) = \hat{f}(x_1, L(x), x_2, \ldots, x_n)$$

and

$$\beta(x) = \hat{f}^{L,1} \circ s_1 \circ v^{L,2}(x) = \hat{f}\left(\frac{1}{x_1}(x_1, L(x), x_2, \ldots, x_n)\right).$$

Thus

$$\beta_r(x) = \left(\frac{1}{x_1}\right)^{d_r} \alpha_r(x)$$

for $r = 1$ to n, and the result follows as in case 1. This completes the proof of Lemma 1.

LEMMA 2.   Let $\gamma = s_k \circ v^{L,j}$ for some k, j, L as above. Let $z \in C^n$, and assume that $L(z) \neq 0$ if $j = k$ or $z_k \neq 0$ if $j \neq k$. Then $d\gamma(z)$ is defined and nonsingular.

Proof.

Case 1.   Suppose that $j = k$. Then

$$\gamma(z) = s_k(v^{L,j}(z)) = \frac{1}{L(z)} z,$$

and thus

$$d\gamma(z) = \frac{-1}{L(z)^2} \begin{bmatrix} a_1 z_1 - L(z) & a_2 z_1 & \cdots & a_n z_1 \\ a_1 z_2 & a_2 z_2 - L(z) & \cdots & a_n z_2 \\ \vdots & \vdots & & \vdots \\ a_1 z_n & a_2 z_n & \cdots & a_n z_n - L(z) \end{bmatrix}.$$

We want to show that det $(d\gamma(z)) \neq 0$. By [Cullen, 1966, Chap. 3] it suffices to show that a matrix derived from $d\gamma(z)$ by elementary row and column operations has a nonzero determinant.

First, assume that $a_r \neq 0$ and $z_r \neq 0$ for $r = 1$ to n. Then $d\gamma(z)$ is equivalent to the following matrix by dividing row i by z_i and column j by a_j:

$$\begin{bmatrix} 1 - \dfrac{L(z)}{a_1 z_1} & 1 & \cdots & 1 \\ 1 & 1 - \dfrac{L(z)}{a_2 z_2} & \cdots & 1 \\ 1 & 1 & \cdots & 1 \\ \vdots & \vdots & & \vdots \\ 1 & 1 & \cdots & 1 - \dfrac{L(z)}{a_n z_n} \end{bmatrix}.$$

Now, subtracting the last row from the other rows yields

$$\begin{bmatrix} \dfrac{-L(z)}{a_1 z_1} & 0 & \cdots & \dfrac{L(z)}{a_n z_n} \\ 0 & \dfrac{-L(z)}{a_2 z_2} & \cdots & \dfrac{L(z)}{a_n z_n} \\ 0 & 0 & \cdots & \dfrac{L(z)}{a_n z_n} \\ \vdots & \vdots & & \vdots \\ 1 & 1 & \cdots & 1 - \dfrac{L(z)}{a_n z_n} \end{bmatrix}.$$

Eliminating the 1's in the last row, we get

$$
\begin{bmatrix}
\dfrac{-L(z)}{a_1 z_1} & 0 & \cdots & 0 & \dfrac{L(z)}{a_n z_n} \\[2ex]
0 & \dfrac{-L(z)}{a_2 z_2} & \cdots & 0 & \dfrac{L(z)}{a_n z_n} \\[2ex]
0 & 0 & \cdots & 0 & \dfrac{L(z)}{a_n z_n} \\[1ex]
\vdots & \vdots & \vdots & \vdots & \vdots \\[1ex]
0 & 0 & \cdots & \dfrac{-L(z)}{a_{n-1} z_{n-1}} & \dfrac{L(z)}{a_n z_n} \\[2ex]
0 & 0 & \cdots & 0 & \dfrac{-a_{n+1}}{a_n z_n}
\end{bmatrix} ,
$$

which is nonsingular because $a_{n+1} \neq 0$.

If some of the a_i or z_j are zero, we can interchange rows and columns to get a matrix that is made up of two blocks, one of which is (nonsingular) diagonal and the other is as above. Case 1 follows.

Case 2.   Suppose that $j \neq k$. This case is similar to case 1 (see Exercise 3-10).

LEMMA 3.   Suppose that z is a solution to $\hat{f}^{L,j} = 0$ and w is the corresponding solution to $\hat{f}^k = 0$ for some j, k, and L. If z is a singular (nonsingular) solution of $\hat{f}^{L,j} = 0$, then w is a singular (nonsingular) solution of $\hat{f}^k = 0$, respectively.

Proof. Using Lemma 1, $\hat{f}^{L,j}(z) = 0$ implies that $\hat{f}^k \circ s_k \circ v^{L,j}(z) = 0$, and if $d\hat{f}^{L,j}(z)$ is singular (nonsingular), then $d(\hat{f}^k \circ s_k \circ v^{L,j})(z)$ is singular (nonsingular), respectively. But

$$
d(\hat{f}^k \circ s_k \circ v^{L,j})(z) = d\hat{f}^k(s_k(v^{L,j}(z))) \circ d(s_k \circ v^{L,j})(z)
$$

by the chain rule, and $d(s_k \circ v^{L,j})(z)$ is nonsingular by Lemma 2. Thus if $d\hat{f}^{L,j}(z) = 0$ is singular (nonsingular), then $d\hat{f}^k(s_k(v^{L,j}(z)))$ is singular (nonsingular), respectively. Now $(s_k \circ v^{L,j})(z)$ corresponds to z and therefore must equal w.

Proof of Proposition 3-5. The solution z corresponds to a solution x of some standard representation $\hat{f}^k = 0$ by Proposition 3-4. Applying Lemma 3 twice, we are done.

3-4 CONTINUATION

Our general continuation system will generalize (2-24). We continue to denote the system to be solved as $f(x) = 0$, although we may be solving the projective transformation, as described in Section 3-3. The basic continuation is the same.

Let's denote the initial system by $g(x) = 0$, where

$$g_j(x) = p_j^{d_j} x_j^{d_j} - q_j^{d_j} \tag{3-23}$$

for $j = 1, \ldots , n$, $d_j = \deg (f_j)$ and p_j, q_j are complex constants to be chosen "independently," as usual. Then $g(x) = 0$ has d solutions, x^0, given by

$$x^0 = \left(u(d_1, k_1) \frac{q_1}{p_1} , u(d_2, k_2) \frac{q_2}{p_2} , \ldots , u(d_n, k_n) \frac{q_n}{p_n} \right) \tag{3-24}$$

for $1 \le k_1 \le d_1, 1 \le k_2 \le d_2, \ldots , 1 \le k_n \le d_n$, where u gives the roots of unity as defined in (2-26). Then the continuation, denoted $h(x, t)$, is defined by

$$h_j(x, t) = (1 - t)g_j (x) + tf_j(x) \tag{3-25}$$

for $j = 1, \ldots , n$. This continuation generates d paths, beginning when $t = 0$ at each of the solutions to $g = 0$. We implement continuation by tracking each path, as in Chapters 1 and 2. In the next chapter, specific implementation schemes are presented in detail. We will refer to the continuation method implementing (3-25) as "CONSOL," with the understanding that many different implementations are possible. Let $p = (p_1, \ldots , p_n)$ and $q = (q_1, \ldots , q_n)$.

THEOREM.   Given f, there are sets of measure zero, A_p and A_q, in C^n so that if $p \notin A_p$ and $q \notin A_q$, then
   a. The solution set $\{(x, t) \in C^n \times [0, 1): h(x, t) = 0\}$ is a collection of d nonoverlapping (smooth) paths.
   b. The paths move from $t = 0$ to $t = 1$ without backtracking in t.
   c. Each geometrically isolated solution of $f = 0$ of multiplicity m has exactly m continuation paths converging to it.
   d. A continuation path can diverge to infinity only as $t \to 1$.
   e. If $f = 0$ has no solutions at infinity, all the paths remain bounded. If $f = 0$ has a solution at infinity, at least one path will diverge to infinity as $t \to 1$. Each geometrically isolated solution at infinity of $f = 0$ of multiplicity m will generate exactly m diverging continuation paths.

Sketch of Proof. a. Choosing $q \notin A_q$ implies that $dh(z)$ is nonsingular whenever $h(z) = 0$ for $\leq t < 1$. It follows that the solutions to $h = 0$ form smooth nonoverlapping paths (see Appendix 4).

b. We can show that $dh(z)$ will have a particular 2×2 block structure, which implies that we may orient paths so that $dt/ds \geq 0$; that is, the partial derivative of t with respect to arc length is nonnegative (see Appendix 3).

c. This follows by applying the definition of multiplicity given in Section 3-2. Let ϵ be a small positive number. The system $h(x, 1 - \epsilon) = 0$ is a perturbation of $h(x, 1) = 0$ with only nonsingular solutions. It follows that exactly m paths must converge to a geometrically isolated solution x^* of multiplicity m. This argument depends, however, on the consistency of the definition of multiplicity, which cannot be established outside the context of algebraic geometry. We can prove some special cases without algebraic geometry. If x^* is nonsingular, we can show that a single path converges to x^* using the Preimage Theorem from Appendix 4. If x^* is singular, we can see that at least one path converges to x^* by arguing that a system of n equations in $n + 1$ unknowns cannot have an isolated solution in complex space. (The technicalities of this, however, seem to require results from several complex variables or results about topological degree. See [Fischer, 1977] or [Ortega and Rheinboldt, 1970], respectively.)

d. If a continuation path diverges to infinity as $t \rightarrow t^0$, then $h(x, t^0) = 0$ has a solution at infinity. The choice of $p \notin A_p$ implies that $h(x, t^0) = 0$ cannot have solutions at infinity for $0 \leq t^0 < 1$ (see Appendix 4).

e. Since $h(x, 1) = f(x)$, we can have diverging paths only if $f = 0$ has solutions at infinity. A geometrically isolated solution at infinity of $f = 0$ cannot be isolated from the solutions of $h = 0$. It must have an associated path. To see this, transform f to $\hat{f}^{L,j} = 0$ (with random L) so that the geometrically isolated solution at infinity is now a solution to $\hat{f}^{L,j} = 0$ and apply the same projective transformation to h. The argument then reduces to part (c) above.

A more complete proof of parts (a) and (d) is given in Appendix 4, and of part (b) in Appendix 3. We consider the issue of how to choose p and q for computations in Chapter 4.

It is important that the solutions to $h(x, t) = 0$ form a set of paths starting at solutions of $h(x, 0) = 0$ and moving across to solutions of $h(x, 1) = 0$ or diverging to infinity as $t \rightarrow 1$. Because of the way we have defined h, this is exactly what happens. If we had not defined h properly, what "bad behavior" might result? First, we might not have paths at all. The schematics in Figs. 3-2 through 3-5 show various forms of "nonpath" behavior.

Figure 3-2 shows a curve "bifurcating" (splitting into two parts). In this book, by definition, at any point on a path, there is one simple curve only; "Y" shapes are not allowed.

Figure 3-3 shows even wilder nonpath behavior. The curve goes into

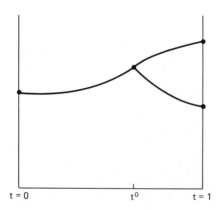

Figure 3-2. Schematic illustrating "nonpath" behavior: one path splitting into two when $t = t^0$.

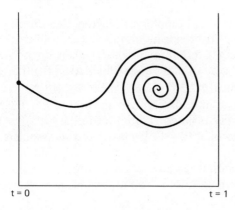

Figure 3-3. Schematic illustrating "nonpath" behavior: a path without one endpoint, spiraling endlessly.

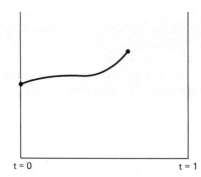

Figure 3-4. Schematic illustrating "nonpath" behavior: a path with endpoint in the middle. It should have its endpoints only when $t = 0$ or $t = 1$.

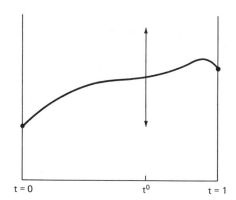

Figure 3-5. Schematic illustrating "nonpath" behavior: the path explodes into a hypersurface at $t = t^0$.

a never-ending spiral. This is not allowed in our definition of path, either. We insist that the path either have two well-defined endpoints, which can occur only when $t = 0$ or $t = 1$, or that it diverge to infinity. Thus Fig. 3-4, with an endpoint in the middle, is ruled out also.

Figure 3-5 shows a continuation "path" that is not even a curve. At $t = t^0$ the curve suddenly becomes an entire hypersurface. (Recall that the x-axis of a schematic stands for all of C^n. Because we rule out such behavior, we are willing to use a graphical representation that does not show it very accurately.)

Now let us suppose that $h(x, t) = 0$ does consist of paths. There are basically three types of behavior that we want to avoid, shown as schematics in Figs. 3-6 through 3-9. In Figs. 3-6 and 3-7, t is not increasing along the path. This in itself we could handle by parameterizing our paths by arc length instead of by t (see Chapter 6 and Appendix 3). But the behavior shown in Fig. 3-7 is a disaster, since the path backtracks all the way to $t = 0$, "neutralizing" two solutions to $g(x) = 0$, messing up the solution count and the method. We avoid backtracking by working in complex rather than real numbers. Consider the continuation:

$$h(x, t) = (1 - t)(x^2 - 1) + t(x^2 + 1).$$

In real arithmetic, it produces a curve like Fig. 3-7. But

$$h(x, t) = (1 - t)(p^2 x^2 - q^2) + t(x^2 + 1),$$

where p and q are (nonreal) complex numbers, generates two well-defined smooth (complex) paths that go from $\pm q/p$ to $\pm i$.

The schematics in Figs. 3-8 and 3-9 show paths that diverge, or more accurately, that pass through points at infinity. The following continuation yields a path similar to the one shown in Fig. 3-9:

$$h(x, t) = (1 - t)(x - 2) + t(-x + 1).$$

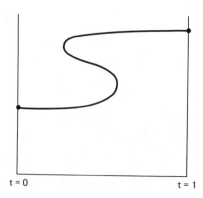

t = 0   t = 1

Figure 3-6. Schematic illustrating "bad path" behavior: backtracking in t.

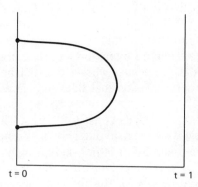

t = 0   t = 1

Figure 3-7. Schematic illustrating "bad path" behavior: backtracking in t.

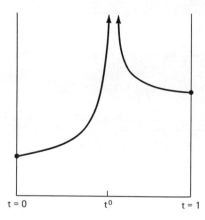

t = 0 t^0 t = 1

Figure 3-8. Schematic illustrating "bad path" behavior: diverging to infinity as $t \rightarrow t^0$.

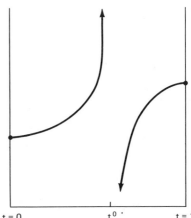

Figure 3-9. Schematic illustrating "bad path" behavior: diverging to infinity as t $\rightarrow t^0$.

If we modify this to

$$h(x, t) = (1 - t)(x - 1) + t(-x + 1),$$

we get something like Fig. 3-5. If we use

$$h(x, t) = (1 - t)(px - q) + t(-x + 1),$$

where p and q are independently chosen complex numbers, we get a path from q/p to 1.

Consider the system $f = 0$ defined by (3-3). We generated the associated $\hat{f}^1 = 0$ as system (3-20) and used this to compute the singularity of the solutions at infinity of $f = 0$. As we observed in Section 3-3, $\hat{f}^1 = 0$ has no solutions at infinity, so it can be used as "the projective transformation" for $f = 0$. Using the theorem above and what we know about the solutions and solutions at infinity of $f = 0$ and $\hat{f}^1 = 0$, we can sketch schematics suggesting the behavior of CONSOL in solving both of these systems. These schematics are given in Figs. 3-10 and 3-11.

Up until now we have defined "continuation" as the following process:

1. Define $h(x, t) = (1 - t)g(x) + tf(x)$, where we can solve $g(x) = 0$ easily and we want to solve $f(x) = 0$.

2. Given a solution x^0 to $g(x) = 0$, solve $h(x, \Delta t) = 0$ using Newton's method with x^0 as a start point. Then solve $h(x, 2\Delta t) = 0$ with the solution to $h(x, \Delta t) = 0$ as a start point, yielding a new solution, and so on until $t = 1$.

3. Repeat step 2 for all solutions to $g(x) = 0$.

I would like to suggest that we develop a more global view of what is going on in the continuation process above. This will give us some added flexibility in Chapter 4, when we consider implementation. The cost of de-

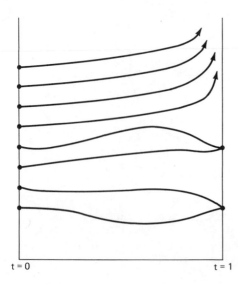

Figure 3-10. Schematic illustrating
CONSOL behavior for system (3-3).

t = 0   t = 1

veloping this global view is that we must encounter some advanced calculus that up until now did not have to be explicit.

Since we have been generating the continuation paths by incrementing t, it is natural to begin by supposing that we have a path parameterized as $x(t)$, where t goes from 0 to 1 and $| x(t) |$ may become unbounded as $t \to 1$.

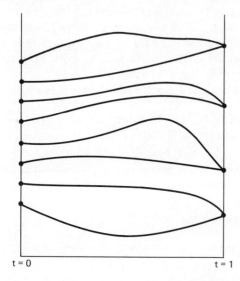

Figure 3-11. Schematic illustrating
CONSOL behavior for system (3-20).

t = 0   t = 1

Then

$$h(x(t), t) = 0 \qquad \text{for } 0 \leq t,$$

since the paths, by definition, are solutions to $h(x, t) = 0$.

What we want to do is find a "formula" for dx/dt, or rather, n formulas for $dx_1/dt, dx_2/dt, \ldots, dx_n/dt$. Knowing dx/dt will tell us precisely how x should change with t. We can "increment t and solve for x" directly from dx/dt, rather than use $h(x, t) = 0$ explicitly. Although we may choose to track continuation paths exactly as before, we will now have alternatives. In fact, having dx/dt defines $x(t)$ as a solution to a system of ordinary differential equations, with $x = x^0$ as an initial condition. So any numerical method for solving ordinary differential equations may be applied to dx/dt to generate the continuation paths. Since there are many such methods, we will have many options.

For what follows it will be useful to establish some notation. Let $I = [0, 1) = \{t : 0 \leq t < 1\}$, $w = (x, t)$, $w(t) = (x(t), t)$, and $h(t) = h(w(t))$. Thus $w_1 = x_1$, $w_2 = x_2, \ldots, w_n = x_n$, $w_{n+1} = t$, and $h(w) = h(x, t)$. Now $h(t) = h(w(t)) = 0$ for $t \in I$ because $x(t)$ is by definition a solution path. Because $h(t)$ is constant, we get

$$\frac{dh}{dt}(t) = 0 \qquad \text{for } t \in I. \tag{3-26}$$

Applying the n-variable chain rule to (3-26) yields

$$dh(w(t)) \frac{dw}{dt} = 0, \tag{3-27}$$

where dh denotes the $n \times (n + 1)$ Jacobian matrix of h with respect to w_1, $\ldots, w_{n+1}$:

$$dh_{r,s} = \frac{\partial h_r}{\partial w_s}(w), \qquad r = 1, \ldots, n; \quad s = 1, \ldots, n + 1.$$

Note that we are doing "polynomial calculus" here. That is, $dh_{r,s}$ is obtained from h by the usual naive interpretation of polynomial derivatives. However, since $x_1, \ldots, x_n$ are complex variables and t is real, this may seem confusing. To be on more solid ground, we can declare $x_1, \ldots, x_n$ to be pairs of real variables, so the calculus is all real-variable calculus. Thus dh would be a $2n \times (2n + 1)$ real matrix. This "more proper" calculus yields formulas equivalent to those obtained by naive differentiation. See Appendix 3 and Exercise 3-7.

Thus (3-27) is the product of an $n \times (n + 1)$ complex matrix $dh(w(t))$ and an $n + 1$ column vector dw/dt set equal to a column of 0's:

$$
\begin{bmatrix}
dh_{1,1} & dh_{1,2} & \cdots & dh_{1,n} & dh_{1,n+1} \\
\vdots & & & \vdots & \vdots \\
dh_{n,1} & dh_{n,2} & \cdots & dh_{n,n} & dh_{n,n+1}
\end{bmatrix}
\begin{bmatrix}
\dfrac{dw_1}{dt} \\[4pt]
\dfrac{dw_2}{dt} \\[4pt]
\vdots \\[4pt]
\dfrac{dw_n}{dt} \\[4pt]
\dfrac{dw_{n+1}}{dt}
\end{bmatrix}
=
\begin{bmatrix}
0 \\
0 \\
\vdots \\
0 \\
0
\end{bmatrix} . \qquad (3\text{-}28)
$$

Let's label the first $n \times n$ block of dh as dh_x and the last column of dh as dh_t. Then, observing that $dw_j/dt = dx_j/dt$ for $j = 1, \ldots, n$ and $dw_{n+1}/dt = dt/dt = 1$, (3-28) can be written in block form

$$
[dh_x \mid dh_t]
\begin{bmatrix}
dx/dt \\
1
\end{bmatrix}
= [0],
$$

which can be rewritten

$$
dh_x \cdot \frac{dx}{dt} + dh_t = 0, \qquad (3\text{-}29)
$$

so our "formula" for dx/dt is

$$
\frac{dx}{dt} = -[dh_x]^{-1} dh_t, \qquad (3\text{-}30)
$$

where the right-hand side of (3-30) has $x_1, \ldots, x_n, t$ as independent variables. Thus (3-30) has the form

$$
\frac{dx}{dt} = P(x, t),
$$

where $P(x, t) = -[dh_x]^{-1} dh_t$, which identifies dx/dt as an ordinary differential equation, as promised. The $n \times n$ matrix dh_x is invertible by the independent choice of the coefficients p_j and q_j of g. (See the footnote to the proof of Lemma A in Appendix 4. See also [Morgan, 1983].) In practice, we will solve (3-29) rather than generate (3-30) (see Section 4-5-3).

EXERCISES

3-1. For the following systems of two second-degree equations, find all solutions and solutions at infinity. Determine whether these are singular. Reconcile your results with Bezout's theorem and draw schematics showing the expected CONSOL behavior. Find a projective representation that has no solutions at infinity for each system, if possible.

(a) $x_1^2 - 2x_1x_2 + x_2^2 + 4 = 0$
    $x_1^2 - 3x_1x_2 + 2x_2^2 \quad = 0$
(b) $100x_1^2 - 200x_1 - 100x_2 + 100 = 0$
    $-x_1^2 + x_1x_2 + 2x_1 - 1 = 0$
(c) $x_1x_2 - x_2 - 2x_1 + 2 = 0$
    $x_1x_2 - 3x_1 - x_2 + 3 = 0$
(d) $x_1^2 - x_1x_2 - 3x_1 + 3x_2 = 0$
    $-x_2^2 + x_1x_2 - 2x_1 + 2x_2 = 0$
(e) $x_1x_2 - x_2^2 - 3x_1 + 3x_2 = 0$
    $x_1x_2 - 3x_1 - x_2 + 3 = 0$
(f) $x_1^2 + x_2^2 - 4 = 0$
    $x_1^2 + x_2^2 - 4 = 0$
(g) $x_1x_2 = 0$
    $x_1x_2 = 0$
(h) $x_1^2 - 21x_1 + 110 = 0$
    $x_1x_2 - x_1 - 10x_2 + 10 = 0$

3-2. For the following systems of equations, find all solutions and solutions at infinity. Determine whether these are singular. Reconcile your results with Bezout's theorem and draw schematics showing the expected CONSOL behavior. Find a projective representation that has no solutions at infinity for each system. Part (c) is discussed in [Brown and Gearhart, 1971], part (d) is cited in [Chien, 1979] and [Chao et al., 1975], and part (e) is "Brown's Almost Linear System," as given in [Moré and Cosnard, 1979].

(a) $x_1^4 - 2x_1^2x_2^2 + x_2^4 + 4 = 0$
    $x_1^4 - 3x_1^2x_2^2 + 2x_2^4 \quad = 0$
(b) $100x_1^4 - 200x_1^2 - 100x_2^2 + 100 = 0$
    $-x_1^4 + x_1^2x_2^2 + 2x_1^2 - 1 = 0$
(c) $4x_1^3 - 3x_1 - x_2 = 0$
    $x_1^2 - x_2 = 0$
(d) $x_1 + x_2 + x_3 + x_4 - 1 = 0$
    $x_1 + x_2 - x_3 + x_4 - 3 = 0$
    $x_1^2 + x_2^2 + x_3^2 + x_4^2 - 4 = 0$
    $x_1^2 + x_2^2 + x_3^2 + x_4^2 - 2x_1 - 3 = 0$

(e) $2x_1 + x_2 + x_3 + x_4 + x_5 - 6 = 0$
    $x_1 + 2x_2 + x_3 + x_4 + x_5 - 6 = 0$
    $x_1 + x_2 + 2x_3 + x_4 + x_5 - 6 = 0$
    $x_1 + x_2 + x_3 + 2x_4 + x_5 - 6 = 0$
    $x_1 \cdot x_2 \cdot x_3 \cdot x_4 \cdot x_5 - 1 = 0$

3-3. Let z, w, and x be solutions or solutions at infinity of three projective transformations of f. Show that if z corresponds to w and w corresponds to x, then z corresponds to x.

3-4. Find specific examples that illustrate the two corollaries to Proposition 3-2.

3-5. Are the solutions at infinity of (2-19) singular? Discuss their multiplicity.

3-6. Consider the continuation

$$h(x, t) = (1 - t)g(x) + t(-x^2 + 1),$$

where $x \in C$, with three choices for the form of g:
(a) $g(x) = px - q$
(b) $g(x) = p^2x^2 - q^2$
(c) $g(x) = p^3x^3 - q^3$
Determine, for each of parts (a), (b), and (c), which values of p, q, x, and t
yield singular solutions of $h = 0$. In each case, most choices of p and q are
OK (even real numbers). Assuming that no solution of $h = 0$ is singular, draw
schematics for each case. [*Hint:* Parts (a) and (b) have two paths, and (c) has
3.] Can you find simple parametric formulas for the paths in each case? (See
[Morgan, 1983].)

3-7. Let

$$h(x, t) = (1 - t)(p^2x^2 - q^2) + t(x^2 + 1),$$

where $(x, t) \in C \times I$. Thus $dh(x, t)$ is a 1×2 complex matrix, as in (3-28),
and dx/dt indicates a direction in complex space. Now substitute $a + bi$ for
x. Consider the system of two equations, "real part of h" and "imaginary part
of h," in the variables a, b, and t. Compute the associated 2×3 Jacobian,
$dh(a, b. t)$. Relate the resulting $(da/dt, db/dt)$ to dx/dt. Find a "formula" for
converting $dh(x, t)$ into $dh(a, b, t)$. (Compare Appendix 3.)

3-8. Complete the proof of Proposition 3-2.

3-9. Show that if z is a solution to $f = 0$, then $\hat{f}^{M,n+1} = 0$ has a corresponding solution, w, exactly when $m(z) \neq 1$, and in this case, $w = cz$ where $c =$
$b_{n+1}/(1 - m(z))$.

3-10. Complete the proof of Lemma 2 by providing an argument for case 2. (*Hint:*
It suffices to let $k = 1$ and $j = 2$.)

EXPERIMENTS

3-1. System (2-3) has a singular solution, $(x, y) = (0, -5)$, of multiplicity 2. By
the definition of multiplicity, we should be able to resolve the multiplicity of
this solution by making small perturbations of the coefficients and solving the
perturbed system. Do this using CONSOL6. Proceed as follows. Add ϵ_a to a,
ϵ'_a to a', ϵ_b to b, ϵ'_b to b', and so on, where the ϵ's are as given below. Take
$k = 4, 8, 16, 32$, and 64. (You might want to modify a copy of CONSOL6 to
generate the perturbed coefficients in the code and save having to do a lot of
arithmetic by hand.) Look at the resulting solution sets. You should see a pair
of nonsingular solutions getting closer and closer to $(0, -5)$ as k gets larger.

$$\epsilon_a = \quad 0.1378 \times 10^{-k} \qquad\qquad \epsilon'_a = \; - \; 0.2974 \times 10^{-k}$$

$$\epsilon_b = -0.8922 \times 10^{-k} \qquad\qquad \epsilon'_b = \quad 0.5233 \times 10^{-k}$$

$$\epsilon_c = \quad 0.1123 \times 10^{-k} \qquad\qquad \epsilon'_c = \quad 0.0581 \times 10^{-k}$$

$$\epsilon_d = -0.8345 \times 10^{-k} \qquad\qquad \epsilon'_d = \; - \; 0.7410 \times 10^{-k}$$

$$\epsilon_e = -0.6603 \times 10^{-k} \qquad \epsilon'_e = 0.3646 \times 10^{-k}$$

$$\epsilon_f = 0.7352 \times 10^{-k} \qquad \epsilon'_f = 0.0431 \times 10^{-k}$$

Compare with the results of Experiment 2-1 for system (2-3). Does solving a system with only nonsingular solutions take less computational work than a comparable system with some singular solutions? What happens if you perturb only some coefficients? That is, if you let most of the ϵ equal 0, does the singularly still resolve? Will perturbing just one coefficient work?

3-2. Solve (3-14) using the projective transformation and CONSOL6. Proceed as follows. Solve $\hat{f}^{L,3} = 0$ where $L = a_1 x_1 + a_2 x_2 + a_3$ and

$$a_1 = 0.1378\ 2974$$

$$a_2 = -0.8922\ 5233$$

$$a_3 = 0.1123\ 0581$$

Then generate the solutions, w, to $f = 0$ from the solutions, z, to $\hat{f}^{L,3} = 0$ by the transformation $z = w/L(z)$ (see Corollary 2 to Proposition 3-2). You have the option of generating the coefficients of $\hat{f}^{L,3}$ directly, by substitution, or numerically, as outlined in Section 3-3 (see Project 3-1).

3-3. Consider (2-19) with all coefficients equal to 1. Solve with CONSOL7 using $\hat{f}^{L,3} = 0$, where L is defined in Experiment 3-2. Compare with Experiment 2-4.

PROJECT

3-1. Modify a copy of the CONSOL6 code so that some projective transformation is evoked "automatically" with a fixed L and j embedded in the code. Thus the user of the code still merely inputs the coefficients of the system to be solved and gets as output the solutions to that system. (Compare Experiments 3-2 and 3-3. See the CONSOL8T code discussed in Chapter 5, especially subroutine FFUNT.)

NOTES

1. For References, see the following note and also Appendix 4, Note 1.

2. The reader familiar with complex projective space, P^n, as a complex analytic manifold will recognize the projective representations of f as restrictions of $\hat{f}$ to various coordinate patches of P^n. The fact that restricting to a randomly chosen patch is sufficient for computations was observed in [Morgan, 1986–2] and confirmed by much computational work (see, e.g., [Watson et al., 1986]). In planning this book, I decided that evoking P^n without developing the concomitant algebraic geometry was to gain little of substance while losing much in the way of language. But see Appendix 4, Note 2.

CHAPTER 4

Implementation

4-1 INTRODUCTION

Once I read a futuristic science fiction story about a society in which resources were so limited the people had to eke out primitive preindustrial lives, even with full descriptions in books of science and technology. The title of the story was ironic, something like "It's the Idea that Counts." The writer of the story was pointing out that there is a difference between knowing and doing. This chapter is about turning the ideas from Chapter 3 into computer codes for solving polynomial systems. We must now bring resources into play that are only suggested by the material in Chapter 3. Our success or failure in setting up codes will depend in an essential way on decisions made without benefit of mathematical theory. Put bluntly, our implementation can be consistent with theory and still not work.

If you are more interested in developing engineering or scientific models than in developing continuation codes, you might want to skip this chapter (and the next two) and go to Part II of the book. I believe that some familiarity with the topics discussed in this chapter will aid your use of polynomial-system solvers, be they continuation or whatever. (For example, the significance of singular solutions and the concept of the "condition" of a matrix are sometimes important.) However, the main focus here will be on software engineering and algorithm design rather than on modeling issues.

It will be helpful to have an outline of the topics we cover:

1. Overview
   - An "elementary" CONSOL: CONSOL-E
   - An "advanced" CONSOL: CONSOL-A

2. Environmental issues
   - What is the anticipated problem environment? Which is more important, speed or reliability? (For example, if the code messes up, will the user just make some adjustments and rerun, or will an entire system crash?)
   - How friendly should the code be?
   - What is the anticipated computing environment? What language features are important? How could alternative architectures be exploited?
   - When should CONSOL be used and when should another method be used?

3. Preprocessing
   - What approaches can be taken to reduce and scale the original system of equations?
   - Would it be worthwhile to generate some solutions or solutions at infinity using another method? How would these be used by CONSOL?

4. Path tracking
   - What methods can be used to follow the continuation paths?
   - Is t just another variable?
   - What methods can be used for the linear algebra?
   - What special consideration should be given to singular solutions?
   - How closely should paths be tracked? What can be done about path crossing?
   - When should a path be declared "converged"? When "diverged"? Should solutions be refined at the end of a continuation run? How do we decide that a path was probably not tracked correctly? Will we rerun paths whose validity is in doubt? What parameters should be changed for a rerun?
   - How do we choose the "independent" coefficients?
   - What sort of "side information" can the method exploit? Examples of such information include a partial solution set computed by another method and the number of solutions at infinity.

5. Performance evaluation
   - How do we judge the success or failure of a particular run?
   - What do we do if a particular run is judged a failure?

6. The experimental laboratory

The "issues" noted above are not independent. For example, creating a more friendly code is likely to be a more complex programming task. The ideal failure-recovery strategy will depend on how friendly the code is intended to be. The development of a method for following the continuation paths will depend on the speed at which the code must run. This is an aspect of the problem environment and the computing environment. All of the "close enough" and "big enough" parameters will be set depending on the trade-off between speed and reliability, which is a function of many of the issues listed above.

Although in Chapter 6 we consider continuation paths that backtrack in t, here we explicitly assume that this cannot occur (see Appendix 3).

4-2 CONSOL-E AND CONSOL-A

Before getting into "issues," let's consider an outline of what a CONSOL code might look like. Recall the following notation from Chapter 3. Let $f(x) = 0$ denote the system to be solved, $g(x) = 0$ the initial system, and $h(x, t) = 0$ the continuation system. I will sometimes refer to $f(x) = 0$ as the "original system."

The solutions to $g(x) = 0$ are the initial x values for the paths, given by formula (3-24). CONSOL will track the paths beginning at these solutions, proceeding in "steps," each of which attempts to increase t toward 1. The simplest way to do this is to begin at $x = x^0$ and $t = 0$, increase t by Δt, solve $h(x, \Delta t) = 0$ to update x, then solve $h(x, 2 \cdot \Delta t) = 0$ to further update x, and so on. Figure 4-1 outlines the overall logic of this "no frills" CONSOL algorithm (called elementary CONSOL or CONSOL-E). The box "Take a Step Along Path" includes evaluating $h(x, t)$ and $dh(x, t)$, incrementing t to $t + \Delta t$, and solving $h(x, t + \Delta t) = 0$ using Newton's method. The codes CONSOL3 through CONSOL7 are variants of CONSOL-E.

An "advanced CONSOL" (CONSOL-A) is outlined in Fig. 4-2. CONSOL-A has PREPROCESSING and POSTPROCESSING, a variable-step-size path tracker, a rerun facility, solution refinement, and a stopping rule that can cut short a run. These features are discussed more fully below. The continuation system is subsumed in the box "Take a Step Along Path." The user must provide a subroutine, FFUN, to generate equation values and partial derivatives for the original system. CONSOL-A allows FFUN to be coded so that its coefficients are read as input, so FFUN itself need not to be modified every time coefficient changes are made. [In Chapter 5 we consider a version of FFUN (labeled "FFUNT") that uses a "tableau" of coefficients and parameters. This FFUN can be used to solve any polynomial system.] There are several valid approaches to step taking. I am focusing on predictor–corrector methods, as discussed in Section 4-5-1.

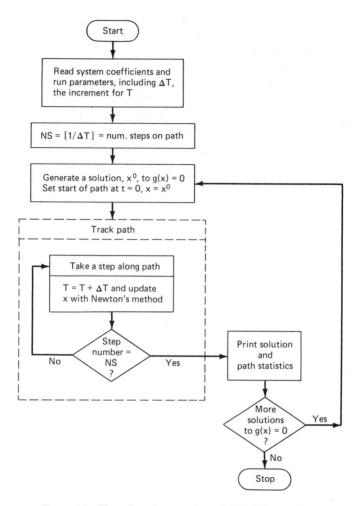

Figure 4-1 Flow chart for overview of CONSOL-E logic.

Preprocessing could include:

1. Automatic system reduction.
2. Automatic scaling.
3. Check for degeneracy (e.g., all coefficients = 0).
4. Evocation of another solution method (say, a fast heuristic†) to compute a partial solution list.

---

† A "heuristic" is a method that cannot be proven to work and that generally fails some of the time, as opposed to an "algorithm," which is proven always to work in exact arithmetic.

5. Evocation of a method to count solutions at infinity.

Postprocessing could include:

1. Transforming solutions of the projective transformation to solutions of the original system.
2. Identifying the physical solutions. If the system has been reduced or scaled, transforming solutions to the units of the original problem.
3. Evaluating the solution list and generating warning and error messages based on solution count, singularity, nonexistence of physical solutions.
4. Deciding, based on the total run, that certain paths should be rerun; setting up the rerun.

Rerunning paths is discussed below. Statistics for a path might include:

1. Diverged or converged? (Based on given criteria.)
2. CPU time.
3. Arc length.
4. Number of steps. Number of successful steps. Number of unsuccessful steps.
5. Number of Newton iterations to track the path. Number of Newton iterations to refine solutions at the end of the path.
6. Number of calls to FFUN, that is, the number of function and Jacobian evaluations (equivalent to the number of linear systems solved in CONSOL).
7. Solution singular? (Based on given criteria.) More detail could be included, such as the condition of the Jacobian at the solution or the singular values of the Jacobian at the solution (see Section 4-5-4).
8. Solution real? (Based on given criteria.)
9. Path tracked correctly? (Based on given criteria.)

Solutions can be refined by evoking Newton's method or some other local solution method, using the endpoint of the continuation path as the start point for the local method. Stopping rules are discussed in Section 4-5-6, and the use of a solution list to save computation time in Section 4-5-8. CONSOL-A, as given in Fig. 4-2, does not, of course, include all desirable features but is presented merely to suggest how CONSOL-E can be extended. Naturally, there are numerous variants. The program that accompanies this chapter, CONSOL8, is an intermediate version, which allows the user to code preprocessing, postprocessing, and FFUN, but omits the rerunning of paths. It will be the fundamental code that we will use for the

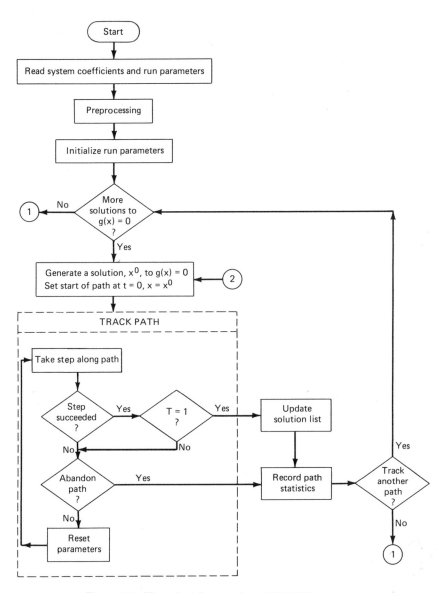

Figure 4-2   Flow chart for overview of CONSOL-A logic.

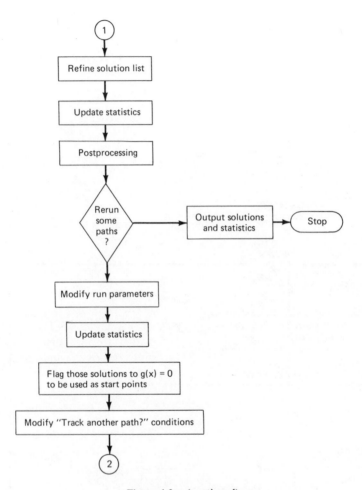

Figure 4-2. (continued)

later chapters. (CONSOL8 uses the TRACK-T predictor, described in Section 4-5-1.)

Within the TRACK PATH part of CONSOL-A, the logic allows for "Abandon Path" conditions. Combinations of various conditions can be used: path too long, endpoint of path too large, maximum number of steps exceeded, minimum step size exceeded. See "Path Tracking" in Section 4-5-1.

In various parts of the CONSOL algorithm we must test if certain numbers are zero or are near zero. Usually, "$z = 0$" will translate in practice into "$|z| \leq \epsilon$," where ϵ is a "zero-epsilon," a small positive number. The effectiveness of zero-epsilons depends on scaling. An inconsistent or clumsy scaling scheme (e.g., none) can make the test unreliable. Ultimately, the only way to feel confident in zero-epsilon choices is by exhaustive testing on a problem set that includes the specific extremes that are relevant to the problem environment. Note that if x_1 and x_2 are solutions, then "$x_1 = x_2$" becomes $| x_1 - x_2 | \leq \epsilon$, and the choice of this ϵ should depend on whether x_1 and x_2 are singular. This is because we can distinguish between close x_1 and x_2 better if they are nonsingular (see Section 4-5-4 and Appendix 1). The use of relative errors can be helpful in some cases, that is, checking $| x_1 - x_2 | / | x_1 | \leq \epsilon$ instead of $| x_1 - x_2 | \leq \epsilon$.

4-3 ENVIRONMENTAL ISSUES

There are basically two types of problem environments: *supervised* and *solitary*. By "supervised" I mean that the code is run, the results are looked over by someone who understands the problem and the method, and the code is rerun if this person judges that better performance is possible and desirable. The code need not be especially "friendly" because the user knows what the code does and can make allowances and adjustments. Error recovery is mostly in the person, not in the code. *You* know enough to run the CONSOL codes this way, having already read Chapters 1 through 3 and worked through the exercises and experiments. In a supervised environment, speed of programming might be emphasized over run speed or reliability in the structure of CONSOL and in the way the basic parameters are set.

A solitary environment is entirely different, because now CONSOL is relied on to produce correct results. Either the user is not familiar with the method or the "user" is a larger computer system that will feed problems to CONSOL and then make use of the results. If CONSOL fails, it must recognize that it has failed, and try to recover. If it cannot recover, it must tell the larger system that it has failed. In designing a version of CONSOL for a solitary environment, great pains are taken to ensure reliability. Using the insights about "solutions at infinity" to get accurate solution counts becomes an important part of CONSOL's logic. Singular solutions are a major source of concern. Making problem-area-specific adaptations of the code for singular solutions is frequently the only way to achieve reasonable reliability. Using preprocessing algorithms to improve scaling and eliminate solutions at infinity becomes worth the effort. In developing a version of CONSOL for a solitary environment, we expend a great deal of energy deciding how closely to track paths the first time, and if path tracking fails, when to track them again more closely and how closely the second time. Chapter 8 discusses developing some automatic system-reduction and solution-at-infinity-counting ideas for systems that arise in computer graphics and geometric modeling. In geometric modelers, the mathematical methods are generally transparent to the user and must work without human supervision. Thus such modelers provide a good example of solitary problem environments.

One major decision, which relates to reliability, is whether to use the projective transformation of the original system (see Section 3-3). It is generally a good idea to use the projective transformation, except when the system has a large number of solutions at infinity, especially if they are singular. In this case, it may be better to avoid the projective transformation and truncate diverging paths. Neither alternative is ideal, because either way we will not be able to generate a reliable "Bezout count" of solutions at a reasonable computational cost.

Naturally, we are limited by the specific computing environments in which we operate. For CONSOL, one of the most convenient features the computing environment can offer is a language with complex arithmetic. At least this makes coding CONSOL easier. Another convenient feature is the availability of good utility programs to do linear algebra or solve ordinary differential equations (see Sections 4-5-1 and 4-5-3).

CONSOL has a natural parallel structure, because the paths are tracked independently. This suggests implementation on computers whose architecture allows for parallel computations. I expect in the near future that parallel implementation will allow a significant improvement in the speed of CONSOL, making medium-sized problems solvable and making certain real-time applications feasible (see Chapter 10).

Here is how to decide whether to use CONSOL or another method. CONSOL is a good first try for small polynomial systems, especially if speed is not an issue. CONSOL computes d continuation paths, where d is the total degree of the system. For systems with d large, you have to decide how painful (or possible) it is to wait. In situations in which only one solution is wanted and a good initial guess is available, Newton's method is usually a winner (see Appendix 1). (There are also many other "local methods," designed to find a single solution and that require an initial solution guess. See [Ortega and Rheinboldt, 1970].)

Newton's method often works even when you cannot get a good initial guess, but then it is less reliable. The principal complaint that I have heard from scientists and engineers is that Newton's method finds a solution but it is the wrong solution! One variant of CONSOL tries a less reliable but fast method on the problem first, and then evokes continuation if the appropriate physical solution is not found. Using CONSOL as a backup to other methods can be a part of CONSOL-A, via "PREPROCESSING."

Aside from local methods, the main alternatives to CONSOL are generalized bisection [Moore and Jones, 1977] and [Kearfott, 1978, 1979] and algebraic techniques based on the method of resultants [Van der Waerden, 1953]. Generalized bisection is not currently an especially developed or tested computational technique. It is an approach that may benefit from parallel implementation. The method of resultants is a systematic way to evoke something like "substitution of variables" for polynomial systems. It is hard to design a numerically reliable general algorithm based on this approach, except for very small systems (say, two quadrics in two unknowns) or for systems in special forms. However, successfully implemented algebraic methods are fast, require no initial guesses, and find all solutions. Thus they may be worth the trouble to develop when the speed of the method justifies a long development time and the system type is appropriate (see Chapter 8).

4-4 PREPROCESSING

The topics of scaling and reduction are so important that they have their own Chapters: 5 and 7, respectively. Actually, much of Part II illustrates reduction. The only aspect that I will note here is that CONSOL-E assumes that the original system has been satisfactorily reduced and scaled, whereas CONSOL-A allows for some reduction and scaling to be done by the code.

Finding solutions at infinity is generally easier than solving the original system, because to find solutions at infinity, all lower-degree terms of the original system are set to 0, and one or more variables are set to 1 or 0. In some contexts we can quickly obtain good counts of solutions at infinity using simple algebra. The best illustration is given in Chapter 8.

Before tracking paths, we can always try to find some solutions by another method. We might do this because we have available a faster but less reliable technique. Suppose that we call this technique FASTSOL. We evoke FASTSOL. If it finds all solutions (or all physical solutions), we need not track any paths and we are done. (In the situation, for example, when we seek a unique positive real solution, we will know if FASTSOL finds it.) If FASTSOL does not find all solutions, it will have generated a partial solution list, and we can use this list when we begin tracking paths. We know how many solutions to expect (from Bezout's theorem), and when the solutions found by path tracking and by FASTSOL add up to the total degree, we can stop. Naturally, singular solutions make maintaining an accurate solution count more difficult, and solutions at infinity, if not accounted for in some way, will prevent the solution count from ever adding up to the total degree. If there are an infinite number of finite solutions, then we will not be able to make use of the FASTSOL solutions like this. If we suspect that there are an infinite number of finite solutions, the best strategy is to let CONSOL run all paths. However, it is sometimes difficult to tell that an infinite solution set has been encountered. The only direct indicator is that solutions on an infinite component will be singular, and of course, a singular solution might also be geometrically isolated. The logic of a version of CONSOL that makes use of precomputed solutions and solutions at infinity is outlined in Section 4-5-8.

4-5 PATH TRACKING

4-5-1 Ways of Taking Steps Along a Path

Let $x(t)$ be a continuation path, defined for $t \in I$, where $I = [0, 1]$ if the path converges and $I = [0, 1)$ if the path diverges. The rather crude method that we have used so far to track $x(t)$ is to fix an increment, Δt, for t, set x^0 equal to a solution to $g(x) = 0$, and solve $h(x, \Delta t) = 0$ with Newton's method, where x^0 is chosen as the start point. This yields a solution x^*. Then we set $x^0 = x^*$ and solve $h(x, 2\Delta t) = 0$ with x^0 as start point, and continue to increment t by Δt and update x until $t = 1$ and x^0 is a solution to $f(x) = 0$.

This approach to tracking a path falls under a general umbrella called "predictor–corrector" methods. (See [Rheinboldt, 1977, 1980], [Heijer and Rheinboldt, 1981], and [Garcia and Zangwill, 1981, Chap. 16].) If (x^0, t^0) is the current point on the path, we *predict* to $(x^1, t^1) = (x^0, t^0 + \Delta t)$ and *correct* (via Newton's method) to (x^*, t^1). Then we set $x^0 = x^*$, $t^0 = t^1$, and iterate the process. The schematic of this particular predictor–corrector has a "staircase" look, as shown in Fig. 4-3. Let's call this basic method "TRACK." TRACK has an obvious limitation: The steps are not oriented along the path. We could take bigger steps if we move in the direction of the path, thereby possibly saving computer time. However, let's explore TRACK before considering improvements.

We have used the "fixed step size" TRACK in Chapters 1 and 2. This is too inflexible to be an efficient path tracker. We can make TRACK a

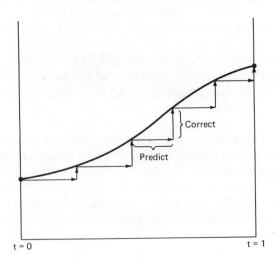

Figure 4-3 Schematic illustrating the TRACK path tracker. This simple method *predicts* by a horizontal shift and *corrects* vertically.

variable-step-size method simply by doubling the step size every time a successful step is taken and halving the step size and restarting the step whenever it fails. Here "success" is signaled by the corrected point being within a specified distance of the path within a specified number of iterations. In other words, the norm of the Newton's method residual must be less than "EPS" within "MAXIT" iterations, where MAXIT and EPS are given parameters. This variable-step-size TRACK works amazingly well considering its simplicity. There are many possible variants on the way the step size is adjusted. We might decide not to increase the step size unless the EPS requirement was attained with some margin: say, keep the same step size unless the corrector residual is less than 10^{-2} EPS. Another idea is to increase the step size only after a fixed number of successive successful steps, say after five in a row. (In fact, this "five rule" is coded into CONSOL8.) Some path trackers try to estimate safe step sizes more analytically (see [Heijer and Rheinboldt, 1981]).

Aside from using a variable-step-size strategy, we can make the path tracker more sophisticated by taking the steps in the direction of the path. We can do this via equation (3-30), which defines dx/dt. It turns out that dx/dt gives a tangent vector to the continuation curve, pointing in the direction of increasing t. More exactly, if (x^0, t^0) is a point on the continuation path, then

$$(x^0, t^0) + \left(\frac{dx}{dt}(t^0), 1\right)$$

gives the coordinates for the tip of the tangent vector.

Thus, instead of solving $h(x, t^0 + \Delta t) = 0$ using $x = x^0$ as a start point, we solve $h(x, t^1) = 0$ with start point x^1, where

$$(x^1, t^1) = (x^0, t^0) + \left(\frac{dx}{dt}(t^0), 1\right)\text{SCL},$$

and the scale factor, SCL, is chosen so that $|(x^1, t^1) - (x^0, t^0)|$ equals the current step size (see Fig. 4-4). The appropriate schematic is given in Fig. 4-5. I call this path tracker TRACK-T since it predicts tangent to the path. TRACK-T clearly respects the geometry of the path better than TRACK. [The actual computation of $(dx/dt)(t^0)$ should be carried out using equation (3-29) and a code for solving linear systems (see Section 4-5-3).] CONSOL8 comes with TRACK-T, implemented in subroutine PREDC. If, for the purpose of comparative testing, you want to run CONSOL8 with TRACK instead, simply replace the call to PREDC in the main routine by the single line of code "T=DMIN1(T+STPSZE,1.D0)."

I should note here that since (3-30) defines a system of ordinary differential equations, it can be solved by any of the available numerical methods for solving ordinary differential equations. See [Boyce and DiPrima,

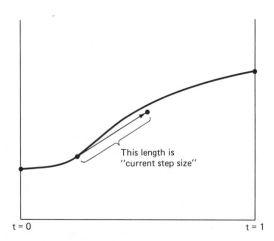

Figure 4-4 Schematic showing a tangent to the continuation path.

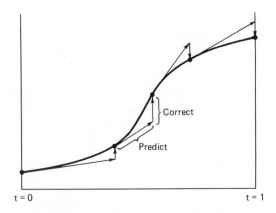

Figure 4-5 Schematic illustrating TRACK-T.

1977], [Hindmarsh, 1974], and [Shampine and Gorden, 1975]. Other path-tracking approaches have been proposed (see [Abbott, 1980], [Kubicek, 1976], and [Rheinboldt, 1977, 1980]). For many additional references, see [Allgower and Georg, 1980] and [Watson et al., 1986].

At this point we have considered all the basic ideas about methods of path tracking that I want to cover. I realize that there are further possible refinements for TRACK-T. For example, the corrector can correct along the hyperplane perpendicular to the predicted tangent. (This allows for more accurate tracking of divergent paths!) The Newton's method for correction can be modified to be faster and less accurate, putting more of the burden for success on the way the step size is managed. These modifications are developed in Section 4-7 as adjuncts to the experimental scheme presented there.

There is a completely different approach to path tracking via simplicial methods (see [Garcia and Zangwill, 1981] and [Allgower and Georg, 1980]). I do not have any experience with it.

4-5-2 Is *t* Just Another Variable?

Continuation paths for complex polynomial systems are increasing in *t*. I have found that incrementing *t* in the predict step and fixing it during the correct step is the most reliable way to track paths. The typical "problem path" looks like Fig. 4-6, with rapid shifts in *t* at the beginning and end of the path. What I observe, in the problem areas of the curve, is that if *t* is allowed to change during correction, then large shifts in *t* may occur that cause the path tracker to lose the path. I do not know if this is due mainly to bad conditioning in the linear algebra or if the path trackers that I have tried have been particularly sensitive for some other reason. This is still a research area. [I have tried schemes that split the implicit system (3-28) differently from (3-29), to better condition the algebra.] Since much of the current research work focuses on tracking paths that can backtrack in *t*, some care must be taken in applying the results of such work to complex polynomial continuation.

There is an additional apparent advantage for treating *t* in this "increment and fix" way. It has to do with singular solutions. Singular solutions are not easy to compute. When a continuation path, $x(t)$, is converging to

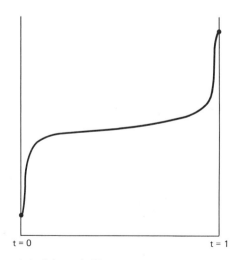

t = 0   t = 1

Figure 4-6   Schematic illustrating typical "problem path."

a singular solution, x^*, we have two concerns:

(a) That the continuation process (say, predict–correct) not break down and abandon the path without generating a reasonable solution estimate
(b) That as accurate a solution estimate as possible be returned

Some path trackers overshoot the solution as a part of their "end game." For singular solutions, I have found it best to respect the singularity and assume that the path is not defined for $t \geq 1$. Thus we are generating the limit to a sequence $\{x^k: k = 1, 2, \ldots\}$, and the process generating the sequence is undefined at the limit.

Consider what happens using the "increment and fix" strategy for t when a path is converging to singular x^*. From point (x^0, t^0), a predicted point (x^1, t^1) is generated; (x^1, t^1) lies on the tangent to the curve and

$$| (x^1, t^1) - (x^0, t^0) |$$

equals the current step size. If $t^1 > 1$, a shorter step is chosen so that $t^1 = 1$. Eventually, we get a predicted point $(x^1, 1)$. Since x^* is singular when $t = 1$, the corrector may fail. Then the step size is shortened and a new prediction is made from (x^0, t^0). We end up with a prediction (x^1, t^1) with $t^0 < t^1 < 1$. Then $h(x, t^1) = 0$ is nonsingular at solutions, and if $dh(x, t^1)$ is not too ill-conditioned, we will get a successful correction. As t converges to 1, $dh(x, t)$ becomes more and more poorly conditioned. Typically, the process terminates in one of two ways:

(a) The predicted point $(x^1, 1)$ meets the convergence criterion without correction and is returned as an endpoint to the curve.
(b) The maximum number of steps or minimum-step-size condition is exceeded, and the path is abandoned.

When the process terminates, we are given an "endpoint," (x^+, t^+), with associated parameters suggesting a singular solution: say, $t^+ = 0.999\,999\,999$, $| f(x^+) | \leq 10^{-12}$, $| \det (df(x^+)) | \leq 10^{-6}$, $C(df(x^+)) \geq 10^5$ [where "$C(df(x^+))$" denotes the "condition" of the matrix $df(x^+)$, as defined in Section 4-5-4]. Now we must decide what to do with x^+. For example, we could decide that x^+ is such a bad estimate of the endpoint that the path will have to be rerun. Or we could decide to refine x^+ with an extended-precision Newton's method (Appendix 1) or some other special local solution method for singular solutions. Perhaps we can judge that x^+ is good enough, and no further refinement is necessary. Whatever decision of this sort is made, I believe that finding such an (x^+, t^+) is the most we can expect from continuation at singular solutions. It makes use of the full power of the generic continuation, where $h(x, 1 - \epsilon) = 0$ is nonsingular at solutions, for $\epsilon > 0$. Pinning down a singular solution in this way costs more computer work, of course, than abandoning the path near the end. Thus, in applications in which singular solutions are numerous and not physically meaningful, we might decide to use that cheaper alternative.

4-5-3 Solving Linear Systems

CONSOL uses a linear-system solver in the Newton's method corrector and in the predictor (for TRACK-T) to find the tangents to the curve. Much of the computer time for CONSOL is spent solving linear systems. Thus it is important to have a fast, reliable linear-system solver.

The problem is to solve a large number of relatively small linear systems quickly. For systems in only two or three variables, using Cramer's rule or using an explicit formula for the matrix inverse may be excellent methods of solution. This is contrary to good practice for larger systems. The approach to solving $Ax = b$ recommended in general is to *factor* (or *decompose*) the square matrix A and then solve the sequence of systems implied by the factorization. For example, using *Gaussian elimination*, we can compute lower and upper triangular matrices, L and U, and a permutation matrix, P, so that

$$LU = PA.$$

Now $Ax = b$ is equivalent to $PAx = Pb$, where Pb is just a permutation of the elements of the column b. Then we can solve $PAx = Pb$ in two steps: Solve $Ly = Pb$ for y and then solve $Ux = y$ for x. These two steps are called *forward elimination* and *back substitution*, respectively. Solving linear systems presented by upper or lower triangular matrices is a relatively cheap computation. Most of the numerical effort goes into factoring the matrix A (see [Forsythe and Moler, 1967, Chaps. 9 and 10]). Part of the advantage of solving $Ax = b$ in this way is the savings in computational expense over some other methods. (For example, by "counting multiplications," we can see that Cramer's rule is more expensive when $n > 3$.) Another advantage has to do with sensitivity to roundoff and errors in coefficients. Also, if we want to solve a sequence of problems, $Ax = b_j$ for $j = 1, \ldots, r$, where A is fixed but the right-hand sides, b_j, change, we need only factor A once. See [Forsythe and Moler, 1967] for a complete discussion of these features.

Other approaches to factoring A yield other numerical methods for solving $Ax = b$. For example, via the *singular value decomposition*, orthogonal square matrices U and V and a diagonal matrix D are computed so that

$$A = UDV^T$$

(see [Forsythe et al., 1977, Chap. 9]). The entries of $D = \text{diag}(\sigma_1, \sigma_2, \ldots, \sigma_m)$

are called the *singular values* of A, and $\sigma_1 \geq \sigma_2 \geq \cdots \geq \sigma_m \geq 0$. This factoring of A has certain advantages, especially if A is badly conditioned (see Section 4-5-4).

I used Cramer's rule to solve the 2×2 systems in CONSOL6 and CONSOL7. In some cases, it may be worthwhile to specialize the linear solver to particular system types. However, it is reasonable in general to use some version of Gaussian elimination that someone else has coded and debugged. For example, the LINPACK code is widely available [Dongarra et al., 1979]. However, if the package you use is highly "modulized" (i.e., if there are a lot of subroutine calls), you should consider "demodulizing" it by copying the code of the subroutines directly in line. The overhead for subroutine calls can sometimes be significant when the linear solver is being called hundreds of times. I have included versions of the Crout variant of Gaussian elimination ([Forsythe and Moler, 1967, Chap. 12]) with the CONSOL codes, but it is easy to replace these by other linear solvers.

Some researchers favor other methods for doing the linear algebra associated with continuation (see [Watson, 1986]). The important point is to relate the method to the problem environment. The considerations of Section 4-5-2 (on t) and of Section 4-5-4 (on singular solutions) are especially relevant.

4-5-4 Singular Solutions

We have two reasons for wanting to know if a solution, x^*, to $f = 0$ is singular. First, if x^* is singular, special numerical techniques may be needed to compute x^* accurately. For example, if we use Newton's method to do a final refinement of x^*, we allow more iterations since the convergence is likely to be slow (see Appendix 1). Second, we "score" the performance of CONSOL by checking the solution list for consistency with Bezout's theorem. This cannot be done unless we can distinguish singular from nonsingular solutions.

Determining exact singularity is generally impossible. The requirement

$$\det\,(df(x^*)) = 0 \tag{4-1}$$

is literally necessary and sufficient for x^* to be singular, but it is unrealistic to expect any computed x^* to obey (4-1). Replacing (4-1) by

$$|\det\,(df(x^*))| < \epsilon \tag{4-2}$$

is logical, but now the condition for singularity is scale dependent. If we are consistently scaling f in some reasonable way, (4-2) may perhaps be acceptable (see Chapter 5). However, the concept of *condition* has been developed to quantify the idea of "nearly singular" independent of scaling.

Let A be a nonsingular $m \times m$ matrix. Then the "condition of A" is defined to be

$$C(A) = \| A \| \| A^{-1} \|, \tag{4-3}$$

where $\| \cdot \|$ denotes a matrix norm. It is not important to our use of (4-3) that any particular matrix norm be used. Thus we have the usual Euclidean norm on R^m, $\| x \| = \sqrt{x_1^2 + \cdots + x_m^2}$, and

$$\| A \| = \left\{ \max \frac{\| Ax \|}{\| x \|} : x \in R^m, x \neq 0 \right\}.$$

Other norms may be more convenient for computations. See [Forsythe and Moler, 1967, Chap. 2] and [Forsythe et al., 1977, pp. 41–42]. Here is a simple approximation that is sometimes useful. Let

$$| A | = \{ \max | a_{k,l} | : k = 1 \text{ to } m, l = 1 \text{ to } m \},$$

where $A = (a_{k,l})$. Then

$$| A | \leq \| A \| \leq m | A |$$

(from [Forsythe and Moler, 1967, p. 4]), and thus

$$| A | | A^{-1} | \leq C(A) \leq m^2 | A | | A^{-1} |. \tag{4-4}$$

Since $\| I \| = 1$, where I denotes the identity matrix, we have

$$1 = \| I \| \leq \| A \| \| A^{-1} \| = C(A)$$

for any A. The idea of "condition" is that the larger $C(A)$ is, the more nearly singular A is. If A is actually singular, we set $C(A) = \infty$.

Let

$$A = \text{diag}(a_1, a_2, \ldots, a_m).$$

Thus A is the matrix with all zero entries, except that a_k appears as the (k, k)th entry. Then

$$A^{-1} = \text{diag}\left(\frac{1}{a_1}, \frac{1}{a_2}, \ldots, \frac{1}{a_m}\right)$$

and

$$C(A) = \frac{\max \{ | a_k | : k = 1 \text{ to } m \}}{\min \{ | a_k | : k = 1 \text{ to } m \}}. \tag{4-5}$$

For example, if

$$A = \begin{bmatrix} 1 & 0 \\ 0 & 10^{-k} \end{bmatrix},$$

then $C(A) = 10^k$. Thus, as $k \to \infty$, $C(A) \to \infty$. Also, if A is any nonsingular square matrix and s is any nonzero number, then $C(sA) = C(A)$. So C is invariant under scaling.

If

$$A = \begin{bmatrix} a & b \\ c & d \end{bmatrix}$$

is any nonsingular 2×2 matrix, then

$$A^{-1} = \frac{1}{ad - bc} \begin{bmatrix} d & -b \\ -c & a \end{bmatrix},$$

and therefore

$$\frac{\max \{a^2, b^2, c^2, d^2\}}{|ad - bc|} \le C(A) \le 4 \frac{\max \{a^2, b^2, c^2, d^2\}}{|ad - bc|} \qquad (4\text{-}6)$$

by (4-4). Thus, if a 2×2 matrix A has its maximum entry of order unity, then

$$1/|\det (A)| \le C(A) \le 4/|\det (A)|.$$

This is a case in which the determinant directly measures the condition of A. But note that generally det (A) is not independent of the scaling of A.

$C(A)$ can be computed from the singular-value decomposition described in Section 4-5-3. Specifically,

$$C(A) = \frac{\max \{\sigma_k : k = 1 \text{ to } m\}}{\min \{\sigma_k : k = 1 \text{ to } m\}}.$$

Computationally inexpensive techniques are available to estimate $C(A)$ [Forsythe and Moler, 1967, p. 51]. Selected codes described in [Dongarra et al., 1979] generate such estimates of condition. For small matrices, however, and in the context of continuation, it is a slight additional expense to compute $C(A)$ directly from (4-3). Specifically, in CONSOL at the end of a path we already have $df(x^+)$ factored into L-U form, where x^+ is the computed estimate of the solution x^* (see Section 4-5-3). We need merely use forward elimination and back substitution n times to generate $[df(x^+)]^{-1}$. Since n is (relatively) small, this is essentially a trivial computation considering the cost of computing the whole path.

How will we use "condition" in continuation? I envision the following sequence:

1. Continuation generates a solution estimate, x^+.
2. $C(df(x^+))$ is evaluated, and based on its value, x^+ is labeled "nonsingular" or "possibly singular."
3. If x^+ is labeled "nonsingular," it is refined by a "nonsingular refinement algorithm."
4. If x^+ is labeled "possibly singular," it is refined by a "singular refinement algorithm." This singular refinement algorithm would (most

likely) be chosen to be effective for both singular and nonsingular solutions, but it would be too computationally expensive to use in general. Now the refined solution estimate is labeled "singular," "nonsingular," or "marginally singular," based on the condition of the Jacobian. These labels will be used to judge the probable accuracy of the computation, and the validity of the total count of solutions.

Note that steps 2 and 4 will require zero-epsilons, which must be chosen based on computational experience to set thresholds for the condition.

What nonsingular and singular refinement algorithms are available? For nonsingular solutions, my experience is that (plain old) Newton's method is excellent. I have found no reason to look for an alternative. In fact, most path-tracking algorithms will generate good solution estimates at nonsingular endpoints, and no special "refinement" need be called for. Singular solutions are a different story. Special algorithms have been developed for singular solutions, but none is clearly preferable to Newton's method for our purposes. Computing singular solutions is considered further in Appendix 1, where I have included a discussion on refining path endpoints. Basically, this is a local problem, not a continuation issue.

However, there is one continuation question associated with singular solutions. Should an apparent singularity be dealt with before $t = 1$? That is, as t approaches 1, if $C(dh_x(x, t))$ is becoming large, the tracker could be changed to a special singular "end game" to complete the continuation. This might be done by checking $C(dh_x(x, t))$, once, say when $t = 0.999$, or more often as t approaches 1. If t is managed as described in Section 4-5-2, then switching to a special end game does not seem to be necessary. However, for other approaches, it is. (For example, it is for the POLSYS driver in HOMPACK. See [Watson et al., 1986].)

Here is a final note. Some systems have a (large) collection of (very) singular solutions at infinity. Rather than solve using the projective transformation and encounter these singularities, it might be better in many cases to solve without the projective transformation and truncate apparently diverging paths.

4-5-5 How Closely Should Paths be Tracked? What Should Be Done About Path Crossing?

Whatever the method used, the actual sequence of points followed in path tracking will not lie on the path but only close to it. We choose parameters in our method to control how close. For example, in CONSOL8 (using TRACK-T) the parameters EPS (zero-epsilon for the corrector residual), MAXIT (maximum number of corrector iterations per step), and SSZBEG (beginning step size) directly control the deviation from the path.

If there were just one path to track, our only consideration would be to stay close enough so that as t approaches 1 we could "tighten up" and get close at the end of the path. In terms of TRACK-T, say, let EPS = EPSBIG (a relatively large number like 10^{-2} or 10^{-3}) until $t \geq$ NEARONE (maybe NEARONE = 0.95), then finish with EPS = EPSSML (something like EPSSML = 10^{-8}). This is done in CONSOL8. A variant sets NEARONE = 1, so that refinement occurs at the end of the path. Although this strategy is a good idea in general, there is another issue. We have several paths and "path crossing" is a real danger (see Fig. 4-7).

In practice, path crossing cannot be ruled out as a possibility. There are several categories of control. The most direct is to set the tracking parameters conservatively; the more conservative, the greater the computational work but the less the possibility of path crossing. The "scaling" of the system influences how close together the paths become. Unfortunately, little is known on how to exploit this. See Chapter 5 for some guidelines. One way to minimize the computational expense of path crossing, however, is to use an adaptive rerun logic.

To see how this works, let's first assume that $f(x) = 0$ has no singular solutions or solutions at infinity. Then it has exactly d nonsingular solutions, where d is the number of continuation paths. If at the end of running all the paths, we get d distinct solutions, we know we are done. In fact, it does not matter if path crossing occurred at all. On the other hand, we may have found fewer than d solutions. This can occur only if two paths converged to the same solution, impossible in exact arithmetic, because the solutions are nonsingular. Thus we know that one of the two paths crossed over to the other. Now we rerun these two, resetting the run parameters to track the paths more closely. The result should be two distinct solutions, although we might have to repeat the tightening-up process. We could, of course,

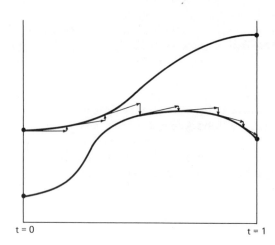

$t = 0$ $t = 1$

Figure 4-7 Schematic illustrating path crossing.

track the paths to machine precision and still have them cross over. This would probably indicate a severe scaling problem, and the system should be scrutinized along the lines of Chapter 5. The rerun strategy I have used is to set EPS $= 10^{-3}$ at first and then EPS $= 10^{-8}$ on rerun.

Now, what if there are singular solutions? We are immediately on less solid ground. If two paths converge to a nonsingular solution, we know that there is crossover. If two (or more) paths converge to a singular solution, we cannot rule out crossover but we cannot say that it has occurred. On the other hand, if only one path converges to a singular solution, we know that some other path has crossed over, but we do not necessarily know which one. There are some "tricks." For real-coefficient systems, complex solutions occur in conjugate pairs and have the same multiplicity. The final results must reflect these known characteristics. If there are only two singular solutions with three paths to one and one to the other, we know that one of the three has crossed over. The logical complexity grows quickly with the total degree. Codes for very small systems can consider all cases. For a general code, we might just shrug and not check crossover for singular solutions. Or we could decide to rerun all paths converging to singular solutions. The appropriate trade-off in computational work, reliability, and algorithm complexity depends on the situation. And, of course, "nearly singular" solutions strain our resources and make a mockery of our zero-epsilons.

If $f(x) = 0$ has solutions at infinity, some paths will diverge. Now we are really at a loss. What if a path that should converge crosses over to a diverging path? How can we detect this and correct it? Unless a complex solution is missing its conjugate or a singular solution its multiple paths, we have no cue. The best cure for this situation is to use the projective transformation of $f = 0$, unless the number of singular solutions thereby generated makes matters worse (see the final note in Section 4-5-4).

4-5-6 Stopping Rules. Detecting "Path Failure"

In exact arithmetic, a path has converged when $t = 1$. This suggests two "stopping rules" for convergent paths: stop when $t = 1$ or stop when $t \geq 1 - \epsilon$ for some small positive ϵ. The danger in using ϵ is that a converging path may be stopped when it is too far from its endpoint or that a diverging path will be labeled "converging." (After all, $t \to 1$ for diverging paths.) Frequently in tests I have observed a significant shift in x as the path converges: say, as t goes from 0.99 to 1. On the other hand, if we merely use the "stop when $t = 1$" rule, then in certain singular cases a path that ought to converge cannot get close enough to make $t = 1$. I tend to use the "stop when $t = 1$" rule.

When I do use the "stop when $t \geq 1 - \epsilon$" rule, I take ϵ near zero. For example,

$$1 - \epsilon = 0.999\ 999\ 999\ 999$$

if the computer arithmetic carries 14 or 15 digits. Then, for each convergent path I check (with Newton's method) if the resulting endpoint of the path is really near a solution to $f(x) = 0$. If not, I label the path "diverging." Checking if the end point of a path is near a solution divides into two cases. If the endpoint is near a *nonsingular* solution, x^*, then Newton's method will converge with high accuracy and in fact

$$| x^* - x^+ | \approx | df^{-1}(x^+)\ f(x^+) |, \tag{4-7}$$

where x^+ denotes the solution estimate. That is, the Newton's method residual is a good estimate of the distance from the estimate x^+ to the actual solution x^* (see Appendix 1). If the endpoint of the path is near a *singular* solution, then generally Newton's method will not converge as quickly or as closely and (4-7) is no longer necessarily valid. However, if the solution is isolated, my experience is that the estimate does not move very far with successive iterations. If you observe such movement, this is a good sign that the path did not converge or converged to an infinite solution set. Frequently, Newton's method will converge if allowed enough iterations from a random start point, so Newton's method convergence per se does not indicate that the path converged. The point is that the Newton's method estimate should be reasonably close to the endpoint of the path: say, within 10 EPS. Note that for paths converging to "nasty" singular solutions (as defined in Appendix 1), the path tracker will most likely fail before any reasonable solution estimate is obtained.

Divergence (in theory) cannot be detected. It means that a path never stops. To really know, we have to wait forever. We would die of old age without being able to say for sure. In practice, we run some examples, see how long convergent paths tend to be, multiply this by some factor (say 10 or 100), and declare "divergent" any path longer than that. This upper bound is problem dependent but does not necessarily vary greatly for a collection of problems with a similar structure. However, there is no relation between the distance between the endpoints of a convergent path and its arc length or the max $\{\ | x(t) |\ :\ t \in I\ \}$. (See Experiment 4-1 at the end of this section.)

Generally, the cleanest resolution of divergent paths is to avoid them by using the projective transformation. However, there are two situations where this is not the case:

1. In supervised environments in which rerunning suspicious CONSOL runs is feasible, it may be more convenient to use $f(x)$.
2. When $f(x) = 0$ has many solutions at infinity (especially if they are singular), it may be more efficient to truncate paths than to use the projective transformation.

A path can "fail" in CONSOL8 in two ways:

- The step size can become less than the minimum step size; in symbols, STPSZE < SSZMIN.

- The number of steps along the path can exceed the maximum allowed; in symbols, NUMSTP > MAXNS.

CONSOL8 halves the step size when the "corrector" part of the path tracker does not converge. If the corrector repeatedly fails to converge after the step size is made smaller and smaller, it is reasonable to abandon the path.

Ideally, a path will never fail. But if $f(x) = 0$ has a solution at infinity, CONSOL8 will generate path failure associated with divergence. The expected failure condition is "NUMSTP > MAXNS." The other condition, "STPSZE < SSZMIN," is more pathological, but does not imply that any isolated solutions have been missed. In practice, crossover is a more significant problem than literal "path failure."

4-5-7 "Independent" Constants

One of the most mysterious aspects of CONSOL is the extraordinary utility of the "independent" coefficients that are a part of the definition of $g(x)$. The function of the independent coefficients is to keep the continuation paths well defined and bounded. You detect that these coefficients are badly chosen by seeing the continuation process "get stuck" or by seeing a path diverge to infinity without $t \to 1$ or by having two paths cross no matter how closely they are followed. Here is an example. Define

$$h(x, t) = (1 - t)(px^2 - q) + t(-x^2 + 1). \qquad (4\text{-}8)$$

If $p = 1$ and $q = 1$, then

$$h(x, t) = (1 - 2t)x^2 - (1 - 2t) = 0.$$

In exact arithmetic both paths "get stuck" at $t = \frac{1}{2}$ because the equation "goes singular." (In computer arithmetic, you might "step over" the singularity and not get stuck.) If $p = 1$, $q \neq 1$, the paths diverge to infinity as $t \to \frac{1}{2}$. We avoid both types of bad behavior by choosing p and q "independently." What this means is that we choose p and q to avoid certain special algebraic relationships implied by the structure of h and the coefficients of f. We see that (4-8) goes singular only when $(1 - t)p - t = 0$, that is, when $p = t/(1 - t)$. Since $t \in [0, 1]$, we avoid this for sure if p is complex with nonzero imaginary part or if p is negative. The point is that given any $f(x)$, there will be "bad" choices of p and q. Generally, we cannot compute these bad choices explicitly, the way we can for (4-8). But if both the real and imaginary parts of p and q are chosen using a random number table, it is unlikely that any special relationship between p, q, and the coefficients of f will hold. This assumes that f's coefficients are not directly derived from p and q. At one time people were saying that p and q should be generated by a random number generator and changed for every problem. I have not found this to be necessary. I choose p and q once and use them over and over (see, e.g., subroutine INPTB in CONSOL8).

It is important to keep in mind that not only do we want p and q to be "independent" but we want them to be "not nearly dependent." Experiment 4-1 shows that if p is near 1 in (4-8), the path length can get arbitrarily long. For efficiency, we want the paths as short as possible.

Aside from being "independent," p and q should be scaled commensurable with the scaling of f. Since I tend to scale f's coefficients about unity (see Chapter 5), I choose the entries of p and q to have absolute value varying between $\frac{1}{10}$ and 1.

The choice of p and q significantly affects the length and shape of the continuation curves, but I know of no way to exploit this systematically. You can, of course, run the same problem with many different choices of p and q to see what happens (as in Experiment 2-2).

4-5-8 Using Solution Counting to Save Computer Time

Let's consider, in the course of running CONSOL, keeping a solution count, say in the variable NUMSOL. At the end of each path, we update NUMSOL. Why would we do this? Because often, by using "side information," NUMSOL can be "ahead" of CONSOL. When NUMSOL equals the correct solution count for the problem, we can terminate CONSOL early, thus saving computational work. First, let's assume that there are neither an infinite number of solutions nor an infinite number of solutions at infinity. Then by Bezout's theorem, the solutions and solutions at infinity of $f(x) = 0$ add up to exactly d, the total degree, counting multiplicities. So when NUMSOL $= d$, we will stop CONSOL. We may be able to precompute the number of solutions at infinity, NUMINF (see Chapters 3 and 8). We may have some (finite) solutions to $f(x) = 0$ that were computed by a different method. Let their number be NUMPRE. We start off with NUMSOL = NUMINF + NUMPRE. We need a list of solutions, LISTSOL, which contains all the solutions, together with flags indicating how many continuation paths have converged to the solution ("0" means it was precomputed), whether the solution is real or nonreal, whether the solution is singular or nonsingular, and the run number of the last path to converge to the solution. Let's call these NUMPAT, REAL, SING, and RUNNUM, respectively. (REAL = 0 means "nonreal"; REAL = 1 means "real." SING = 0 means "nonsingular"; SING = 1 means "singular.")

If $f(x)$ has real coefficients, the complex solutions occur in conjugate pairs. That is, if $x^0 = a + bi$ is a solution, $\bar{x}^0 = a - bi$ is also a solution. Since in my work, $f(x)$ always has real coefficients, I will assume that solutions occur in conjugate pairs. (The modifications if this is not the case will be clear.)

Now, each time we find a solution, x^0, we check to see if x^0 is in LISTSOL. If it is not, we enter x^0 in LISTSOL and set the associated flags.

If REAL = 0, we put $\bar{x}^0$ in the list also. Then we increase NUMSOL by 1 if REAL = 1 and SING = 0, by 2 if either REAL = 1 and SING = 1 or REAL = 0 and SING = 0, and by 4 if REAL = 0 and SING = 1.

If x^0 is already in LISTSOL and NUMPAT = 0, we set NUMPAT = 1 and we do not change NUMSOL. If x^0 is in the list and NUMPAT > 0, we check SING. If SING = 0, this is an error. CONSOL should not have found the solution twice. We will want to rerun the two paths that led to this solution (the current run and RUNNUM) with a smaller EPS to see if they "separate out." If they do not after rerunning, we will set an error flag and proceed. If SING = 1, we check NUMPAT. If NUMPAT = 1, we set NUMPAT = 2 and proceed. If NUMPAT > 1, we increase NUMPAT by one and increase NUMSOL by one also.

When NUMSOL equals d, we stop CONSOL, confident that LISTSOL is complete. I have seen CONSOL's performance on small systems improve by a factor of 2 using such a scheme.

There are some dangers. To decide singularity, realness, and identity (with a solution already in the list) requires three zero-epsilons that must be chosen with care. It is especially delicate to decide realness and identity for singular solutions, and zero-epsilons different from the nonsingular cases should be used. (The singular zero-epsilons should be larger.) Note that we are trusting CONSOL to properly count the multiplicity of nonsingular solutions.

If the number of solutions at infinity is underestimated, no errors will (necessarily) be committed by CONSOL, but it will compute all paths, because NUMSOL will never add up to d. Thus this approach is most useful in cases in which (often) we can get an accurate count of solutions at infinity or in which we can be (reasonably) sure that there are no solutions at infinity (e.g., when we use the projective transformation).

This method is also useful when CONSOL is being used to back up a faster, less reliable method, FASTSOL. In this case, FASTSOL may find all solutions and NUMSOL will start out equal to d. Then no paths will be run. Or perhaps FASTSOL will miss one or two solutions, so that NUMSOL is not quite d. Then CONSOL will run paths until it finds the missing solutions. On the average, less computational work will be spent than if CONSOL were being run alone. Side information can be very useful: If we know there is only one positive real solution and the positive real solution is the one we want, we can stop as soon as we have found it (see Chapter 9).

EXPERIMENT

4-1. Using CONSOL8, solve equation (4-8) with

$$p = 1 + \epsilon(1 + i)$$
$$q = 2 + \epsilon(1 + i),$$

where $i^2 = -1$, for $\epsilon = 10^{-2}, 10^{-4}, 10^{-6}, 10^{-8}, 10^{-16}$, and 10^{-32}. Note the way the arc length, the maximum $|x|$, and the number of steps grow. (You will have to modify subroutine INPTB by temporarily changing the definitions of arrays P and Q.) If you can, plot t against $|x|$ to generate schematics, for each value of ϵ.

4-6 PERFORMANCE EVALUATION

How do you judge the success or failure of a particular CONSOL run? Our most powerful tool for deciding if a run of CONSOL has been successful is Bezout's theorem. If the count of solutions and solutions at infinity equals the total degree, we have succeeded. (Qualifications will be given below.) The primary value of this test is that it is independent of CONSOL. We know that CONSOL *in theory* will generate all solutions, but every numerical method is a heuristic in implementation.

Singular solutions make the count less certain, because we must rely on CONSOL for computing multiplicities greater than 2. If there are an infinite number of solutions, we can no longer make accurate assessments of success or failure. The ideal situation is not to submit such problems to CONSOL. If there are an infinite number of solutions, CONSOL can be proved (in *exact arithmetic*) to find all *isolated* solutions (Section 3-4).

In my experience it tends also to find at least two representative points on each of the connected components of the nonisolated solutions. These representative points are always found as singular solutions.

The case that $f(x) = 0$ has a finite number of solutions but an infinite number of solutions at infinity is particularly unpleasant. If we encounter such a system (as in Chapter 10), we may have to judge success or failure without Bezout's theorem, generally by using side information specific to the problem.

Naturally, in scrutinizing a CONSOL run, we look for basic consistency. If x^+ is a solution estimate, $|f(x^+)|$ should be small. We judge the condition of $df(x^+)$ and decide if x^* is singular or not, where x^+ is an estimate of x^* (see Section 4-5-4). Each singular solution will be found by more than one path, and each nonsingular solution will be found by exactly one path if CONSOL is working properly. If f has real coefficients, complex solutions should occur in conjugate pairs. If some paths diverge, a quick back-of-the-envelope calculation of $f^0(x) = 0$ (the homogeneous part of f) may tell if some solutions at infinity exist, even if computing them carefully is not possible.

What do you do if a run is judged a failure? Actually, there are different levels of success and failure. For example, we can use the following categories:

- "Successful" if the solution count equals the total degree, and there are no singular solutions
- "Probably successful" if the count of solutions by multiplicities equals the total degree, and the singular solutions have only multiplicity 2
- "Maybe successful" if the count of multiplicities equals the total degree, with some higher multiplicities computed by CONSOL and used in the count
- "Unknown" if the "solutions at infinity" count is not available or known to be infinite, or if the projective transformation has solutions at infinity (suggesting an infinite number of solutions at infinity), or if $f(x) = 0$ is known to have infinite numbers of solutions where CONSOL is being used to find isolated solutions
- "Failure" if two paths converged to a nonsingular solution, or if only one path converged to a singular solution, or if only one of a pair of complex conjugate solutions was found (and f has real coefficients), or if a path is abandoned that was not diverging, or if a solution that is known to exist is not found (say, the positive real one)

When failure is detected or suspected, there two things to do:

1. Rerun with "path closeness" zero epsilons set smaller and/or the maximum number of steps set larger. For example, in CONSOL8 make EPSBIG (the initial bound for the corrector residual), SSZBEG (the beginning step size), and/or MAXIT (the maximum number of iterations allowed for a correction step) smaller and/or MAXNS (the maximum number of steps allowed on a path) larger.
2. Revise the model.

The usual reason for running closer to the paths is that we believe some of the paths to have crossed numerically. By tracking them more closely, we hope to uncross them. (Recall that paths do not cross in exact arithmetic.) The usual reason for increasing the maximum number of steps is that we believe some path might have converged if more steps had been allowed. Note, however, that singular solutions tend to use up a lot of steps at the end. A path may have converged as close as it can to a singular solution and finally generate a message such as "maximum number of steps exceeded."

The idea of revising the model is that we may have encountered difficulties because of the presentation of $f(x)$ and we can possibly improve it. For example, f might be reducible or f might be poorly scaled (see Chapters 5 and 7). Or the physical basis of the model may be weak. This is typical of situations in which CONSOL produces a nice solution set but no physically meaningful solutions.

4-7 THE EXPERIMENTAL LABORATORY

Here is the heart and soul of this book. In this section we set up a laboratory for mathematical experiments. We learn how CONSOL really behaves. We see if "improvements" really improve.

The theory of CONSOL is qualitative. This theory is as true and as vague as the statement that we must eat to live. It is perfectly consistent with theory that all continuation paths have arc length greater than 10^{15} with hideous curvature every inch of the way. This would make CONSOL useless for computations, just as we would die if we had only a slice of bread a year to eat. We must bring CONSOL down to earth. We start a problem, and while it runs we will wait, or get a cup of coffee, or read a newspaper, or go to sleep, or perhaps grow old and die. To confront CONSOL's quantitative character is to learn what we cannot prove: Will it do us any good? And how can we make it serve us best?

In the experiments described below you will try out and compare the various main options we have noted for solving systems with CONSOL:

- Solve $f = 0$ (the original system) or the projective transformation of $f = 0$.
- Use path tracker TRACK, or TRACK-T, or TRACK-P (defined below). Note that CONSOL8 comes with TRACK-T (see Section 4-5-1).
- Use full Newton or modified Newton (defined below).

For each main option, you will explore the effects of various choices of the run parameters, in terms of selected run statistics that summarize the performance of CONSOL8. Although doing the total series of experiments will require many runs, the value will be in developing your sense of the effect of these choices on performance, something difficult to discover any other way. The total series of experiments generates thousands of tables (each summarizing a run), and this total collection of tables will quantify whatever sense of "best" choices you develop. Naturally, you may run a subset of the experiments, omitting some options, and you will (I hope) explore some extra options and systems not presented here. For example, the number of valid path tracking schemes is large, and we have explicitly considered only the simplest. You might try HOMPACK as an alternative. See [Watson et al., 1986]. Scaling is another factor that can have a profound effect on the performance of CONSOL and is not explicitly included here (see Chapter 5).

CONSOL8 is coded to print out the following statistics for each path:

- Number of steps
- Number of successful steps
- Number of unsuccessful steps
- Total corrector iterations on path (Newton iterations)
- Total number of linear systems solved
- Arc length of path (approximate)
- $C(df(x))$ at the endpoint (the condition of the Jacobian)
- $|\operatorname{resid}| = |df(x)^{-1}f(x)|$ at endpoint (Newton residual)

In addition, for each run, you should record the following information:

- Total number of corrector iterations over all paths
- Total number of linear systems solved over all paths
- Total arc length over all paths
- CPU time (total for run) if available; otherwise, clock time
- Number of solutions found
- Number of solutions not found
- Number of crossed paths (best estimate)

For specific problems, note the parameter choices that yield the shortest run times and still generate all solutions. Over all problems, note the parameter choices that tend to work best.

Other statistics you might collect and study include:

- Distance from start point to endpoint for each path
- $\max |x|$ on path

Use the following problem sets:

- Set 1. Systems (2-2) through (2-9), (2-18), and (2-19) with all coefficients set equal to 1. You solved these using CONSOL6 and CONSOL7 in Experiments 2-1, 2-3, and 2-4. Compare the CONSOL8 statistics with what you observed before. (Note the efficiency of a variable-step-size method.)
- Set 2. System (3-3) and the systems in Exercises 3-1 and 3-2.
- Set 3. The following three systems are taken from physical applications. Note the scaling extremes and the solutions at infinity.

$$a_{1,1}x^2 + a_{1,2}y^2 + a_{1,3}xy + a_{1,4}x + a_{1,5}y + a_{1,6} = 0 \quad (4\text{-}9)$$
$$a_{2,1}x^2 + a_{2,2}y^2 + a_{2,3}xy + a_{2,4}x + a_{2,5}y + a_{2,6} = 0,$$

where

$$a_{1,1} = \quad -0.00098$$
$$a_{1,2} = 978000.$$
$$a_{1,3} = \quad -9.8$$
$$a_{1,4} = \quad -235.$$
$$a_{1,5} = \quad 88900.$$
$$a_{1,6} = \quad -1.$$
$$a_{2,1} = \quad -0.01$$
$$a_{2,2} = \quad -0.984$$
$$a_{2,3} = \quad -29.7$$
$$a_{2,4} = \quad 0.00987$$
$$a_{2,5} = \quad -0.124$$
$$a_{2,6} = \quad -0.25$$

$$b_{1,1}xy^2 + b_{1,2}xy + b_{1,3}y^2 + b_{1,4}y + b_{1,5} = 0$$
$$b_{2,1}x^3 + b_{2,2}x^2y + b_{2,3}x^2 + b_{2,4}xy^2 \qquad (4\text{-}10)$$
$$+ b_{2,5}xy + b_{2,6}x + b_{2,7} = 0,$$

where

$$b_{1,1} = \quad 0.124D+08$$
$$b_{1,2} = 3480.$$
$$b_{1,3} = 3720.$$
$$b_{1,4} = \quad 1.$$
$$b_{1,5} = \quad -0.00003$$
$$b_{2,1} = \quad 0.186D+07$$
$$b_{2,2} = \quad 0.126D+07$$
$$b_{2,3} = 5850.$$
$$b_{2,4} = \quad 0.110D+08$$
$$b_{2,5} = 6590.$$
$$b_{2,6} = \quad 1.99$$
$$b_{2,7} = \quad -0.00008$$

and "D" denotes "power of 10," so that $0.124D+08 = 12400000$.

$$c_{1,1}x^2z + c_{1,2}y^6 + c_{1,3}y^5 + c_{1,4}y^4 + c_{1,5}y^3$$
$$+ c_{1,6}y^2 + c_{1,7}y + c_{1,8} = 0$$
$$c_{2,1}x^2y + c_{2,2}x^2z + c_{2,3}xy + c_{2,4}y^5 + c_{2,5}y^4 \qquad (4\text{-}11)$$
$$+ c_{2,6}y^3 + c_{2,7}y^2 + c_{2,8}y + c_{2,9} = 0$$
$$c_{3,1}x^2 + c_{3,2}xz + c_{3,3}y + c_{3,4} = 0,$$

where

$$c_{1,1} = -0.625D+14$$
$$c_{1,2} = 0.538D+09$$
$$c_{1,3} = 0.503D+09$$
$$c_{1,4} = 0.895D+08$$
$$c_{1,5} = 0.578D+07$$
$$c_{1,6} = 0.107D+06$$
$$c_{1,7} = 0.617D+03$$
$$c_{1,8} = 1.00$$
$$c_{2,1} = 0.625D+14$$
$$c_{2,2} = 0.188D+15$$
$$c_{2,3} = 0.202D+08$$
$$c_{2,4} = -0.503D+09$$
$$c_{2,5} = -0.179D+09$$
$$c_{2,6} = -0.173D+08$$
$$c_{2,7} = -0.429D+06$$
$$c_{2,8} = -0.308D+04$$
$$c_{2,9} = -6.00$$
$$c_{3,1} = -0.556D+16$$
$$c_{3,2} = 0.111D+17$$
$$c_{3,3} = 0.180D+10$$
$$c_{3,4} = 1.00$$

- Set 4. The following 10 systems present singular or near singular solutions.

$$2ax^3 - 2xy = 0$$

$$y - x^2 = 0$$

for a = 33/32, 161/160, 19/20, 53/56, and 29/32 (see [Griewank and Osborne, 1983].

$$x^2 + y^2 - r = 0$$

$$x^2 - y^2 = 0$$

for $r = 2 \cdot 10^{-4}, 2 \cdot 10^{-8}, 2 \cdot 10^{-16}, 2 \cdot 10^{-32}$, and 0.

Use the following choices of input parameters:

- EPSBIG: $10^{-1}, 10^{-2}, 10^{-3}, 10^{-6}$
- MAXIT: 1, 2, 3, 6
- MAXNS: 10, 25, 100, 1000
- SSZBEG: $10^{-1}, 10^{-2}, 10^{-4}, 10^{-6}$

We are missing only the definitions of "TRACK-P" and "modified Newton" for you to be able to run the full range of options for this series of experiments. I will describe these missing options now, but if you are impatient to begin, by all means, plunge in! Just be aware that these options exist, are relatively easy to implement (by modifying CONSOL8), and may provide improvements.

TRACK-T (as defined in Section 4-5-1) always corrects by holding t fixed. Let (x^1, t^1) be the predicted point. Then TRACK-T solves $h(x, t^1) = 0$, using Newton's method with start point $x = x^1$. A possible improvement would be to correct perpendicular to the tangent used for prediction (see Fig. 4-8). The idea is to shorten the distance from the predicted point to the curve. It is, in fact, easy to modify TRACK-T to do this. Let z denote "(x, t)," so that $z^1 = (x^1, t^1)$ and $z^0 = (x^0, t^0)$. Here z^0 is the last computed point on the curve and z^1 is the predicted point. Then a point z is on the hyperplane perpendicular to the vector from z^0 to z^1 and passing through z^1 whenever

$$\langle z - z^1, z^0 - z^1 \rangle = 0 \tag{4-12}$$

where "$\langle \cdot, \cdot \rangle$" denotes the dot product. Therefore, z is on this hyperplane and also on the continuation path if z satisfies (4-12) and $h(z) = 0$ at the same time. But this defines a system of $n + 1$ equations in $n + 1$ unknowns; say, $H(z) = 0$. We now (try to) solve with Newton's method, using start

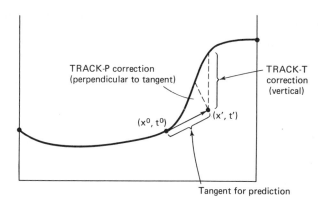

Figure 4-8   Schematic illustrating TRACK-P in comparison with TRACK-T.

point z^1. This Newton's method would replace the TRACK-T correction. In fact, the Newton formula to update z^k to z^{k+1} (starting at z^1) is

$$dH(z)(z^{k+1} - z^k) = -H(z)$$

or, in terms of h,

$$\begin{bmatrix} dh(z^k) \\ w \end{bmatrix} (z^{k+1} - z^k) = - \begin{bmatrix} h(z^k) \\ 0 \end{bmatrix}, \tag{4-13}$$

where $w = z^0 - z^1$ and $[dh] = [dh_x \mid dh_t]$, the (complex) $n \times (n + 1)$ Jacobian of h with dh_x denoting the first $n \times n$ block of partials with respect to $x_1, x_2, \ldots, x_n$ and dh_t denoting the last column of partials with respect to t. Note that the successive iterates, z^k, do stay on the desired hyperplane and that now t is not fixed but changes with the iteration. Compare (4-13) with the following corrector formula used in TRACK-T:

$$dh_x(x^k, t^1)(x^{k+1} - x^k) = -h(x^k, t^1). \tag{4-14}$$

[Compare (3-28) and (3-29).]

To implement TRACK-P, you should modify CORCT, the Newton's method corrector subroutine, to implement (4-13). The vector w will be additional input. If you want to use this new CORCT to refine a path endpoint, $(x^+, 1)$, merely set $w = (0, 0, \ldots, 0, 1)$ so that t will not be changed.

The ultimate in a "close to the curve" corrector would correct by solving the constrained minimization problem:

$$\text{minimize } (z - z^1) \tag{4-15}$$
$$\text{subject to } h(z) = 0.$$

But the most natural way of evoking Lagrange multipliers and linearization reduces solving (4-15) to solving (4-13), gaining us nothing (see [Fletcher, 1980, 1981]).

The modified Newton's method you might like to try (in TRACK, TRACK-T, or TRACK-P) is a common variant. It evaluates the Jacobian only once for a step. Thus

$$dh_x(z^0)(x^{k+1} - x^k) = -h(x^k, t^1) \tag{4-16}$$

replaces (4-14), where z^1 is taken as the initial point for the iterations. [Similarly, "$dh(z^0)$" can replace "$dh(z^k)$" in (4-13).] The *justification* for doing this is that the modified method is known to be a valid alternative to Newton's method, although its convergence tends to be slower (see [Rheinboldt, 1974, p. 26]). The *reason* for using this modification is that the bulk of the work in solving $Az = b$ (for any square matrix A and column b) is in factoring A (see Section 4-5-3). After A is factored, solving several systems of the form $Az = b$ (with different b) is cheap. We *predict*, finding a solution to

$$dh(z^0) \frac{dz}{dt} = 0, \tag{4-17}$$

by solving

$$dh_x(z^0) \frac{dx}{dt} = -dh_t(z^0), \tag{4-18}$$

which yields the same results as

$$\begin{bmatrix} dh(z^0) \\ v \end{bmatrix} \begin{bmatrix} dx/dt \\ 1 \end{bmatrix} = v^T \tag{4-19}$$

with $v = (0, 0, \ldots, 0, 1)$. Compare (4-19) with (4-13). We see that we can predict and correct for a moderate increase in the previous cost merely of predicting if we use $dh(z^0)$ instead of $dh(z^k)$ in (4-13) or instead of $dh(x^k, t^1)$ in (4-14), depending on whether we use TRACK-P or TRACK-T, respectively. The advantages and disadvantages of using a less accurate but cheaper corrector is, of course, what you will study in doing the experiments.

These modifications suggest many variants and extensions. I am sure you will be delighted that I am not going to say anything about them, save one modest exercise below. Happy experimenting!

EXERCISE

4-1. *Modifying the Predictor.* To solve the predictor equation (4-17), we have been solving (4-18), which yields the same results as solving (4-19) with $v = (0, 0, \ldots, 0, 1)$. However, we could replace v in (4-19) by any nonzero vector; say, we could use any vector of the form $(0, \ldots, 0, 1, 0, \ldots, 0)$. Why might we want to solve (4-17) this way? Is it easy to decide what v to use? Would any v work? Can we (easily) reduce such (4-19) from $n + 1$ equations to n equations, for efficiency?

CHAPTER 5

Scaling

5-1 INTRODUCTION

Scaling means transforming the coefficients of a system so that they do not have extreme values. The purpose of scaling is to reduce the possibility of catastrophic arithmetic problems when a solution method is evoked on a computer. Scaling is critically important for CONSOL. Aside from the obvious issues of being able to evaluate polynomial expressions effectively and solve linear systems using finite-precision arithmetic, scaling affects quantities such as arc length and curvature for paths.

   In this chapter two algorithms for scaling are developed, "SCLGEN" and "SCLCEN." I routinely use SCLGEN for all systems, with SCLCEN being an option for special situations. A bonus of using these algorithms is that the coefficient tableau they require for input (defined below) can also be used by a generic FFUN subroutine. (FFUN is the subroutine in CONSOL that returns system and Jacobian values.) In other words, instead of rewriting FFUN for each system type, we can use this generic FFUN for all systems. The drawback is that this FFUN is usually significantly slower than a specialized FFUN. Still, it is convenient to use and allows us quickly to set up a new system for CONSOL. The code for this tableau FFUN is given in Appendix 6, Part A, Section 4. (Subroutines INPTAT, INPTBT, FFUNT, and OTPUTT are included.) The version of CONSOL I use when CPU time is not an issue is CONSOL8 with these subroutines, referred to hereafter as CONSOL8T (see Section 5-3).

   The projective transformation functions as a scaling transformation. Its effect is to shorten arc lengths and bring solutions closer to unity. (This is true even if $f = 0$ has an infinite number of solutions at infinity, in which case Proposition 3-7 does not hold.) The scaling featured in this chapter is different, in that it directly addresses extreme values in the system coefficients. The two scaling schemes work well together. This is illustrated in Section 5-3.

Although there is general agreement that some attention to scaling is better than none, there is no practical theory that specifies how to scale nonlinear systems. We must proceed by "common sense." To illustrate scaling, let's consider some simple examples. The equation

$$10^{20}x^2 + 3 \cdot 10^{20}x + 2 \cdot 10^{20} = 0 \tag{5-1}$$

obviously ought to be multiplied through by 10^{-20}. Then (5-1) becomes

$$x^2 + 3x + 2 = 0, \tag{5-2}$$

which is as numerically tame as we could wish. Now

$$10^{-20}x^2 + 3x + 2 \cdot 10^{20} = 0 \tag{5-3}$$

clearly cannot be tamed by multiplying each coefficient by the same number, but the change of variables

$$x = 10^{20}z \tag{5-4}$$

yields (5-1) (with x replaced by z) and thus yields (5-2). This sort of change of variables is called "variable scaling" as opposed to "equation scaling." Thus (5-3) becomes (5-1) via variable scaling and (5-1) becomes (5-2) via equation scaling.

To summarize, we have distinguished two types of scaling:

- Equation scaling
- Variable scaling

An equation is scaled by multiplying it by a nonzero constant. A variable is scaled by making a change of variables of the form

$$x = \text{constant} \cdot z.$$

The effect of scaling the variables and equations in a system is to redefine the coefficients of the system. The structure of the system and its solutions will not change, except that the solutions will be expressed in new units.

It is a common practice to scale equations and variables by various schemes specific to the application; say, by dividing variables through by known upper bounds so that the scaled physical solutions will be in the unit box. These sorts of specific scalers often do the job, but they must be customized to each application. The two scalers presented here capture the essence of the successful scaling schemes I have seen, which is not to say that they are cure-alls.

SCLGEN is a general-purpose scaler to be used when you have little knowledge of the solutions. It uses variable and equation scaling to center the coefficients about unity and minimize the variation within equations.

SCLCEN assumes that you have mean values (or upper bounds) for each variable. It "centers" the variable scaling around the given values and then does minimum variance equation scaling.

Let me illustrate the approach before giving details. Although we have seen a satisfactory scaling for (5-3), let's try to find an "ideal" scaling systematically. Let $x = 10^{c_1}z$ and scale the equation by 10^{c_2}. Thus (5-3) becomes

$$10^{c_2}[10^{-20}(10^{2c_1}z^2) + 3 \cdot 10^{c_1}z + 2 \cdot 10^{20}] = 0,$$

yielding

$$10^{E_1}z^2 + 10^{E_2}z + 10^{E_3} = 0,$$

where

$$E_1 = -20 + 2c_1 + c_2$$

$$E_2 = c_1 + c_2 + \log(3)$$

$$E_3 = c_2 + 20 + \log(2)$$

and "log" denotes "$\log_{10}$." We seek c_1 and c_2 so that

$$r_1 \equiv E_1^2 + E_2^2 + E_3^2 \tag{5-5}$$

is minimized, to center the coefficients about unity, and at the same time

$$r_2 \equiv (E_1 - E_2)^2 + (E_1 - E_3)^2 + (E_2 - E_3)^2 \tag{5-6}$$

is minimized, to reduce the variability of coefficients.

Now (5-5) and (5-6) together yield a system of two second-degree equations in the two unknowns, c_1 and c_2. We can minimize this system by finding a solution to it. We might, in some contexts, be willing to solve a system to scale a system. We will discuss this in Section 5-2. But let's take a simpler approach for now. Thus let's define

$$r(c_1, c_2) = r_1(c_1, c_2) + r_2(c_1, c_2) \tag{5-7}$$

and find a (c_1, c_2) pair that minimizes r. A glance at r_1 and r_2 convinces us that r has no maximum, so if we find a place where all the partial derivatives of r are zero, this will be a minimum. In fact, the system

$$\frac{\partial r}{\partial c_1} = 0$$

$$\frac{\partial r}{\partial c_2} = 0 \tag{5-8}$$

is a linear system of two equations in two unknowns and generically will have a unique solution. Thus we can minimize r by solving a single linear system. This is a trivial computational price to pay for scaling. System (5-8) is

$$\frac{\partial r}{\partial c_1} = 22c_1 + 6c_2 - 320 - 6 \log(2) + 2 \log(3) = 0$$

$$\frac{\partial r}{\partial c_2} = 6c_1 + 6c_2 + 2 \log(2) + 2 \log(3) = 0,$$

(5-9)

which has solution $(c_1, c_2) = (19.956, -20.409)$. If we ignore the log terms ("set them equal to zero"), the solution to (5-9) is $(c_1, c_2) = (20, -20)$, the scaling used at the beginning of this chapter.

5-2 TWO SCALING ALGORITHMS: SCLGEN AND SCLCEN

We scale variables by

$$x_j = 10^{c_j} z_j \qquad \text{for } j = 1, \ldots, n \qquad (5\text{-}10)$$

and scale the ith equation, $f_i(x) = 0$, by $10^{c_{n+i}}$ for $i = 1, \ldots, n$. Then the scaled $f_i(x_1, \ldots, x_n)$ becomes

$$10^{c_{n+i}} f_i(10^{c_1} z_1, 10^{c_2} z_2, \ldots, 10^{c_n} z_n).$$

Letting

$$a_{i,j} x_1^{d_{i,1,j}} x_2^{d_{i,2,j}} \cdots x_n^{d_{i,n,j}}$$

denote the jth term of the unscaled f_i (thus d is triple subscripted), we get the associated scaled term

$$10^{c_{n+i}} a_{i,j} (10^{c_1} z_1)^{d_{i,1,j}} \cdots (10^{c_n} z_n)^{d_{i,n,j}}$$

$$= 10^{c_{n+i}} 10^{c_1 d_{i,1,j}} \cdots 10^{c_n d_{i,n,j}} a_{i,j} z_1^{d_{i,1,j}} \cdots z_n^{d_{i,n,j}}$$

$$= \exp_{10} \left(c_{n+i} + \sum_{k=1}^{n} c_k d_{i,k,j} \right) a_{i,j} z_1^{d_{i,1,j}} \cdots z_n^{d_{i,n,j}},$$

where "$\exp_{10}(\text{expression})$" denotes "$10^{\text{expression}}$."

After we have found acceptable values for $c_1, \ldots, c_{2n}$, the new (scaled) coefficients will be

$$b_{i,j} = \left[\exp_{10} \left(c_{n+i} + \sum_{k=1}^{n} c_k d_{i,k,j} \right) \right] a_{i,j}. \qquad (5\text{-}11)$$

After we solve $f(z) = 0$ using $b_{i,j}$ in place of $a_{i,j}$, we recover x from z via equation (5-10).

To find the c_j, we set up a minimization problem just as we did for the illustrative example. Define a quadratic objective function[†] in analogy with (5-7):

$$r = r_1 + r_2 = \sum_{i=1}^{n} r_{1,i} + \sum_{i=1}^{n} r_{2,i}, \qquad (5\text{-}12)$$

where

$$r_{1,i} = \sum_{j=1}^{l_i} [\sigma_{1,i,j}]^2 \qquad (5\text{-}13)$$

with

$$\sigma_{1,i,j} = c_{n+i} + \left(\sum_{k=1}^{n} c_k d_{i,k,j} \right) + \log \left(| a_{i,j} | \right)$$

to penalize the deviation from zero of the exponents of terms in f_i, and

$$r_{2,i} = \sum_{1 \le j' < j'' \le l_i} [\sigma_{2,i,j',j''}]^2 \qquad (5\text{-}14)$$

with

$$\sigma_{2,i,j',j''} = \left(c_{n+i} + \sum_{k=1}^{n} c_k d_{i,k,j'} + \log \left(| a_{i,j'} | \right) \right)$$
$$- \left(c_{n+i} + \sum_{k=1}^{n} c_k d_{i,k,j''} + \log \left(| a_{i,j''} | \right) \right)$$

to penalize the variation in magnitude between coefficients of f_i. Here l_i denotes the number of terms in f_i. Note that $\sigma_{2,i,j',j''}$ can be simplified:

$$\sigma_{2,i,j',j''} = \sum_{k=1}^{n} c_k (d_{i,k,j'} - d_{i,k,j''}) + \log \left(\frac{| a_{i,j'} |}{| a_{i,j''} |} \right).$$

However, if $| a_{i,j} | = 0$, then $\sigma_{1,i,j}$ is omitted from (5-13), and if either $| a_{i,j'} | = 0$ or $| a_{i,j''} | = 0$, then $\sigma_{2,i,j',j''}$ is omitted from (5-14). In essence, we do not scale zero coefficients.

The scaling algorithm based on minimizing (5-12) is called SCLGEN. SCLGEN is described in [Meintjes and Morgan, 1987], where the extreme scaling of chemical equilibrium systems is addressed (see also chapter 9). A computer code implementing SCLGEN is included in Appendix 6, Part A, Section 4, and in HOMPACK (where it is called SCLGNP) [Watson et al., 1986].

[†] An "objective function" is an equation we are going to minimize. See [Fletcher, 1980, 1981] for more background on optimization.

The minimization of (5-12) turns out to be a relatively mild linear least-squares problem. The most straightforward approach is via the calculus: namely, to solve the $2n \times 2n$ linear system

$$\frac{\partial r}{\partial c_j} = 0 \qquad \text{for } j = 1, \ldots, 2n. \tag{5-15}$$

We can rewrite (5-15) as $\alpha \cdot c = \beta$, where α is a $2n \times 2n$ matrix and c and β are $2n \times 1$ column matrices. Note that α is a matrix of integers whose entries are algebraic combinations of the various degrees that define f. The system $\alpha \cdot c = \beta$ always has a solution, even if α is singular. (Consider that the minimization always has a solution, although perhaps an infinite number.) If α is singular, this simply means that we can choose some of the c_i arbitrarily. For the small systems of polynomials of low degree that I tend to solve, using Gaussian elimination on (5-15) seems to work well. (In the Gaussian elimination of singular α, diagonal entries reduce to zero for the unconstrained variables.)

However, if we anticipated working with singular or ill-conditioned α, we would approach the linear least-squares problem not via (5-15) but rather by a different approach. We would make a rectangular system, $A \cdot c = b$, with rows derived from the $\sigma_{1,i,j}$ and $\sigma_{2,i,j',j''}$, and then minimize $|A \cdot c - b|$ using the singular value decomposition (see [Forsythe et al., 1977, Chap. 9]). This is a more numerically stable approach than dealing with a singular α. In fact, a singular α is an indication that f is poorly presented. The "cure" for singular α is not a better least-squares method but rather the reduction of f. The f that I have found to yield a singular (or badly conditioned) α have a degenerate structure. In particular, they have either an infinite number of solutions or a subset of variables that can be set identically to zero. Consider

$$x_1^2 x_2^2 - 2x_1 x_2 + 1 = 0 \tag{5-16}$$
$$x_1 x_2 - 1 = 0$$

and

$$x_1^2 + x_2^2 = 0 \tag{5-17}$$
$$x_1^2 - x_2^2 = 0.$$

See Section 7-3 for a fuller discussion of the implications of α's being singular. Just to keep the SCLGEN code from breaking when α *is* singular, I have implemented a patch to assign values to the unconstrained variables and generate a warning message. But do not solve $f = 0$ if α is singular.

I said that we would omit terms in (5-13) and (5-14) if some $|a_{i,j}| = 0$. There are significant delicacies in implementing this. If we omit a term only if $|a_{i,j}| = 0$ literally, we may be allowing an "outlyer" coefficient to completely unbalance our scaling. On the other hand, if we choose a zero-epsilon, ϵ, so that the term will be omitted whenever $| a_{i,j} | \leq \epsilon$, what value shall we pick for ϵ? The system $f = 0$ with its coefficients $a_{i,j}$ comes to us unscaled. It might happen, for example, that $10^{-30} \leq | a_{i,j} | \leq 10^{-20}$ for all i, j. [Chemical problems have a way of yielding such systems (see Chapter 9).] If we set $\epsilon = 10^{-15}$, we would wipe out the whole system. In a supervised environment, we might examine the coefficient list to try to pick out obvious outlyers. We could also run SCLGEN with $\epsilon = 0$ and then examine the resulting $b_{i,j}$ for lack of balance, perhaps omitting apparent outlyers and repeating SCLGEN on the new set of coefficients. We might even want to solve the system with several scalings to see if significant differences in solutions are obtained. Sometimes

$$10^{-10}x^2 + 2x - 8 = 0$$

might as well be

$$2x - 8 = 0,$$

but the decision to replace the one by the other is a modeling decision. The big solution may be the one we want.

In a solitary environment, a prior decision must be made on how to set ϵ. The default $\epsilon = 0$ is perhaps the best you can do without any knowledge of the problem environment (which is rarely the case). If you want a code that does not need information on the problem environment, a scheme that uses two passes of SCLGEN might be better. The first pass takes $\epsilon = 0$, and the second sets ϵ to be a small multiple (say, 10^{-15}) of the largest $| b_{i,j} |$ produced by the first pass. Working out a reliable scaling algorithm for a solitary problem environment requires a lot of patience, testing with as representative a problem set as possible.

The main variation on scaling I want to present involves centering x_1, $\ldots, x_n$ about nominal values. This is useful when we have reasonable ranges on $x_1, \ldots, x_n$ for the physically meaningful solutions or, alternatively, reasonable upper bounds for these quantities. Thus if the range for x_1 in the natural units of the problem is $10^{-38} \leq x_1 \leq 10^{-32}$, we might want to "center" x_1 via the scaling $x_j = 10^{-35}z_j$. On the other hand, if all we know is that $0 \leq x_1 \leq 10^{-35}$, the same scaling will at least give us the advantage of working in the unit box. It is reasonable to take the nominal values either as *mean values* or *upper bounds*, depending on the modeling context. The scaling algorithm based on centering variables is called "SCLCEN." It is discussed further in Section 5-3, and a computer code is given in Appendix 6, Part A, Section 4.

For SCLCEN, we choose typical (nonzero) values x_j^0 for $x_j, j = 1$, $\ldots, n$, and set

$$c_j = \log (| x_j^0 |)   \text{for } j = 1, \ldots, n.$$

Then we use r_1 from (5-12) as our objective function, where r_1 is now a function of c_j for $j > n$ alone, since $c_1, \ldots, c_n$ are constant. To minimize r_1 we solve

$$\frac{\partial r_1}{\partial c_{n+j}} = 0 \qquad \text{for } j = 1, \ldots, n, \tag{5-18}$$

an $n \times n$ linear system. Equation (5-18) is in fact diagonal, and therefore trivial to solve.

I hate math books that overgeneralize or catalog alternatives that do not help the general reader and are obvious to experts. However, I feel compelled here to note some variations on SCLGEN and SCLCEN.

Equation (5-14) does not involve c_j for $j > n$, so we could minimize r_2 first to find $c_1, \ldots, c_n$, then minimize r_1 with $c_1, \ldots, c_n$ fixed (compare with SCLCEN). This might, of course, yield a different scaling than the one generated by SCLGEN.

Equation (5-14) does not penalize variability in coefficients between equations. We can achieve this by substituting for r_2 in (5-12) the equation

$$r_2 = \sum_{\substack{i',j' \\ i'',j''}} [(c_{n+i'} + c_1 d_{i',1,j'} + \cdots + c_n d_{i',n,j'} + \log(|\,a_{i',j'}\,|))$$

$$- (c_{n+i''} + c_1 d_{i'',1,j''} + \cdots + c_n d_{i'',n,j''} + \log(|\,a_{i'',j''}\,|))]^2$$

$$= \sum_{1 \le i' \le i'' \le n} \sum_{\substack{1 \le j' \le l_{i'} \\ 1 \le j'' \le l_{i''} \\ j' \le j''}} \left[(c_{n+i'} - c_{n+i''}) + \sum_{k=1}^{n} c_k (d_{i',k,j'} - d_{i'',k,j''}) \right.$$

$$\left. + \log\left(\frac{|\,a_{i',j'}\,|}{|\,a_{i'',j''}\,|}\right) \right]^2 .$$

We could decide, instead of minimizing (5-12), to solve the system of $2n$ second-degree equations in $2n$ unknowns

$$r_{1,i} = 0 \qquad \text{for } i = 1, n \tag{5-19}$$
$$r_{2,i} = 0 \qquad \text{for } i = 1, n.$$

Of course, (5-19) may not have real solutions. However, we can use complex scaling factors in CONSOL; it is already defined in a context of complex arithmetic. We could use a complex Newton's method to find a solution to (5-19). I have not tried this approach, so am not sure what its advantages and disadvantages might be.

5-3 COMPUTING WITH SCLGEN AND SCLCEN; THE COEFFICIENT TABLEAU

We prepare input for SCLGEN by listing coefficients and degrees by term for each equation. SCLGEN generates matrices α and β and solves $\alpha \cdot c = \beta$, returning the scale factors c. [The linear equation $\alpha \cdot c = \beta$ is (5-15).] We generate the new coefficients $b_{i,j}$ using (5-11). (It is convenient to include the generation of the $b_{i,j}$ in the SCLGEN code.) Then we solve the system $f = 0$ using $b_{i,j}$ instead of $a_{i,j}$. For each solution z to this $b_{i,j}$ system, we generate a solution x to the original system via (5-10). If SCLGEN and CONSOL are put together in a single package, the operation of scaling can be transparent to the user.

The CONSOL8T code (CONSOL8 with SCLGEN and the tableau versions of FFUN, INPTA, INPTB, and OTPUT) uses the coefficient tableau to set up the FFUN subroutine. No special FFUN code need be written. With the input flags IFLGPT and IFLGSC the user specifies whether the projective transformation is to be used and whether SCLGEN is to be used. It is a very convenient package for exercising the most significant CONSOL options. (The POLSYS driver for HOMPACK [Watson et al., 1986] uses the same tableau as input, although the path tracker is different.)

To illustrate the effect of SCLGEN on solving a system of equations, let's consider an example from Chapter 9, which I will take out of context for use here. The example consists of two cubics:

$$f_1 = a_{1,1}x_1x_2^2 + a_{1,2}x_1x_2 + a_{1,3}x_2^2 + a_{1,4}x_2 + a_{1,5} = 0$$

$$f_2 = a_{2,1}x_2^3 + a_{2,2}x_1^2x_2 + a_{2,3}x_1^2 + a_{2,4}x_1x_2^2 + a_{2,5}x_1x_2 \qquad (5\text{-}20)$$

$$+ a_{2,6}x_1 + a_{2,7} = 0.$$

The basic SCLGEN input values, term by term, for f_1 and f_2 are given in Table 5-1.

The input to SCLGEN is n, l_1, l_2, $a_{1,1}$, $\ldots$, $a_{1,5}$, $a_{2,1}$, $\ldots$, $a_{2,7}$, and the $d_{i,k,j}$. The output will be c_1, c_2, c_3, and c_4, from which we will generate $b_{1,1}$, $\ldots$, $b_{1,5}$, $b_{2,1}$, $\ldots$, $b_{2,7}$ using (5-11). Thus, for (5-20), α is a 4×4 matrix and β a 4×1 matrix. Taking the values given in Table 5-1, $\alpha \cdot c = \beta$ becomes

$$
\begin{bmatrix}
5 & 0 & 2 & 6 \\
0 & 7 & 10 & 4 \\
2 & 10 & 22 & 8 \\
6 & 4 & 8 & 16
\end{bmatrix}
\begin{bmatrix}
c_1 \\
c_2 \\
c_3 \\
c_4
\end{bmatrix}
=
\begin{bmatrix}
-87.4 \\
-211.0 \\
-457.0 \\
-328.0
\end{bmatrix},
$$

TABLE 5–1   SCLGEN input for system (5–20)[a]

Input	Comments
N = 2	Number of equations
NUMT(1) = 5	Equation 1 has five terms
DEG(1, 1, 1) = 1	The degree of x_1 in the first term of equation 1
DEG(1, 2, 1) = 2	The degree of x_2 in the first term of equation 1
A(1, 1) = 0.194 997 2 D + 48	The coefficient for the first term of equation 1
DEG(1, 1, 2) = 1	
DEG(1, 2, 2) = 1	
A(1, 2) = 0.131 824 2 D + 23	
DEG(1, 1, 3) = 0	
DEG(1, 2, 3) = 2	
A(1, 3) = 0.321 383 6 D + 23	
DEG(1, 1, 4) = 0	
DEG(1, 2, 4) = 1	
A(1, 4) = 0.100 000 0 D + 01	
DEG(1, 1, 5) = 0	
DEG(1, 2, 5) = 0	
A(1, 5) = −0.300 000 0 D − 04	
NUMT(2) = 7	
DEG(2, 1, 1) = 3	
DEG(2, 2, 1) = 0	
A(2, 1) = 0.118 048 7 D + 51	
DEG(2, 1, 2) = 2	
DEG(2, 2, 2) = 1	
A(2, 2) = 0.115 344 4 D + 48	
DEG(2, 1, 3) = 2	
DEG(2, 2, 3) = 0	
A(2, 3) = 0.309 912 2 D + 26	
DEG(2, 1, 4) = 1	
DEG(2, 2, 4) = 2	
A(2, 4) = −0.862 090 5 D + 47	
DEG(2, 1, 5) = 1	
DEG(2, 2, 5) = 1	
A(2, 5) = −0.872 350 0 D + 25	
DEG(2, 1, 6) = 1	
DEG(2, 2, 6) = 0	
A(2, 6) = −0.262 495 9 D + 21	
DEG(2, 1, 7) = 0	
DEG(2, 2, 7) = 0	
A(2, 7) = −0.800 000 0 D − 04	

[a] The actual input to a SCLGEN program includes several additional parameters. See the listing in Appendix 6. "DEG" is indexed as DEG(equation, variable, term). "A" is indexed as A(equation, term). "D" denotes powers of 10. Thus "0.194 997 2 D + 48" means "0.194 997 2 $\times$ 10^{48}."

yielding the scale factors

$$c_1 = 8.829$$

$$c_2 = 4.385$$

$$c_3 = -17.75$$

$$c_4 = -16.01$$

By (5-11) we have

$$b_{i,j} = [\exp_{10}(c_{2+i} + c_1 d_{i,1,j} + c_2 d_{i,2,j})] a_{i,j}.$$

Thus we get the $b_{i,j}$ values listed in Table 5-2. Observe that these coefficients

TABLE 5-2 Coefficients for scaled system, generated by SCLGEN from Table 5-1 input

B(1, 1) =	0.227 128 0 D + 07
B(1, 2) =	0.155 956 8 D − 02
B(1, 3) =	0.210 111 7 D + 00
B(1, 4) =	0.664 038 7 D − 07
B(1, 5) =	− 0.202 339 9 D + 05
B(2, 1) =	0.162 005 9 D + 02
B(2, 2) =	0.874 749 2 D + 00
B(2, 3) =	0.238 722 1 D − 05
B(2, 4) =	− 0.361 291 2 D + 02
B(2, 5) =	− 0.371 331 8 D − 04
B(2, 6) =	− 0.113 490 9 D + 08
B(2, 7) =	− 0.194 138 9 D + 01

TABLE 5-3 CONSOL8T input for system (5–20) using coefficients from Table 5–2

Input	Comments
IFLGPT = 0	Do not use projective transformation
IFLGSC = 1	Use SCLGEN
IFLGCR = 0	Do not print out corrector residuals
IFLGST = 0	Do not print out step summaries
MAXNS = 5000	Maximum number of steps per path
MAXIT = 3	Maximum number of corrector iterations per step
EPSBIG = 1.D − 06	Most of the path will be tracked this closely
SSZBEG = 1.D − 10	Beginning step size

Part 2

The Table 5–1 tableau, but with coefficients "B" from Table 5–2 instead of the "A."

TABLE 5–4 Scaled solutions to (5–20)[a]

Path	Re(x_1)	Im(x_1)	Re(x_2)	Im(x_2)
1	$-0.719D-07$	0.0	$-0.658D+03$	$-0.114D-30$
2	$-0.219D-06$	$0.201D-15$	$-0.944D-07$	$0.264D+03$
3	$0.834D+03$	$-0.279D-28$	$-0.326D-02$	$0.103D-26$
4	$-0.925D-07$	$0.629D-11$	$0.286D+05$	$-0.241D+05$
5	$-0.837D+03$	$0.881D-04$	$-0.172D-09$	$-0.326D-02$
6	$-0.220D-06$	$-0.201D-15$	$-0.944D-07$	$-0.264D+03$
7	$0.837D+03$	$0.440D-28$	$0.326D-02$	$-0.163D-26$
8	$-0.837D+03$	$-0.881D-04$	$-0.172D-09$	$0.326D-02$
9	$-0.719D-07$	$0.175D-45$	$0.658D+03$	$0.329D-36$

[a] No solution was generated for path 4. The given (x_1, x_2) is the endpoint of the path. The final t value is 0.999 999 999 964 968 for path 4.

TABLE 5–5 Unscaled solutions to (5–20)[a]

Path	Re(x_1)	Im(x_1)	Re(x_2)	Im(x_2)
1	$-0.128D-24$	0.0	$-0.648D-13$	$-0.112D-46$
2	$-0.392D-24$	$0.358D-33$	$-0.929D-23$	$0.260D-13$
3	$0.149D-14$	$-0.498D-46$	$-0.321D-18$	$0.102D-42$
4	$-0.165D-24$	$0.112D-28$	$0.281D-11$	$-0.236D-11$
5	$-0.149D-14$	$0.157D-21$	$-0.169D-25$	$-0.321D-18$
6	$-0.392D-24$	$-0.358D-33$	$-0.929D-23$	$-0.260D-13$
7	$0.149D-14$	$0.783D-46$	$0.321D-18$	$-0.160D-42$
8	$-0.149D-14$	$-0.157D-21$	$-0.169D-25$	$0.321D-18$
9	$-0.128D-24$	$0.312D-63$	$0.648D-13$	$0.324D-52$

[a] The "physical" solution is the positive real one, generated by path 7. The symmetry in the real solutions seems characteristic of chemical equilibrium problems. (See Chapter 9 and [Meintjes and Morgan, 1987].) No solution was generated for path 4. The given (x_1, x_2) is the unscaled endpoint of the path.

TABLE 5–6 Summary of four CONSOL8T runs using the input in Table 5–3, with and without SCLGEN and with and without the projective transformation[a]

Run	Norm of Endpoint (Exponents of 10)	Total Arc Length	Total Number of Linear Systems
Scaled; no trans.	3, 3, 3, 5*, 3 3, 3, 3, 3	45,000	36,727
Scaled; trans.	0, 0, 1, 1, 0, 1, 0*, 1, 0	39	12,168
Not scaled; trans.	-12, -15, -14, -15, -12, -14, -15, -14, 0*	50	16,823
Not scaled; no trans.	-12, -13, -13, -13, 9*, -14, -14, -14, -12	$0.1125D+09$	17,184

[a] The "*" indicates the endpoint of the path that did not converge (in the "no trans" cases) or that converged to points at infinity (in the "trans" cases). Also included are the sum of all the arc lengths of all continuation paths and the total number of linear systems solved, for both prediction and correction.

range by many less orders of magnitude than the original coefficients and that they are much better centered about unity. Now substituting $b_{i,j}$ for $a_{i,j}$ in (5-20), we solve this system (via CONSOL8T whose input is listed in Table 5-3), yielding the eight scaled solutions given in Table 5-4. (There is a solution at infinity.) Then, evoking (5-10), we get the unscaled solutions in Table 5-5.

The main thing to observe from this run is that the unscaled solutions are so small that it would be difficult to compute them without scaling and retain any resolution. Also note the t value for the divergent path.

To demonstrate how SCLGEN and the projective transformation compare with each other and how they work together, I have run three variants on this problem, as follows.

1. The coefficients were scaled by SCLGEN and then the system was solved using the projective transformation.
2. The coefficients were not scaled, and the system was solved using the projective transformation.
3. The system was solved without scaling or the projective transformation.

In Table 5-6 the four runs (including scaling without the projective transformation) are summarized. Note the regularizing effect of scaling and/or the projective transformation. Especially note the total arc length and the total number of linear equations solved. The first measures the geometry of the paths and the second measures how much work CONSOL had to do to compute the geometry. However, these statistics do not tell the whole story. It turns out that only the two scaled runs produced results sufficiently accurate for the original chemical equilibrium problem that generated it. (See Chapter 9, model A, case 1.)

EXPERIMENTS

5-1. Solve problem set 3 from Section 4-7 using CONSOL8T with IFLGPT = 0 and 1 and with IFLGSC = 0 and 1. Note the effects of scaling and the projective transformation. Generate a table like Table 5-6 for each problem.

5-2. Solve problem set 3 from Section 4-7 using CONSOL8T as in Experiment 5-1, but now with SCLCEN rather than SCLGEN. Try using the positive real solution as a nominal value. Try changing this by orders of magnitude.

CHAPTER 6

Other
Continuation Methods

6-1 INTRODUCTION

This chapter focuses on methods for finding *one* solution to a nonlinear system. Why do we want such methods when CONSOL8 guarantees finding all solutions? There are two reasons:

1. CONSOL8 is limited to polynomial systems.
2. CONSOL8 tracks d paths, where d is the total degree of the system, and the time to compute many paths can be prohibitive.

Although the methods in this chapter are applicable to nonpolynomial systems and are relatively fast, they have a serious drawback. They do not always work. First, they may not converge to any solution. Second, they may converge to an unwanted solution. For example, a negative solution might be found in a context in which only the positive solutions have physical meaning. Thus we must view these methods as being less reliable in practice than CONSOL8. However, because their convergence characteristics tend to be *global* rather than *local*, they may converge under circumstances in which Newton's method does not. This is especially significant when we use *real* rather than *complex* arithmetic.

In problem environments in which we seek real solutions, it is natural to want to limit our algorithms to real arithmetic. If we compute a full solution list, then (of course) we get whatever solutions have physical meaning. But seeking one solution in complex space will often yield only an unwanted complex solution.

Thus continuation in real space will be a theme of this chapter. However, we will also consider complex continuations. In fact, I will make a point of directing your attention to comparisons between real and complex continuations, to illustrate what the complex context does for the solution process.

In the next section I am going to outline some general principles for defining continuation systems. We will have to consider paths that "turn back in t" and develop a CONSOL code, CONSOL9, to track such paths. In Section 6-3 we look at a specific type of continuation. The examples, exercises, and experiments for this chapter not only demonstrate CONSOL9's ability to solve larger polynomial systems and nonpolynomial systems, but include cases for which the method fails.

6-2 GENERAL PRINCIPLES

The basic continuation idea is very simple. We obtain by some means a continuation system

$$h(x, t) = 0 \qquad\qquad (6\text{-}1)$$

for which we can solve the initial system

$$h(x, 0) = 0, \qquad\qquad (6\text{-}2)$$

where now "solve" means that we can find one or more solutions but not necessarily all of them. We want solutions to the system

$$h(x, 1) = 0. \qquad\qquad (6\text{-}3)$$

We track a continuation path defined by (6-1) with start point a solution to (6-2), as in CONSOL8. When $t = 1$ we get a solution to (6-3). Often, in engineering applications, the t is a physical parameter (e.g., temperature or pressure) and continuation is a straightforward way of using numerical techniques to extend what we can easily compute to more difficult cases.

The approach we will take is much more structured than is implied by this loose description of continuation. First, we base our choice of (6-1) on theorems that guarantee success, within the limitations of certain hypotheses. Second, we draw on experience and experimentation, which suggest some generally successful rules of thumb.

An approach to constructing general continuation systems is given in [Chow et al., 1978] and in [Watson, 1979, 1980]. From this work we can extract the following recipe.

Define a continuation system

$$h(q, x, t) = 0 \qquad\qquad (6\text{-}4)$$

with $q \in V^n$, $x \in V^n$, $t \in [0, 1]$, and $h(q, x, t) \in V^n$, where V^n denotes either

R^n or C^n. The system (6-4) need not consist of polynomial equations, but it should be smooth (twice continuously differentiable). Now, the following conditions guarantee that we can track paths to find solutions.

1. $h(q, x, 1) = 0$ is the system we want to solve (for x, given q).
2. $h(q, x, 0) = 0$ is easy to solve for x, given q.
3. The matrix $dh_{q,x,t}(q, x, t)$ has rank n whenever $h(q, x, t) = 0$, where $dh_{q,x,t}$ denotes the partial derivative matrix (Jacobian) of h with respect to the variables $q_1, q_2, \ldots, q_n, x_1, \ldots, x_n, t$, an $n \times (2n + 1)$ matrix. (If V^n denotes C^n, the "rank" is the complex rank. But see Appendix 3.) This need hold only for $t \in [0, 1)$.
4. One of the following holds:
   (a) The continuation curves are increasing in t.
   (b) $h(q, x, 0) = 0$ has a unique solution, x, for each q.
5. The continuation curves for (6-4) are bounded. We want this to hold for $t \in [0, 1]$.

Condition 3 implies that if we choose q "independently" (e.g., at random), we will have well-defined continuation paths. (This is in analogy with the "Independence Principles" cited in previous chapters. It is formally a consequence of the Preimage and Transversality Theorems referenced in Appendix 4.) Condition 4 implies that the paths do not turn around and end up with $t = 0$ (see Fig. 3-7). Condition 5 implies that a path will eventually progress until $t = 1$, and in particular it will not diverge to infinity.

   If conditions 1 through 5 hold, then we fix q in V^n, and attempt to track the path from a solution of $h(q, x, 0) = 0$ to a solution of $h(q, x, 1) = 0$. If we are relying on condition 4b rather than 4a, we must allow for the continuation path to "turn back" in the t parameter, a situation that we have not discussed before since it cannot occur in CONSOL8. The 4b case is more a nuisance than a stumbling block, but it does rule out the simplest path-tracking schemes, those that merely increment t. We consider path tracking below.

   What is easy and what is difficult about this recipe? There are several "standard" ways of constructing h. If these do not work out, then finding h can be difficult, in the sense that it requires experience and persistence. (See below. Also see [Watson, 1981].) Once we have h, conditions 1 and 2 are taken care of. Condition 3 is usually easy to satisfy. Merely adding "$(1 - t)q$" to h will often be satisfactory. If $V^n = C^n$ and h is complex analytic, 4a will hold as a consequence of the Cauchy–Riemann equations. This is the case, for example, for the CONSOL8 continuation (see Appendix 3). Otherwise, 4a or 4b will have to be established by a separate argument. Condition 5 is generally the most difficult to establish rigorously. If h is polynomial and $h = 0$ has no solutions at infinity for any fixed t, then paths are bounded as in Chapters 2 and 3, even in real arithmetic. But often h will have solutions at infinity (see Exercises 6-2 through 6-4). In some cases a plausibility argument based on physical reasoning will suggest that we can assume 5 when we cannot prove it. Of course, there is no law against trying the continuation to see if the path diverges. If your guesses are informed by the conditions above, at least you will have a better sense of why things are failing, if they fail.

What is involved in tracking paths that are not increasing in t? Figure 3-6 shows a path that "backtracks" in t. It is labeled "bad path behavior" because in the context of CONSOL8 it is undesirable and avoidable. In our current context it is still undesirable, but it is (sometimes) unavoidable also. The good news is that we can handle such paths. However, we must work harder, to develop a more sophisticated path tracker, and the computer will have to work harder too, expending CPU time to make sure that everything works out.

It clearly makes no sense to proceed by incrementing t, as we did in the simple predictor–corrector path trackers for CONSOL described in Chapters 1 through 4. However, a straightforward extension of these approaches will work. Imagine that we are at the point (x^0, t^0) on the path and we want to move along the path to a new point (see Fig. 4-8). First, we compute a tangent vector to the path at (x^0, t^0). Since we cannot now count on t increasing, we must take some care to "predict" along the path in a consistent direction. Then we "correct" back to the path using Newton's method. Let's call the version of TRACK-T we are about to develop TRACK-B, since it can backtrack.

However, we should backtrack ourselves a bit here and reconsider the way that we have been *parametrizing* the continuation paths. Up until now, we have used t as our path parameter. This was reasonable, since we could count on each t corresponding to a unique point on the curve. We inched along the curve by incrementing t bit by bit. But t is no longer a satisfactory path parameter. If the path backtracks in t, there will be at least one value of t that corresponds to several points on the path (consider Fig. 3-6). It makes† more sense now to parametrize the path by its arc length.† That is, as we inch along the path, we mark where we are by how far we have come. Let's denote *arc length* by s and use s as our path parameter. For each s there is a unique path point, $(x(s), t(s))$, the point on the path that is distance s along the path in the direction we are moving. Although it may seem at first that arc length will be harder to deal with computationally than t, we will see that the formal algebraic manipulations we must carry out are similar to those needed to implement TRACK-T.

Let's reconsider the material at the end of Chapter 3 and rederive equations (3-26) through (3-28) using s as the path parameter instead of t.

---

† See Appendix 3 for a discussion of parametrizations and arc length.

As before, $I = [0, 1) = \{t: 0 \leq t < 1\}$, $w = (x, t)$, but now we take $w(s) = (x(s), t(s))$ to parametrize the path. Then $h(s) = h(w(s)) = 0$ because $w(s)$ is by definition a solution path. Because $h(s)$ is *constant*, we get

$$\frac{dh}{ds}(s) \equiv 0. \tag{6-5}$$

Applying the n-variable chain rule to (6-5) yields

$$dh(w(s)) \frac{dw}{ds} = 0, \tag{6-6}$$

where dh denotes the $n \times (n + 1)$ Jacobian matrix of h with respect to w_1, ..., w_{n+1}:

$$dh_{i,j} = \frac{\partial h_i}{\partial w_j}(w) \qquad \text{for } i = 1, \ldots, n; \quad j = 1, \ldots, n + 1.$$

This dh is identical with the dh in equation (3-27), and the comments given in Chapter 3 on real and complex calculus are still valid (as clarified in Appendix 3).

Now (6-6) is the product of an $n \times (n + 1)$ matrix $dh(w(s))$ and an $n + 1$ column vector dw/ds set equal to a column of 0's:

$$\begin{bmatrix} dh_{1,1} & dh_{1,2} & \cdots & dh_{1,n} & dh_{1,n+1} \\ \vdots & & & \vdots & \vdots \\ dh_{n,1} & dh_{n,2} & \cdots & dh_{n,n} & dh_{n,n+1} \end{bmatrix} \begin{bmatrix} \dfrac{dw_1}{ds} \\[2mm] \dfrac{dw_2}{ds} \\[2mm] \vdots \\[2mm] \dfrac{dw_n}{ds} \\[2mm] \dfrac{dw_{n+1}}{ds} \end{bmatrix} = \begin{bmatrix} 0 \\ 0 \\ \vdots \\ 0 \\ 0 \end{bmatrix}. \tag{6-7}$$

At this point we reach our first significant deviation from Chapter 3. We will not factor the dh matrix into dh_x and dh_t. The t no longer has the special status it had in Chapter 3. We can no longer count on dh_x being nonsingular. We must proceed with more caution, as follows. We know that dh has rank n, even though dh_x may not have rank n. (This is a consequence of condition 3 above.) Thus there is (at least) one column of dh, say the kth, with the property that if we omit this column from dh, the resulting square matrix is nonsingular. When we were developing CONSOL in the previous chapters, we could always take $k = n + 1$. But now we will have to choose k as we go along, changing the choice as necessary. Once we choose k, we proceed in much the same way as before. It will be convenient to have a name for this kth column. Let's call it the "balance column" of dh and k the "balance index." Note that the balance index is not uniquely determined by the criterion that dh with the balance column omitted is nonsingular. (Indeed, every column may have this property.) Thus we will need to develop a method for picking out a balance index, and different methods will yield different choices. When we reference the balance index, it is understood implicity that such a method has been selected.

We need to establish some notation for breaking equation (6-7) into blocks. Let dh_k denote the kth column of dh and $d\hat{h}_k$ the $n \times n$ matrix formed by omitting the kth column from dh. Let $dh_{\langle -k \rangle}$ denote the $n \times (k-1)$ matrix consisting of the first $(k-1)$ columns of dh, and $dh_{\langle +k \rangle}$ the $n \times (n-k+1)$ matrix consisting of the last $(n-k+1)$ columns of dh. Thus we have

$$dh = [dh_{\langle -k \rangle} \mid dh_k \mid dh_{\langle +k \rangle}],$$

and

$$d\hat{h}_k = [dh_{\langle -k \rangle} \mid dh_{\langle +k \rangle}].$$

Similarly, dw_k/ds denotes the kth element of dw/ds, $d\hat{w}_k/ds$ denotes dw/ds with the kth element omitted, $dw_{\langle -k \rangle}/ds$ the first $(k-1)$ elements of dw/ds, and $dw_{\langle +k \rangle}/ds$ the last $(n-k+1)$ elements. Then equation (6-7) can be written

$$[dh_{\langle -k \rangle} \mid dh_k \mid dh_{\langle +k \rangle}] \cdot \begin{bmatrix} \dfrac{dw_{\langle -k \rangle}}{ds} \\[1.5ex] \dfrac{dw_k}{ds} \\[1.5ex] \dfrac{dw_{\langle +k \rangle}}{ds} \end{bmatrix} = 0, \qquad (6\text{-}8)$$

where the "0" now denotes a column of 0's. This gives us

$$d\hat{h}_k \frac{d\hat{w}_k}{ds} + dh_k \frac{dw_k}{ds} = 0, \qquad (6\text{-}9)$$

in analogy with equation (3-29).

Note that (6-9) requires a choice of the scalar dw_k/ds. (Usually, I take $dw_k/ds = 1$.) Equation (6-9), being the equation of the tangent line to the path, must have a *line* of solutions. By choosing dw_k/ds, we fix a point on the line. After we solve (6-9), obtaining dw/ds, any scalar multiple of this dw/ds will also satisfy (6-9), yielding another point on the tangent line. It is important, however, that our new tangent be oriented along the path in the same direction as the last tangent (see Fig. 6-1). Thus fixing the *sign* of dw/ds, or more precisely, deciding when to multiply dw/ds as obtained from (6-9) by -1, is an important numerical decision which we must make for

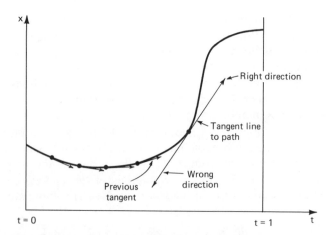

Figure 6-1 A tangent line to a continuation curve always has two directions.

TRACK-B and which we did not have to make for TRACK-T. (If we know that t is always increasing along the path, then the appropriate tangent orientation is predetermined.)

As in Chapter 3, we see that dw/ds can be viewed as an ordinary differential equation. Thus we can track the paths computationally with any numerical ordinary differential equation solver, as noted in Chapter 4. This is always an alternative to the predictor–corrector methods described here. Further alternatives are available in HOMPACK ([Watson et al., 1986]) and as cited in the other references to Section 4-5-1.

Now, having rederived the basic equations, let's consider the new numerical issues we must settle:

1. At each step, how do we choose the balance index for equation (6-9)?
2. At each step, after solving (6-9), how do we decide whether to change the sign of dw/ds?
3. To carry out the correction part of the step, what do we need to do differently from TRACK-T?

There are no unique or obviously best ways to settle the issues listed above. As always, I will suggest simple approaches that have worked for me. But see [Allgower, 1981] for a further discussion of these ideas, or see any of the references on path tracking cited in Chapter 4.

As above, we denote the balance index by k. For every step, take $dw_k/ds = 1$. After computing dw/ds from equation (6-9) (and changing its sign, if necessary), we will multiply the tangent by a scale factor consistent with the current step size, just as in TRACK-T. For the initial step of a path, choose $k = n + 1$. For any later step along the path, we will choose k by examining dw/ds from the previous step. Let k correspond to the component of dw/ds of greatest absolute value. That is, let k be the index of the $|dw_i/ds|$ that is the greatest for $i = 1$ to $n + 1$. Basically, what we want to avoid is choosing for k an index of a component of the tangent vector that is approaching zero. Thus we choose the index of the largest entry from the previous step. Clearly, this will work if we are not taking too large a step.

Let (x^0, t^0) be a point on the path. The tangent line to the path at (x^0, t^0) has two directions. The solution to (6-9) always yields a unique tangent direction, but not necessarily the direction we want to go. Let dw^0/ds denote the previous tangent and dw/ds the (current) solution to (6-9). We will change the sign of dw/ds if the angle between dw^0/ds and dw/ds is not acute (see Figs. 6-1 and 6-2). The logic of this rule is that the next tangent should, in fact, make a small angle with the previous one, but an erroneous switch in direction will change the small angle θ into the large angle $180° - \theta$. This test reduces simply to checking the sign of the dot product of dw^0/ds and dw/ds. The rule is:

$$\text{If } \left\langle \frac{dw^0}{ds}, \frac{dw}{ds} \right\rangle < 0, \quad \text{then } \frac{dw}{ds} = -\frac{dw}{ds},$$

where $\langle \cdot, \cdot \rangle$ denotes the dot product. (Recall that if $a \in R^m$ and $b \in R^m$, then $\langle a, b \rangle = a_1b_1 + a_2b_2 + \cdots + a_mb_m$.) This rule follows from the familiar formula for the cosine of the angle, θ, between two vectors, v_1 and v_2:

$$\cos \theta = \frac{\langle v_1, v_2 \rangle}{|v_1||v_2|}.$$

After we have "predicted" our next path point (the tip of the chosen tangent vector, suitably scaled by the step length), we need to "correct" it back toward the path. In strict analogy with TRACK-T, we fix the kth com-

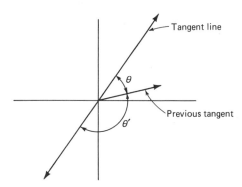

Figure 6-2 Previous tangent and tangent line at (x^0, t^0), translated to the origin. The acute angle θ indicates the correct direction for the new tangent. Note that $\theta + \theta' = 180$ degrees.

ponent of the predicted point, and do an $n \times n$ Newton's method correction on the other components. TRACK-B is implemented in CONSOL9, the code I have included for continuation on general nonlinear systems (Appendix 6, Part A, Section 5).

It will be useful to define *turning points* with respect to t of continuation curves. Given a continuation curve, $w(s) = (x(s), t(s))$, t is a smooth function of the real variable s. We call $w(s^0)$ a *turning point* of the curve (with respect to t) if $t(s^0)$ is either a relative maximum or relative minimum of $t(s)$. Thus $(dt/ds)(s^0) = 0$ at a turning point, $w(s^0)$. (Naturally, $(dt/ds)(s^0) = 0$ does not imply that $w(s^0)$ is a turning point. See [Kaplan, 1973, pp. 176–177] for a review of the relevant material from calculus.) We could consider turning points with respect to the other components of $w(s)$, but we are interested here only in $t(s)$. For complex-polynomial continuations, the continuation paths have no turning points (see Appendix 3).

To illustrate the new features of TRACK-B, let's develop several examples. Consider continuations of the form

$$h(x, t) = f^a(x) + f^b(t), \tag{6-10}$$

where $x \in R^1$,

$$f^a(x) = x(x - 1)(x - 2) = x^3 - 3x^2 + 2x,$$

and f^b depends on t alone. For the first example, we take

$$f^b(t) = 0.45(1 - 2t). \tag{6-11}$$

We will be tracking a real continuation path from $t = 0$ to $t = 1$. In fact, Fig. 6-3 shows the path. Its simple cubic character is, of course, to be expected from the form of f^a. Let us consider how TRACK-B will track this path. First, observe that

$$dh = [dh_x, dh_t] = [df^a_x, df^b_t]$$

$$= [3x^2 - 6x + 2, -0.9],$$

and according to the TRACK-B logic we will choose

$$\frac{dw}{ds} = \begin{bmatrix} \dfrac{dx}{ds} \\ \dfrac{dt}{ds} \end{bmatrix}$$

to be

$$\frac{dw}{ds} = \begin{bmatrix} -dh_t \\ dh_x \\ 1 \end{bmatrix} \tag{6-12}$$

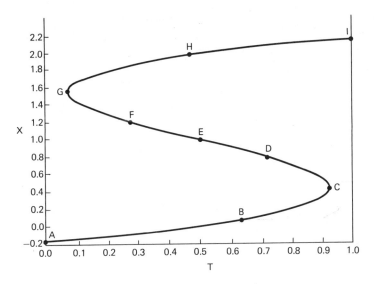

Figure 6-3   Path for continuation (6-10) with $f^b(t) = 0.45(1 - 2t)$.

if $k = 2$ (where k is the balance index), and

$$\frac{dw}{ds} = \begin{bmatrix} 1 \\ -dh_x \\ \overline{dh_t} \end{bmatrix} \tag{6-13}$$

if $k = 1$. Then we will change the sign of dw/ds if necessary. Our rule for choosing k is: For step 1, take $k = 2$. For step i, consider the $dw/ds = (dx/ds, dt/ds)$ used in step $i - 1$. Take $k = 1$ if $| dx/ds | > | dt/ds |$, take $k = 2$ if $| dx/ds | < | dt/ds |$, and leave k unchanged if $| dx/ds | = | dt/ds |$. Our rule for changing the sign of dw/ds is: Replace dw/ds by $-dw/ds$ if $\langle dw^0/ds, dw/ds \rangle < 0$, where dw^0/ds denotes the tangent used in the previous step, dw/ds is the vector that we have just obtained using (6-12) or (6-13), and $\langle \cdot, \cdot \rangle$ denotes the dot product.

Note in Fig. 6-3 that nine points on the path are designated by letters: A through I. Point A is the start point for the path; $A = (x, t) = (-0.1759, 0)$. At point C, $dh_x = df_x^a = 0$. (Points C and G are turning points of the path.) Therefore, at some point between A and C, k must change from 2 to 1. For future reference, note that $dh_x > 0$ between A and C, $dh_x < 0$ between C and G, and $dh_x > 0$ after G. We begin with $dw/ds = (0.9/dh_x, 1)$, switching to $(1, dh_x/0.9)$ as we approach and move past C. Note that the signs are correct in terms of increasing and decreasing x and t, and they do not need to be changed. The TRACK-B "dot product test" confirms this. However, at point $E = (1, 0.5)$, we have $dh = (-1, -0.9)$. We see that $dw/ds = (-0.9, 1)$ if $k = 2$ and $dw/ds = (1, -1/0.9)$ if $k = 1$. Thus in either case, we choose $k = 2$. (We must assume, for this discussion, a reasonably small step size.) Note also that if $k = 2$, the sign of dw/ds must be changed. From C to G, t is decreasing and x is increasing. Thus $dx/ds > 0$ and $dt/ds < 0$. When we take the dot product we get

$$\left\langle \frac{dw^0}{ds}, \frac{dw}{ds} \right\rangle = \left\langle (-0.9, 1), \left(\frac{dx}{ds}, \frac{dt}{ds} \right) \right\rangle$$

$$= -0.9 \frac{dx}{ds} + \frac{dt}{ds} < 0,$$

which tells TRACK-B to multiply dw/ds by -1.

At G, $dh_x = 0$, which means that k must have changed to 1 between E and G. Past G, dh_x will grow, until at some point (designated H), k must change back to 2. At the endpoint, $I = (2.176, 1)$, $dh = [3.144, -0.9]$, and $dw/ds = (0.2863, 1)$. This confirms that k changed back to 2 before the path reaches its endpoint.

Now consider a second example, where

$$f^b(t) = (1 - 2t)^3. \tag{6-14}$$

The associated continuation path is graphed in Fig. 6-4. The characteristics of this path are similar to those of the first example. One key difference is that at point $E = (1, 0.5)$, $dh = [-1, 0]$, forcing $k = 2$ in the vicinity of E. Although $k = 2$ at E in the previous example, dh_t never equaled zero. Thus, although TRACK-B called for changing to $k = 2$, this change was not required numerically. Had we decided, in defining TRACK-B, to change k

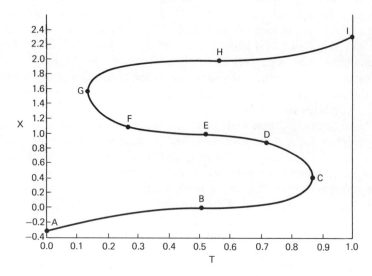

Figure 6-4   Path for continuation (6-10) with $f^b(t) = (1 - 2t)^3$.

only when the absolute value of some component of dw/ds exceeded twice that of dw_k/ds (a reasonable alternative rule), then k would not have changed at E in the first example. However, in the second example k must be 2, and no reasonable rule would allow k to remain 1 at E.

There are two natural ways to set up continuations, both of which often do not work very well:

(a) Let the continuation parameter t be a physically meaningful parameter: temperature, pressure, or such.
(b) Put t's in front of the complicated terms in the system so that, when $t = 0$, it is so simple that it can be solved easily.

Obviously, (a) is natural in some contexts and (b) is always tempting. I am not saying that these never work or even not to try them. I am saying don't be surprised when they fail, and when they fail, consider some "unnatural" alternatives.

The issue here is that when $h(x, 0)$ is *structurally* simpler than $h(x, t)$ for $t > 0$, the *path structure* may also be qualitatively different as $t \to 0$. Bifurcation and divergence may be the result. Although we can sometimes live with such path anomalies, we would rather not have to. And sometimes in these cases we cannot get started from our start point. The $t = \Delta t$ jump is too big.

One way of constructing continuations with some favorable generic properties is

$$h(x, t) = f(x) - (1 - t)f(q) = 0 \tag{6-15}$$

for a randomly chosen $q \in V^n$. When $t = 0$, $x = q$ is a solution. Another general-purpose continuation is

$$h(x, t) = (1 - t)(x - q) + tf(x) = 0, \tag{6-16}$$

where $q \in V^n$ is chosen at random. In both of these standard continuations we try to "make them more generic" by the choice of a random vector q. In [Garcia and Zangwill, 1981, Chap. 1], (6-15) is called the "Newton homotopy" and (6-16) the "fixed-point homotopy," where here "homotopy" and "continuation" are essentially synonyms. Both continuations have been considered by a number of investigators. For example, see [Watson, 1979].

I have found that adding a small imaginary part to q sometimes "smooths out" the path without leading to a complex solution.† This will not work, though, when there is a powerfully attractive complex solution nearby and no real solution in the neighborhood. But it does give me the

---

† But the continuation does have to be implemented in complex arithmetic. This means that the HFNN subroutine must acknowledge the complex structure, although the rest of CONSOL9 need not. See HFNN3, included with CONSOL9.

alternative of converging to a solution I do not want, rather than diverging. The former is usually a lot cheaper (in computer time) than the latter.

The exercises and experiments for this section develop some of these ideas further. In Section 6-3 we will focus on a variant of (6-16).

EXERCISES

6-1. Consider (6-8) and (6-9). Assume that dh is nonsingular and dw/ds is a solution to (6-8). Show that if $dw_k/ds \neq 0$, then $d\hat{h}_k$ is nonsingular.

6-2. Assume that $h(x, t) = (1 - t)g(x) + tf(x)$, where g and f are polynomial systems and $0 < \deg (g_j) < \deg (f_j)$ for all j. Show that
   (a) $h(x, 0) = 0$ has a solution at infinity. (*Hint:* Recall the comments on "declared degree" in Chapter 3.)
   (b) $h(x, t) = 0$ has a solution at infinity for $t > 0$ if and only if $f(x) = 0$ has a solution at infinity.

6-3. Define

$$f(x, y) = \begin{bmatrix} x^2 + xy - 1 \\ y^2 + x - 5 \end{bmatrix} = \begin{bmatrix} 0 \\ 0 \end{bmatrix} \tag{6-17}$$

and

$$h(x, y, t) = \begin{bmatrix} x^2 + txy - 1 \\ y^2 + tx - 5 \end{bmatrix} = \begin{bmatrix} 0 \\ 0 \end{bmatrix}. \tag{6-18}$$

Note that $h(x, y, 0) = 0$ has the four real solutions $(x, y) = (\pm 1, \pm\sqrt{5})$, generating four real continuation paths.
   (a) Show that for any fixed $t \in [0, 1]$, (6-18) has no solutions at infinity.
   (b) Show that if a continuation curve to (6-18) has a turning point in t at (x_0, y_0, t_0), then

$$4x_0y_0 + 2t_0y_0^2 - t_0^2x_0 = 0.$$

Hint: At a turning point in t, $dt/ds = 0$. But $w = [dx/ds, dy/ds, dt/ds]^T$ satisfies $dh \cdot w = 0$, where

$$dh = \begin{bmatrix} 2x + ty & tx & xy \\ t & 2y & x \end{bmatrix}.$$

   (c) Show that the system

$$x^2 + txy \quad - 1 \quad = 0$$
$$y^2 + tx \quad - 5 \quad = 0$$
$$4xy + 2ty^2 - t^2x = 0$$

has no real solution with $0 \leq t \leq 1$. (*Suggestion:* Why not use CONSOL8T?)
   (d) Is it fair to conclude from parts (a), (b), and (c) that the four real continuation curves for (6-18) will converge to four real solutions of (6-17)? (Compare Experiment 6-4.)

6-4. Define

$$h(x, y, t) = \begin{bmatrix} tx^2 + xy - 1 + 10(1 - t) \\ ty^2 + x - 5 \end{bmatrix} = \begin{bmatrix} 0 \\ 0 \end{bmatrix}. \qquad (6\text{-}19)$$

(a) Accepting that the *declared degree* of each equation in (6-19) is 2, show that for any fixed $t \in [0, 1]$, (6-19) has solutions at infinity only when $t = 0$, and in this case the solutions at infinity are $(x, y) = (1, 0)$ and $(0, 1)$. What does this imply about the continuation paths?

(b) Show that continuation paths for (6-19) have turning points only if the system

$$tx^2 + xy - 1 + 10(1 - t) = 0$$
$$ty^2 + x - 5 = 0 \qquad (6\text{-}20)$$
$$4t^2xy + 2ty^2 - x = 0$$

has a real solution with $0 \le t \le 1$. Does (6-20) have such a solution?

(c) Describe all possible types of behavior for continuation paths allowed by parts (a) and (b). (Compare Experiment 6-5.)

(d) Define

$$\bar{h}(x,y,z,t) = \begin{bmatrix} tx^2 + xy - z^2 + 10(1 - t)z^2 \\ ty^2 + xz - 5z^2 \\ r_1x + r_2y + r_3z \end{bmatrix} = \begin{bmatrix} 0 \\ 0 \\ 1 \end{bmatrix}, \qquad (6\text{-}21)$$

with r_1, r_2, and r_3 real numbers, as the projective transformation of h. Find r_1, r_2, r_3 so that (6-21) has no real solutions at infinity for any fixed $t \in [0, 1]$. (Compare Experiment 6-6.)

(e) Can you show that (6-21) has no real solutions at infinity for any fixed $t \in [0, 1]$ if the r_j are randomly chosen?

EXPERIMENTS

Note: For the experiments below, run CONSOL9 with the IFLGST = 1 option so that the full continuation path will be printed out (see subroutine INPA1). In some cases it will be enlightening to plot the paths, t versus one of the other variables or t versus $|x|$. In some cases, there is more than one solution. In some cases, one continuation works much better than the other.

6-1. Run the example defined by (6-10) and (6-11). Confirm its "backtracking" characteristics as described in the text.

6-2. Run the example defined by (6-10) and (6-14). Confirm its "backtracking" characteristics as described in the text.

6-3. Let $f^b = (1 - 2t)^2$ and run CONSOL9 with (6-10). What does the path look like? Does it converge? What if "2" is replaced by another positive integer?

6-4. Try to solve (6-17) using the continuation (6-18). (Compare Exercise 6-3.)

6-5. Try to solve (6-17) using the continuation (6-19). (Compare Exercise 6-4.)

6-6. Try to solve (6-17) using the continuation (6-21) with

$$r_1 = \quad 1.762\ 537\ 462\ 980$$

$$r_2 = -0.993\ 758\ 928\ 100$$

$$r_3 = \quad 0.393\ 839\ 300\ 090$$

(Compare Exercise 6-4.)

6-7.(a) Using the (6-15) continuation, try to solve $f = 0$ with f defined as follows (from [Boggs, 1971]):

$$f_1 = x_1^2 - x_2 + 1$$

$$f_2 = x_1 - \cos(0.5\pi x_2)$$

(b) Try the (6-16) continuation.

6-8. (a) Using the (6-15) continuation, try to solve $f = 0$ with f defined as follows (from [Branin, 1972]):

$$f_1 = 1 - 2x_2 + 0.13 \sin(4\pi x_2) - x_1$$

$$f_2 = x_2 - 0.5 \sin(2\pi x_1)$$

(b) Try the (6-16) continuation.

6-9. (a) Using the (6-15) continuation, try to solve $f = 0$ with f defined as follows (from [Broyden, 1969] as cited by [Abbott and Brent, 1975]):

$$f_1 = 0.5 \sin(x_1 x_2) - 0.25 \frac{x_2}{\pi} - 0.5x_1$$

$$f_2 = \left(1 - \frac{1}{4\pi}\right)(e^{2x_1} - e) + \frac{ex_2}{\pi} - 2ex_1,$$

where e denotes the base of the natural logarithms.

(b) Try the (6-16) continuation.

6-10. (a) Using the (6-15) continuation, try to solve $f = 0$ with f defined as follows (from [Fletcher and Powell, 1963] as cited by [Incerti et al., 1981]):

$$f_i = \sum_{j=1}^{5} (a_{ij} \sin x_j - b_{ij} \cos x_j) - c_i \qquad \text{for } i = 1, 5,$$

where

$$(a_{ij}) = \begin{bmatrix} 52 & 33 & 39 & 68 & 41 \\ 15 & 92 & 72 & 52 & 31 \\ 50 & 63 & 26 & 41 & 64 \\ 95 & 14 & 61 & 39 & 72 \\ 32 & 3 & 98 & 69 & 82 \end{bmatrix}$$

and

$$(b_{ij}) = \begin{bmatrix} 45 & 14 & 74 & 10 & 54 \\ 9 & 8 & 46 & 67 & 40 \\ 69 & 76 & 73 & 97 & 44 \\ 23 & 54 & 63 & 99 & 72 \\ 29 & 70 & 42 & 30 & 87 \end{bmatrix}$$

(b) Try the (6-16) continuation.

6-3 THE CONVEX–LINEAR CONTINUATION

In this section we consider the continuation

$$h(x, t) = (1 - t)(Px - q) + tf(x) \qquad (6-22)$$

for finding a solution to $f(x) = 0$, where P is an $n \times n$ matrix with entries in V and $q \in V^n$. P is assumed to be invertible, and generally q is chosen at random. I call this the convex–linear continuation because it is the convex combination of f and a linear system. It is a variant of (6-16). Conditions 1 and 2 (from Section 6-2) are trivial. Condition 3 also follows easily, since

$$dh_{q,x,t} = [-(1 - t)I_n, (1 - t)P + t\, df(x), -(Px - q) + f(x)],$$

where I_n denotes the $n \times n$ identity matrix. Condition 4b is immediate, although we cannot assume that paths are monotonic in t (without some further argument). However, if $q \in C^n$ and f is complex analytic, condition 4a holds, as discussed in Section 6-2. (We must choose q at random in C^n. See the discussion that closes this section.)

Condition 5 may be false in even the simplest cases. For example, assume that f consists of polynomials and all the solutions of $f = 0$ are at infinity. Or, assume that f consists of polynomials, $V^n = R^n$, and all the solutions are complex (nonreal). In either case, condition 5 cannot hold.

Let's consider what happens when f is polynomial in a little more detail. First, assume that $V^n = C^n$, q is chosen at random in C^n, and the continuation takes place in complex arithmetic. Thus paths are increasing in t. Now, if a path diverges to infinity, then this implies that there is a t^0, with $0 < t^0 \leq 1$, such that $h(x, t^0) = 0$ has a solution at infinity. (The argument is similar to that given at the end of Chapter 2. See also Appendix 4, the proof of Lemma B.) This implies that $f = 0$ has a solution at infinity. [Here we must assume that deg $(f_j) > 1$ for all j.] Thus if $f = 0$ has no solutions at infinity, condition 5 must hold. We have already observed that if $f = 0$ has all its solutions at infinity, condition 5 cannot hold. If $f = 0$ has some solutions at infinity, condition 5 may hold or not, depending on the choice of P and q.

Now suppose that f is polynomial with $V^n = R^n$ and $q \in R^n$. The path is no longer necessarily increasing in t. It can backtrack, although it cannot return to intersect the $t = 0$ hyperplane. Again, we cannot count on the path being bounded. Interestingly, there are now two distinct ways that a path can become unbounded. If the path becomes unbounded, this implies that there is a t^0 with $0 \le t^0 \le 1$ such that the system $h(x, t^0) = 0$ has a solution at infinity. If $t^0 > 0$, this implies that $f = 0$ has a solution at infinity, as above, but if $t^0 = 0$, no such conclusion can be made. In fact, consider the case that all the solutions of $f = 0$ are nonreal (i.e., have nonzero imaginary parts) and $f = 0$ has no solutions at infinity. Then no solutions can be found in real arithmetic, so the continuation must fail. The only way for it to fail is for the path to become unbounded. But because $f = 0$ has no solutions at infinity, the only way the path can become unbounded is for $t \to 0$ as the path diverges (see Fig. 6-5).

The function of q in (6-22) is clear from theory; it makes the continuation path well defined and smooth. (See Appendix 4, the proof of Lemma A.) The utility of P is not as straightforward. In CONSOL8, the existence of p implies that $h(x, t) = 0$ will not have solutions at infinity for $t < 1$, so that paths can diverge only as $t \to 1$ (Appendix 4, the proof of Lemma B). Here P cannot do this (unless f is linear). What P does is control the algebraic independence of the x term in the continuation, thereby making h "more generic." It is an element of the Jacobian of h and influences overall qualitative characteristics of paths, such as arc length (see Exercises 6-5 through 6-8).

Let's consider several examples. Let $f(x) = x^2 - 1$, with $x \in R^1$,

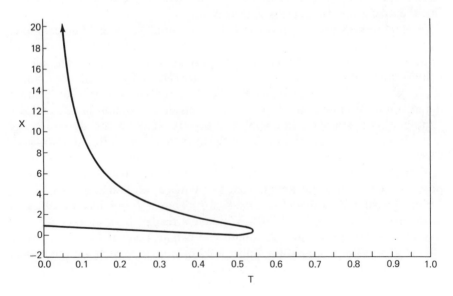

Figure 6-5   Continuation path for (6-22) diverging as $t \to 0$.

$P = 1$, and $q \in R^1$. Then we can find the continuation path for (6-22) explicitly, using the quadratic formula. Thus we have the continuation equation

$$h(x, t) = tx^2 + (1 - t)x - [(1 - t)q + t] = 0, \qquad (6\text{-}23)$$

which is equivalent to

$$x^2 + Tx - (Tq + 1) = 0, \qquad (6\text{-}24)$$

where $T = (1 - t)/t$, except when $t = 0$. Therefore,

$$x = \frac{-T \pm \sqrt{T^2 + 4(Tq + 1)}}{2}. \qquad (6\text{-}25)$$

Observe that (6-25) generates two "branches," one given by "$+\sqrt{}$" and the other by "$-\sqrt{}$."

Now we have two cases:

(a) The two branches do not meet.

(b) The two branches do meet.

From (6-25), the condition that they meet is

$$T^2 + 4(Tq + 1) = 0,$$

or

$$T = 2(-q \pm \sqrt{q^2 - 1})$$

for positive real T.

If the branches do not meet, they form two distinct paths. The path given by $-\sqrt{}$ goes to infinity as $t \to 0$ and is not the path we are associating with the continuation. The path given by $+\sqrt{}$ converges to $x = 1$ as $t \to 1$ and converges to $x = q$ as $t \to 0$. This is the path we are associating with the continuation (see Fig. 6-6). Notice that no matter what q is, the continuation "works" and that no matter what q is, the continuation always finds the solution $x = 1$.

Now let us consider the other case. Suppose that the branches do meet. If we presume that the branches match up smoothly, Fig. 6-7 is the only possible schematic. Continuation from $x = q$ diverges to infinity, while the solutions $x = 1$ and $x = -1$ are linked by a path we cannot reach using (6-23). There are several ways to argue that the branches must match up smoothly for a randomly chosen q. A formal argument can proceed along the lines of "Independence Principles." But observing from (6-25) that for any t, we cannot have more than two x values, the type of overlapping given in Fig. 6-8 is ruled out. The "just touching" case (Fig. 6-9) is the only reason q must be chosen at random. There is a (single) choice of q that allows this to happen.

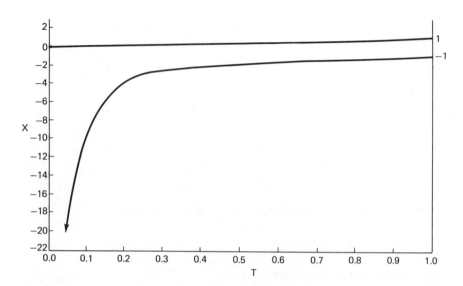

Figure 6-6 Paths for continuation (6-23), case a.

The second example I would like us to consider is $f(x) = x^2 + 1$, with $x \in R^1$, $P = 1$, and $q \in R^1$. The solutions to $f = 0$ are pure imaginary, namely $x = \pm i$, so the continuation must fail. Let's find out exactly what happens.

We have

$$h(x, t) = tx^2 + (1 - t)x - [(1 - t)q - t] = 0, \tag{6-26}$$

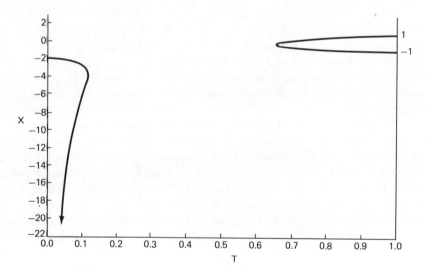

Figure 6-7 Paths for continuation (6-23), case b.

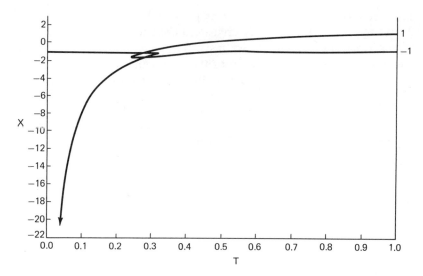

Figure 6-8 Paths for continuation (6-23) cannot overlap like this.

which is equivalent to

$$x^2 + Tx - (Tq - 1) = 0, \tag{6-27}$$

where $T = (1 - t)/t$, except when $t = 0$. Therefore,

$$x = \frac{-T \pm \sqrt{T^2 + 4(Tq - 1)}}{2}. \tag{6-28}$$

Now, as t goes to 1, T goes to zero. Thus we see that the expression under the square-root sign goes negative. Since we cannot enter complex space, the curve must "turn around" at that point and, in fact, the $+\sqrt{}$ and $-\sqrt{}$ paths must join up smoothly by the random choice of q (or by more direct arguments, as in the example above). Thus the path will look as shown in Fig. 6-5. The path becomes unbounded, not as $t \to 1$, because t never goes to 1, but as $t \to 0$.

Now let's consider a "fine point." I said at the beginning of this section that if $q \in C^n$ and f is complex analytic, paths are monotonic in t. In the example just completed, f is a polynomial, and polynomials are complex analytic. Why did the path turn around? Let me make this a more general question. If we take f to be complex analytic but with pure real constants (like a polynomial with real coefficients) and we start at a pure real start point, the continuation will behave just like a real continuation. Being complex analytic does not change anything.

The resolution of this apparent paradox is that the path goes singular at the "turning point" when we view the continuation as being carried out in complex space. In fact, the single real path that turns back at the turning point is seen to be two distinct complex paths that cross at the turning point and pass into complex space (increasing in t) and continue. However, to carry out smooth continuation, we insist that paths not cross.† Choosing q at random (in *complex* space) assures that the paths will not cross. Even a small random q will lift the crossing paths off each other, although to make

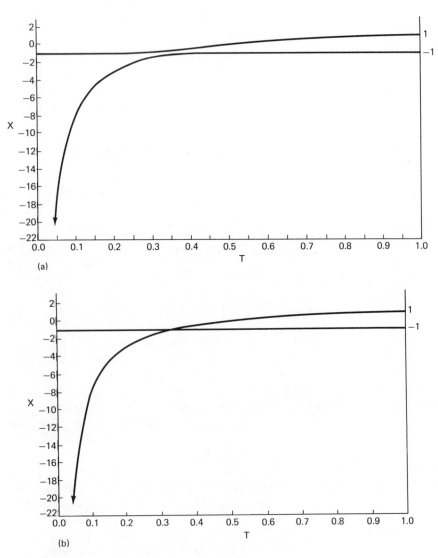

(a)

(b)

Figure 6-9   Paths for continuation (6-23) can "just touch," but only for one value of q. Shown are curves for (a) $q = -0.99$, (b) $q = -1.0$, and (c) $q = -1.001$.

† Some researchers are willing to accept paths that cross or bifurcate (see, e.g., [Allgower, 1981]).

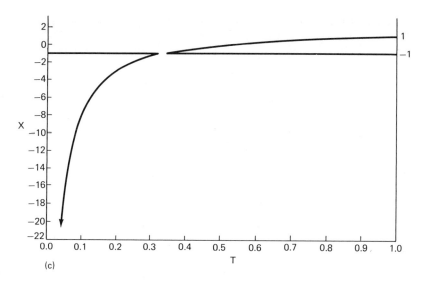

(c)

Figure 6-9   (continued)

numerical path tracking easier, we want the imaginary parts of q to be a reasonable size.

Consider dh [for h defined by (6-26) with q real] as a complex Jacobian:

$$dh = \begin{bmatrix} 2t\,\mathrm{Re}(x) + (1 - t) & -2t\,\mathrm{Im}(x) & \mathrm{Re}(x^2 - x) + (q + 1) \\ 2t\,\mathrm{Im}(x) & 2t\,\mathrm{Re}(x) + (1 - t) & \mathrm{Im}(x^2 - x) \end{bmatrix}$$

with $\mathrm{Re}(z) \equiv$ the real part of z and $\mathrm{Im}(z) \equiv$ the imaginary part of z. (See Appendix 3 for a description of how to set up the complex Jacobian.) Thus

$$\frac{1}{t}\,dh = \begin{bmatrix} 2\,\mathrm{Re}(x) + T & -2\,\mathrm{Im}(x) & \dfrac{\mathrm{Re}(x^2 - x) + (q + 1)}{t} \\ 2\,\mathrm{Im}(x) & 2\,\mathrm{Re}(x) + T & \dfrac{\mathrm{Im}(x^2 - x)}{t} \end{bmatrix}.$$

If x is real and at a turning point, then (6-28) implies that $x = -T/2$, giving

$$\frac{1}{t}\,dh = \begin{bmatrix} 0 & 0 & \dfrac{T^2 + 2T + 4q + 4}{4t} \\ 0 & 0 & 0 \end{bmatrix}.$$

Thus the real rank of dh is less than 2, and the path at the turning point has gone singular when viewed from the perspective of complex space. Note, however, that it does not go singular as a real continuation. We have

$$\frac{1}{t} dh = \frac{1}{t} [dh_x, dh_t] = \left[2x + T, \frac{x^2 - x + q + 1}{t} \right]$$

$$= \left[0, \frac{T^2 + 2T + 4q + 4}{4t} \right]$$

and $dh_t \neq 0$.

EXERCISES

6-5. Consider the continuation (6-26).
    (a) When q is real, find a turning point, (x', t'), for the continuation path that begins at q. Confirm that the path is nonsingular at that point. Confirm that $t < t^0$ on this path for some $t^0 < 1$.
    (b) Suppose that q is complex with nonzero imaginary part. Show that there are two continuation paths that do not cross. Using the formula (6-28), show that the path associated with $+\sqrt{}$ connects $x = i$ to $x = q$ and the path associated with $-\sqrt{}$ begins at $x = -i$ and diverges to $-\infty$.
6-6. Consider the continuation (6-23), and assume that the continuation paths are nonsingular.
    (a) Show that the path always finds the solution $x = 1$ (independent of q).
    (b) Make a simple change in (6-23) so that the continuation always finds the solution $x = -1$.
    (c) Show that the sign of $dh_x(x_0, 0)$ is equal to the sign of $dh_x(x^*, 1)$, where $(x_0, 0)$ is the start point and $(x^*, 1)$ is the endpoint of the continuation path. (*Hint:* What happens to the path when dh_x changes sign?)
6-7. Analyze the behavior of the continuation paths for (6-23) with P real but not necessarily equal to 1. In other words, consider $h(x, t) = (1 - t) (Px - q) + t(x^2 - 1)$ for all real choices of P and q. For which P and q will the continuation "work"?
6-8. Let $h(x, t)$ be a continuation in real space with a well-defined smooth continuation path from $(x_0, 0)$ to $(x^*, 1)$. Show that the sign of det $(dh_x(x_0, 0))$ is equal to the sign of det $(dh_x(x^*, 1))$. [*Hint:* Express dt/ds in terms of det (dh_x) as in Appendix 3.] What does this result suggest about the utility of P in (6-22)?

EXPERIMENTS

6-11. Track the continuation (6-26) in complex space using $q = 1 + 10^{-k}i$ where $i = \sqrt{-1}$ for $k = 0, 1, 2, 4, 8, 16,$ and 32. What should happen? What does happen? Why?

Note: In the experiments that follow, solve the given system using the continuation

(6-22). Use various choices for P and q. Try to find extra solutions by varying P as well as q.

6-12. The following system has a real solution, a solution at infinity, and a complex conjugate pair. Can you detect how the presence of these nonreal and nonfinite solutions influences the behavior of (6-22)?

$$xy + x = 0$$
$$y - x^2 = 0$$

6-13. The following system has no finite solutions:

$$x^2 + y^2 - 1 = 0$$
$$x^2 + y^2 - 2 = 0$$

6-14. What happens when there are an infinite number of well-spaced solutions?

$$\cos(x) - 1 = 0$$

6-15. What happens when there are an infinite number of solutions that are not well spaced?

$$\sin(e^x) - x = 0$$

6-16. Two-point boundary value problem from [Moré and Cosnard, 1979]:

$$f_k = 2x_k - x_{k+1} - x_{k-1} + 0.5h^2(x_k + t_k + 1)^3 \qquad \text{for } k = 1, n$$

with $x_0 = x_{n+1} = 0$, $t_k = k \cdot h$, $h = 1/(n + 1)$. Solve for $n = 5$ and $n = 10$. There is a unique real solution x and $-0.5 \le x_k \le 0$ for $k = 1, n$.

6-17. Brown's almost linear system:

$$f_k = x_k + \left(\sum_{j=1}^{n} x_j \right) + (n + 1) \qquad \text{for } k = 1, n - 1$$

$$f_n = \prod_{j=1}^{n} x_j - 1.$$

Solve for $n = 5$ and $n = 10$. There are three and two real roots, respectively, for these two cases. See [Moré and Cosnard, 1979] for details. Compare with Exercise 3-2 part (e).

6-18. From [Watson, 1979]:

$$f_k = \frac{1}{2n} \left[\left(\sum_{i=1}^{n} x_i^3 \right) + k \right] - x_k \qquad \text{for } k = 1, n.$$

Solve for $n = 5$ and $n = 10$.

6-19. From [Watson, 1979]:

$$f_k = \exp \left[\cos \left(k \sum_{i=1}^{n} x_i \right) \right] - x_k \qquad \text{for } k = 1, n.$$

Solve for $n = 2$ and $n = 5$.

PART II

Applying the Method

The following chapters contain systematic case histories for various applications, "systematic" because they present general approaches but "case histories" because they each focus on a single problem. Chapter 7 on "reduction" is preliminary to the others. Chapters 8, 9, and 10 deal with applications from geometric modeling, chemical equilibrium equations, and the kinematics of mechanisms, respectively. Workers in these areas may find the specific techniques of interest, but the general purpose of these case histories is to illustrate how certain commonly encountered difficulties are overcome. Specifically:

- Using on-line reduction to minimize the numerical difficulties caused by nongenericity† (Chapter 8)
- Dealing with extreme scaling using both on-line and off-line methods (Chapter 9)
- Using off-line reduction to make a system "small enough" to solve (Chapter 10)

The exposition is organized in such a way that you need not be familiar with the application areas to follow the solution methodology.

Let me finish this introduction with a final, important observation. You should never solve a system of polynomial equations in the form in which it arises. A system should always be "reduced" and "scaled" before it is "solved." The chapters that follow—the "systematic case histories"—merely illustrate this basic reduction–scaling–solution process for specific application areas. When you understand the goals and approaches of this process, you will be able to apply it to your own problems.

Reduction changes the original system into a system with a new structure. *Scaling* preserves the structure of the system but modifies the coefficients. *Solution* generates solutions to the resulting system. Reduction does not involve approximations, but rather, simplifies the original system by generating a strictly derived system that preserves the basic information of the original. Scaling improves the numerical characteristics of the system without altering the solution structure at all. We will assume here that solving will be effected by CONSOL, but the concepts described are useful for almost every general solution method I have used.

---

† I call a system "generic" if it has no singular solutions or solutions at infinity. Most systems that arise in practice are nongeneric.

CHAPTER 7

Reduction

7-1 INTRODUCTION

The purpose of reduction is to reduce the total degree of a system. A reduced system is generally easier to solve. By "easier" I mean that the solution method will be faster and more reliable. The advantage for CONSOL is immediate: fewer paths to track. Here is a simple illustration of reduction: the system

$$x^2 + y^2 = a^2$$
$$(x - b)^2 + y^2 = c^2 \qquad (7\text{-}1)$$
$$x^2 + y^2 + z^2 = d^2,$$

where a, b, c, and d are constants, has total degree 8. (This is two cylinders and a sphere.) Observing that the second equation of (7-1) can be rewritten

$$x^2 + y^2 - 2bx + b^2 = c^2,$$

we see that (7-1) can be reduced by exploiting the first equation to substitute for "$x^2 + y^2$." We get

$$a^2 - 2bx + b^2 = c^2 \qquad (7\text{-}2)$$
$$a^2 + z^2 = d^2,$$

a new system that has total degree 2. It is geometrically the intersection of lines in the xz-plane; y does not appear. After (7-2) is solved, we will have solutions (x, z) for (7-2), and the associated (x, y, z) solutions to (7-1) are obtained (say) by the relation $y = \pm\sqrt{a^2 - x^2}$ from the first equation.

Usually, the process of reduction will not simplify a system to this extent. Consider the following example, which came up in economics modeling:

$$(x_1 + x_1x_2 + x_2x_3 + x_3x_4)x_5 = c_1$$

$$(x_2 + x_1x_3 + x_2x_4)x_5 = c_2$$

$$(x_3 + x_1x_4)x_5 = c_3 \qquad (7\text{-}3)$$

$$x_4x_5 = c_4$$

$$x_1 + x_2 + x_3 + x_4 + 1 = 0,$$

where c_1, c_2, c_3 and c_4 are constants. This system of three third-degree, one second-degree, and one linear equation in five variables has total degree 54. The fourth equation implies that $x_5 = c_4/x_4$. Thus we can substitute into the other equations to eliminate x_5, yielding the new system

$$(x_1 + x_1x_2 + x_2x_3 + x_3x_4) = \frac{c_1x_4}{c_4}$$

$$(x_2 + x_1x_3 + x_2x_4) = \frac{c_2x_4}{c_4}$$

$$(x_3 + x_1x_4) = \frac{c_3x_4}{c_4} \qquad (7\text{-}4)$$

$$x_1 + x_2 + x_3 + x_4 + 1 = 0.$$

This is a system of three second-degree equations and one linear equation in four variables. It has total degree 8. The system can be reduced further simply by eliminating x_4 using the linear equation. The resulting system of three second-degree equations in three unknowns is

$$x_1x_2 - x_1x_3 - x_3^2 + \left(\frac{c_1}{c_4} + 1\right)x_1 + \frac{c_1}{c_4}x_2 + \left(\frac{c_1}{c_4} - 1\right)x_3 + \frac{c_1}{c_4} = 0$$

$$x_1x_3 - x_2x_3 - x_1x_2 + \frac{c_2}{c_4}x_1 + \frac{c_2}{c_4}x_2 + \frac{c_2}{c_4}x_3 + \frac{c_2}{c_4} = 0 \qquad (7\text{-}5)$$

$$-x_1^2 - x_1x_2 - x_1x_3 + \left(\frac{c_3}{c_4} - 1\right)x_1 + \frac{c_3}{c_4}x_2 + \left(\frac{c_3}{c_4} + 1\right)x_3 + \frac{c_3}{c_4} = 0$$

Like (7-4), system (7-5) has total degree 8.

Reduction, as illustrated by the examples, may include eliminating variables as well as decreasing the total degree. By eliminating variables a reduced system will generate (in CONSOL) lower-order linear algebra: (7-5) generates 3×3 linear algebra, while (7-4) generates 4×4 and (7-3) generates 5×5.

I want our focus to be on *systematic* reduction. In the following chapters, we will see the pattern: An application area produces a consistent structure in the systems that arise, and this structure leads to a consistent reduction strategy. All the systematic reduction described in this book involves eliminating solutions at infinity or reducing the number of variables or both.

Suppose, for example, in a particular application area systems of the form

$$f^a(x) + f^b(x) = 0$$
$$f^b(x) = 0$$
(7-6)

arise where f^a and f^b are systems, and the total degree of f^a is d_a, the total degree of f^b is d_b, and $d_b > d_a$. Then (7-6) has total degree d_b^2. However, we may always replace (7-6) by

$$f^a(x) = 0$$
$$f^b(x) = 0$$
(7-7)

which has total degree $d_a \cdot d_b < d_b^2$, and (7-7) has the same finite solutions as (7-6). The extra solutions of (7-6) are all at infinity.

7-2 REDUCIBILITY

When we first encounter a system, we try to judge its potential for reduction, its "reducibility." Typically, discovering a specific reduction strategy for an application area takes time and energy. Before trying to find such a strategy, we want to be able to estimate the potential for reduction.

I have in mind two indicators of reducibility. The first is the ratio, RINF, of the number of solutions at infinity (counted by multiplicity) to the total degree. (See Chapter 3 for definitions.) Thus, if there are only a finite number of solutions at infinity, then $0 \le \text{RINF} \le 1$. (This follows from a variant of Bezout's theorem.) Otherwise, RINF $= \infty$. We use RINF as an indicator of reducibility via: the larger RINF, the greater the system's potential for reduction. My rule of thumb is to *consider* reduction when RINF > 0, and to *try hard* for reduction when RINF $> \frac{1}{2}$. We would especially like to avoid RINF $= \infty$.

Assuming that there are not an infinite number of solutions or an infinite number of solutions at infinity, RINF measures the potential for reduction in the following exact sense. The theory of resultants guarantees that there is a reduction of $f = 0$ to a single polynomial whose solutions give the finite solutions of $f = 0$ [Van der Waerden, 1953]. Generating the resultant polynomial is rarely feasible in practice. Still, the existence of this reduction shows that when RINF is nonzero, $f = 0$ is reducible. I do not mean to imply that RINF is always easy to estimate. Naturally, counting the solutions at infinity may be a nontrivial calculation. Unfortunately, solutions at infinity with high multiplicity seem to come up a lot, making the count more difficult.

Let's look back at the examples. The homogeneous part of system (7-1) is

$$x^2 + y^2 = 0$$
$$x^2 + y^2 = 0$$
$$x^2 + y^2 + z^2 = 0,$$

which is equivalent to

$$x^2 + y^2 = 0$$
$$z^2 = 0,$$

which has solutions $(x, y, z) = (1, \pm i, 0)$ when $x = 1$, and no solutions if $x = 0$ and $y = 1$ or if $x = 0$, $y = 0$, and $z = 1$. Thus it has two solutions at infinity and RINF $= \frac{2}{8} = \frac{1}{4}$. The homogeneous part of (7-3) is

$$(x_1x_2 + x_2x_3 + x_3x_4)x_5 = 0$$
$$(x_1x_3 + x_2x_4)x_5 = 0$$
$$x_1x_4x_5 = 0$$
$$x_4x_5 = 0$$
$$x_1 + x_2 + x_3 + x_4 = 0,$$

which has the isolated solution at infinity $x = (0, 0, 0, 0, 1)$ and the infinite set of solutions at infinity

$$\{x \in C^5 : x_5 = 0, x_1 + x_2 + x_3 + x_4 = 0\}.$$

Thus RINF $= \infty$. For both examples RINF suggests reduction is possible, and in the second example that it is strongly indicated.

The second indicator of reducibility is the "sparsity" of the system, which I will now define. Given the degree d and number of variables n of an equation, the possible terms consist of a constant term and one term for each way of multiplying powers of $x_1, \ldots, x_n$ together to get a degree less than or equal to d. For example, if $n = 3$ and $d = 2$, we have x_1^2, x_2^2, x_3^2, $x_1x_2, x_1x_3, x_2x_3, x_1, x_2$, and x_3. Thus a general equation with $n = 3$ and $d = 2$ looks like

$$a_1x_1^2 + a_2x_2^2 + a_3x_3^2 + a_4x_1x_2 + a_5x_1x_3 + a_6x_2x_3$$
$$+ a_7x_1 + a_8x_2 + a_9x_3 + a_{10} = 0.$$

Let's denote by NCOEF(n, d) the number coefficients of a general dth degree equation with n variables. For example, NCOEF(3, 2) = 10, NCOEF(3, 3) = 20, NCOEF(5, 3) = 56, and NCOEF(n, 1) = n + 1. Now define the sparsity of f by

$$\text{sparsity of } f = \frac{\text{number of nonzero coefficients of } f}{\sum_{i=1}^{n} \text{NCOEF}(n, d_i)},$$

where d_i = deg (f_i), as usual. Note that the sparsity of f is less than or equal to 1. Now, sparsity as an indicator for reduction works like this: The smaller the sparsity, the more reduction is indicated.

Table 7-1 gives the sparsity for systems (7-1) through (7-5). We see that in the process of reduction we have moved from smaller to larger sparsity. The idea of using sparsity to measure reducibility relates to the concept of *genericity*. A *generic* system is one that has no solutions at infinity or singular solutions. Given n and $d_1, \ldots, d_n$, consider a system whose coefficients are chosen at random. Then the ith equation will have NCOEF(n, d_i) terms and these terms will have nonzero coefficients. It can be shown that such a "random" system is generic, using ideas along the lines of the "Independence Principles" cited in previous chapters (see Appendix 4). Systems with random coefficients rarely arise in practice. However, they provide us with a model for genericity beyond the (not particularly helpful) definition of genericity.

A system that arises in practice may deviate in two significant ways from a system whose coefficients are random:

(a) It may have some zero coefficients.

(b) There may be some particular algebraic relationships among its nonzero coefficients.

In some application areas (a) and (b) are both a concern, as in geometric modeling (see Chapter 8). In other areas, only (a) seems to be significant, as in chemical equilibrium equations (see Chapter 9). In the latter areas, a

TABLE 7-1 Sparsity for systems (7–1) through (7–5)

Equation	Number of Coefficients	$\sum_{i=1}^{n} \text{NCOEF}(n, d_i)$	Sparsity
(7–1)	12	30	0.40
(7–2)	4	9	0.44
(7–3)	19	195	0.01
(7–4)	17	53	0.32
(7–5)	21	30	0.70

system's character deviates from that of a random (and generic) system only in its zero coefficients.

In summary, we have two indicators of reducibility:

1. RINF (the ratio of the number of solutions at infinity to the total degree)
2. Sparsity (the ratio of the number of coefficients to the generic number of coefficients)

Reducibility is indicated when RINF is large and when the sparsity is small.

7-3 THE SINGULARITY OF THE SCLGEN MATRIX

We do have a third indicator of reducibility: the singularity of the SCLGEN scaling matrix, α, as defined in Chapter 5 [the discussion after (5-15)]. That is, if α is singular, the system should be examined for reducibility. Actually, I do not consider this an especially useful indicator, because the only systems I can find that make α singular are so obviously degenerate. Consider the following result. Recall that $d_{i,k,j}$ denotes the degree of x_k in the jth term of f_i (see Section 5-2).

PROPOSITION.   If either of the following two conditions on the degrees of f hold, then the scaling matrix, α, is singular. Further, each geometrically isolated solution of $f = 0$ can be obtained by setting some of the variables of $f = 0$ identically to zero and solving the resulting reduced system, as indicated.

Condition 1.   There are integers m_1 and m_2 with $1 \le m_1 < m_2 \le n$ such that

$$\sum_{k=1}^{m_1} d_{i,k,j} = \sum_{k=m_1+1}^{m_2} d_{i,k,j}$$

for all i and j. In words: The sums of the degrees of the variables in $\{x_1, \ldots, x_{m_1}\}$ and in $\{x_{m_1+1}, \ldots, x_{m_2}\}$ are always the same for each term of each equation. (The sum can change from term to term.) The geometrically isolated solutions of $f = 0$ are all solutions to the reduced system

$$f(0, \ldots, 0, x_{m_2+1}, \ldots, x_n) = 0$$

$$x_1 = 0$$

$$\vdots$$

$$x_{m_2} = 0.$$

Condition 2.   There is an integer m with $1 \le m \le n$ and, for each i, an integer δ_i such that

$$\sum_{k=1}^{m} d_{i,k,j} = \delta_i$$

for all j. In words: With $x_{m+1}, \ldots, x_n$ fixed, f is homogeneous in $x_1, \ldots, x_m$. (If $m = n$, f is homogeneous. If $m = 1$, then for each equation, f_i, there is a degree, δ_i, such that x_1 appears in every term raised to the δith power.) The geometrically isolated solutions of $f = 0$ are all solutions to the reduced system

$$f(0, \ldots, 0, x_{m+1}, \ldots, x_n) = 0$$

$$x_1 = 0$$

$$\vdots$$

$$x_m = 0.$$

The following examples illustrate the proposition.

Example 1.   System (5-16) obeys condition 1, with $n = 2$, $m_1 = 1$, and $m_2 = 2$. The solutions are $\{(x_1, x_2) : x_1 x_2 = 1\}$.

Example 2.   System (5-17) obeys condition 2, with $n = 2$ and $m = 2$. The only solution is $(x_1, x_2) = (0, 0)$.

Example 3.   The following two equations could be a part of a system obeying condition 1, with $n = 6$, $m_1 = 2$, and $m_2 = 4$:

$$x_1^3 x_2 x_3^2 x_4^2 x_5^6 x_6 + x_1 x_2^2 x_3^3 x_5 x_6 + x_5^2 x_6^5 + x_5 + 1 = 0$$

$$x_5^6 x_6^2 + x_5 x_6 + x_5 + x_6 + 1 = 0.$$

Example 4.   The following two equations could be part of a system obeying condition 2, with $n = 4$ and $m = 2$:

$$x_1^2 x_2 x_3^6 x_4 + x_1 x_2^2 x_3 + x_1^3 x_3^3 x_4^5 + x_1^3 x_3 = 0$$

$$x_3^6 x_4^2 + x_3 x_4 + x_3 + x_4 + 1 = 0.$$

It would be interesting to have a full characterization of those f for which α is singular. In particular, we would like to know if there exists an f with a singular α where $f = 0$ is not obviously degenerate or reducible.

Proof of the Proposition.   For each of the two conditions, we shall see that if c is a solution to $\alpha \cdot c = \beta$, we can generate a second solution, $\bar{c}$. In fact, we show that $\sigma_{1,i,j}$ and $\sigma_{2,i,j',j''}$ in (5-13) and (5-14) are the same for c and $\bar{c}$. Then we show that if $x = (x_1, \ldots, x_n)$ is a solution to $f = 0$ and $\lambda \in C$, $\lambda \neq 0$, then

$$\bar{x} = \left(\lambda x_1, \ldots, \lambda x_{m_1}, \frac{x_{m_1+1}}{\lambda}, \ldots, \frac{x_{m_2}}{\lambda}, x_{m_2+1}, \ldots, x_n \right)$$

is also a solution (under condition 1) or

$$\bar{x} = (\lambda x_1, \ldots, \lambda x_m, x_{m+1}, \ldots, x_n)$$

is also a solution (under condition 2). This will suffice to prove the proposition.

Assume condition 1. Let $\bar{c}_k = c_k + 1$ for $k = 1$ to m_1, $\bar{c}_k = c_k - 1$ for $k = m_1 + 1$ to m_2, and $\bar{c}_k = c_k$ for $k > m_2$. Then

$$\sigma_{1,i,j}(\bar{c}) = \bar{c}_{n+i} + \left(\sum_{k=1}^{n} \bar{c}_k d_{i,k,j} \right) + \log (|\, a_{i,j}\,|)$$

$$= c_{n+i} + \left(\sum_{k=1}^{n} c_k d_{i,k,j} \right) + \sum_{k=1}^{m_1} d_{i,k,j}$$

$$- \sum_{k=m_1+1}^{m_2} d_{i,k,j} + \log (|\, a_{i,j}\,|)$$

$$= c_{n+i} + \left(\sum_{k=1}^{n} c_k d_{i,k,j} \right) + \log (|\, a_{i,j}\,|)$$

$$= \sigma_{1,i,j}(c)$$

and

$$\sigma_{2,i,j',j''}(\bar{c}) = \sum_{k=1}^{n} \bar{c}_k (d_{i,k,j'} - d_{i,k,j''}) + \log \left(\frac{|\, a_{i,j'}\,|}{|\, a_{i,j''}\,|} \right)$$

$$= \sum_{k=1}^{n} c_k (d_{i,k,j'} - d_{i,k,j''}) + \sum_{k=1}^{m_1} (d_{i,k,j'} - d_{i,k,j''})$$

$$- \sum_{k=m_1+1}^{m_2} (d_{i,k,j'} - d_{i,k,j''}) + \log \left(\frac{|\, a_{i,j'}\,|}{|\, a_{i,j''}\,|} \right)$$

$$= \sum_{k=1}^{n} c_k (d_{i,k,j'} - d_{i,k,j''}) + \log \left(\frac{|\, a_{i,j'}\,|}{|\, a_{i,j''}\,|} \right)$$

$$= \sigma_{2,i,j',j''}(c).$$

This shows that $\alpha \cdot c = \beta$ has two distinct solutions, so α is singular.

Now we will see that $f(\bar{x}) = f(x) = 0$ by comparing terms. The jth term of $f_i(\bar{x})$ is

$$a_{i,j} \prod_{k=1}^{m_1} (\lambda x_k)^{d_{i,k,j}} \prod_{k=m_1+1}^{m_2} \left(\frac{x_k}{\lambda}\right)^{d_{i,k,j}} \prod_{k=m_2+1}^{n} (x_k)^{d_{i,k,j}}$$

$$= a_{i,j} \prod_{k=1}^{n} (x_k)^{d_{i,k,j}} \prod_{k=1}^{m_1} (\lambda)^{d_{i,k,j}} \prod_{k=m_1+1}^{m_2} \left(\frac{1}{\lambda}\right)^{d_{i,k,j}}$$

$$= a_{i,j} \prod_{k=1}^{n} (x_k)^{d_{i,k,j}} \exp_\lambda \left(\sum_{k=1}^{m_1} d_{i,k,j} - \sum_{k=m_1+1}^{m_2} d_{i,k,j} \right)$$

$$= a_{i,j} \prod_{k=1}^{n} (x_k)^{d_{i,k,j}},$$

where "$\exp_\lambda$(expression)" denotes "$\lambda^{\text{expression}}$." This completes the proof of the condition 1 part of the proposition.

Assume condition 2. Let $\bar{c}_k = c_k + 1$ for $k = 1$ to m, $\bar{c}_k = c_k$ for $k = m + 1$ to n, and $\bar{c}_{n+i} = c_{n+i} - \delta_i$ for $i = 1$ to n. Then

$$\sigma_{1,i,j}(\bar{c}) = \sigma_{1,i,j}(c) + \sum_{k=1}^{m} d_{i,k,j} - \delta_i$$

$$= \sigma_{1,i,j}(c)$$

and

$$\sigma_{2,i,j',j''}(\bar{c}) = \sigma_{2,i,j',j''}(c) + \sum_{k=1}^{m} (d_{i,k,j'} - d_{i,k,j''})$$

$$= \sigma_{2,i,j',j''}(c) + \delta_i - \delta_i$$

$$= \sigma_{2,i,j',j''}(c).$$

The jth term of $f_i(\bar{x})$ is

$$a_{i,j} \prod_{k=1}^{m} (\lambda x_k)^{d_{i,k,j}} \prod_{k=m+1}^{n} (x_k)^{d_{i,k,j}}$$

$$= a_{i,j} \prod_{k=1}^{n} (x_k)^{d_{i,k,j}} \exp_\lambda \left(\sum_{k=1}^{m} d_{i,k,j} \right)$$

$$= a_{i,j} \prod_{k=1}^{n} (x_k)^{d_{i,k,j}} \lambda^{\delta_i},$$

so $f_i(\bar{x}) = \lambda^{\delta_i} f_i(x) = 0$.

This completes the proof of the proposition.

7-4 ROW REDUCING THE COEFFICIENT MATRIX

Aside from the examples, little has been said here about how to do reduction. This is intentional, since I can say little about it in general. The following chapters illustrate various approaches that have been successful. The best I can do is show you these in the expectation that they may be inspirational.

However, there is one "trick" I would like to present here. Sometimes a system is not sparse but has a "hidden" sparse structure. This is especially the case when "(b)" cited in Section 7-2 holds. A simple technique that has often been helpful is to "row reduce" the system coefficient matrix. This creates (or uncovers) sparsity in the system and may lead to reduction.

For example, consider

$$x^2 + y^2 + z^2 + xy + yz + zx + x + y + z + 1 = 0$$

$$2x^2 + 2y^2 + z^2 + xy + yz + zx + x + y + z - 3 = 0 \qquad (7\text{-}8)$$

$$x^2 + y^2 + z^2 + xy + yz + zx + 2x + 2y + 2z + 2 = 0.$$

The sparsity of this system is 1; thus it is as unsparse as it can be. However, RINF = 4/8 = 0.5, because the system has four solutions at infinity. Sparsity is a poor indicator when the system has characteristic (b). A glance at (7-8) shows (outrageous) relationships in the coefficients of the equations. Denote the first, second, and third equations of (7-8) by (7-8-1), (7-8-2), and (7-8-3), respectively. If we subtract (7-8-1) from (7-8-2), we get

$$x^2 + y^2 - 4 = 0, \qquad (7\text{-}9\text{-}2)$$

and if we subtract (7-8-1) from (7-8-3), we get

$$x + y + z + 1 = 0. \qquad (7\text{-}9\text{-}3)$$

Now, letting (7-9-1) $\equiv$ (7-8-1), the system (7-9), defined by (7-9-1), (7-9-2), and (7-9-3), has sparsity 17/24 $\approx$ 0.71. The (finite) solutions of (7-8) are identical with the solutions of (7-9). The reduction in total degree from 8 to 4 indicates that we have eliminated the four solutions at infinity.

To make this sort of adding and subtracting of equations more systematic, we can use row reduction of the coefficient matrix. (The row reduction algorithm, called "GEL," is sketched in Appendix 5. It is essentially the same Gaussian elimination algorithm as that used for solving linear systems.) In generating a coefficient matrix, I write equations down the rows and terms across the columns. For (7-8) the coefficient matrix is

$$
\begin{array}{l}
(7\text{-}8\text{-}1) \rightarrow \\
(7\text{-}8\text{-}2) \rightarrow \\
(7\text{-}8\text{-}3) \rightarrow
\end{array}
\begin{bmatrix}
1 & 1 & 1 & 1 & 1 & 1 & 1 & 1 & 1 & 1 \\
2 & 2 & 1 & 1 & 1 & 1 & 1 & 1 & 1 & -3 \\
1 & 1 & 1 & 1 & 1 & 1 & 2 & 2 & 2 & 2
\end{bmatrix}.
$$

$$
\begin{array}{cccccccccc}
\uparrow & \uparrow & \uparrow & \uparrow & \uparrow & \uparrow & \uparrow & \uparrow & \uparrow & \uparrow \\
x^2 & y^2 & z^2 & xy & yz & zx & x & y & z & 1
\end{array}
$$

I always write the highest-degree terms to the left, because GEL starts with the first column and proceeds by columns to the right. Since there will generally be more columns than rows, the choice of the ordering of the coefficients (left to right) will affect the reduction, sometimes significantly. I want reduction to eliminate high-degree terms, so I order by degree from the left.

The result of GEL is

$$
\begin{bmatrix}
1 & 1 & 0 & 0 & 0 & 0 & 0 & 0 & 0 & -4 \\
0 & 0 & 1 & 1 & 1 & 1 & 0 & 0 & 0 & 4 \\
0 & 0 & 0 & 0 & 0 & 0 & 1 & 1 & 1 & 1
\end{bmatrix},
$$

which corresponds to the system

$$
x^2 + y^2 - 4 = 0
$$
$$
z^2 + xy + yz + z + 4 = 0 \tag{7-10}
$$
$$
x + y + z + 1 = 0,
$$

which has sparsity $12/24 = 0.5$.

Systems (7-8) and (7-10) are equivalent, in the sense that they have exactly the same (finite) solutions. However, the sparsity of (7-10)) suggests reduction: say, substitute $z = -(1 + x + y)$. Perhaps more significantly for this example, the solutions at infinity of (7-8) have been essentially eliminated in (7-10), since we can easily recognize that the third equation is linear. Row reduction often makes simpler the checking of conditions associated with nongenericity (solutions at infinity, infinite number of solutions, multiple solutions). For example, a zero row would immediately signal an infinite number of solutions, while a row with a single nonzero element signifies a multiple solution or no (finite) solutions or a "splitting" of the system into subsystems. Row reduction is useful especially in "on-line" reduction schemes (see Chapter 8).

EXERCISES

7-1. Find the solutions at infinity for (7-1) through (7-5).

7-2. What indicator(s) suggest that row reduction might be useful for (7-1) and (7-3)?

7-3. I said (7-1) represented two cylinders and a sphere, but this is not true for all values of a, b, c, and d. For what values is this true? When does the system have real solutions? An infinite number of solutions? Singular solutions? (Note that in using CONSOL we do not have to consider these cases.)

7-4. Design an algorithm to generate NCOEF(n, d), given n and d.

7-5. Find the solutions at infinity of (7-8). Find the solutions at infinity of (7-10) when the third equation is declared to have deg = 1 and then when it is declared to have deg = 2.

7-6. Discuss the behavior of CONSOL that you expect on systems (7-1) through (7-5).

7-7. Show that if x^0 is a finite solution of (7-6), x^0 is a solution of (7-7). Conclude that (7-7) has $d_b^2 - d_a \cdot d_b$ solutions at infinity, counting multiplicities.

7-8. Characterize the systems of two second-degree equations in two unknowns that generate a singular scaling matrix, α.

EXPERIMENTS

7-1. Write a code to generate the sparsity of a system, given n, d_1, . . . , d_n, and the total number of nonzero coefficients. (*Hint:* See Exercise 7-4.)

7-2. Solve systems (7-1) through (7-5) using CONSOL. Compare the results with Exercise 7-6.

CHAPTER 8

Geometric
Intersection Problems

8-1 INTRODUCTION

The first time I saw a computer-generated shaded picture, it was a chess piece, a rook, which had been created on the General Motors solid modeling system, GMSOLID (see Fig. 8-1, taken from [Roth, 1982]). I was struck by the thought: "It looks like a photograph, but the object represented by the image doesn't exist." That seemed eerie, to make a photograph of something that did not exist. I have had the experience a number of times since, with more and more sophisticated images. And that sense of "being there" persists. These ghosts have a reality, even though I know they are not real.

Solid modeling and computer graphics systems depend on programs that compute geometric intersections. (I am going to call such systems "modelers" from now on.) Basic shapes are created: lines, curves, planes, surfaces, solids. To carry out the business of the modeler, these shapes must be intersected in various combinations: three-surface intersections, line–surface intersections, two-curve intersections. This chapter is devoted to solving such intersection problems when the geometrical shapes are defined by polynomial equations. In this case, a geometric intersection corresponds to the simultaneous solution of the associated equations. These systems of equations are generally simpler than those that arise in chemistry, kinematics, nonlinear circuits, and other fields. However, what makes geometric intersection problems challenging is that they must be solved very quickly and reliably in a solitary problem environment. Further, they are commonly nongeneric. This is because designers align surfaces and curves in special ways (just touching, say, or perpendicular) that force the associated poly-

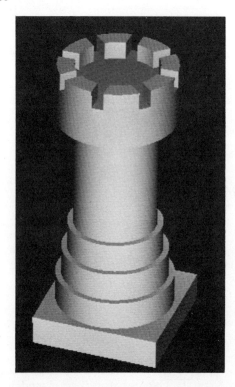

Figure 8-1   Picture of a chess rook produced by GMSOLID. (From [Roth,
1982].)

nomials to have special relations among their coefficients. Nongeneric systems strain the robustness of CONSOL (and other methods), which is particularly unpleasant in a modeler, since the solvers in a modeler must run
without supervision, transparent to the user.

Although cast in a general context, most of this chapter will focus on
quadrics (second-degree equations.)† There are two reasons for this. First,
many geometric modeling systems are based on quadrics. Either all the
shapes in the system are quadric or the most used shapes are. (GMSOLID
[Boyse and Gilchrist, 1982] is an example of such a system, as is PADL
[Voelcker and Requicha, 1977]. See [Requicha and Voelcker, 1983] for other
examples.) Second, we can consider a full development for quadrics of some
reduction ideas that are applicable in general but would be tedious to present
generally. Solving systems of two and three quadrics is harder than it might
seem; a number of general issues connected with nongenericity and reduction arise (as noted, e.g., in [Goldman, 1983]). We will consider some higher-
degree equations: bicubics in Section 8-2, the torus (a fourth-degree surface)
in Section 8-3, and tubes in space in Section 8-4.

---

† If the equation has only one variable, it is called a *quadratic*.

If you are not particularly interested in modelers, this chapter offers:

* Examples of nongeneric sytems of equations and their effects on solvers
* The idea of using on-line reduction to diminish the effects of nongenericity (subsequent chapters concentrate on off-line reduction)
* Specific algorithms for the on-line reduction of systems of two and three quadric equations

The systems of equations that arise in modelers are either emphatically generic or emphatically nongeneric. A designer wants two mechanical parts to just touch, or not touch, or fuse together, but will not want two parts to nearly touch without touching. Thus the equations are scaled at a human level, as opposed to, say, chemical systems. As noted in Chapter 7, any polynomial system can in theory be reduced to a single polynomial (via the "method of resultants" [Van der Waerden, 1953]), thereby eliminating solutions at infinity and (perhaps) making the singularity structure of the system clearer. Although the method of resultants is sometimes usable, it is hard to implement into a stable numerical method, except in the simplest cases. The reduction scheme I am going to present is numerically conservative and less universal than the method of resultants. It will simplify some nongeneric systems and not others. It is, however, especially suited to the quadric systems that come up in modelers.

CONSOL will not always be the method of choice in modelers. The main alternatives are solvers based on algebraic methods, which are harder to make numerically stable and are inherently less general and logically more complex, but are (comparatively) fast. (I do not consider local methods, like Newton's method, because they cannot be relied on to find all solutions.) In GMSOLID, three-quadric intersections are solved with a method similar to CONSOL and quadric–quadric–plane intersections with an algebraic method. The factors that determine which approach to take are as follows. CONSOL can be coded in general and used in many contexts. It is inherently reliable. This makes it ideal for system development and debugging, even when a specialized method will be used in production. CONSOL will allow you to begin to study a proposed new modeler before a specialized solver is ready. Or if the specialized solver is giving trouble, CONSOL can be used to verify that the solver is failing and to help with debugging. CONSOL will not be as fast as an algebraic method. On-line reduction can improve its speed and enhance its reliability, but it cannot be competitive in speed. On the other hand, the total time used by the solver may be so small relative to the time used by the whole modeler that "speed" is irrelevant. This is the case for the three-quadric solver in GMSOLID. It might be evoked, say, 100 times in a half-hour design session and average half a second of CPU time per call. (This is on an IBM 3033.) The quadric–quadric–plane code, however, might be called thousands of times in the same session. Thus it is worthwhile to have a solver that takes only a few milliseconds per call. Many hours of algorithm development and programming were spent to produce a stable algebraic method for this simple problem. And we used CONSOL in verifying the fast code.

This chapter deals with solving systems of two and three variables, with the focus on quadrics. Why, you may ask, do we need such a chapter? We already have a continuation method, CONSOL, that will solve such systems readily. Can't we just plug it in and go? The answer is "yes and no," as you have probably already guessed, but it is more "no" than you may think. I have seen some royal failures of system solvers plugged in off-the-shelf in geometric modelers, especially solvers that rely on algebraic methods alone. In this chapter an enhancement of CONSOL will be developed that in essence puts an algebraic preprocessor in the solver and exploits all the information that can be thereby obtained to make CONSOL faster and more efficient, without sacrificing reliability. The cost will be in algorithm complexity.

8-2 INTERSECTION OF TWO CURVES

8-2-1 Quadric Curves

Let's consider a typical "easy" two quadric intersection problem, an intersection of two circles. Here are the equations

$$x^2 + y^2 - 1 = 0 \tag{8-1}$$
$$(x - 2)^2 + y^2 - 1 = 0$$

and the associated picture is Fig. 8-2.

We see at once (8-1) has one (finite, real) solution: $(x, y) = (1, 0)$. Now there is something about system (8-1) typical of the way geometric modelers are used, and I would like to discuss this before we go any further. The

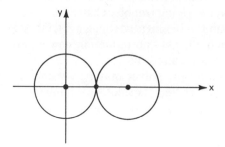

Figure 8-2 Graph for system (8-1).

circles are lined up to just touch. In a design context, this is very natural. Designers tend to use lines that are perpendicular or parallel, curves that just touch (with coincident tangents at the point of intersection), shapes that mesh and match. The intersection shown in Fig. 8-3 is less typical in that the circles overlap, and such a picture would arise from simple perturbations of (8-1) such as

$$x^2 + y^2 - 1 = 0 \qquad\qquad (8\text{-}2)$$
$$[x - (2 - \epsilon_1)]^2 + y^2 - 1 = 0,$$

where $\epsilon_1 > 0$, and

$$x^2 + y^2 - 1 = 0 \qquad\qquad (8\text{-}3)$$
$$[(1 + \epsilon_2)x - (2 - \epsilon_1)]^2 + y^2 - 1 = 0$$

where $\epsilon_1 > 0$ and $\epsilon_2 > 0$.

Now (8-1) is less generic than (8-2) and (8-2) is less generic than (8-3). (I'll explain this in a moment.) The situation I am trying to communicate is that the more "neatly aligned" a geometric intersection, the less generic the associated algebraic system. And designers tend to align their shapes neatly. A further implication is that "nearly neatly aligned" systems can often be assumed to be intended to be "neatly aligned." Thus we are justified in erring on the side of assuming that nearly nongeneric systems are nongeneric. This has implications about the way we choose zero-epsilons (as defined at the end of Section 4-2). Further, nongenericity is both a curse and an opportunity. It is a curse because it gives solvers trouble. It is an opportunity because it signals the possibility of reducing the system to a simpler (and easier to solve) form. These intuitive remarks will be clarified throughout the chapter.

Observe the following progression from nongeneric to generic:

- Equation (8-1) has a singular solution at $(1, 0)$ and two nonsingular solutions at infinity.
- Equation (8-2) has two nonsingular solutions and two nonsingular solutions at infinity when ϵ_1 is not too large.

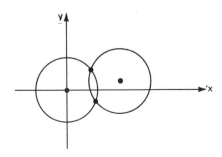

Figure 8-3 Graph for systems (8-2) and (8-3).

- Equation (8-3) has four nonsingular (finite) solutions when ϵ_1 and ϵ_2 are not too large (see Experiments 8-2 and 8-5).

One form of degeneracy is not represented above: having an infinite number of solutions. If a system of two nonzero quadrics has an infinite number of solutions, either the solution curves coincide [i.e., the coefficients of the first equation are a (nonzero) constant multiple of the coefficients of the second] or the equations have a common straight line in their solution sets. For example, the equations

$$xy - y - x + 1 = 0 \tag{8-4}$$
$$xy - y - 2x + 2 = 0$$

have the line $x = 1$ in common in their solution sets, even though the equations' coefficient sets are not multiples of each other.

Actually, identifying that there are an infinite number of solutions in a two-quadric intersection is not difficult. However, identifying that this is *almost* true and the intent of the designer was that it *be* true is delicate. It is a matter of choosing appropriate zero-epsilons. I recall a case in which the inputs to a double-precision solver were being generated by a single-precision modeler. Geometric shapes that were single-precision coincident were judged by the double-precision code to be distinct. It was chaos until the solver's zero-epsilons were loosened up.

What do we do about degenerate systems? We could, of course, make no special provisions. After all, CONSOL is supposed to solve any system. But in a modeling environment, this approach will rarely be as reliable or as fast as it could be. I want us to consider "on-line reduction" as a cure for degeneracy. This is not without dangers, as it risks exchanging one type of numerical instability for another. In reducing the original system to a more generic one, our reduction method may itself generate instabilities.

The reduction method we shall consider is called RED2 and it is in two steps:

1. Change the form of the system into certain standard "types" via the GEL algorithm (see Chapter 7 and Appendix 5).
2. Within each "type," classify the system as to degrees of equations, number and multiplicity of solutions at infinity, and other special information on degeneracies, as appropriate.

With the information provided by RED2, we can solve the (reduced) system more efficiently. For example, if we know that there are exactly two solutions at infinity, we know there are exactly two finite solutions and we can stop CONSOL when we find them. I should emphasize that RED2 is independent of CONSOL and can be used with other solvers.

Here is how RED2 works for (8-1). First we create the matrix A of coefficients as in Chapter 7,

$$A: \quad \begin{matrix} f_1 \to \\ f_2 \to \end{matrix} \begin{bmatrix} 1 & 1 & 0 & 0 & 0 & -1 \\ 1 & 1 & 0 & -4 & 0 & 3 \end{bmatrix}.$$

$$\begin{matrix} \uparrow & \uparrow & \uparrow & \uparrow & \uparrow & \uparrow \\ x^2 & y^2 & xy & x & y & \text{constant} \end{matrix}$$

Then, as in Chapter 7, we apply the GEL algorithm to A to generate A_r, a row-reduced matrix. A_r will be in one of the following four standard forms:

$$\text{type } (1, 2): \begin{bmatrix} 1 & 0 & * & * & * & * \\ 0 & 1 & * & * & * & * \end{bmatrix}$$

$$\text{type } (1, 3): \begin{bmatrix} 1 & * & 0 & * & * & * \\ 0 & 0 & 1 & * & * & * \end{bmatrix}$$

$$\text{type } (2, 3): \begin{bmatrix} 0 & 1 & 0 & * & * & * \\ 0 & 0 & 1 & * & * & * \end{bmatrix}$$

$$\text{type } (i, 0); \begin{bmatrix} * & * & * & * & * & * \\ 0 & 0 & 0 & * & * & * \end{bmatrix},$$

where "$*$" indicates that any numbers, including 0, can appear. The "type" notation works like this: Type (i, j) indicates that the first nonzero entry of row 1 occurs in the ith column and the first nonzero entry of row 2 in the jth column, except that $i = 0$ if the first three components of row 1 are zero and $j = 0$ if the first three components of row 2 are zero.

Once A has been reduced to a standard type, we can classify the system easily. For example, the A for system (8-1) is reduced by GEL to

$$A_r = \begin{bmatrix} 1 & 1 & 0 & 0 & 0 & -1 \\ 0 & 0 & 0 & -4 & 0 & 4 \end{bmatrix}.$$

This is type $(1, 0)$. The associated system is

$$x^2 + y^2 - 1 = 0 \tag{8-5}$$
$$-4x + 4 = 0.$$

We see that the total degree of the A_r-system is 2, and that it has two finite solutions and no solutions at infinity. (We identify the second equation as linear.)

After we have determined the type of the reduced system, there are generally some subtypes that affect classification. For example, in a type $(i, 0)$ system, the first equation might be linear also. Thus, for each type we check for a collection of subtypes. Further, the reduced system will be equivalent to a single polynomial of degree 4 or less. In some contexts, it will be reasonable and desirable to generate these polynomials explicitly and solve them using an appropriate numerical method.† Otherwise, we can solve

---

† I have found available codes based on the cubic and quartic formulas to be unreliable. Although not especially customized to small problems, the method described in [Jenkins and Traub, 1970] seems to be reliable and tolerably fast.

these reduced systems using CONSOL without generating the single-polynomial form. But either way, the basic reduction is unchanged.

For each type we would like to have a count of the solutions at infinity
(by their multiplicities). Then, using Bezout's theorem, we would know how
many finite solutions there are (counted by multiplicities) and how many
convergent continuation paths there are. However, we will content ourselves
here with something less. We will compute the solutions at infinity and
determine whether each is singular. Thus, when a solution at infinity is
singular, we will know that its multiplicity is at least 2 but not necessarily
what the multiplicity is. Letting NINF denote the number of solutions at
infinity (counted by multiplicities), we will find NINF in some cases and a
lower bound for NINF in other cases. The main part of "subtype classification" is this computation of NINF.

We proceed using material from Section 3-3, as follows. For each sys-
tem $f = 0$, we generate $\hat{f}$ and its Jacobian $\hat{J}$, from which we get f^0, $\hat{J}^1$, and
$\hat{J}^2$; that is, we generate the homogenization of f and its Jacobian, from which
we get the homogeneous part of f and the Jacobians of the standard representations $\hat{f}^1$ and $\hat{f}^2$, respectively. Then a solution at infinity of $f = 0$
must be of the form $(x, y) = (1, y)$ or it must be the point $(x, y) = (0, 1)$.
Such (x, y) are solutions at infinity if they satisfy $f^0(x, y) = 0$. A solution
at infinity $(1, y)$ is singular exactly when it satisfies $\det(\hat{J}^1(y, 0)) = 0$. A
solution at infinity $(0, 1)$ is singular exactly when it satisfies $\det(\hat{J}^2(0, 0))$
$= 0$.

For future reference, here are the key equations. If $\deg(f_1) = 2$ and
$\deg(f_2) = 2$, then:

$$\hat{f}_1 = a_{1,1}x^2 + a_{1,2}y^2 + a_{1,3}xy + a_{1,4}xz + a_{1,5}yz + a_{1,6}z^2$$

$$\hat{f}_2 = a_{2,1}x^2 + a_{2,2}y^2 + a_{2,3}xy + a_{2,4}xz + a_{2,5}yz + a_{2,6}z^2,$$

where here we have homogenized f via $x/z \to x$ and $y/z \to y$. Then

$$\hat{J} = \begin{bmatrix} 2a_{1,1}x + a_{1,3}y + a_{1,4}z & 2a_{1,2}y + a_{1,3}x + a_{1,5}z & a_{1,4}x + a_{1,5}y + 2a_{1,6}z \\ 2a_{2,1}x + a_{2,3}y + a_{2,4}z & 2a_{2,2}y + a_{2,3}x + a_{2,5}z & a_{2,4}x + a_{2,5}y + 2a_{2,6}z \end{bmatrix}.$$

Now

$$f_1^0 = a_{1,1}x^2 + a_{1,2}y^2 + a_{1,3}xy$$

$$f_2^0 = a_{2,1}x^2 + a_{2,2}y^2 + a_{2,3}xy,$$

$$\det(\hat{J}^1(y, 0)) = \det \begin{bmatrix} 2a_{1,2}y + a_{1,3} & a_{1,5}y + a_{1,4} \\ 2a_{2,2}y + a_{2,3} & a_{2,5}y + a_{2,4} \end{bmatrix}$$

$$= (2a_{1,2}y + a_{1,3})(a_{2,5}y + a_{2,4})$$

$$- (2a_{2,2}y + a_{2,3})(a_{1,5}y + a_{1,4}),$$

and

$$\det (\hat{J}^2(0, 0)) = \det \begin{bmatrix} a_{1,3} & a_{1,5} \\ a_{2,3} & a_{2,5} \end{bmatrix} = a_{1,3}a_{2,5} - a_{2,3}a_{1,5}.$$

For comparison, let's consider another case. If deg $(f_1) = 2$ and deg $(f_2) = 1$, then

$$\hat{f}_1 = a_{1,1}x^2 + a_{1,2}y^2 + a_{1,3}xy + a_{1,4}xz + a_{1,5}yz + a_{1,6}z^2$$

$$\hat{f}_2 = a_{2,4}x + a_{2,5}y + a_{2,6}z,$$

and therefore

$$\hat{J} = \begin{bmatrix} 2a_{1,1}x + a_{1,3}y + a_{1,4}z & 2a_{1,2}y + a_{1,3}x + a_{1,5}z & a_{1,4}x + a_{1,5}y + 2a_{1,6}z \\ a_{2,4} & a_{2,5} & a_{2,6} \end{bmatrix}.$$

Thus we get

$$f_1^0 = a_{1,1}x^2 + a_{1,2}y^2 + a_{1,3}xy$$

$$f_2^0 = a_{2,4}x + a_{2,5}y,$$

$$\det (\hat{J}^1(y, 0)) = \det \begin{bmatrix} 2a_{1,2}y + a_{1,3} & a_{1,5}y + a_{1,4} \\ a_{2,5} & a_{2,6} \end{bmatrix}$$

$$= (2a_{1,2}y + a_{1,3})a_{2,6} - (a_{1,5}y + a_{1,4})a_{2,5},$$

and

$$\det (\hat{J}^2(0, 0)) = \det \begin{bmatrix} a_{1,3} & a_{1,5} \\ a_{2,4} & a_{2,6} \end{bmatrix} = a_{1,3}a_{2,6} - a_{1,5}a_{2,4}.$$

If we use the deg $(f_2) = 2$ formulas when deg $(f_2) = 1$, no "error" will be committed, but we will always have a pair of solutions at infinity (counted by multiplicity) that we could have avoided by assigning to f_2 its actual degree. Note the remarks in Chapter 3 on "declared degree."

Let's develop a full subtype classification for type $= (1, 3)$. The reduced system is

$$f_1 = x^2 + a_{1,2}y^2 + a_{1,4}x + a_{1,5}y + a_{1,6} = 0 \tag{8-6}$$

$$f_2 = \qquad\qquad xy + a_{2,4}x + a_{2,5}y + a_{2,6} = 0$$

Let's check for solutions at infinity. The homogeneous part of (8-6) is

$$f_1^0 = x^2 + a_{1,2}y^2 = 0$$

$$f_2^0 = xy = 0.$$

By $f_2^0 = 0$, either $x = 0$ or $y = 0$. This yields two possible solutions at infinity, $(1, 0)$ and $(0, 1)$. Now $f_1^0 = 0$ is not satisfied by $(1, 0)$. But it is satisfied by $(0, 1)$ when $a_{1,2} = 0$. We conclude that the system has a solution at infinity when $a_{1,2} = 0$, and it does not have a solution at infinity when $a_{1,2} \neq 0$. We can test for the singularity of the solution at infinity by checking $\det (\hat{J}^2(0, 0)) = 0$. In fact, $\det (\hat{J}^2(0, 0)) = -a_{1,5}$. So the solution at infinity is singular whenever $a_{1,5} = 0$.

When $a_{1,2} = 0$ and $a_{1,5} = 0$, however, $f_1 = x^2 + a_{1,4}x + a_{1,6}$. Thus we can solve $f_1 = 0$ for x and then get y from f_2. (From further analysis, we see that one or two finite solutions, no solutions, or an infinite number of solutions can result.)

Here is a summary of type $= (1,3)$: $\deg (f_1) = \deg (f_2) = 2$, and

1. If $a_{1,2} \neq 0$, then NINF $= 0$.
2. If $a_{1,2} = 0$ and $a_{1,5} \neq 0$, then NINF $= 1$.
3. If $a_{1,2} = 0$ and $a_{1,5} = 0$, then NINF ≥ 2. Also, $f_1 = x^2 + a_{1,4}x + a_{1,6} = 0$.

We will do types $(1, 2)$ and $(i, 0)$ shortly. But let's consider two observations now.

First, note the importance of "zero tests" on the methodology we are developing. Such tests make us scale dependent and scale limited.

Second, we can use the information generated by RED2 in two ways. After RED2 produces its "typed" system, we can proceed to solve this system with CONSOL, using the "NINF" value to double check the CONSOL behavior. Or we can solve some cases using alternative methods, saving CONSOL only for the "tough" ones. For example, we could easily solve the $a_{1,2} = 0$, $a_{1,5} = 0$ case of type $(1, 3)$ with the quadratic formula, and the other cases with CONSOL. We can even further develop the RED2 output to reduce each case to a single polynomial in a single variable. In essense, we are "halfway there" already. I am not going to do this here, but it can be done. If done carefully, it produces a pretty method, fast and stable. For maximum speed, this is the way to go.

Type $(1, 2)$ is settled in exactly the same spirit as type $(1, 3)$. The homogeneous part is

$$x^2 + a_{1,3}xy = 0$$

$$y^2 + a_{2,3}xy = 0.$$

We seek solutions at infinity. If $x = 0$, then $y = 0$; if $y = 0$, then $x = 0$. Therefore, we may as well assume that $x \neq 0$ and $y \neq 0$. Then

$$x + a_{1,3}y = 0$$

$$y + a_{2,3}x = 0,$$

which has a solution exactly when $a_{1,3}a_{2,3} = 1$. Thus the unique solution at infinity is $(x, y) = (1, -1/a_{1,3})$. It is a multiple solution whenever

$\det(\hat{J}^1(-1/a_{1,3}, 0)) = 0$. This yields the condition

$$a_{2,4}a_{1,3}^3 - a_{2,5}a_{1,3}^2 + a_{1,4}a_{1,3} - a_{1,5} = 0.$$

Here is the summary for type $(1, 2)$: $\deg(f_1) = \deg(f_2) = 2$, and

1. If $a_{1,3}a_{2,3} \neq 1$, then NINF $= 0$.
2. If $a_{1,3}a_{2,3} = 1$ but $a_{2,4}a_{1,3}^3 - a_{2,5}a_{1,3}^2 + a_{1,4}a_{1,3} - a_{1,5} \neq 0$, then NINF $= 1$.
3. If $a_{1,3}a_{2,3} = 1$ and $a_{2,4}a_{1,3}^3 - a_{2,5}a_{1,3}^2 + a_{1,4}a_{1,3} - a_{1,5} = 0$, then NINF ≥ 2.

If case 3 holds, NINF might equal 2, 3, or 4. For example,

$$x^2 + xy + x + y + 1 = 0 \tag{8-7}$$
$$y^2 + xy - x - y - 1 = 0$$

has a quadruple root at infinity and therefore no finite solutions.

Type $(2, 3)$ is similar to type $(1, 3)$. In fact, we can convert type $(2, 3)$ into type $(1, 3)$ by relabeling the variables, that is, by exchanging x and y. In matrix terms, this is the same as exchanging the first and second and fourth and fifth columns of A. Having an extra reduction operation suggests a refinement of the "typing" strategy. If we allow the operation "relabeling variables" as well as the elementary row operations of GEL, we can reduce matrix A to fewer types. For two quadrics, we would have three rather than four types, because type $(2, 3)$ would reduce to type $(1, 3)$. (In the computer code that implements RED2 we would have to set a flag to remember that we exchanged x and y.) Now, having fewer types is good, but we have lost some information: "type $(2, 3)$" is not exactly the same as "type $(1, 3)$ with x and y exchanged." I have chosen not to reduce using this column operation. However, sometimes it would be useful to have fewer types, so you should keep it in mind as an option.

Let's consider type $(i, 0)$. In a sense, this is an easy case. Since f_2 is linear, we can reduce the system to a single quadratic equation in one variable. However, because the actual degrees of f_1 and f_2 are not fixed by the type, there are more cases than for the other types. We have the following possibilities:

$$\deg(f_1) = 2, \quad \deg(f_1) = 1, \quad \deg(f_1) = 0,$$

and

$$\deg(f_2) = 1, \quad \deg(f_2) = 0.$$

The system $f = 0$ in reduced form is

$$f_1 = a_{1,1}x^2 + a_{1,2}y^2 + a_{1,3}xy + a_{1,4}x + a_{1,5}y + a_{1,6} = 0,$$
$$f_2 = \qquad\qquad\qquad\qquad\qquad a_{2,4}x + a_{2,5}y + a_{2,6} = 0.$$

We have

$$f_1^0 = a_{1,1}x^2 + a_{1,2}y^2 + a_{1,3}xy$$

$$f_2^0 = a_{2,4}x + a_{2,5}y,$$

and we need to check $(x,y) = (1, y)$ and $(x, y) = (0, 1)$. The second possibility is easy to settle: $(0, 1)$ is a solution at infinity whenever $a_{1,2} = 0$ and $a_{2,5} = 0$. It is a singular solution if, in addition,

$$\det (\hat{J}^2(0, 0)) = a_{1,3}a_{2,6} - a_{1,5}a_{2,4} = 0.$$

Now, suppose that $(x, y) = (1, y)$. If $a_{2,5} \neq 0$, then $f_2^0 = 0$ implies that $y = -a_{2,4}/a_{2,5}$. Substituting into $f_1^0 = 0$ yields the condition for $(1, y)$ to be a solution at infinity:

$$f_1^0 \left(1, \frac{-a_{2,4}}{a_{2,5}} \right) = a_{1,1}a_{2,5}^2 + a_{1,2}a_{2,4}^2 - a_{1,3}a_{2,4}a_{2,5} = 0.$$

It is a singular when

$$\det \left(\hat{J}^1 \left(\frac{-a_{2,4}}{a_{2,5}}, 0 \right) \right)$$

$$= a_{2,6} \left(-2a_{1,2} \frac{a_{2,4}}{a_{2,5}} + a_{1,3} \right) - a_{2,5} \left(-a_{1,5} \frac{a_{2,4}}{a_{2,5}} + a_{1,4} \right) = 0.$$

If $a_{2,5} = 0$ and $(1, y)$ is a solution, $f_2^0 = 0$ implies that $a_{2,4} = 0$. Thus deg $(f_2) = 0$.

The "deg $= 0$" cases are degenerate and should not be submitted to CONSOL. For example, deg $(f_2) = 0$ means either that $f_2 = 0$ cannot be satisfied because it is in the form "nonzero constant $= 0$" or it is always satisfied because it is in the form "$0 = 0$." In the first case, the message "no solutions" can be returned, while in the second case, the solutions to $f_1 = 0$ are the solutions to the system. In a supervised environment, a simple expedient is to treat all "deg $= 0$" cases as "fatal errors." In a solitary environment, if a "deg $= 0$" case arises, the solver should continue the analysis of subcases and take appropriate action (which will depend on the application).

These details are essentially "obvious," but the point I want to make is: It is the *unrecognized easy degenerate* cases that mess up solvers. RED2 gives you control by letting you know when such a case has occurred.

Here is a summary of type $(i, 0)$: deg $(f_1) = 2$ or 1 or 0, and deg $(f_2) = 1$ or 0. We omit the analysis of the cases when deg $(f_2) = 0$.

1. If $a_{2,5} \neq 0$ and $f_1^0(1, -a_{2,4}/a_{2,5}) \neq 0$, then NINF $= 0$.

2. If $a_{2,5} \neq 0$ and $f_1^0(1, -a_{2,4}/a_{2,5}) = 0$, but $\det (\hat{J}^1(-a_{2,4}/a_{2,5}, 0)) \neq 0$, then NINF $= 1$.

3. If $a_{2,5} \neq 0$, $f_1^0(1, -a_{2,4}/a_{2,5}) = 0$, and det $(\hat{J}^1(-a_{2,4}/a_{2,5}, 0)) = 0$, then NINF = 2.

4. If $a_{2,5} = 0$, $a_{2,4} \neq 0$, and $f_1^0(0, 1) \neq 0$, then NINF = 0.

5. If $a_{2,5} = 0$, $a_{2,4} \neq 0$, $f_1^0(0, 1) = 0$, and det $(\hat{J}^1(0, 0)) \neq 0$, then NINF = 1.

6. If $a_{2,5} = 0$, $a_{2,4} \neq 0$, $f_1^0(0, 1) \neq 0$, and det $(\hat{J}^1(0, 0)) = 0$, then NINF = 2.

7. If $a_{2,5} = 0$ and $a_{2,4} = 0$, then deg $(f_2) = 0$.

As noted above, in the cases in which the reduced system has a singular solution at infinity, we may be left in doubt on the exact value of NINF. When type = (1, 3), this is not much of a concern, since the system collapses to a quadratic. When type = (1, 2), it is more of a problem. NINF might be 2, 3, or 4. We know that two paths will diverge (or, using the projective transformation, two paths will converge to a point at infinity). Without further analysis, we know no more. If CONSOL finds two finite solutions, the uncertainty is eliminated. If CONSOL does not, we must trust the continuation or do further algebraic analysis (see Project 8-3).

A summary of the RED2 logic is given in Table 8-1 at the end of this chapter.

8-2-2 General Curves

The on-line reduction ideas of RED2 can be applied to higher-degree equations. Let's consider an equation type common in graphics and modeling applications, the bicubic. Bicubics are sixth-degree equations of the form

$$f(x, y) = a_1 x^3 y^3 + a_2 x^3 y^2 + a_3 x^2 y^3 + a_4 x^3 y + a_5 x^2 y^2 + a_6 x y^3$$

$$+ a_7 x^3 + a_8 x^2 y + a_9 x y^2 + a_{10} y^3 + a_{11} x^2 + a_{12} x y + a_{13} y^2$$

$$+ a_{14} x + a_{15} y + a_{16}$$

$$= 0.$$

Naturally, CONSOL can be used to solve systems of two such curves. However, since each equation has (declared) degree 6, the system will have total degree 36. This means, that we must track 36 curves unless the system is reduced. Can we construct a "RED2" for bicubic curves? The answer is "yes." We can row reduce the coefficient matrix of the system, beginning with the highest-degree term. There will be, in essence, many cases, depending on the actual degrees of the equations, but we will generally get a significant reduction. For example, consider

$$f_1 = x^3 y^3 + 2x^3 y^2 + 7x^2 y + 6x + y - 8 = 0$$

$$f_2 = 2x^3 y^3 + 4xy^2 + 16x^2 + 2x + 1 = 0.$$

I have let a number of the coefficients be zero to make the example less bulky, but it is still a total degree 36 system. The associated coefficient matrix is

$$\begin{bmatrix} 1 & 2 & 7 & 0 & 0 & 6 & 1 & -8 \\ 2 & 0 & 0 & 4 & 16 & 2 & 0 & 1 \end{bmatrix}.$$

$$\begin{array}{cccccccc} \uparrow & \uparrow & \uparrow & \uparrow & \uparrow & \uparrow & \uparrow & \uparrow \\ x^3y^3 & x^3y^2 & x^2y & xy^2 & x^2 & x & y & \text{constant} \end{array}$$

Now row reducing "from the left" we get

$$\begin{bmatrix} 1 & 0 & 0 & 2 & 8 & 1 & 0 & 0.5 \\ 0 & 1 & 3.5 & -1 & -4 & 2.5 & 0.5 & -4.25 \end{bmatrix}.$$

The reduced system has total degree 30. In general, a bicubic system can be reduced to have no more than degree 30, as this example illustrates. Thus a total degree 36 bicubic system always has six solutions at infinity. The reduced system is

$$\begin{aligned} x^3y^3 \quad\quad\quad + 2xy^2 + 8x^2 + \quad x \quad\quad + 0.5 &= 0 \\ x^3y^2 + 3.5x^2y - \quad xy^2 - 4x^2 + 2.5x + 0.5y - 4.25 &= 0. \end{aligned} \tag{8-8}$$

This system still has solutions at infinity, namely (1, 0) and (0, 1).

We can find the solutions at infinity for a general bicubic system in the usual way: Create the homogenization of the system by setting the lower-degree terms to zero, then solve for $x = 1$, then check if $x = 0$, $y = 1$ is a solution. This is relatively easy to do for bicubics, since it reduces to solving a cubic polynomial in one variable. Thus we can find the solutions at infinity and their singularity status (via a Jacobian evaluation) and use this information to guide CONSOL as before. [Both solutions at infinity of (8-8) are singular.]

To summarize, we can row reduce the bicubic system coefficient matrix (resulting in a reduction in total degree at least to 30) and count nonsingular solutions at infinity and find a lower bound on the number of singular solutions at infinity, just as for quadrics. Because the bicubic system is structurally more complex than the quadric, this reduction will often be less complete. The resulting reduced system will frequently not be as reduced as we would like. Clearly, (8-8) can be further reduced easily. Merely replace the first equation by the first less y times the second, yielding the system

$$-3.5x^2y^2 + xy^3 + 4x^2y + 2xy^2 + 8x^2$$

$$- 2.5\,xy - 0.5y^2 + 4.25y + x + 0.5 = 0 \tag{8-9}$$

$$x^3y^2 + 3.5x^2y - xy^2 - 4x^2 + 2.5x + 0.5y - 4.25 = 0$$

Now (8-9) has total degree 20. The "auxiliary reduction" principle here is to continue to reduce until one equation has highest-degree terms that are not a factor of the other equation's terms. But note that the system still has singular solutions at infinity, suggesting that further reduction is possible (see Project 8-2).

EXERCISES

8-1. Solve (8-4) by simple algebra and confirm that it has a line of solutions.

8-2. For TYPE = (1, 3), case 3, when does NINF = 4? When does NINF = 3?

8-3. (a) Confirm the Table 8-1 summary of type (2, 3).
   (b) Show how type (2, 3) reduces to type (1, 3) if we interchange x and y. Which subtype cases of type (1, 3) apply?

8-4. (a) Confirm that (8-7) is type (1, 2), case 3.
   (b) Show (by an elementary argument) that (8-7) has a quadruple root at infinity. (*Hint:* Show that it has no finite solution.)
   (c) Predict the behavior of CONSOL in Solving (8-7).
   (d) Predict the behavior of CONSOL in solving the projective transformation of (8-7).

EXPERIMENTS

Note: I recommend that you use CONSOL8T for these experiments, as described in Chapter 5. Then you can turn on the projective transformation with the IFLGPT flag.

8-1. Solve (8-4) using CONSOL. Is the result of the CONSOL run consistent with its theoretically proven behavior?

8-2. Solve (8-1), (8-2), (8-3), and (8-5) using CONSOL. (Let $\epsilon_1 = 0.00321$ and $\epsilon_2 = 0.00123$.) Compare the run statistics for these four. Do this for the projective transformations of these systems (see Experiment 8-5).

8-3. Solve (8-7) and the projective transformation of (8-7) using CONSOL.

8-4. Solve (8-8) and (8-9) using CONSOL. The ratio of total degrees is 20/30 = 2/3. Does (8-9) get solved two-thirds times faster than (8-8)? Why? What do these runs reveal about the structure of the solution sets at infinity? How can you sort out this structure further?

8-5. This experiment is a quantitative study of transitions from generic to nongeneric. Let us consider system (8-3), where ϵ_1 and ϵ_2 are restrained for convenience by $1 > \epsilon_1 \geq 0$ and $1 > \epsilon_2 \geq 0$. Then (8-3) has solutions at infinity exactly when $\epsilon_2 = 0$, and singular solutions exactly when $\epsilon_1 = 0$. Thus when $\epsilon_1 > 0$ and $\epsilon_2 > 0$, (8-3) is generic. However, as $\epsilon_1 \to 0$ it becomes nongeneric in one way, as $\epsilon_2 \to 0$ it becomes nongeneric in another way, and as $\epsilon_1 \to 0$, $\epsilon_2 \to 0$, it becomes nongeneric both ways at once. This experiment is designed to study these phenomena, to gain quantitative insights into the generic to nongeneric evolution and to study CONSOL's behavior under this evolution. In particular, this experiment will build your confidence in CONSOL's robustness with respect to degeneracy. In part (b) you will see how the computational situation changes when you use the projective transformation. Compare with experiment 8-2.

(a) Let $\epsilon_1^0 = 0.00321$ and $\epsilon_2^0 = 0.00123$. Solve (8-3) with CONSOL for $\epsilon_j = 10^{-k}\epsilon_j^0$ for $j = 1, 2$ and $k = 2, 4, 8, 16$, and 32. Make a table showing ϵ_1, ϵ_2, and "total number of linear systems solved" as entries. Set MAXNS $= 1000$ (the maximum number of steps). Note how the accuracy of the solution degrades as $\epsilon_1 \to 0$. Note also cases where you do not get all solutions and watch the two big solutions grow as $\epsilon_2 \to 0$.

(b) Rerun part (a) using the projective transformation. Compare the computational work (total number of linear systems solved) needed by CONSOL for parts (a) and (b).

PROJECTS

8-1. Develop an algorithm based on RED2 for solving systems of two equations in two unknowns using algebraic methods: that is, reducing the system to either a quadratic, a cubic, or a quartic polynomial in one variable and then solving this polynomial. Code and test your algorithm.

8-2. Develop a complete on-line reduction scheme for systems of two bicubics in two unknowns. Incorporate the row operations "multiply a row by x" and "multiply a row by y" to improve the range of the scheme.

8-3. For an algebraic geometer: Develop a numerically stable and efficient algorithm to compute the exact multiplicity of singular solutions at infinity for a system of two quadric equations in two unknowns; for a system of three quadrics in three unknowns.

8-3 INTERSECTION OF THREE SURFACES

8-3-1 Introduction

The solutions to a single nonzero polynomial equation in three variables form a surface in three-dimensional space. For example,

$$ax + by + cz + d = 0$$

forms a plane,

$$x^2 + y^2 - r^2 = 0$$

forms a cylinder of radius r along the z axis,

$$x^2 + y^2 + z^2 - r^2 = 0$$

forms a sphere with center at the origin and radius r,

$$x^2 + y^2 - z^2 = 0$$

forms a double cone along the z-axis, and

$$x^4 + y^4 + z^4 + 2x^2y^2 + 2x^2z^2 + 2y^2z^2 - 2(R_1^2 + R_2^2)x^2$$
$$- 2(R_1^2 + R_2^2)y^2 + 2(R_1^2 - R_2^2)z^2 + (R_1^2 - R_2^2)^2 = 0$$

forms a torus whose center circle is in the (x, y)-plane and has radius R_1, and whose other radius is R_2 (see Fig. 8-4 through 8-8).

The geometric intersection of three polynomial surfaces corresponds to the simultaneous solution of three polynomial equations in three unknowns, the subject of this section. Naturally, we are extending the ideas from Section 8-2, most of which carry over to this case in some form. The degeneracies we discussed in Section 8-2 all can occur: singular solutions, solutions at infinity, and an infinite number of solutions. We have in addition a new potential degeneracy: an infinite number of solutions at infinity. (This can occur with two equations only if the degree of one of the equations has been declared too large or is zero.)

8-3-2. Quadric Surfaces

A general system of three quadric equations in three variables looks like

$$a_{i,1}x^2 + a_{i,2}y^2 + a_{i,3}z^2 + a_{i,4}xy + a_{i,5}xz + a_{i,6}yz$$
$$+ a_{i,7}x + a_{i,8}y + a_{i,9}z + a_{i,10} = 0$$

for $i = 1, 2, 3$, where $a_{i,1}, \ldots, a_{i,10}$ are constants and x, y, z are the variables. We are going to develop an on-line reduction algorithm, RED3, based on row reduction via GEL, in analogy with RED2 in Section 8-2.

RED3 will take as input the 3×10 matrix of coefficients, $A = (a_{i,j})$,

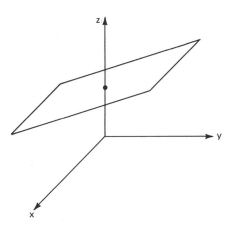

Figure 8-4 Plane in space, the solution set of a linear equation.

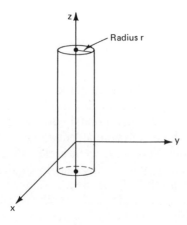

Figure 8-5 Cylinder of radius r given by $x^2 + y^2 - r^2 = 0$. The true solution cylinder extends infinitely along the z-axis.

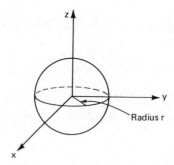

Figure 8-6 The solutions to $x^2 + y^2 + z^2 - r^2 = 0$ form a sphere with center the origin and radius r.

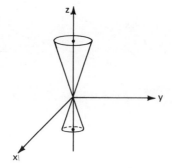

Figure 8-7 Solutions to $x^2 + y^2 - z^2 = 0$ form a double cone along the z-axis. The true solution cone extends infinitely along the z-axis.

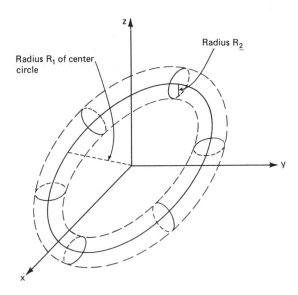

Figure 8-8 The torus has the characteristic "donut shape." Note the center line is in the x-y plane with center the origin and radius R_1.

and a zero-epsilon, ϵ_0, and it will produce as output:

- A 3×10 matrix, A_r, of coefficients for a reduced system, equivalent to the input A-system. That is, the finite solutions to the A_r-system are identical to those of the A-system.
- Either NINF, the number of solutions of the A_r-system at infinity, or a lower bound on NINF.
- Messages, including flags, indicating:
  - Whether NINF has been calculated or a lower bound
  - If the A_r-system has an infinite number of finite solutions
  - If the A_r-system has an infinite number of solutions at infinity
  - If the third equation of the A_r-system is linear

Depending on the messages, we may:

- Solve the A_r-system,
- Convert the A_r-system to two quadrics in two unknowns (if the third equation is linear) and solve, or
- Not submit the A_r-system to a solver.

The advantages of running RED3 are:

- The A_r-system may have lower total degree than the A-system.
- In many cases we will know exactly how many CONSOL paths diverge (or converge to infinity). In the other cases, we will know that some paths diverge, and we will have a lower bound on the number.
- Certain nongeneric cases will be revealed either for *action*, as when the third equation is linear, or for *caution*, as when there are an infinite number of solutions at infinity. Other nongeneric cases will be flagged as errors, such as when there are an infinite number of finite solutions.

When the third equation of the A_r-system is linear, this equation looks like

$$a_{3,7}x + a_{3,8}y + a_{3,9}z + a_{3,10} = 0. \tag{8-10}$$

[In this chapter I denote the elements of A_r as $(a_{i,j})$. The context will always distinguish these from the coefficients of A.] Generally, we have various possibilities and options in this case. If $a_{3,7} = a_{3,8} = a_{3,9} = 0$ and $a_{3,10} \neq 0$, the system has no finite solutions and we are done. If $a_{3,7} = a_{3,8} = a_{3,9} = a_{3,10} = 0$, we would generally expect the system to have an infinite number of finite solutions, but to be sure we must examine the first two equations, which may similarly be nongeneric. If one of $a_{3,7}$, $a_{3,8}$, $a_{3,9}$ is nonzero, (8-10) is "nondegenerate." In this case, we may solve the system as three equations in three unknowns, noting that deg $((8–10)) = 1$. Or, we may use (8–10) to eliminate one of the unknowns in the system, either by literally substituting, say

$$z = \frac{-(a_{3,10} + a_{3,7}x + a_{3,8}y)}{a_{3,9}}$$

(if $a_{3,9} \neq 0$) into equations 1 and 2, or by treating z as an implicitly defined function of x and y in the subroutine FFUN, which evaluates the equations and their partial derivatives. Then we can solve the system as two equations in two unknowns, replacing the 3×3 linear algebra by 2×2 linear algebra. In a solitary environment, where the fact that (8-10) is linear is "discovered" only after on-line reduction, merely noting that deg $((8–10)) = 1$ and running in three variables is the simplest-to-code option. The two "elimination" options call for extra coding, essentially requiring a special routine to convert the three-equation coefficients into two-equation coefficients or a special equation evaluation routine that treats z as implicitly defined by (8–10). Naturally, since $a_{3,9} = 0$ is possible for a nondegenerate (8–10), provision must be made for eliminating x or y instead of z. None of this is "hard," but the more branches the logic must take, the more difficult it becomes to generate bug-free code.

The most complicated algebra in RED3 is solving cubic equations in one variable. It is possible to augment the reduction by further algebra, making the reduction logic more sophisticated, but as in Section 8-2, I will not pursue this.

After row reducing A to A_r, A_r will fall into one of 21 different "types." Each type will be easy to analyze; the principal ambiguity will be counting solutions at infinity if they are singular, especially since we may encounter an infinite number of solutions at infinity.

If you go through the details of this three-variable case with me, you will get a feeling for what you would be up against extending to more variables or higher degree. There is no particular barrier to such extensions. However, counting the solutions at infinity can quickly become a nontrivial calculation. Let's also recall that row reduction is valuable in a modeling context because of the frequency of occurrence of nongeneric forms which reduce especially nicely with this type of reduction. In other contexts, such reduction might be worthless.

We will benefit from some notational conventions. As in Section 8-2, we will row reduce the entire matrix A, but the "typing" of A will be in reference to the first six columns, the "nonlinear" part of A. I will denote this A^0. After row reduction, A will have become A_r, and A^0 will have become A_r^0. A_r^0 is "type (p, q, r)" if the first nonzero entry of row 1 occurs in the pth column, of row 2 the qth column, and of row 3 the rth column, except that a zero is assigned when a row of A_r^0 is zero. By the process of reduction, $p < q < r$, unless one of these is zero. In this latter case, $r = 0$, or $q = r = 0$, or $p = q = r = 0$, depending on how many rows are zero. The first nonzero entry of each nonzero row of A_r^0 equals 1. If p, q, or r is zero, the first six columns of the corresponding row of A_r are zero and the corresponding equation is linear. If $r = 0$, we can reduce the system to two quadrics in two unknowns and solve as in Section 8-2. The 21 types are

$(1, 2, 3)$, $(1, 2, 4)$, $(1, 2, 5)$, $(1, 2, 6)$

$(1, 3, 4)$, $(1, 3, 5)$, $(1, 3, 6)$

$(1, 4, 5)$, $(1, 4, 6)$

$(1, 5, 6)$

$(2, 3, 4)$, $(2, 3, 5)$, $(2, 3, 6)$

$(2, 4, 5)$ $(2, 4, 6)$

$(2, 5, 6)$

$(3, 4, 5)$, $(3, 4, 6)$

$(3, 5, 6)$

$(4, 5, 6)$

$(p, q, 0)$, where p and q may be any values from 0 to 6.

Each type must be separately analyzed to count the solutions at infinity. To do this, we will sequentially set $(x, y, z) = (1, y, z)$, $(0, 1, z)$, and $(0, 0, 1)$, and solve in each case the homogeneous part of the A_r-system, the A_r^0-system. To determine if solutions at infinity are singular, we must check if the appropriate Jacobians of the homogenization have zero determinants. Thus we check the relations

$$\det (\hat{J}^1(y, z)) = 0$$

$$\det (\hat{J}^2(0, z)) = 0$$

$$\det (\hat{J}^3(0, 0)) = 0,$$

where $\hat{J}^i$ denotes the Jacobian of $\hat{f}^i$, as described in Chapter 3. Since we do not determine the exact multiplicity when a solution is singular, the computed NINF is a lower bound whenever a solution at infinity is singular. But if you are an algebraic geometer, see Project 8-3.

The analysis of three types are presented below in detail, and that of all types is summarized in Table 8-2 at the end of this chapter [except type $(p, q, 0)$]. To check for solutions at infinity, we will be solving systems of three equations in two or one unknown(s). Always, we will choose a sub-system of two or one equation(s), respectively, convert the subsystem to a single cubic or quadratic in one variable, solve the single polynomial, and then check if the resulting solutions solve all three equations. (I call the solutions generated by solving the single polynomial "candidates." We do not know if they are solutions to the entire system until they have been tested in each equation.) We will generally have several ways of choosing the subsystem, and the logic of RED3 could vary depending on these choices. I make choices to simplify the logic of RED3 rather than to squeeze out the last drop of CPU efficiency. Also, I try to choose the subsystems so that the possibility of an infinite number of solutions is eliminated. That is, if possible, I would like the single polynomial to have at least one nonzero coefficient for sure. If this choice can be made, I make it. Otherwise, the RED3 logic must test several polynomials for zero coefficients and if all are zero, report that there are an infinite number of solutions at infinity.

RED3 needs a method for solving cubic equations in one variable. In the following I let "CUBE" denote such a method. CUBE solves cubic, quadratic, or linear polynomials. Writing a reliable code to implement CUBE is not trivial. See the footnote in Section 8-2.

The types we will look at in detail are $(1, 2, 3)$, $(1, 4, 5)$, and $(2, 4, 6)$. I will also describe the shorthand used in Table 8-2 and make some general comments on all types.

If TYPE $= (1, 2, 3)$, then

$$A_r^0 = \begin{bmatrix} 1 & 0 & 0 & a_{1,4} & a_{1,5} & a_{1,6} \\ 0 & 1 & 0 & a_{2,4} & a_{2,5} & a_{2,6} \\ 0 & 0 & 1 & a_{3,4} & a_{3,5} & a_{3,6} \end{bmatrix}. \tag{8-11}$$

Thus the A_r^0-system is

$$x^2 + a_{1,4}xy + a_{1,5}xz + a_{1,6}yz = 0$$

$$y^2 + a_{2,4}xy + a_{2,5}xz + a_{2,6}yz = 0 \qquad (8\text{-}12)$$

$$z^2 + a_{3,4}xy + a_{3,5}xz + a_{3,6}yz = 0.$$

First we solve with $x = 1$. Thus (8-12) becomes

$$(a_{1,4} + a_{1,6}z)y = -(1 + a_{1,5}z)$$

$$y^2 + a_{2,4}y + a_{2,5}z + a_{2,6}yz = 0 \qquad (8\text{-}13)$$

$$z^2 + a_{3,5}z + a_{3,6}yz + a_{3,4}y = 0.$$

Since (8-13-1) [denoting the first equation of (8-13)] is bilinear, it allows us to define either variable as a ratio of linear expressions in the other. Substituting such a linear ratio in (8-13-2) and (8-13-3) will generate cubics in one variable. Solving these cubics will give us candidates for solutions to (8-13). We must, however, be wary of the various special (nongeneric) cases that may arise.

More specifically, consider

$$y = -\frac{1 + a_{1,5}z}{a_{1,4} + a_{1,6}z} \qquad (8\text{-}14)$$

derived from (8-13-1). Equation (8-14) is not well defined for choices of z making $a_{1,4} + a_{1,6}z = 0$. Thus, let us first consider the case $a_{1,4} + a_{1,6}z = 0$. Then, by (8-13-1), $1 + a_{1,5}z = 0$. If $a_{1,5} = 0$, this is impossible and the case we are considering has no solutions. If $a_{1,5} \neq 0$, then $z = -1/a_{1,5}$. Since z is now a constant value, we may substitute this for z in (8-13-2) and (8-13-3), getting two equations in the one unknown y:

$$y^2 + \left[a_{2,4} - \frac{a_{2,6}}{a_{1,5}} \right] y - \left[\frac{a_{2,5}}{a_{1,5}} \right] = 0 \qquad (8\text{-}15\text{-}2)$$

$$\left[-\frac{a_{3,6}}{a_{1,5}} + a_{3,4} \right] y + \left[\frac{1}{a_{1,5}^2} - \frac{a_{3,5}}{a_{1,5}} \right] = 0 \qquad (8\text{-}15\text{-}3)$$

Any triple $(x, y, z) = (1, y, -1/a_{1,5})$ satisfying both (8-15-2) and (8-15-3) is a solution at infinity to the A_r-system. Further, any $(x, y, z) = (1, y, -1/a_{1,5})$ satisfying *either* (8-15-2) or (8-15-3) is a candidate solution. That is, we can solve either of (8-15-2) or (8-15-3) to generate the resulting $(1, y, -1/a_{1,5})$, and check if this triple solves (8-13). *All* solutions at infinity of the A_r-system (under the case $a_{1,4} + a_{1,6}z = 0$) will be found. The point is, we will have to solve only one equation, not two, to get a complete candidate list. I choose (8-15-2) to solve because, even though it has higher degree than (8-15-3), there is no possibility of its having an infinite number of solutions (because the coefficient of y^2 is 1), while (8-15-3) might be identically zero. Rather than check if (8-15-3) is zero and solve (8-15-2) if it is, I would prefer to solve the equation that I know is nonzero to start with.

To summarize the "$a_{1,4} + a_{1,6}z = 0$" case, we have the following logic:

IF $a_{1,5} = 0$,

THEN "NO SOLUTIONS"; skip to next case.

ELSE use CUBE to solve (8-15-2) for y with $x = 1$, $z = -1/a_{1,5}$ and TEST resulting $(1, y, -1/a_{1,5})$ triples in (8-13). Those that solve (8-13) are solutions at infinity.

Note that the case "infinite number of solutions at infinity" cannot occur.

The other case is "$a_{1,4} + a_{1,6}z \neq 0$." First we make sure that either $a_{1,4} \neq 0$ or $a_{1,6} \neq 0$. If one of these coefficients is nonzero, we can substitute (8-14) into (8-13-2) and (8-13-3), generating the two cubics

$$[a_{2,5}a_{1,6}^2 - a_{2,6}a_{1,5}a_{1,6}]z^3 + [2a_{2,5}a_{1,4}a_{1,6} - a_{2,6}(a_{1,4}a_{1,5} + a_{1,6})$$

$$- a_{2,4}a_{1,5}a_{1,6} + a_{1,5}^2]z^2 + [a_{2,5}a_{1,4}^2 - a_{2,6}a_{1,4}$$

$$- a_{2,4}(a_{1,4}a_{1,5} + a_{1,6}) + 2a_{1,5}]z + [1 - a_{2,4}a_{1,4}] = 0 \quad (8\text{-}16\text{-}2)$$

$$a_{1,6}z^3 + [a_{1,4} + a_{3,5}a_{1,6} - a_{3,6}a_{1,5}]z^2$$

$$+ [a_{3,5}a_{1,4} - a_{3,6} - a_{3,4}a_{1,5}]z - a_{3,4} = 0. \quad (8\text{-}16\text{-}3)$$

Any solution z to (8-16-2) and (8-16-3) generates a solution at infinity $(1, y, z)$ to the A_r-system, where y is defined from z via (8-14). As before, we need only solve *one* of (8-16-2) and (8-16-3) to generate candidate $(1, y, z)$ to test in (8-13). But here our logic must include the possibility of solving either because there is no guarantee that either (or both) will not be identically zero.

If both have all coefficients equal to zero, then any $(1, y, z)$ solves (8-13), where y is generated by (8-14) [for values of z where (8-14) is well defined]. Thus NINF $= \infty$. Otherwise, one or the other might have all coefficients zero. We then solve the nonzero one (or discover that it has no solutions), and for each solution z, $(1, y, z)$ is a solution to (8-13) where y is generated by (8-14). If both (8-16-2) and (8-16-3) are nonzero, we can solve either as noted above. This exhausts all possibilities when $x = 1$.

When $x = 0$ and $y = 1$, (8-12) becomes

$$a_{1,6}z = 0$$

$$1 + a_{2,6}z = 0 \quad (8\text{-}17)$$

$$z^2 + a_{3,6}z = 0.$$

Observe that (8-17-2) implies that $z \neq 0$, and then (8-17-3) implies that $z = -a_{3,6}$. So our only candidate is $(x, y, z) = (0, 1, -a_{3,6})$. Finally, we must always test $(0, 0, 1)$. It will be a candidate for every type.

Let's summarize the entire type $= (1, 2, 3)$ logic. Define a flag "SNGFLG" so that if SNGFLG $= 1$, some solution at infinity is singular.

- Step 0. Set NINF $= 0$. Set SNGFLG $= 0$.
- Step I. IF ($a_{1,5} \neq 0$)
    THEN: use CUBE to solve (8-15-2) for y with $x = 1$, $z = -1/a_{1,5}$. Call TEST, which tests the resulting (x, y, z) triples in (8-12). For each solution to (8-12), TEST increments NINF by 1. If

$$\det[\hat{J}^1(y, -1/a_{1,5})] = 0,$$

    increment NINF by 1 (again) and set SNGFLG $= 1$.
- Step II. IF $a_{1,4} = 0$ and $a_{1,6} = 0$, THEN go to step III.
    IF (8-16-3) is not identically zero,
    THEN: use CUBE to solve (8-16-3) for z. For each solution z with $a_{1,4} + a_{1,6}z \neq 0$, generate a candidate (x, y, z) with $x = 1$ and y given by (8-14). Evoke the TEST described above.
    ELSE   IF (8-16-2) is not identically zero,
        THEN: use CUBE to solve (8-16-2) for z and continue as above.
        ELSE   NINF $= -1$ (i.e., NINF $= \infty$) and return.
- Step III. TEST candidates $(0, 1, -a_{3,6})$ and $(0, 0, 1)$. Check singularity via $\hat{J}^2$ and $\hat{J}^3$, respectively.

The RED3 logic for each type will be essentially parallel (but not *strictly* parallel) to the type $= (1, 2, 3)$ case, and I will not go into this level of detail in considering the other cases. However, I would like to introduce a shorthand that describes the preceding logic and that is used in Table 8-2. For type $= (1, 2, 3)$, we have:

I. $x = 1$, $z = -1/a_{1,5}$, [2]
II. $x = 1$, $y = -(1 + a_{1,5}z)/(a_{1,4} + a_{1,6}z)$, [3], [2]
III. $x = 0$, $y = 1$, $z = -a_{3,6}$

where "I" indicates step I, "II" step II, and "III" step III. The shorthand shows which variables are fixed and which equations will be solved (the equation numbers in square brackets), with II describing the substitution that defines the cubics (8-16-2) and (8-16-3). If, in the equations for I, II, or III, all coefficients are zero, NINF $= \infty$, but since two equations are given for II, both must be zero before NINF $= \infty$. Otherwise, we may choose either (nonzero) equation to solve, but the indicated order is recommended. The triple $(0, 0, 1)$ is not mentioned because it is *always* a candidate. In the shorthand, if an indicated division is invalid, the step is skipped. In other words, if $a_{1,5} = 0$, step I is skipped. If $a_{1,4} = 0$ and $a_{1,6} = 0$, step II is skipped.

Let's consider type $= (1, 4, 5)$. Then

$$A_r^0 = \begin{bmatrix} 1 & a_{1,2} & a_{1,3} & 0 & 0 & a_{1,6} \\ 0 & 0 & 0 & 1 & 0 & a_{2,6} \\ 0 & 0 & 0 & 0 & 1 & a_{3,6} \end{bmatrix}.$$

Thus the A_r^0-system is

$$x^2 + a_{1,2}y^2 + a_{1,3}z^2 + a_{1,6}yz = 0$$

$$xy + a_{2,6}yz = 0 \qquad\qquad (8\text{-}18)$$

$$xz + a_{3,6}yz = 0.$$

Now let $x = 1$. If $y = 0$, then $z = 0$ by (8-18-3), yielding candidate $(1, 0, 0)$. If $z = 0$, then $y = 0$ by (8-18-2). Otherwise, (8-18-2) and (8-18-3) imply that $y = -1/a_{3,6}$ and $z = -1/a_{2,6}$ yielding $(1, -1/a_{3,6}, -1/a_{2,6})$. (If either $a_{3,6} = 0$ or $a_{2,6} = 0$, we omit this candidate). When $x = 0$ and $y = 1$, (8-18) becomes

$$a_{1,2} + a_{1,3}z^2 + a_{1,6}z = 0$$

$$a_{2,6}z = 0 \qquad\qquad (8\text{-}19)$$

$$a_{3,6}z = 0.$$

If either $a_{2,6} \neq 0$ or $a_{3,6} \neq 0$, then $z = 0$, yielding candidate $(0, 1, 0)$. If both $a_{2,6} = 0$ and $a_{3,6} = 0$, we get candidates $(0, 1, z)$, where z satisfies (8-19-1). If there are an infinite number of such z (i.e., if $a_{1,2} = a_{1,3} = a_{1,6} = 0$), then NINF $= \infty$. We can organize this to be more formally parallel to the logic of the type $= (1, 2, 3)$ case via:

    I. $x = 1, y = 0, z = 0$
    II. $x = 1, y = -1/a_{3,6}, z = -1/a_{2,6}$
    III. $x = 0, y = 1, [2], [3], [1]$

We note in III the possible need to check all three equations (if some are zero) and also the possibility, if all are zero, that NINF $= \infty$.

The type $= (2, 4, 6)$ case is similar. We have

$$A_r^0 = \begin{bmatrix} 0 & 1 & a_{1,3} & 0 & a_{1,5} & 0 \\ 0 & 0 & 0 & 1 & a_{2,5} & 0 \\ 0 & 0 & 0 & 0 & 0 & 1 \end{bmatrix},$$

yielding

$$y^2 + a_{1,3}z^2 + a_{1,5}xz = 0$$

$$xy + a_{2,5}xz = 0 \qquad (8\text{-}20)$$

$$yz = 0.$$

Let $x = 1$. Then (8-20-3) implies either $y = 0$ or $z = 0$. But (8-20-2) implies that if $z = 0$, then $y = 0$, yielding the candidate $(1, 0, 0)$. If $z \neq 0$, then $y = 0$, and (8-20-1) and (8-20-2) become

$$a_{1,3}z + a_{1,5} = 0 \qquad (8\text{-}21)$$

$$a_{2,5}z = 0.$$

There is a nonzero solution, z, to (8-21) only if $a_{2,5} = 0$, $a_{1,3} \neq 0$ and $a_{1,5} \neq 0$ or if $a_{2,5} = a_{1,3} = a_{1,5} = 0$. In the latter case, any z is a solution and NINF $= \infty$. If $x = 0$ and $y = 1$, then (8-20-3) implies that $z = 0$, contradicting (8-20-1). The formal summary of this case is:

$$x = 1, y = 0, [1], [2],$$

which includes $x = 1$, $y = 0$, $z = 0$ implicitly.

Table 8-2 lists all the types with the RED3 logic indicated in the shorthand described above, except type $(p, q, 0)$. Only four of the 20 types generate cubic equations, while 11 other types require some equation solving (quadratic or linear). In five cases, we need merely check two or three specific candidates [as well as $(0, 0, 1)$]. Only seven types can yield NINF $= \infty$, those showing two or three equations in one line.

8-3-3 General Surfaces

Reducing higher-degree systems with GEL before solving is reasonable, although the number of cases tends to increase with the total degree of the system. If you need to include higher-degree surfaces in a modeler, the more manageable situation is to be able to assume special forms. For example, perhaps it will be possible to assume that only one torus will occur, with the two other surfaces quadric. Then you could plan to transform coordinates so that the torus is in a standard position and row reduce only the quadric part of the coefficient matrix. Then the number of cases that must be dealt with will be decreased.

Sometimes an application calls for the intersection of *two*, not *three*, polynomial surfaces. Generally, in this case the expected solution set will consist of curves. Tracking each curve is relatively easy once we find one point on it.† How can we find a point on each of the curves that are in the intersection of two surfaces? Let P_1 and P_2 denote the two defining polynomials. The system of three equations in three unknowns consisting of $P_1 = 0$, $P_2 = 0$, and $P_3 = 0$ (defined below) contains in its solution set at least one point on each curve. The extra equation is

$$P_3(x, y, z) = \det \begin{bmatrix} \dfrac{\partial P_1}{\partial x} & \dfrac{\partial P_1}{\partial y} & \dfrac{\partial P_1}{\partial z} \\[2mm] \dfrac{\partial P_2}{\partial x} & \dfrac{\partial P_2}{\partial y} & \dfrac{\partial P_2}{\partial z} \\[2mm] x - r_1 & y - r_2 & z - r_3 \end{bmatrix},$$

where "det" denotes the determinant and constants r_1, r_2, and r_3 are chosen independently of P_1 and P_2, as in the definition of the projective transformation. Now if a point is on an intersection curve, it satisfies $P_1 = 0$ and $P_2 = 0$. If, in addition, it is either a minimum or a maximum distance on the curve from the point $r = (r_1, r_2, r_3)$, it satisfies $P_3 = 0$. Each curve must have a point that is a minimum distance from r, so each curve will have a representative point in the solution set of $P_j = 0$ for $j = 1, 2, 3$. (Compare with the normality equation in Section 8-4.) Note that if P_1 and P_2 are quadrics, P_3 is a cubic. A poor choice of r might yield a system with an infinite number of solutions. The independent choice of r rules out this possibility, as long as the solutions to $P_1 = 0$ and $P_2 = 0$ have a one-dimensional intersection.

I do not have a lot to add, except that it might be fun to try some of Exercises 8-5 through 8-9. Also see the following section.

EXERCISES

8-5. A torus looks as shown in Fig. 8-8, but it need not be oriented along the (x, y) plane. It is interesting to work out the maximum number of nonsingular intersection points three tori can have. The total degree of a three-torus system is 64 but that is not the answer. What does this imply about solutions at infinity?

8-6. The following system gives the intersection of a torus (with $R_1 = 2$, $R_2 = 1$), a cylinder of radius $\frac{1}{2}$, and a sphere with x offset a and radius R:

$$x^4 + y^4 + z^4 + 2x^2y^2 + 2x^2z^2 + 2y^2z^2 - 10x^2 - 10y^2 + 6z^2 + 9 = 0$$

$$x^2 + z^2 - \tfrac{1}{4} = 0$$

$$(x - a)^2 + y^2 + z^2 - R^2 = 0.$$

† The cross product of the gradients of the two defining equations gives a tangent vector to the curve, which can be tracked by a path tracker.

Discuss reduction and the number of solutions at infinity for various values of a and R.

8-7. Assume that you have a fast reliable solver for systems of two kth degree polynomials in two unknowns. Describe an algorithm using this solver for counting the solutions at infinity of a system consisting of two kth-degree polynomials in three unknowns and an arbitrary degree polynomial in three unknowns.

8-8. The sphere–cylinder intersection

$$P_1 = x^2 + y^2 + z^2 - 4 = 0$$

$$P_2 = x^2 + y^2 - 1 = 0$$

consists of two circles. (Make a sketch.) Generate P_3 as in Section 8-3-3. Show that any point on the z-axis would be a poor choice for r. Is any choice off the z-axis OK? Solve the system $P_1 = 0$, $P_2 = 0$, $P_3 = 0$. Confirm that you get points on both intersection circles.

8-9. The cylinder and torus from Exercise 8-6 intersect in four closed arcs. Generate $P_3 = 0$ as above. Show that $r = (0, 0, 0)$ is a poor choice. How about $r = (0, 0, 1)$? Solve the appropriate system to find points on all four intersection arcs.

8.4 TUBES IN SPACE

We finish this chapter by considering a type of algebraic surface called a "tube." It is a generalization of a surface of revolution. I do not know of any current modeling system that uses algebraic tubes, so this material will have to be labeled "potentially useful."

A tube is the "fattening" of a curve. For example, a torus is the fattening of a circle. Since a tube is a surface, we are talking about the surface of the fat curve. Tubes can be defined as the simultaneous solution of several polynomial equations. Thus even though we will not generally have a single equation whose solutions give the tube, we can proceed as if we did. (In some cases, such as the torus, it is fairly easy to put the several equations that define the tube together to get a single equation.)

Let's take a first cut at the definition of a tube. Denote by C a curve in space. Let r be a positive number, the "radius" of the tube. Define

$$T = T(C, r) = \text{all points in space of distance } r \text{ from } C$$

(see Fig. 8-9). Note that the radius r must be "reasonably scaled" to C, or the geometry of the tube may be unusual, as shown in Fig. 8-10. We want to rule out this sort of "self-intersecting" tube. Intuitively, r should be small relative to C, and C should not fold back on itself, getting arbitrarily close without touching. If C is a circle, T is a torus, where C is the center circle.

We must clarify some things before we can identify T as a polynomial

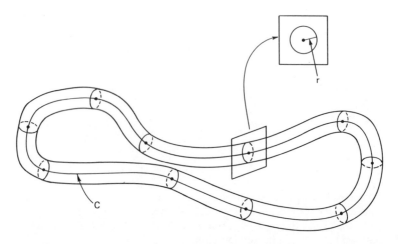

Figure 8-9   Tube in space of radius r generated by curve C, showing circular cross section.

surface. The most important clarification is the meaning of "distance"; in particular, distance of a point from a curve. An easier matter to clarify is the appropriate definition of C. For now, we take C to be the intersection of two polynomial surfaces:

$$C = \{(x', y', z') : P_1(x', y', z') = 0 \text{ and } P_2(x', y', z') = 0\},$$

where P_1 and P_2 are two polynomials. (Later, we will take a look at the case that C is defined, instead, by a parameterization.) For example, if $P_1 = x'^2 + y'^2 - 1$ and $P_2 = z'$, defining a cylinder and a plane, respectively, C is a circle in the (x', y')-plane.

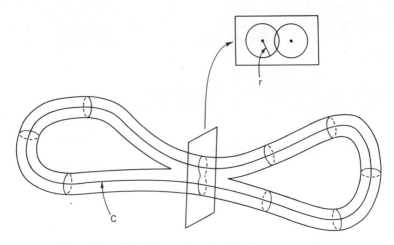

Figure 8-10   Self-intersecting tube, showing cross section.

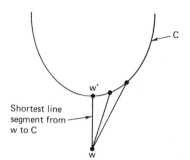

Figure 8-11 Distance of point w from curve C is given by shortest line segment.

Now let's consider "distance." If $w = (x, y, z)$ is a point in space, the distance of w from C ought to be the length of some line segment from w to C. Since there are generally lots of such line segments, we have to choose one. It makes sense to take the shortest (see Fig. 8-11). Thus let $d(w, C)$ denote the distance from w to C. Then

$$d(w, C) = \min \{| w - w' | : w' \in C\},$$

where "$| \cdot |$" denotes some norm (say, $| (a, b, c) | = \sqrt{a^2 + b^2 + c^2}$). To turn this into algebra, we need the following.

PROPOSITION.   There is a w' in C such that the straight line from w to w' is perpendicular to the tangent to C at w', and $d(w, C) = | w - w' |$.

Figure 8-12 makes the proposition reasonable. Here is a proof.

Proof. We may assume, for the purposes of the proof, that the curve C has a smooth parameterization, $p(\theta)$. Thus if $w' \in C$, then there is a θ such that $w' = p(\theta)$. Then $| w - w' |^2 = | w - p(\theta) |^2$, which attains its minima only when the derivative with respect to θ equals zero. But

$$0 = \frac{d(| w - p(\theta) |^2)}{d\theta} = 2 \left\langle w - p(\theta), \frac{dp}{d\theta} \right\rangle ,$$

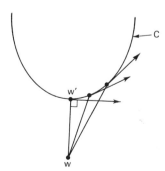

Figure 8-12 The shortest line segment from w to C is perpendicular to the tangent to C at the intersection point, w'.

where "$\langle \, , \, \rangle$" denotes the dot product: $\langle (a, b, c), (d, e, f) \rangle = a \cdot d + b \cdot e + c \cdot f$. [Recall that two vectors $u = (a, b, c)$ and $v = (d, e, f)$ are perpendicular exactly when $\langle u, v \rangle = 0$.] Since $dp/d\theta$ gives the tangent of C at $p(\theta)$, we see that the line from w to $w' = p(\theta)$ is perpendicular to the tangent to C at w'. This proves the proposition.

Geometrically, the sphere centered at w and with radius less than $|w - w'|$ does not contain any point of C, while the sphere with radius $= |w - w'|$ contains w'. The tangent line to C at w' cannot pierce this sphere (since points of C cannot be closer to w than w'). Hence the tangent to C is tangent to the sphere and therefore perpendicular to the radius vector $w - w'$.

This proposition leads us to the two fundamental tube equations:

$$|w - w'|^2 = r^2 \qquad \text{(radius equation)}$$

$$\langle \tan_C(w'), w - w' \rangle = 0 \qquad \text{(perpendicularity equation)},$$

where "$\tan_C(w')$" denotes a nonzero tangent to C at the point w'. The radius equation expresses the relation $|w - w'| = r$, squared so that it will be a polynomial equation. The perpendicularity equation says the tangent to C at w' is perpendicular to $w - w'$. If w is in $T(C, r)$, there is a $w' \in C$ such that (w, w') satisfies these two equations. If (w, w') satisfies these two equations, then w is in $T(C, r)$, if r is small enough. If r is not small relative to C, some "extra" w points may be included. However, in this case, the tube is self-intersecting and, as noted above, it is no longer what we have in mind by a tube.

It is convenient, when C is defined as the intersection of two surfaces, to replace the perpendicularity equation by an equivalent "normality equation." The normal space to C at w' is a plane in space spanned by the gradient vectors $dP_1(w')$ and $dP_2(w')$, where

$$dP_i = \left(\frac{\partial P_i}{\partial x}, \frac{\partial P_i}{\partial y}, \frac{\partial P_i}{\partial z} \right).$$

This is because $dP_1(w')$ is normal to the surface given by $P_1(w') = 0$, and $dP_2(w')$ is normal to the surface given by $P_2(w') = 0$, and both are normal to the intersection of these surfaces, C (see Fig. 8-13). Thus $w - w'$ is perpendicular to $\tan_C(w')$ exactly when $w - w'$ is in the normal space to C at w'; in other words, when the three vectors $dP_1(w')$, $dP_2(w')$, and $w - w'$ are linearly dependent; that is, when

$$\det \begin{bmatrix} dP_1(w') \\ dP_2(w') \\ w - w' \end{bmatrix} = 0 \qquad \text{(normality equation)}.$$

Thus the normality equation is a restatement of the perpendicularity equation

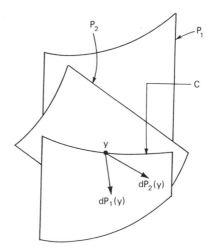

Figure 8-13 Surfaces given by $P_1 = 0$ and $P_2 = 0$ with normal vectors $dP_1(y)$ and $dP_2(y)$ at point y on the intersection curve C. The normal vectors are both normal to C, the curve of intersection.

when C is given as the intersection of two surfaces. It will be convenient to have a notation for the left-hand side of the normality equation. Let's use

$$N(P_1, P_2, w, w') = \det \begin{bmatrix} dP_1(w') \\ dP_2(w') \\ w - w' \end{bmatrix}.$$

Thus $N(P_1, P_2, w, w') = 0$ is a polynomial equation in w and w' that expresses exactly the requirement that the line from w to w' be normal to the curve C at the point w', where C is defined to be the intersection of $P_1 = 0$ and $P_2 = 0$.

So we have accomplished our goal of defining $T(C, r)$ algebraically, although we have had to introduce a supplemental variable, w', that is not really of interest to us. Sometimes, we can combine $P_1(w') = 0, P_2(w') = 0$, the radius equation, and the normality equation in such a way as to eliminate w'. Otherwise, it will just have to go along for the ride.

To summarize, we have the following definition of a tube:

$T(C, r) = \{w \in R^3:$ there is a $w' \in R^3$ such that $P_1(w') = 0,$

$$P_2(w') = 0, \mid w - w' \mid^2 - r^2 = 0, N(P_1, P_2, w, w') = 0\}.$$

Here R^3 denotes the points in three-dimensional space. Let's see what this means for a torus in standard position. We have

$$P_1(w') = x'^2 + y'^2 - R_1^2 = 0 \qquad (8\text{-}22\text{-}1)$$

$$P_2(w') = z' = 0 \qquad (8\text{-}22\text{-}2)$$

$$\mid w - w' \mid^2 - R_2^2 = 0 \qquad (8\text{-}22\text{-}3)$$

and

$$N(P_1, P_2, w, w') = \det \begin{bmatrix} 2x' & 2y' & 0 \\ 0 & 0 & 1 \\ x - x' & y - y' & z - z' \end{bmatrix}$$ (8-22-4)

$$= 2(y' \cdot x - x' \cdot y) = 0.$$

These four equations define T, a torus. If we want to intersect T with two other surfaces, say defined by $S_1(w) = 0$ and $S_2(w) = 0$, we must solve the system consisting of the six equations (8-22-1) through (8-22-4), $S_1(w) = 0$, and $S_2(w) = 0$. This system has six unknowns: x, y, z, x', y', and z', where we have been using the notation $w = (x, y, z)$ and $w' = (x', y', z')$. A solution (w, w') to the system consists of a common intersection point, w, and the point w' on C closest to w. In fact, the distance from w to w' is exactly R_2. This system has total degree 32 if S_1 and S_2 are quadrics. Although not outrageously large, we can reduce this total degree by simplifying the subsystem (8-22-1) through (8-22-4). In general, such a simplification may be hard to find, but in this case it is relatively easy.

We will eliminate w' from the subsystem. First, (8-22-4) implies that

$$y' = \frac{x'y}{x}$$ (8-23)

and substituting into (8-22-1) and (8-22-3) yields

$$x' = \frac{\pm R_1 x}{\sqrt{x^2 + y^2}}$$ (8-24)

$$x^2 + y^2 + z^2 + (R_1^2 - R_2^2) = \frac{2x'(x^2 + y^2)}{x},$$ (8-25)

where we have also taken advantage of $x'^2 + y'^2 = R_1^2$ and $z' = 0$ in generating (8-25). Now squaring both sides of (8-24) and substituting the resulting x'^2 into the equation obtained by squaring both sides of (8-25) yields the fourth-degree equation for the torus given at the beginning of Section 8-3.

Thus the subsystem has degree 4 and the full system degree 16, which is as small as it can be unless S_1 or S_2 are in some special form. I would like to emphasize that the reduction of the system from total degree 32 to 16 is not necessary to solve the intersection problem. But, of course, this type of reduction is a good idea if it can be done.

It may be more natural to define C by a parameterization. In this case, we proceed much as before, except that the parameterization must be transformable into algebraic relations if we want to use a polynomial solver. For example, if C is the circle, a natural parameterization would be

$$C(\theta) = (\cos \theta, \sin \theta, 0). \tag{8-26}$$

Then the tangent line to the curve at the point $c(\theta)$ is given by

$$\frac{dC}{d\theta}(\theta) = (-\sin \theta, \cos \theta, 0),$$

and the tube equations become

$$(w - C(\theta))^2 = r^2 \quad \text{(radius equation)}$$

and

$$\left\langle \frac{dC}{d\theta}(\theta), w - C(\theta) \right\rangle = 0 \quad \text{(perpendicularity equation)}.$$

Now the radius and perpendicularity equations form a system of two equations in the four unknowns x, y, z, and θ and define a surface (in fact, a torus). This system is not polynomial. However, we can make it polynomial by applying a transformation to eliminate the sines and cosines. In this case we have an easy transformation:

$$\cos \theta = x', \sin \theta = y' \tag{8-27}$$

with the additional relation

$$x'^2 + y'^2 = 1. \tag{8-28}$$

This yields a system identical to (8-22).

In general, we will have other expressions for $C(\theta)$ besides (8-26). The radius equation and perpendicularity equations will be as above. However, if $C(\theta)$ is naturally defined in terms of transcendental elements, it may be difficult (or impossible) to find a satisfactory transformation of the system into algebraic form. For example, the simple spiral

$$C(\theta) = (\cos \theta, \sin \theta, \theta)$$

defines this type of tranformation. However, the spiral

$$C(\theta) = (\cos \theta, \sin \theta, \tan \theta)$$

can be transformed satisfactorily via (8-27).

EXERCISES

8-10. Display systems for an elliptical torus (i.e., a tube with centerline an ellipse) and a parabolic torus (centerline a parabola). Can you reduce these systems?

8-11. Define two tori and a plane so that the common intersection gives eight real solutions. Confirm your choices by running CONSOL. (*Hint:* Make a sketch with one torus represented by its center line.)

8-12. Define two elliptical tori and a plane so that the common intersection gives 16 real solutions. Confirm your choices by running CONSOL.

8-13. You cannot define a sphere or a cone as a tube. But if you generalize the definition of tube so that r is not constant, you can. Find systems that express the cone and sphere as "tubes with nonconstant radius."

8-14. Find a value for a so that

$$x^2 + y^2 + z^2 - a^2 = 0$$

$$(x - 1)^2 + y^2 - 1 = 0$$

$$T((\cos \theta, \sin \theta, \tan \theta), 0.5) = 0$$

has at least one real solution. Solve using CONSOL to confirm your answer. (*Hint:* Make a sketch.)

8-15. Find a value for a so that

$$x^2 + y^2 + z^2 - a^2 = 0$$

$$(x - 1)^2 + y^2 - 1 = 0$$

$$T((\theta, \theta^2, \theta^3), 0.5) = 0$$

has at least one real solution. Solve using CONSOL to confirm your answer.

TABLE 8–1   RED2 logic summary[a]

---

1. Type $= (1, 2)$:
    $\deg(f_1) = \deg(f_2) = 2$
    IF $a_{1,3}a_{2,3} \neq 1$
    THEN NINF $= 0$
    ELSE IF $a_{2,4}a_{1,3}^3 - a_{2,5}a_{1,3}^2 + a_{1,4}a_{1,3} - a_{1,5} \neq 0$
         THEN NINF $= 1$
         ELSE NINF ≥ 2.

2. Type $= (1, 3)$:
    $\deg(f_1) = \deg(f_2) = 2$
    IF $a_{12} \neq 0$
    THEN NINF $= 0$
    ELSE IF $a_{1,5} \neq 0$
         THEN NINF $= 1$
         ELSE NINF ≥ 2.
    IF (NINF ≥ 2) THEN $f_1 = x^2 + a_{1,4}x + a_{1,6} = 0$

3. Type $= (2, 3)$:
    $\deg(f_1) = \deg(f_2) = 2$
    IF $a_{1,4} \neq 0$,
    THEN NINF $= 1$
    ELSE NINF ≥ 2.
    IF (NINF ≥ 2) THEN $f_1 = y^2 + a_{1,5}y + a_{1,6} = 0$

---

[a] If the second equation is identically a constant, this version of RED2 returns a "fatal error."

TABLE 8–1   (continued)

4. Type = $(i, 0)$:
    IF $(a_{1,1} \neq 0$ or $a_{1,2} \neq 0$ or $a_{1,3} \neq 0)$
    THEN $\deg(f_1) = 2$
    ELSE IF $(a_{1,4} \neq 0$ or $a_{1,5} \neq 0)$
          THEN $\deg(f_1) = 1$
          ELSE $\deg(f_1) = 0$.
    IF $(a_{2,4} \neq 0$ or $a_{2,5} \neq 0)$
    THEN $\deg(f_2) = 1$
    ELSE $\deg(f_2) = 0$
    IF $\deg(f_2) = 0$ THEN return "fatal error"
    IF $a_{2,5} \neq 0$
    THEN IF $a_{1,1}a_{2,5}^2 + a_{1,2}a_{2,4}^2 - a_{1,3}a_{2,4}a_{2,5} \neq 0$
          THEN NINF = 0
          ELSE IF $a_{2,6}(-2a_{1,2}(a_{2,4}/a_{2,5}) + a_{1,3}) - a_{2,5}(-a_{1,5}(a_{2,4}/a_{2,5}) + a_{1,4}) \neq 0$
                THEN NINF = 1
                ELSE NINF = 2
    ELSE IF $a_{2,4} \neq 0$
          THEN IF $a_{1,2} \neq 0$
                THEN NINF = 0
                ELSE IF $a_{1,3}a_{2,6} - a_{1,5}a_{2,4} \neq 0$
                      THEN NINF = 1
                      ELSE NINF = 2

TABLE 8–2   List of types for RED3[a]

1. $(1, 2, 3)$: $x = 1, z = -1/a_{1,5}$ [2]
                  $x = 1, y = -(1 + a_{1,5}z)/(a_{1,4} + a_{1,6}z)$ [3], [2]
                  $x = 0, y = 1, z = -a_{3,6}$
2. $(1, 2, 4)$: $x = 1, z = -1/a_{3,6}$ [2]
                  $x = 1, y = -a_{3,5}z/(1 + a_{3,6}z)$ [1], [2]
                  $x = 0, y = 1,$ [2]
3. $(1, 2, 5)$: $x = 1, z = -a_{2,4}, z = 0$
                  $x = 1, y = -1/a_{3,6}$ [1], [2]
                  $x = 0, y = 1,$ [2]
4. $(1, 2, 6)$: $x = 1, y = -1/a_{1,4}, z = 0$
                  $x = 1, y = 0$ [1]
5. $(1, 3, 4)$: $x = 1, z = -1/a_{3,6}$ [1], [2], [3]
                  $x = 1, y = -a_{3,5}z/(1 + a_{3,6}z)$ [1], [2]
                  $x = 0, y = 1, z = 0$
                  $x = 0, y = 1, z = -a_{2,6}$

[a] The numbers in square brackets refer to rows in the (reduced) matrix. The associated equations must be solved with the indicated substitutions to find all solutions at infinity. (The $(p, q, 0)$ case can be reduced to the two-variable problem and is omitted from this table.) See Section 8-3-2 for explanation.

TABLE 8–2   (continued)

6. (1, 3, 5): $x = 1, z = 0$ [1]
    $x = 1, y = -1/a_{3,6}$ [2]
    $x = 0, y = 1, z = 0$
    $x = 1, y = 1, z = -a_{2,6}$

7. (1, 3, 6): $x = 1, y = 0, z = -1/a_{1,5}$
    $x = 1, z = 0$ [1]
    $x = 0, y = 1, z = 0$

8. (1, 4, 5): $x = 1, y = 0, z = 0$
    $x = 1, y = -1/a_{3,6}, z = -1/a_{2,6}$
    $x = 0, y = 1$ [2], [3], [1]

9. (1, 4, 6): $x = 1, y = 0, z = 0$
    $x = 1, y = 0$ [1]
    $x = 0, y = 1, z = 0$

10. (1, 5, 6): $x = 1, z = 0,$ [1]
    $x = 0, y = 1, z = 0$

11. (2, 3, 4): $x = 1, z = -1/a_{3,6}$ [1]
    $x = 1, y = -a_{3,5}z/(1 + a_{3,6}z)$ [1], [2]
    $x = 0, y = 1, z = -1/a_{1,6}$

12. (2, 3, 5): $x = 1, y = 0, z = 0$
    $x = 1, y = -a_{1,4}, z = 0$
    $x = 1, y = -1/a_{3,6},$ [2]
    $x = 0, y = 1, z = -a_{2,6}$

13. (2, 3, 6): $x = 1, y = 0, z = -a_{2,5}$
    $x = 1, y = 0, z = 0$
    $x = 1, y = -a_{1,4}, z = 0$

14. (2, 4, 5): $x = 1, y = 0, z = 0$
    $x = 1, y = -1/a_{3,6}, z = -1/a_{2,6}$
    $x = 0, y = 1,$ [1]

15. (2, 4, 6): $x = 1, y = 0,$ [1], [2]

16. (2, 5, 6): $x = 1, y = 0, z = 0$
    $x = 1, y = -a_{1,4}, z = 0$

17. (3, 4, 5): $x = 1, y = 0, z = 0$
    $x = 1, y = -1/a_{3,6}, z = -1/a_{2,6}$
    $x = 0, y = 1, z = -a_{1,6}$

18. (3, 4, 6): $x = 1, y = 0, z = 0$
    $x = 1, y = 0, z = -a_{1,5}$
    $x = 0, y = 1, z = 0$

19. (3, 5, 6): $x = 1, z = 0,$ [1]
    $x = 0, y = 1, z = 0$

20. (4, 5, 6): $x = 1, y = 0, z = 0$
    $x = 0, y = 1, z = 0$

CHAPTER 9

Chemical Equilibrium Systems

9-1 INTRODUCTION

"The plane almost crashed on its maiden flight because they left out the chemistry." We were a little astonished, but before anyone could say anything he went on. "The flight-test computer model was just aerodynamic. It didn't include the dissociation effects. Frankly, they didn't know how to solve the chemistry, and they didn't think it was that important." We were glad to hear that nobody got hurt. And glad to hear that they were going to add chemistry to their model. Sometimes solving polynomial systems can be a matter of life or death.†

The principal focus here is on standard chemical equilibrium problems. These problems are defined by reaction and conservation equations. The resulting polynomial systems have a structure that can be systematically reduced and scaled. It is critically important that they be reduced and scaled, since in the wrong form they are virtually impossible to solve.

For the general reader this chapter illustrates two ideas:

- Exploiting system structure to carry out preliminary off-line reduction
- Dealing with problems whose initial scaling is extreme

The chapter is organized as follows. We consider a simple illustrative problem in Section 9-2, and the general concepts follow in Section 9-3. More

---

† I was told this story by an aircraft designer who works for a large company not associated with the General Motors Corporation. He would rather I did not give details.

complicated examples are given in Section 9-4, where we see the full reduction–scaling–solution process carried out on a "real" problem. Section
9-5 discusses a maverick example that is not strictly an equilibrium system.
This section illustrates how the spirit of the approach can be used when the
letter cannot.

The material in this chapter is based on joint work with Keith Meintjes
of General Motors Research Laboratories, and I would like to thank Dr.
Meintjes for his assistance in its preparation.

9-2 SIMPLE ILLUSTRATIVE PROBLEM

Imagine a container into which hydrogen (H), oxygen (O), and water (H_2O)
are placed and heated at a constant temperature. The amounts of hydrogen,
oxygen, and water vapor in the sealed container will initially change but
eventually remain constant. Then exact relations among these amounts will
hold. For example, water can exchange itself with hydrogen and oxygen
atoms in a fixed ratio: two hydrogens and one oxygen can make one water,
and vice versa; in chemical shorthand,

$$H_2O \rightleftarrows 2H + O.$$

This is called an "equilibrium reaction equation." Reaction equations tell
how some molecules can transform into others. (We focus on "reversible"
reactions, in which the transformation can go both ways—thus the double
arrows.) Another type of relation among the chemicals in the container is
"conservation." The elements we put into the pot cannot go anywhere else.
No matter how the compounds change around, the amount of each element
remains the same. My mental image is of discrete ingredients; that is, little
pellets for H and O and little clusters of pellets for water. We are counting
the pellets in all the molecules. Conservation of hydrogen is expressed as

$$2[H_2O] + [H] = \text{total hydrogen},$$

where the square brackets indicate quantity, according to the prescription
that each compound containing H appears in the "conservation of hydrogen" equation with a number in front for how many hydrogen atoms are in
the compound (2 for H_2O, 1 for H). The "total hydrogen" is a number
expressing how much hydrogen was put in at the beginning in the form of
hydrogen atoms and in the form of water. Similarly, we have conservation
of oxygen:

$$[H_2O] + [O] = \text{total oxygen}.$$

The basic chemistry for the equilibrium system is defined by the three equations above.† However, we have not derived an algebraic system yet. We are still in the world of chemistry. To get to the world of algebra, we proceed as follows.

1. First, we name algebraic variables that stand for the quantities of each of the chemical elements and compounds. (We will use molar concentrations here [Benson, 1971, p. 25].) Thus

$$x_1 = [\text{H}]$$

$$x_2 = [\text{O}]$$

$$x_3 = [\text{H}_2\text{O}].$$

It is convenient to have the first named variables represent the elements and the later ones the compounds.

2. The conservation equations express total quantities and translate naturally into linear algebraic equations. Conservation of hydrogen yields

$$2x_3 + x_1 = T_{\text{H}},$$

where T_{H} stands for total hydrogen, and conservation of oxygen

$$x_3 + x_2 = T_{\text{O}},$$

where T_{O} stands for total oxygen. (T_{H} and T_{O} are constants that must be part of the problem definition.)

3. So far, we have two equations in three unknowns, and we need another relation to have a solvable system. This other equation will come from the reaction equation. Here is how chemical reaction equations yield algebraic equations: The "$\rightleftarrows$" becomes "$=$," the element and compound names are replaced by the associated algebraic variables, "$+$" becomes "$\cdot$" (multiplication), and "muliplication by numbers" becomes "raising to powers." (Formally, this transformation is identical with applying the exponential and replacing e^{H} by x_1, e^{O} by x_2, and $e^{\text{H}_2\text{O}}$ by x_3.) Finally, we must include an "equilibrium constant," K. The result is

$$Kx_3 = x_1^2 x_2.$$

The numerical value of the equilibrium constant will depend on the conditions of the reaction (in particular, the temperature). In the context of a particular chemical problem, the appropriate values for K

---

† Those with some chemistry experience might observe that molecular oxygen and hydrogen, O_2 and H_2, are likely to be present. They have been left out to clarify the essentials of the problem.

would typically be derived from standard tables. (See [Benson, 1971, Chap. 18] for more background.)

We have arrived at a mathematical system:

$$2x_3 + x_1 = T_H$$

$$x_3 + x_2 = T_O \qquad (9\text{-}1)$$

$$x_3 = Rx_1^2 x_2,$$

where $R = 1/K$. The constants T_H, T_O, and R are defined by the chemical context. Given T_H, T_O, and R, we want to find all the solutions (x_1, x_2, x_3) to (9-1) that are positive and real. (From chemical considerations, we in fact know that there is always one and only one positive real solution.) From this point forward, we have a purely mathematical problem. No chemical reasoning will play a part in our reduction–scaling–solution methodology. However, the way (9-1) has arisen has given it a particular structure that we will exploit. (In Section 9-3 more is said about this in general.) For now, observe that we can use the third equation in (9-1) to eliminate x_3 from the system by substitution, yielding a new system:

$$2(Rx_1^2 x_2) + x_1 = T_H \qquad (9\text{-}2)$$

$$Rx_1^2 x_2 + x_2 = T_O.$$

This is the system we will solve. Once we solve (9-2) for x_1 and x_2, we can generate the x_3 values from the third equation of (9-1). Note that if x_1 and x_2 are real and positive, so is x_3. Observe that (9-2) is a system with only the elements as variables and with one equation for each conservation equation.

Generating (9-2) is our "basic reduction." The system (9-2) consists of two cubic equations and therefore has total degree 9. At this point we must decide whether to continue with an "auxiliary reduction" or proceed to scaling and solution.

Although the problem is artificially simple, let us proceed systematically. We can test the likelihood of reduction by computing RINF and sparsity (see Chapter 7). Let $f = 0$ denote the system (9-2). To generate the solutions at infinity and compute their singularity, we create $\hat{f}$, the homogenization of f, and its Jacobian, $\hat{J}$. Then we consider f^0, $\hat{f}^1$, and $\hat{f}^2$ as required (see Section 3-3). Thus

$$\hat{f}_1(y) = 2Ry_1^2 y_2 + y_1 y_3^2 - T_H y_3^3 = 0 \qquad (9\text{-}3)$$

$$\hat{f}_2(y) = Ry_1^2 y_2 + y_2 y_3^2 - T_O y_3^3 = 0$$

and

$$\hat{J} = \begin{bmatrix} 4Ry_1 y_2 + y_3^2 & 2Ry_1^2 & 2y_1 y_3 - 3T_H y_3^2 \\ 2Ry_1 y_2 & Ry_1^2 + y_3^2 & 2y_2 y_3 - 3T_O y_3^2 \end{bmatrix}.$$

Now $f^0 \equiv \hat{f}(x_1, x_2, 0)$, so

$$f_1^0 = 2Ry_1^2 y_2 = 0 \tag{9-4}$$
$$f_2^0 = Ry_1^2 y_2 = 0.$$

We see that $(x_1, x_2) = (1, 0)$ and $(0, 1)$ are solutions of $f^0 = 0$ and therefore solutions at infinity of $f = 0$. These two solutions at infinity are singular, if $\det(\hat{J}^1(0, 0)) = 0$ or $\det(\hat{J}^2(0, 0)) = 0$, respectively, where $\hat{J}^i$ denotes the Jacobian of the standard representation $\hat{f}^i$ for $i = 1, 2$. Thus

$$\hat{J}^1(0, 0) = \begin{bmatrix} 2R & 0 \\ R & 0 \end{bmatrix}$$

and

$$\hat{J}^2(0, 0) = \begin{bmatrix} 0 & 0 \\ 0 & 0 \end{bmatrix}$$

shows that both are singular. This implies that RINF $\geq \frac{4}{9}$. Since NCOEF(2, 3) = 9, the sparsity of (9-2) is $\frac{6}{18} = \frac{1}{3}$. Thus both indicators from Chapter 7 suggest that reduction is possible.

What to do at this point to reduce (9-2) is not predetermined in the way the reduction that generated (9-2) was. We must try approaches that have worked before on other systems or try whatever we can devise. It turns out that we can reduce (9-2) further by using the row reduction technique, GEL, as illustrated in Chapters 7 and 8 (see Appendix 5). We have the coefficient matrix for (9-2):

$$\begin{array}{c} \text{first eq.} \rightarrow \\ \text{second eq.} \rightarrow \end{array} \begin{bmatrix} 2R & 1 & 0 & T_H \\ R & 0 & 1 & T_O \end{bmatrix}$$

$$\uparrow \quad \uparrow \quad \uparrow \qquad \uparrow$$

$$x_1^2 x_2 \quad x_1 \quad x_2 \quad \text{constant}$$

and row reduction yields

$$\begin{bmatrix} 0 & 1 & -2 & T_H - 2T_O \\ 1 & 0 & \dfrac{1}{R} & \dfrac{T_O}{R} \end{bmatrix},$$

giving the equivalent system

$$x_1 - 2x_2 = T_H - 2T_O \tag{9-5}$$
$$x_1^2 x_2 + \frac{1}{R} x_2 = \frac{T_O}{R},$$

which has total degree 3. Since the first equation of (9-5) is linear, we can use it to eliminate one variable. In this case we get a single cubic equation:

$$Rx_1^3 - (T_H - 2T_O)Rx_1^2 + x_1 - T_H = 0. \tag{9-6}$$

We cannot hope for a further reduction, so we have completed the reduction part of solving (9-1).

It is time to consider scaling. To get a sense of what is involved, take $T_H = 10^{-2}$, $T_O = 10^2$, and $R = 10^{-16}$. (These values are plausible.) Then our equation is

$$x_1^3 + 1.9999 \cdot 10^2 x_1^2 + 10^{16}x_1 - 10^{14} = 0. \tag{9-7}$$

This range of coefficients (and worse) is typical of chemical problems and indicates why scaling is important. I know of no problem area where scaling is more critical to the success of solution methods. This will be illustrated further by the examples developed in later sections of this chapter. From Chapter 5 we have two general approaches to scaling, SCLGEN and SCLCEN.

With

$$a_1 = 1, \quad a_2 = 1.9999 \cdot 10^2, \quad a_3 = 10^{16}, \quad a_4 = 10^{14}$$

for input coefficients, SCLGEN yields scale factors $c_1 = 5.57$ and $c_2 = -16.43$, which produce the scaled coefficients

$$b_1 = 1.91$$

$$b_2 = 1.03 \cdot 10^{-3}$$

$$b_3 = 1.38 \cdot 10^5$$

$$b_4 = 3.72 \cdot 10^{-3}$$

in the notation of Chapter 5.

For SCLCEN, we need to input a nominal value for x_1, denoted x_1^0. Since x_1 represents hydrogen and $T_H = 10^{-2}$, we'll choose $x_1^0 = 10^{-2}$. Thus $c_1 = -2$, and SCLCEN computes $c_2 = -5.2$, which yields

$$b_1 = 6.31 \cdot 10^{-12}$$

$$b_2 = 1.26 \cdot 10^{-7}$$

$$b_3 = 6.30 \cdot 10^8$$

$$b_4 = 6.30 \cdot 10^8.$$

The final step is, of course, to solve the equation.

This completes the discussion of the illustrative example. In the next section we describe the general case, which will follow the steps outlined here point by point.

EXERCISES

9-1. Which of the two scaling given for (9-7) is best? (Define "best.")

9-2. Solve (9-7) with CONSOL using both scalings.

9-3 THE METHODOLOGY IN GENERAL

Assume that we have a system $f(x) = 0$ of n polynomial equations in the n unknowns $x_1, \ldots, x_n$ as follows: $x_1, \ldots, x_m$ are called "components" and $x_{m+1}, \ldots, x_n$ are called "compounds." The first m equations in the system $f_1(x) = 0, \ldots, f_m(x) = 0$ are linear and involve all variables. These are the "conservation equations." The last $n - m$ equations are in the form

$$f_{m+1} = x_{m+1} - s_1(x_1, \ldots, x_m) = 0$$

$$\vdots \qquad \qquad \vdots \qquad \qquad \qquad \qquad (9\text{-}8)$$

$$f_n \quad = x_n - s_{n-m}(x_1, \ldots, x_m) = 0,$$

where the s_i are polynomial equations with only the components as variables. These are the "equilibrium reaction equations."

Chemical equilibrium systems can always be put in this form.† (See [Brinkley, 1946] and [Meintjes and Morgan, 1987].) Most often, the "components" will be individual atoms of each element: for example, H or O. In some cases it will be convenient to take simple compounds as the components: for example, CO instead of C. This is OK as long as the resulting $f(x) = 0$ can be put in the form of (9-8). Now let's consider how to reduce–scale–solve this type of system. There is no chemistry in this section of the chapter.

The first reduction step—the *basic reduction*—substitutes for the compounds in the conservation equations using the expressions given by the equilibrium reaction equations. In other words, we generate the new system made up of equations:

$$f_i(x_1, \ldots, x_m, s_1(x_1, \ldots, x_m), \ldots, s_{n-m}(x_1, \ldots, x_m)) = 0 \ (9\text{-}9)$$

for $i = 1, 2, \ldots, m$. System (9-9) consists of m equations in m unknowns. Often, m is considerably smaller than n and consequently (9-9) can already be a significant reduction over (9-8). The basic reduction can always be accomplished because of the fundamental structure of chemical equilibrium problems. It is not a stroke of luck.

The second reduction step—the *auxiliary reduction*—is less cut-and-

---

† General equilibrium problems always include the equations above and may involve additional linear equations. These additional equations do not significantly affect the analysis.

dried. We must proceed as in Chapter 7 to test for reducibility and then to try previously successful approaches, such as GEL, or substitution, or change of variables. The examples in this chapter illustrate both the basic and the auxiliary reductions. Even though (9-9) might be further reduced, I will continue to denote the system as "(9-9)."

SCLGEN seems to scale chemical problems well. As with all polynomial systems, I use SCLGEN and the projective transformation on chemical problems unless I have a reason not to. However, a stock scaling for chemistry problems is to evoke SCLCEN with the "mean values," x_i^0, equal to the component totals used in the conservation equations. Then all physical solutions will be scaled to the unit box. Because of the extremely small solution values that can result, I am not sure that this is generally a good idea, except for certain special studies (e.g., mapping basins of attraction for Newton's method; see [Meintjes and Morgan, 1987]).

To choose a solution method, I use the following rules of thumb. Newton's method is good if I have a reasonable way of getting start points, if CPU time is an issue, and if I am expecting only a single physically meaningful solution. (Using total concentrations for each component as a start point can work well. In the SCLCEN "stock scaling," this is $x_j = 1$ for $j = 1, \ldots, m$. We have had success on chemical problems with the "absolute Newton's method." In this variant of Newton's method, after each iterate is computed, any negative coordinates have their signs changed and then the method is allowed to continue. See [Meintjes and Morgan, 1987].) I use CONSOL if I want to look at the full solution set and if I can spare the extra CPU time. Note that sometimes a problem has *no* or *several* physical solutions.† Only CONSOL can convince me of this. Newton's failure to converge does not prove that there is no solution, and Newton's failure to find several solutions, even with several start points, does not imply anything about how many solutions there are. Naturally, these general comments on Newton's method and CONSOL assume that the problem has not been reduced to one of those special cases for which an algebraic or other nongeneral method applies.

Finally, after the scaled–reduced system is solved, the solutions to the original system must be generated. There is one trap to look out for here. After the components have been generated, always use the (9-8) equations to get the compounds. These equations are product formulas, involving no addition or subtraction. Do not use the conservation equations, which are sums of components and compounds. The numbers involved typically differ from each other by many orders of magnitude and "catastrophic cancellation" in summing is likely (see [Forsythe et al., 1977, p. 15]).

---

† Not the basic chemical equilibrium problem. But my comments here include more general models.

9-4 TWO COMBUSTION CHEMISTRY EXAMPLES

I have chosen two combustion chemistry examples to illustrate the methodology described in Section 9-3. I will not focus on the chemistry but will spend a good deal of time on the mathematics. We will consider the examples in detail, from initial analysis through reduction, scaling, and solution. This is partly to give you a sense of the magnitude of the undertaking, as well as to show you the "technique" per se. Finally, I will suggest that you go through some experiments to get hands-on experience with the difficulties that can be encountered.

Here is a brief discussion of the type of combustion model that we will be dealing with. See [Meintjes and Morgan, 1985, 1987] for a more technical and detailed description. The phenomenon modeled is the burning of fuel in a combustion chamber (as in an automobile engine). We want to know what chemical components and compounds are in the chamber at various stages of burning (so that, for example, we can estimate the quantity of various gases in the exhaust). We start with air and fuel, but after burning begins, product gases (e.g., carbon monoxide) are generated. Although the burning will be fast, we envision a "snapshot" of the chamber between the time when the burning begins and when it ends. Also, we envision a very small piece of the combustion chamber that, for brief instants, will behave like a chemically closed system. With these modeling assumptions, the composition of gases can be described by a chemical equation system of the sort we have been considering. The coefficients of the system will be derived from reaction constants and component conservation totals as before, although now these will be related to temperature and other boundary conditions. The two models below describe the same phenomenon but one is more complicated, as it includes a term deemed "negligible" in the other. The reduced forms are significantly different. Therefore, the neglected term in mathematically significant. However, deciding which model to choose is a chemical, not a mathematical, issue.

Table 9-1 lists the chemical components and compounds we will be considering. Thus the fourth component is nitrogen, and we take $x_4 = [N]$.

Model A: (Omits NO).
Conservation equations:†

$$\text{hydrogen:} [H] + 2[H_2] + [OH] + 2[H_2O] = T_H \qquad (9\text{-}10\text{-}1)$$

$$\text{carbon:}\quad [CO] + [CO_2] = T_C \qquad (9\text{-}10\text{-}2)$$

† The brackets around the components and compounds indicate "molar concentrations," which are our variables. By contrast, the "reaction equations" are more symbolic, indicating "atom exchange mechanisms" rather than equations in the sense that we have been using the concept in this book.

TABLE 9–1 Variable indexing for models A and B

Subscript of x		Components
1	O	atomic oxygen
2	H	atomic hydrogen
3	CO	carbon monoxide[a]
4	N	atomic nitrogen
		Compounds
5	O_2	molecular oxygen
6	H_2	molecular hydrogen
7	N_2	molecular nitrogen
8	CO_2	carbon dioxide
9	OH	hydroxyl radical
10	H_2O	water vapor
11	NO	nitric oxide

[a] In these models, C (pure carbon) does not enter into the reactions. Thus CO, the most elementary carbon compound, is used as the carbon "component."

oxygen: $[O] + [CO] + 2[O_2] + 2[CO_2] + [OH] + [H_2O] = T_O$ (9-10-3)

nitrogen: $[N] + 2[N_2] = T_N$ (9-10-4)

Equilibrium reaction equations:

$$O_2 \rightleftarrows 2O \qquad (9\text{-}10\text{-}5)$$

$$H_2 \rightleftarrows 2H \qquad (9\text{-}10\text{-}6)$$

$$N_2 \rightleftarrows 2N \qquad (9\text{-}10\text{-}7)$$

$$CO_2 \rightleftarrows O + CO \qquad (9\text{-}10\text{-}8)$$

$$OH \rightleftarrows O + H \qquad (9\text{-}10\text{-}9)$$

$$H_2O \rightleftarrows O + 2H \qquad (9\text{-}10\text{-}10)$$

Model B: (Includes NO).
The same as model A, except: replace conservation equations (9-10-3) and (9-10-4) by

oxygen: $[O] + [CO] + 2[O_2] + 2[CO_2]$

$$+ [OH] + [H_2O] + [NO] = T_O \qquad (9\text{-}10\text{-}3\text{-}1)$$

nitrogen: $[N] + 2[N_2] + [NO] = T_N$ (9-10-4-1)

and append the additional equilibrium reaction equation:

$$NO \rightleftarrows O + N \qquad (9\text{-}10\text{-}11)$$

The totals T_O, T_H, T_C, and T_N are part of the initial conditions of the problem. The reaction coefficients, denoted K_1, K_2, . . . , K_7 below, are determined from standard thermochemical tables, and are functions of the temperature. See [Meintjes and Morgan, 1985] for details.

Let's focus first on model A. We can change the chemical system into a mathematical system by rote, as described in Section 9-3. This yields

$$x_2 + 2x_6 + x_9 + 2x_{10} = T_H \qquad\qquad (9\text{-}11\text{-}1)$$

$$x_3 + x_8 = T_C \qquad\qquad (9\text{-}11\text{-}2)$$

$$x_1 + x_3 + 2x_5 + 2x_8 + x_9 + x_{10} = T_O \qquad\qquad (9\text{-}11\text{-}3)$$

$$x_4 + 2x_7 = T_N \qquad\qquad (9\text{-}11\text{-}4)$$

$$K_1 x_5 = x_1^2 \qquad\qquad (9\text{-}11\text{-}5)$$

$$K_2 x_6 = x_2^2 \qquad\qquad (9\text{-}11\text{-}6)$$

$$K_3 x_7 = x_4^2 \qquad\qquad (9\text{-}11\text{-}7)$$

$$K_4 x_8 = x_1 x_3 \qquad\qquad (9\text{-}11\text{-}8)$$

$$K_5 x_9 = x_1 x_2 \qquad\qquad (9\text{-}11\text{-}9)$$

$$K_6 x_{10} = x_1 x_2^2, \qquad\qquad (9\text{-}11\text{-}10)$$

where equation (9-11-j) corresponds to equation (9-10-j) for $j = 1$ to 10 and K_1 through K_6 are the equilibrium reaction constants. The basic reduction is obtained by substituting (9-11-5) through (9-11-10) into (9-11-1) through (9-11-4), yielding

$$2R_6 x_1 x_2^2 + R_5 x_1 x_2 + 2R_2 x_2^2 + x_2 - T_H = 0 \qquad\qquad (9\text{-}12\text{-}1)$$

$$R_4 x_1 x_3 + x_3 - T_C = 0 \qquad\qquad (9\text{-}12\text{-}2)$$

$$2R_1 x_1^2 + R_6 x_1 x_2^2 + R_5 x_1 x_2 + 2R_4 x_1 x_3 + x_1 + x_3 - T_O = 0 \quad (9\text{-}12\text{-}3)$$

$$2R_3 x_4^2 + x_4 - T_N = 0, \qquad\qquad (9\text{-}12\text{-}4)$$

where $R_j = 1/K_j$ for $j = 1, \ldots , 6$.

Before we go on, let's stop to note an important point about our original model A, equations (9-10-1) through (9-10-10). Equations (9-10-5) through (9-10-10) are all of the form

compound = sum of components.

This yields equations (9-11-5) through (9-11-10), which are in exactly the right form for substitution into (9-11-1) through (9-11-4) (to yield the basic reduction). Sometimes, it will be natural for the reaction equations to arise in a different form: say,

$$2CO_2 \rightleftarrows O_2 + 2CO \qquad (9\text{-}10\text{-}8\text{-}1)$$

instead of (9-10-8). This yields the mathematical equation

$$Kx_8^2 = x_5 x_3^2, \qquad (9\text{-}11\text{-}8\text{-}1)$$

which is not in a form convenient for the basic reduction. If our original system arises this way, we must make appropriate substitutions so that we end up with equations in the form of (9-11-5) through (9-11-10). The simplest way to assure this is to formulate the model as in (9-10-5) through (9-10-10). Otherwise, we can find algebraic substitutions to accomplish the same thing. For example, (9-11-5) substituted into (9-11-8-1) yields

$$Kx_8^2 = x_1^2 x_3^2, \qquad (9\text{-}11\text{-}8\text{-}2)$$

whose square root is in the form of (9-11-8). The key observation is that we can always get the system in the right form if it arises from a pure equilibrium problem. (See the comments in Section 9-3.)

We have reduced system (9-11-1) through (9-11-10), whose total degree is $2^5 \cdot 3 = 96$, to (9-12-1) through (9-12-4), whose total degree is $2^2 \cdot 3^2 = 36$. This is already a significant reduction, but with only a little insight, we can do much better via an auxiliary reduction. Observe that (9-12-4) is completely independent of (9-12-1) through (9-12-3), so we can split the system into two parts. Since (9-12-4) is merely a quadratic equation, let's focus on (9-12-1) through (9-12-3), whose total degree is 18.

Let's check for solutions at infinity. The homogeneous part of (9-12) yields

$$2x_1 x_2^2 = 0$$

$$x_1 x_3 = 0 \qquad (9\text{-}13)$$

$$x_1 x_2^2 = 0.$$

If $x_1 = 1$, then $x_2 = 0$, $x_3 = 0$ gives one solution at infinity. If $x_1 = 0$, $x_2 = 1$, then (9-13) is satisfied for any x_3. If $x_1 = 0$, $x_2 = 0$, and $x_3 = 1$, this is another solution. Thus (9-13) has an infinite number of solutions at infinity, a strong sign of reducibility. To check sparsity, we note that NCOEF(3, 3) = 20 and NCOEF(3, 2) = 10, as given in Chapter 7. Thus

$$\text{sparsity} = \frac{5 + 3 + 7}{20 + 10 + 20} = \frac{15}{50} = 0.3,$$

which also suggests reducibility. We especially do not want to have to deal with a system having an infinite number of solutions at infinity.

Examining (9-12-1) through (9-12-3), there are two clear reduction ideas:

(a) Since (9-12-1) and (9-12-3) have as their only cubic terms scalar multiples of "$x_1 x_2^2$," row reduction of the system will make one of (9-12-1) or (9-12-3) second degree.

(b) Equation (9-12-2) can be used to eliminate x_1 or x_3.

I would like to make several points:

- We see from (a) that we can easily get the system to have total degree $3 \cdot 2 \cdot 2 = 12$.

- It turns out that there is a simple reduction to two cubics, total degree $3 \cdot 3 = 9$. The difference between two equations with total degree 9 and three equations with total degree 12 is considerable: one-fourth fewer paths to compute using CONSOL and 2×2 linear algebra rather than 3×3. (I need to acknowledge here, with much appreciation, the symbolic manipulation program REDUCE, which has made looking for alternative reductions a reasonable venture (see [Hearn, 1983]).)

- There is a further reduction to a single eighth-degree polynomial, total degree = 8. I favor the two-cubic reduction, because:
  □ The coefficients of the eighth-degree polynomial are complicated expressions in the original coefficients (over 60 terms), possibly difficult to compute accurately and surely a potential source of the sort of numerical instability discussed in [Wilkinson, 1963].
  □ The computer arithmetic involved in evaluating an eighth-degree polynomial is inherently more subject to roundoff error than that in evaluating a cubic.
  □ The difference between total degree 8 and 9 is less compelling.
  In most cases in chemical systems (and kinematics systems, and other applications areas I have seen) such single polynomial reductions are much harder to derive than in this case and are much more complicated. Generally, I seek reduction to a system with minimal total degree, in which the maximum degrees of terms are still relatively small and the expressions for the coefficients are still reasonable.

I will now present a reduction to two cubics, omitting the false starts and dead ends we encountered finding it. For example, row reduction turns out to be unnecessary. Also, there are at least two different ways to reduce the system to two cubics. I want to make a clear distinction between the basic and auxiliary reductions: the basic reduction is achieved by a purely mechanical procedure; the auxiliary reduction requires insight and trial and error.

By (9-12-2), $x_3 = T_C/(1 + R_4 x_1)$. Substituting into (9-12-3) and clearing the denominator yields

$$2R_1R_4x_1^3 + R_4R_6x_1^2x_2^2 + R_4R_5x_1^2x_2$$

$$+ (2R_1 + R_4)x_1^2 + R_6x_1x_2^2 + R_5x_1x_2$$

$$+ (-R_4T_O + 2R_4T_C + 1)x_1 - T_O + T_C = 0. \quad (9\text{-}12\text{-}3\text{-}1)$$

We can eliminate the fourth-degree term by replacing (9-12-3-1) by the algebraic combination of equations "(9-12-3-1) − [R_4x_1 (9-12-1)]/2," yielding

$$4R_1R_4x_1^3 + R_4R_5x_1^2x_2 + 2(2R_1 + R_4)x_1^2$$

$$+ 2(-R_2R_4 + R_6)x_1x_2^2$$

$$+ (-R_4 + 2R_5)x_1x_2$$

$$+ (-2R_4T_O + 4R_4T_C + R_4T_H + 2)x_1$$

$$+ 2(-T_O + T_C) = 0. \quad (9\text{-}12\text{-}3\text{-}2)$$

The final two-cubic system is (9-12-1) with (9-12-3-2).

To find solutions at infinity for this system, we solve

$$2R_6x_1x_2^2 = 0$$

$$4R_1R_4x_1^3 + R_4R_5x_1^2x_2 + 2(-R_2R_4 + R_6)x_1x_2^2 = 0,$$

which yields a single solution $(x_1, x_2) = (0, 1)$. This solution is nonsingular, so we have exactly one solution at infinity and therefore eight finite solutions.

From equation (9-12-1) we can derive the relation

$$x_1 = \frac{-2x_2^2R_2 + x_2 - T_H}{2x_2^2R_6 + x_2R_5}. \quad (9\text{-}14)$$

Using (9-14) to substitute for x_1 in equation (9-12-3-2) yields the eighth-degree polynomial noted above. This polynomial is also a valid reduction of the original system. However, I prefer the two-cubic model, for the reasons cited above.

It will be convenient to rename the equations (9-12-1) and (9-12-3-2) to be (9-15) and (9-16) respectively, and to denote the coefficients systematically as follows:

$$a_{1,1}x_1x_2^2 + a_{1,2}x_1x_2 + a_{1,3}x_2^2 + a_{1,4}x_2 + a_{1,5} = 0 \quad (9\text{-}15)$$

with

$$a_{1,1} = 2R_6$$

$$a_{1,2} = R_5$$

$$a_{1,3} = 2R_2$$

$$a_{1,4} = 1$$

$$a_{1,5} = -T_H$$

and

$$a_{2,1}x_1^3 + a_{2,2}x_1^2x_2 + a_{2,3}x_1^2 + a_{2,4}x_1x_2^2$$
$$+ a_{2,5}x_1x_2 + a_{2,6}x_1 + a_{2,7} = 0 \quad (9\text{-}16)$$

with

$$a_{2,1} = 4R_1R_4$$

$$a_{2,2} = R_4R_5$$

$$a_{2,3} = 2(2R_1 + R_4)$$

$$a_{2,4} = 2(-R_2R_4 + R_6)$$

$$a_{2,5} = (-R_4 + 2R_5)$$

$$a_{2,6} = (-2R_4T_O + 4R_4T_C + R_4T_H + 2)$$

$$a_{2,7} = 2(-T_O + T_C).$$

Let's call the system consisting of (9-15) and (9-16) "system A." We will use the three sets of values for $R_1, \ldots, R_6$, T_O, T_C, and T_H, given in Table 9-2 and labeled cases 1 through 3. Actually, case 1 was the key example for Chapter 5. I will reference that chapter for results when appropriate. The coefficients for system A are given in Table 9-3.

I used CONSOL8T to solve system A with the case 1 coefficients in Chapter 5. The solutions are listed in Tables 5-4 and 5-5. Table 5-3 gives the input parameters and Table 5-6 the run statistics. Finally, for complete-

TABLE 9–2   Constants for model A and model B[a]

Constant	Case 1 (1000°)	Case 2 (3000°)	Case 3 (6000°)
R_1	24.528	7.289	3.108
R_2	22.206	6.997	3.270
R_3	47.970	15.107	6.942
R_4	24.942	6.825	2.559
R_5	22.120	7.208	3.541
R_6	46.989	14.680	6.791
R_7	32.187	10.285	4.878

Constant	All Cases
T_O	5.D−05
T_C	3.D−05
T_H	1.D−05
T_N	1.D−05

[a] Given as $\log_{10}(R_j)$ for $j = 1, 7$, and component total concentrations ($R_j = 1/K_j$, where K_j is the equilibrium constant).

TABLE 9-3 Coefficients for system A, derived from Table 9-2[a]

Coefficient	Case 2	Case 3
$a_{1,1}$	0.957 260 8 D + 15	0.123 603 6 D + 08
$a_{1,2}$	0.161 435 9 D + 08	0.347 536 4 D + 04
$a_{1,3}$	0.198 623 0 D + 08	0.372 417 8 D + 04
$a_{1,4}$	0.100 000 0 D + 01	0.100 000 0 D + 01
$a_{1,5}$	− 0.300 000 0 D − 04	− 0.300 000 0 D − 04
$a_{2,1}$	0.520 067 0 D + 15	0.185 806 0 D + 07
$a_{2,2}$	0.107 894 6 D + 15	0.125 892 6 D + 07
$a_{2,3}$	0.911 811 9 D + 08	0.585 380 6 D + 04
$a_{2,4}$	0.824 512 3 D + 15	0.110 112 8 D + 08
$a_{2,5}$	0.256 037 4 D + 08	0.658 848 6 D + 04
$a_{2,6}$	− 0.198 503 0 D + 03	0.198 913 2 D + 01
$a_{2,7}$	− 0.800 000 0 D − 04	− 0.800 000 0 D − 04

[a] For case 1, see Table 5-1.

ness, Table 9-4 gives the physical solution, obtained from the positive real solution in Table 5-5 and equations (9-12-2) and (9-11-5) through (9-11-10). ([N] and [NO] are included for comparison with model B, but were obtained from (9-12-4) and (9-11-7) using the quadratic formula.) Note in Tables 5-4 and 5-5 that there is exactly one physical (i.e., positive real) solution. Also note the characteristic symmetry in the set of real solutions.

Model B, as you might expect, can be handled partly in parallel to Model A. However, there are some differences. Equations (9-10-3-1) and (9-10-4-1) yield

$$x_1 + x_3 + 2x_5 + 2x_8 + x_9 + x_{10} + x_{11} = T_O \qquad (9\text{-}11\text{-}3\text{-}1)$$

$$x_4 + 2x_7 + x_{11} = T_N \qquad (9\text{-}11\text{-}4\text{-}1)$$

and (9-10-11) yields

$$K_7 x_{11} = x_1 x_4. \qquad (9\text{-}11\text{-}11)$$

Substituting (9-11-5) through (9-11-11) into (9-11-1), (9-11-2), (9-11-3-1), and (9-11-4-1) yields the basic reduction for model B: equations (9-12-3-1), (9-12-2), and

$$2R_1 x_1^2 + R_6 x_1 x_2^2 + R_5 x_1 x_2 + 2R_4 x_1 x_3 + R_7 x_1 x_4$$
$$+ x_1 + x_3 - T_O = 0 \qquad (9\text{-}12\text{-}3\text{-}1)$$

$$R_7 x_1 x_4 + 2R_3 x_4^2 + x_4 - T_N = 0, \qquad (9\text{-}12\text{-}4\text{-}1)$$

where $R_j = 1/K_j$ for $j = 1, 7$. This system (the "basic reduction") has total degree $2^2 \cdot 3^2 = 36$. We can immediately reduce its total degree to 24 by

TABLE 9–4   Physical solution for model A, case 1

Subscript of x	Components	Value
1	O	0.149 118 D $-$ 14
2	H	0.321 204 D $-$ 18
3	CO	0.766 422 D $-$ 15
4	N	0.231 464 D $-$ 26
	Compounds	
5	O_2	0.750 000 D $-$ 05
6	H_2	0.165 789 D $-$ 14
7	N_2	0.500 000 D $-$ 05
8	CO_2	0.100 000 D $-$ 04
9	OH	0.631 403 D–11
10	H_2O	0.150 000 D $-$ 04
11	NO	0.530 899 D $-$ 09

replacing equation (9-12-3-1) by the algebraic combination of equations "$2 \cdot (9\text{-}12\text{-}3\text{-}1) - (9\text{-}12\text{-}1)$," yielding

$$4R_1x_1^2 + R_5x_1x_2 + 4R_4x_1x_3 + 2R_7x_1x_4$$

$$+ 2x_1 - 2R_2x_2^2 - x_2 + 2x_3 - 2T_O + T_H = 0. \quad (9\text{-}12\text{-}3\text{-}3)$$

Now, using (9-12-2) to justify $x_3 = T_C/(R_4x_1 + 1)$ and substituting into (9-12-3-3), as in model A, yields

$$4R_1R_4x_1^3 + R_4R_5x_1^2x_2 + 2R_4R_7x_1^2x_4$$

$$- 2R_2R_4x_1x_2^2 + 2(2R_1 + R_4)x_1^2$$

$$+ (-R_4 + R_5)x_1x_2 + 2R_7x_1x_4 - 2R_2x_2^2$$

$$+ (-2R_4T_O + 4R_4T_C + R_4T_H + 2)x_1$$

$$- x_2 + (-2T_O + 2T_C + T_H) = 0. \quad (9\text{-}12\text{-}3\text{-}4)$$

The system consisting of (9-12-1), (9-12-3-4), and (9-12-4-1) has total degree $3 \cdot 3 \cdot 2 = 18$. It is the best reduction of model B I can find.

Let's rename (9-12-3-4) and (9-12-4-1) to be (9-17) and (9-18), respectively. Then the system for model B will consist of equations (9-15), (9-17), and (9-18). We get

$$a_{3,1}x_1^3 + a_{3,2}x_1^2x_2 + a_{3,3}x_1^2x_4 + a_{3,4}x_1x_2^2$$

$$+ a_{3,5}x_1^2 + a_{3,6}x_1x_2 + a_{3,7}x_1x_4$$

$$+ a_{3,8}x_2^2 + a_{3,9}x_1 + a_{3,10}x_2 + a_{3,11} = 0 \quad (9\text{-}17)$$

with

$$a_{3,1} = 4R_1R_4$$

$$a_{3,2} = R_4R_5$$

$$a_{3,3} = 2R_4R_7$$

$$a_{3,4} = -2R_2R_4$$

$$a_{3,5} = 2(2R_1 + R_4)$$

$$a_{3,6} = -R_4 + R_5$$

$$a_{3,7} = 2R_7$$

$$a_{3,8} = -2R_2$$

$$a_{3,9} = -2R_4T_O + 4R_4T_C + R_4T_H + 2$$

$$a_{3,10} = -1$$

$$a_{3,11} = -2T_O + 2T_C + T_H$$

and

$$a_{4,1}x_1x_4 + a_{4,2}x_4^2 + a_{4,3}x_4 + a_{4,4} = 0 \qquad (9\text{-}18)$$

TABLE 9–5 Coefficients for the second and third equations of system B, derived from the values given in Table 9–2

Coefficient	Case 1	Case 2	Case 3
$a_{3,1}$	0.118 048 7 D+51	0.520 067 0 D+15	0.185 806 0 D+07
$a_{3,2}$	0.115 344 3 D+48	0.107 894 7 D+15	0.125 892 7 D+07
$a_{3,3}$	0.269 170 7 D+58	0.257 649 7 D+18	0.547 054 1 D+08
$a_{3,4}$	−0.281 206 3 D+48	−0.132 748 5 D+15	−0.134 905 8 D+07
$a_{3,5}$	0.309 912 2 D+26	0.911 812 0 D+08	0.585 380 6 D+04
$a_{3,6}$	−0.873 668 2 D+25	0.946 015 6 D+07	0.311 312 1 D+04
$a_{3,7}$	0.307 628 4 D+33	0.385 504 8 D+11	0.151 018 5 D+06
$a_{3,8}$	−0.321 383 6 D+23	−0.198 623 1 D+08	−0.372 417 8 D+04
$a_{3,9}$	−0.262 495 9 D+21	−0.198 503 1 D+03	0.198 913 3 D+01
$a_{3,10}$	−0.100 000 0 D+01	−0.100 000 0 D+01	−0.100 000 0 D+01
$a_{3,11}$	−0.500 000 0 D−04	−0.500 000 0 D−04	−0.500 000 0 D−04
$a_{4,1}$	0.153 814 2 D+33	0.192 752 4 D+11	0.755 092 7 D+05
$a_{4,2}$	0.186 651 4 D+49	0.255 876 5 D+16	0.174 996 9 D+08
$a_{4,3}$	0.100 000 0 D+01	0.100 000 0 D+01	0.100 000 0 D+01
$a_{4,4}$	−0.100 000 0 D−04	−0.100 000 0 D−04	−0.100 000 0 D−04

with

$$a_{4,1} = 4R_1R_4$$

$$a_{4,2} = R_4R_5$$

$$a_{4,3} = 2R_4R_7$$

$$a_{4,4} = -2R_2R_4.$$

TABLE 9–6   Input parameters and run statistics for CONSOL8T solution of model B, case 1

Parameter	Comments
Input Part 1	
IFLGPT = 1	Use the projective transformation
IFLGSC = 1	Use SCLGEN
IFLGCR = 0	Do not print out corrector residuals
IFLGST = 0	Do not print out step summaries
MAXNS = 5000	Maximum number of steps per path
MAXIT = 3	Maximum number of corrector iterations per step
EPSBIG = 1.D − 06	Most of the path will be tracked this closely
SSZBEG = 1.D − 10	Beginning step size
Input Part 2	
A tableau based on system B with coefficients from Tables 9–3 and 9–5. (Compare Table 5–1.)	
Run Statistics	
Total arc length: 257.0	
Total number of linear systems solved: 28,312	

TABLE 9–7   Physical solution for model B, case 1

Subscript of x	Components	Value
1	O	0.149 116 D − 14
2	H	0.321 207 D − 18
3	CO	0.766 436 D − 15
4	N	0.231 458 D − 26
	Compounds	
5	O_2	0.749 973 D − 05
6	H_2	0.165 792 D − 14
7	N_2	0.499 973 D − 05
8	CO_2	0.100 000 D − 04
9	OH	0.631 398 D − 11
10	H_2O	0.150 000 D − 04
11	NO	0.530 888 D − 09

Let's call the system consisting of (9-15), (9-17), and (9-18) "system B." Table 9-5 gives coefficients for (9-17) and (9-18) from the R_j values in Table 9-2. I used CONSOL8T to solve system B with the case 1 coefficients. Table 9-6 gives the input parameters and the run statistics. The physical solution is given in Table 9-7.

EXERCISES

9-3. Show that system A has a nonsingular solution at infinity.

9-4. Show that system B has a finite number of solutions at infinity. Give as full a description of the solutions at infinity as you can.

EXPERIMENTS

9-1. Solve system A with the case 2 coefficients given in Table 9-3, as was done for case 1 in the text.

9-2. Solve system A with the case 3 coefficients given in Table 9-3. (This is (4-10) from problem set 3 in section 4-7.)

9-3. Solve system B with the case 1 coefficients given in Tables 9-3 and 9-5. Compare your results with the summary in Table 9-6 and the result in Table 9-7.

9-4. Solve system B with the case 2 coefficients.

9-5. Solve system B with the case 3 coefficients.

9-6. Investigate the effects of alternate scalings on solving models A and B in cases 1 through 3. Try, for example, no scaling and SCLCEN. (Compare Table 5-6.)

9-5 A FINAL EXAMPLE

The chemical model presented in this section does not yield a mathematical system obeying the conditions given in Section 9-3. It is not a true equilibrium system. Because it attempts to capture certain precipitation–saturation behavior with a particular kind of model, some new types of equations are generated. However, the spirit, if not the letter, of our reduction–scaling–solution methodology is effective. This model illustrates the flexibility of the approach.†

Consider a chemical bath used for electroplating nickel onto metal plates. We want to compute the bath composition at various times, as this is significant for understanding and managing the electroplating process. Table 9-8 lists the relevant components and compounds, most of which are

---

† I am indebted to Patrick Ng and Dexter Snyder of General Motors Research Laboratories for bringing this class of models to my attention.

TABLE 9–8   Indexing scheme for electrochemistry
model

Subscript of x	Components and Compounds
1	N_i^{+2}
2	OH^-
3	NH_4^+
4	NH_3
5	$N_i(NH_3)^{+2}$
6	$N_i(NH_3)_2^{+2}$
7	$N_i(NH_3)_3^{+2}$
8	$N_i(NH_3)_4^{+2}$
9	$N_i(NH_3)_5^{+2}$
10	$N_i(NH_3)_6^{+2}$
11	H^+
12	H_2O
13	$1/\epsilon$ (porosity)

expressed in ionic forms, due to the nature of the process. We imagine the
plate as having been impregnated with pores in which the basic chemistry
takes place. As an approximation we take a pore to be a closed system. The
volume of the pore changes as the plating process progresses, and the change
in volume is accounted for via the "porosity," which we treat mathematically as another variable. The model is as follows:

Mass Balance:

For N_i: $[N_i^{+2}] + [N_i(NH_3)^{+2}] + [N_i(NH_3)_2^{+2}]$

$$+ [N_i(NH_3)_3^{+2}] + [N_i(NH_3)_4^{+2}]$$

$$+ [N_i(NH_3)_5^{+2}] + [N_i(NH_3)_6^{+2}]$$

$$= P_1 \frac{1}{\epsilon} + Q_1, \tag{9-19}$$

where P_1 and Q_1 are constants reflecting the total nickel in solution at a
previous time in the process and the amount of nickel precipitated since that
time. The $1/\epsilon$ is a variable, the "porosity."

For N: $[NH_4^+] + [NH_3] + [N_i(NH_3)^{+2}] + 2[N_i(NH_3)_2^{+2}]$

$$+ 3[N_i(NH_3)_3^{+2}] + 4[N_i(NH_3)_4^{+2}]$$

$$+ 5[N_i(NH_3)_5^{+2}] + 6[N_i(NH_3)_6^{+2}]$$

$$= P_2, \tag{9-20}$$

where P_2 is a constant reflecting the total NH_3 at a previous time in the process and the amount of NH_4^+ generated since that time.

$$\text{For } OH^-: \quad [OH^-] + [H_2O] = P_3 \frac{1}{\epsilon} + Q_3, \tag{9-21}$$

where P_3 and Q_3 are constants, as above.

Electrochemistry:

$$2[N_i^{+2}] + [H^+] + [NH_4^+] + 2([N_i(NH_3)^{+2}]$$
$$+ [N_i(NH_3)_2^{+2}] + [N_i(NH_3)_3^{+2}]$$
$$+ [N_i(NH_3)_4^{+2}] + [N_i(NH_3)_5^{+2}]$$
$$+ [N_i(NH_3)_6^{+2}])$$
$$= [OH^-] + P_4, \tag{9-22}$$

where P_4 is a constant representing $NO_3{}^-$ in solution and the extra negative charge from outside sources being input into the system.

Equilibrium Reactions:

$$NH_4^+ \rightleftarrows NH_3 + H^+ \tag{9-23-1}$$
$$N_i(NH_3)^{+2} \rightleftarrows N_i^{+2} + NH_3 \tag{9-23-2}$$
$$N_i(NH_3)_2^{+2} \rightleftarrows N_i(NH_3)^{+2} + NH_3 \tag{9-23-3}$$
$$N_i(NH_3)_3^{+2} \rightleftarrows N_i(NH_3)_2^{+2} + NH_3 \tag{9-23-4}$$
$$N_i(NH_3)_4^{+2} \rightleftarrows N_i(NH_3)_3^{+2} + NH_3 \tag{9-23-5}$$
$$N_i(NH_3)_5^{+2} \rightleftarrows N_i(NH_3)_4^{+2} + NH_3 \tag{9-23-6}$$
$$N_i(NH_3)_6^{+2} \rightleftarrows N_i(NH_3)_5^{+2} + NH_3 \tag{9-23-7}$$
$$H_2O \rightleftarrows H^+ + OH^- \tag{9-23-8}$$

Precipitation:

$$[N_i^{+2}][OH^-]^2 = P_5, \tag{9-24}$$

where P_5 is a constant.

We will proceed to the derivation and analysis of the resulting mathematical system. The main feature of this model which distinguishes it from equilibrium models (as in Section 9-4) is equation (9-24), which leads to a unique mathematical relationship unlike either the reaction or conservation equations that we have encountered. The mass balance and electrochemistry equations, although chemically distinct, function mathematically the same as the conservation equations. In fact, we will proceed to substitute the equilibrium reactions and precipitation equation into the mass balance and electrochemical equations, in analogy with the basic reduction. Then we will continue with an auxiliary reduction. First, though, let's look at the mathematical system:

$$x_1 + x_5 + x_6 + x_7 + x_8 + x_9 + x_{10} = P_1 x_{13} + Q_1 \qquad (9\text{-}25)$$

$$x_3 + x_4 + x_5 + 2x_6 + 3x_7 + 4x_8 + 5x_9 + 6x_{10} = P_2 \qquad (9\text{-}26)$$

$$x_2 + x_{12} = P_3 x_{13} + Q_3 \qquad (9\text{-}27)$$

$$2x_1 + x_{11} + x_3 + 2(x_5 + x_6 + x_7 + x_8 + x_9 + x_{10}) = x_2 + P_4 \qquad (9\text{-}28)$$

$$K_1 x_3 = x_4 x_{11} \qquad (9\text{-}29\text{-}1)$$

$$K_2 x_5 = x_1 x_4 \qquad (9\text{-}29\text{-}2)$$

$$K_3 x_6 = x_5 x_4 \qquad (9\text{-}29\text{-}3)$$

$$K_4 x_7 = x_6 x_4 \qquad (9\text{-}29\text{-}4)$$

$$K_5 x_8 = x_7 x_4 \qquad (9\text{-}29\text{-}5)$$

$$K_6 x_9 = x_8 x_4 \qquad (9\text{-}29\text{-}6)$$

$$K_7 x_{10} = x_9 x_4 \qquad (9\text{-}29\text{-}7)$$

$$K_8 x_{12} = x_{11} x_2 \qquad (9\text{-}29\text{-}8)$$

$$x_1 x_2^2 = P_5. \qquad (9\text{-}30)$$

Note the recursive nature of equations (9-29-3) through (9-29-7) and the "unbalanced" form of equation (9-30). We can use (9-29) to define x_3, x_5 through x_{10}, and x_{12} in terms of x_1, x_2, x_4, x_{11}, and x_{13}. Further, (9-30) implies that $x_1 = P_5/x_2^2$, eliminating one more variable. Substituting into (9-25) through (9-28) yields our "basic reduction" (where $R_j = 1/K_j$, as before):

$$-(Q_1 + P_1 x_{13})x_2^2$$

$$+ P_5 R_2 R_3 R_4 R_5 R_6 R_7 x_4^6$$

$$+ P_5 R_2 R_3 R_4 R_5 R_6 x_4^5$$

$$+ P_5 R_2 R_3 R_4 R_5 x_4^4$$

$$+ P_5 R_2 R_3 R_4 x_4^3$$

$$+ P_5 R_2 R_3 x_4^2$$

$$+ P_5 R_2 x_4$$

$$+ P_5 = 0 \tag{9-31}$$

$$R_1 x_2^2 x_4 x_{11} + x_2^2 x_4 - P_2 x_2^2$$

$$+ 6 P_5 R_2 R_3 R_4 R_5 R_6 R_7 x_4^6$$

$$+ 5 P_5 R_2 R_3 R_4 R_5 R_6 x_4^5$$

$$+ 4 P_5 R_2 R_3 R_4 R_5 x_4^4$$

$$+ 3 P_5 R_2 R_3 R_4 x_4^3$$

$$+ 2 P_5 R_2 R_3 x_4^2$$

$$+ P_5 R_2 x_4 = 0 \tag{9-32}$$

$$R_8 x_2 x_{11} + x_2 - (Q_3 + P_3 x_{13}) = 0 \tag{9-33}$$

$$-x_2^3 + R_1 x_2^2 x_4 x_{11} + x_2^2 x_{11} - P_4 x_2^2$$

$$+ 2 P_5 R_2 R_3 R_4 R_5 R_6 R_7 x_4^6$$

$$+ 2 P_5 R_2 R_3 R_4 R_5 R_6 x_4^5$$

$$+ 2 P_5 R_2 R_3 R_4 R_5 x_4^4$$

$$+ 2 P_5 R_2 R_3 R_4 x_4^3$$

$$+ 2 P_5 R_2 R_3 x_4^2$$

$$+ 2 P_5 R_2 x_4$$

$$+ 2 P_5 = 0. \tag{9-34}$$

The system (9-31) through (9-34) has total degree $6 \cdot 6 \cdot 2 \cdot 6 = 432$, which is higher than the original system, (9-25) through (9-30). Basically, the recursive relations (9-29-3) through (9-29-7) have been "unrolled" by the substitution, and the implied sixth-degree relationship between x_5 and x_{12} has been transformed into powers of x_4 up to x_4^6.

Looking over the system for auxiliary reduction ideas, we note:

(a) Equation (9-33) can be used to eliminate either x_{11} or x_2.

(b) The only sixth-degree term in any equation is x_4^6. Further, equations (9-31) and (9-34) share the terms

$$x_4^6 P_5 R_2 R_3 R_4 R_5 R_6 R_7 + x_4^5 P_5 R_2 R_3 R_4 R_5 R_6$$

$$+ x_4^4 P_5 R_2 R_3 R_4 R_5 + x_3^4 P_5 R_2 R_3 R_4$$

$$+ x_4^2 P_5 R_2 R_3 + x_4 P_5 R_2 + P_5.$$

[In equation (9-34) this expression is multiplied by 2.] The remaining terms of both equations have degree no more than 3. These facts suggest that we can reduce one equation's degree from 6 to 3 and another's from 6 to 5, via "row reduction."

(c) Since only a single variable, x_4, accounts for the high-degree terms in all equations, introducing a new relationship and a new variable, such as $z = x_4^2$, may be worthwhile. Appending this new equation to the system doubles its total degree, but the substitution $z = x_4^2$ may more than offset this increase in the rest of the system.

It is difficult to predict the effect of (a) on total degree without examining the whole system, since it will involve division. But (b) promises to reduce the total degree by a factor of $3 \cdot (5/6) \cdot 6 \approx 0.42$.

Let's take

$$x_{11} = \frac{-(x_2 - R_4 - x_{13}P_3)}{x_2 R_8}$$

from (9-33) and substitute into (9-32) and (9-34), yielding

$$(-R_1 + R_8)x_2^2 x_4 - R_8 P_2 x_2^2 + R_1(Q_3 + P_3 x_{13})x_2 x_4$$

$$+ 6P_5 R_2 R_3 R_4 R_5 R_6 R_7 R_8 x_4^6$$

$$+ 5P_5 R_2 R_3 R_4 R_5 R_6 R_8 x_4^5$$

$$+ 4P_5 R_2 R_3 R_4 R_5 R_8 x_4^4$$

$$+ 3P_5 R_2 R_3 R_4 R_8 x_4^3$$

$$+ 2P_5 R_2 R_3 R_8 x_4^2$$

$$+ P_5 R_2 R_8 x_4 = 0 \qquad\qquad (9\text{-}32\text{-}1)$$

$$-R_8x_2^3 - R_1x_2^2x_4 - (R_8P_4 + 1)x_2^2$$

$$+ (Q_3 + P_3x_{13})R_1x_2x_4 + (Q_3 + P_3x_{13})x_2$$

$$+ 2P_5R_2R_3R_4R_5R_6R_7R_8x_4^6$$

$$+ 2P_5R_2R_3R_4R_5R_6R_8x_4^5$$

$$+ 2P_5R_2R_3R_4R_5R_8x_4^4$$

$$+ 2P_5R_2R_3R_4R_8x_4^3$$

$$+ 2P_5R_2R_3R_8x_4^2$$

$$+ 2P_5R_2R_8x_4$$

$$+ 2P_5R_8 = 0. \tag{9-34-1}$$

Thus our system now consists of the three equations (9-31), (9-32-1), and (9-34-1). We see that (9-32-1) and (9-34-1) both have degree 6, so our substitution has been "for free" and we have halved the total degree.

Now let's substitute the algebraic combination of equations: "(9-34-1) $-$ 2(9-31)" for (9-34-1) to implement the first part of (b), and reduce (9-34-1)'s degree from 6 to 3. We get

$$-R_8x_2^3 - R_1x_2x_4$$

$$+ (2R_8Q_1 + 2R_8P_1x_{13} - R_8P_4 - 1)x_2$$

$$+ R_1(Q_3 + P_3x_{13})x_4 + Q_3 + P_3x_{13} = 0. \tag{9-34-2}$$

Happily, (9-34-2) has degree 2 rather than 3 because (9-34-1) $-$ 2(9-31) had a common factor of x_2, which I have eliminated.

Our system now is (9-31), (9-32-1), and (9-34-2) with total degree $6 \cdot 6 \cdot 2 = 72$. We have done quite well to reduce the total degree from 432 to 72 without increasing the degree of any equation or generating complicated expressions for the coefficients. However, we can reduce further. We can definitely combine (9-31) and (9-32-1) together to reduce one of them to fifth-degree [say, replace (9-32-1) by "(9-32-1) $-$ 6 $\cdot$ R_8(9-31)"]. Then the total degree would be $6 \cdot 5 \cdot 2 = 60$. However, the idea from (c) is better. Let's introduce a new variable, x_4^2, and turn the sixth-degree terms in x_4 into third-degree terms in x_4^2. It is convenient at this point to rename all the variables, via

$$z_1 = x_2$$

$$z_2 = x_4$$

$$z_3 = x_{13}$$

$$z_4 = x_4^2.$$

Then the system becomes

$$P_5 R_2 R_3 R_4 R_5 R_6 R_7 z_4^3$$
$$+ P_5 R_2 R_3 R_4 R_5 R_6 z_2 z_4^2$$
$$- P_1 z_1^2 z_3$$
$$+ P_5 R_2 R_3 R_4 R_5 z_4^2$$
$$+ P_5 R_2 R_3 R_4 z_2 z_4$$
$$- Q_1 z_1^2$$
$$+ P_5 R_2 R_3 z_4$$
$$+ P_5 R_2 z_2$$
$$+ P_5 = 0 \qquad\qquad (9\text{-}31\text{-}1)$$

$$6 P_5 R_2 R_3 R_4 R_5 R_6 R_7 R_8 z_4^3$$
$$+ 5 P_5 R_2 R_3 R_4 R_5 R_6 R_8 z_2 z_4^2$$
$$+ (-R_1 + R_8) z_1^2 z_2$$
$$- R_8 P_2 z_1^2$$
$$+ R_1 P_3 z_1 z_2 z_3$$
$$+ 4 P_5 R_2 R_3 R_4 R_5 R_8 z_4^2$$
$$+ 3 P_5 R_2 R_3 R_4 R_8 z_2 z_4$$
$$+ R_1 Q_3 z_1 z_2$$
$$+ 2 P_5 R_2 R_3 R_8 z_4$$
$$+ P_5 R_2 R_8 z_2 = 0 \qquad\qquad (9\text{-}32\text{-}2)$$

$$-R_8 z_1^2 - R_1 z_1 z_2$$
$$+ 2 R_8 P_1 z_1 z_3 + R_1 P_3 z_2 z_3$$
$$+ (2 R_8 Q_1 - R_8 P_4 - 1) z_1 + R_1 Q_3 z_2 + P_3 z_3 + Q_3 = 0 \qquad (9\text{-}34\text{-}3)$$

$$z_2^2 - z_4 = 0. \qquad\qquad (9\text{-}35)$$

This system [consisting of (9-31-1), (9-32-2), (9-34-3), and (9-35)] has total degree $3 \cdot 3 \cdot 2 \cdot 2 = 36$. It is the best auxiliary reduction of (9-31) through (9-34) that I can find.

As in Section 9-4, it will be convenient to rename the equations and coefficients before running tests. Denote (9-31-1), (9-32-2), (9-34-3), and (9-35) by (9-36), (9-37), (9-38), and (9-39), respectively, as follows.

$$a_{5,1}z_4^3 + a_{5,2}z_2z_4^2 + a_{5,3}z_1^2z_3 + a_{5,4}z_4^2$$

$$+ \; a_{5,5}z_4z_2 + a_{5,6}z_1^2 + a_{5,7}z_4$$

$$+ \; a_{5,8}z_2 + a_{5,9} = 0, \tag{9-36}$$

where

$$a_{5,1} = P_5R_2R_3R_4R_5R_6R_7$$

$$a_{5,2} = P_1R_2R_3R_4R_5R_6$$

$$a_{5,3} = -P_1$$

$$a_{5,4} = P_5R_2R_3R_4R_5$$

$$a_{5,5} = P_5R_2R_3R_4$$

$$a_{5,6} = -Q_1$$

$$a_{5,7} = P_5R_2R_3$$

$$a_{5,8} = P_5R_2$$

$$a_{5,9} = P_5$$

$$a_{6,1}z_4^3 + a_{6,2}z_2z_4^2 + a_{6,3}z_1^2z_2 + a_{6,4}z_1^2$$

$$+ \; a_{6,5}z_1z_2z_3 + a_{6,6}z_4^2 + a_{6,7}z_2z_4$$

$$+ \; a_{6,8}z_1z_2 + a_{6,9}z_4 + a_{6,10}z_2 = 0, \tag{9-37}$$

where

$$a_{6,1} = 6P_5R_2R_3R_4R_5R_6R_7R_8$$

$$a_{6,2} = 5P_5R_2R_3R_4R_5R_6R_8$$

$$a_{6,3} = -R_1 + R_8$$

$$a_{6,4} = R_8P_2$$

$$a_{6,5} = R_1P_3$$

$$a_{6,6} = 4P_5R_2R_3R_4R_5R_8$$

$$a_{6,7} = 3P_5R_2R_3R_4R_8$$

$$a_{6,8} = R_1Q_3$$

$$a_{6,9} = 2P_5R_2R_3R_8$$

$$a_{6,10} = P_5R_2R_8$$

$$a_{7,1}z_1^2 + a_{7,2}z_1z_2 + a_{7,3}z_1z_3 + a_{7,4}z_2z_3$$
$$+ a_{7,5}z_1 + a_{7,6}z_2 + a_{7,7}z_3 + a_{7,8} = 0, \qquad (9\text{-}38)$$

where

$$a_{7,1} = -R_8$$

$$a_{7,2} = -R_1$$

$$a_{7,3} = 2P_1R_8$$

$$a_{7,4} = R_1P_3$$

$$a_{7,5} = 2R_8Q_1 - R_8P_4 - 1$$

$$a_{7,6} = R_1Q_3$$

$$a_{7,7} = P_3$$

$$a_{7,8} = Q_3$$

and

$$a_{8,1}z_2^2 + a_{8,2}z_4 = 0, \qquad (9\text{-}39)$$

TABLE 9–9   Constants for electrochemistry
model

Constant	
R_1	5.832 D + 09
R_2	617.0
R_3	174.0
R_4	53.8
R_5	15.5
R_6	5.62
R_7	1.07
R_8	1.8 D + 16
P_1	− 0.043 149 9
P_2	0.028 125 0
P_3	− 0.086 299 9
P_4	4.746 875
P_5	0.16 D − 13
Q_1	2.535 383
Q_3	55.670 77

TABLE 9–10   Coefficients for
electrochemistry model, defined from Table
9–9 constants

Coefficient	
$a_{5,1}$	0.861 367 5 D $-$ 05
$a_{5,2}$	0.805 016 3 D $-$ 05
$a_{5,3}$	0.431 499 5 D $-$ 01
$a_{5,4}$	0.143 241 3 D $-$ 05
$a_{5,5}$	0.924 137 7 D $-$ 07
$a_{5,6}$	$-$ 0.253 538 3 D $+$ 01
$a_{5,7}$	0.171 772 8 D $-$ 08
$a_{5,8}$	0.987 200 0 D $-$ 11
$a_{5,9}$	0.160 000 0 D $-$ 13
$a_{6,1}$	0.930 276 9 D $+$ 12
$a_{6,2}$	0.724 514 7 D $+$ 12
$a_{6,3}$	$-$ 0.180 000 1 D $+$ 17
$a_{6,4}$	$-$ 0.506 250 0 D $+$ 15
$a_{6,5}$	$-$ 0.503 301 0 D $+$ 09
$a_{6,6}$	0.103 133 8 D $+$ 12
$a_{6,7}$	0.499 034 3 D $+$ 10
$a_{6,8}$	0.324 671 9 D $+$ 12
$a_{6,9}$	0.618 382 1 D $+$ 08
$a_{6,10}$	0.177 696 0 D $+$ 06
$a_{7,1}$	$-$ 0.180 000 0 D $+$ 17
$a_{7,2}$	$-$ 0.583 200 0 D $+$ 10
$a_{7,3}$	$-$ 0.155 339 8 D $+$ 16
$a_{7,4}$	$-$ 0.503 301 0 D $+$ 09
$a_{7,5}$	0.583 003 6 D $+$ 16
$a_{7,6}$	0.324 671 9 D $+$ 12
$a_{7,7}$	$-$ 0.862 998 9 D $-$ 01
$a_{7,8}$	0.556 707 7 D $+$ 02
$a_{8,1}$	0.100 000 0 D $+$ 01
$a_{8,2}$	$-$ 0.100 000 0 D $+$ 01

TABLE 9–11   Input parameters and run statistics for CONSOL8T solution of
electrochemistry problem

Input Part 1
   Identical to Table 9–6.
Input Part 2
   A table based on system C with coefficients from Table 9–10. (Compare Table 5–1.)
Run Statistics
   Total arc length: 11,190
   Total number of linear systems solved: 55,860

TABLE 9–12   Physical solution for electrochemistry model

Subscript of x	Components and Compounds	Value
1	N_i^{+2}	0.234 719 D+01
2	OH^-	0.825 630 D−07
3	NH_4^+	0.365 132 D−02
4	NH_3	0.168 123 D−04
5	$N_i(NH_3)^{+2}$	0.243 478 D−01
6	$N_i(NH_3)_2^{+2}$	0.712 256 D−04
7	$N_i(NH_3)_3^{+2}$	0.644 236 D−07
8	$N_i(NH_3)_4^{+2}$	0.167 882 D−10
9	$N_i(NH_3)_5^{+2}$	0.158 623 D−14
10	$N_i(NH_3)_6^{+2}$	0.285 349 D−19
11	H^+	0.372 397 D−07
12	H_2O	0.553 432 D+02
13	$1/\epsilon$ (porosity)	0.379 540 D+01

where

$$a_{8,1} = \quad 1$$

$$a_{8,2} = -1.$$

Call the system consisting of the four equations (9-36) through (9-39) "system C." Table 9-10 gives a set of coefficients for system C derived from the values for $R_1, \ldots, R_8, P_1, \ldots, P_5, Q_1,$ and Q_3 given in Table 9-9. I have run CONSOL8T to solve model C with these coefficients. The run statistics are given in Table 9-11, and the physical solution in Table 9-12.

EXERCISES

9-5. Does the system consisting of equations (9-31) through (9-34) have only a finite number of solutions at infinity?

9-6. What is the total degree of the system consisting of equations (9-25) through (9-30)? Calculate (or estimate) its sparsity and RINF (as described in Chapter 7). If each path can be tracked in 1 second of CPU time, how long would the run take? If $\frac{1}{10}$ second?

9-7. Do Exercise 9-6 for the two systems: "(9-31) through (9-34)" and system C.

9-8. A supercomputer might implement the CONSOL code 10 times faster than a standard mainframe and 10,000 times faster than a PC. Compare the performance of the supercomputer, the mainframe, and the PC on the three systems "(9-25) through (9-30)," "(9-31) through (9-34)," and system C. (*Moral:* Reduction is worth a supercomputer.)

CHAPTER 10

Kinematics
of Mechanisms

10-1 INTRODUCTION

I have a fondness for robots from the science fiction films of my childhood: *Tobor the Great*, *Gog*, and especially *Forbidden Planet*. Robbie the robot was an evocative creation and grabbed my imagination long before Anne Francis could. Robbie's superhuman manufacturing abilities were combined with humanoid looks and talking. You could tell him your troubles. He would listen, do what he could. Until the monster from the id got him.

Current industrial robots are disappointing in comparison, dumb fake arms doing the same repetitive tasks all day. And they have to be told exactly what to do: "Now move your third joint 73 degrees counterclockwise with the following velocity profile. . . ." However, we will consider one small step to make robots more like Robbie. We will teach the robot how to figure out where to put his own joints when we tell him where we want his hand. Since we will develop a general robot model, the technique can be used to study the kinematic behavior of various robot designs. Until recently, designs have been limited to the special cases where this "inverse position problem" is easy to solve. Thus this chapter documents a new and practical application of the reduction–scaling–solution ideas that we are developing.

The scope of this chapter is actually more general than robot kinematics. We will consider polynomial systems that result from questions about mechanisms, and our focus will be on solving the inverse-position problem for manipulators (denoted the "IPP" from now on). However, the approach is applicable to a much broader class of kinematic problems: those that are formulated in terms of algebraic relations based on local coordinate systems and the rigid-motion coordinate transformations between them. Because of this broad applicability, I will develop the material a little more fully than would be strictly necessary to solve the IPP.

I would like to thank Charles Wampler of the General Motors Research Laboratories for reading this chapter and making valuable suggestions to improve its exposition. Here, as in Chapter 9, I acknowledge the value of the REDUCE symbolic manipulation code to this work [Hearn, 1983].

If you are already interested in the kinematics of mechanisms, this chapter will be useful for its solution of the IPP and the potential applicability of the approach to other kinematics problems. If this area is new to you, Sections 10-2 and 10-3 are especially written to ease you into the rather complex notation used in this field and to present some simple problems that suggest the flavor of the area. If you decide to work your way through the entire chapter (and it is X-rated for beginners), the benefit will be that you will see reduction in its most challenging form. Imagine the thrill of reducing a system of total degree 2,073,600 to 256. Solving the reduced system would almost be an anticlimax, except that it has an infinite number of solutions at infinity. The IPP for spatial mechanisms is the most colossal reduction–scaling–solution project I have ever worked on. But don't let that scare you. At least read the next section. Otherwise, Dr. Morbius will be displeased.

10-2 REVOLUTE JOINT PLANAR MANIPULATORS

Before giving any general definitions or methods, let's look at some simple examples. My favorite simple manipulator consists of two links and two joints. It is shown in Fig. 10-1. I like to think of the links as long, narrow pieces of cardboard, and the two joint axes as thumbtacks, one stuck into the table and the other sticking up through the two links. (Use an old table.) The end of the second link, the one without a thumbtack, is called "the hand." The basic question about this mechanism is: Given a position for the hand, what angles should the links form to put the hand in that specified position?

Let's make this more explicit. Measure angles counterclockwise from the horizontal. Then given two angles, θ_1 and θ_2, the hand position is determined (see Fig. 10-2). Given a hand position, several sets of axis angles might put the hand there, as shown in Fig. 10-3. The question is: How do we construct a computer program so that if we input the hand position, the associated sets of axis angles are output?

To see how to accomplish this (by turning the problem into a polynomial system to solve), let's make the question more exact. We will preserve the essence of the question if we let the width of the links shrink to zero, and thus model them by line segments. We'll put the first joint axis at the origin, as in Fig. 10-4.

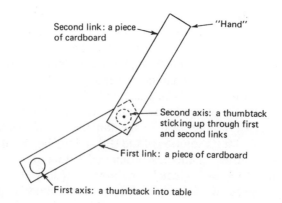

Second link: a piece of cardboard

"Hand"

Second axis: a thumbtack sticking up through first and second links

First link: a piece of cardboard

First axis: a thumbtack into table

Figure 10-1 Thumbtack and cardboard manipulator.

Here a_1 and a_2 are the *link lengths*. Our problem is: We want the hand at (x, y). Find all angle pairs (θ_1, θ_2) that put it there. Trigonometry generates some basic relations (see Fig. 10-5).

Since we want to deal with *polynomial*, not *trigonometric* equations, we make the substitutions

$$s_1 = \sin \theta_1 \qquad s_2 = \sin \theta_2$$
$$c_1 = \cos \theta_1 \qquad c_2 = \cos \theta_2$$

and append the relations

$$\cos^2 \theta_1 + \sin^2 \theta_1 = 1$$
$$\cos^2 \theta_2 + \sin^2 \theta_2 = 1,$$

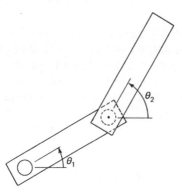

θ_2

θ_1

Figure 10-2 Axis angles on the thumbtack and cardboard manipulator are measured from the horizontal counterclockwise.

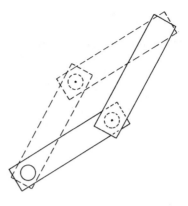

Figure 10-3 Two sets of axis angles may put the hand in the same position.

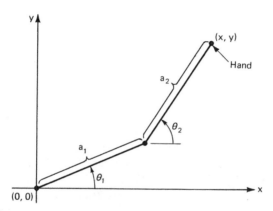

Figure 10-4 Abstract manipulator showing axis angles θ_1 and θ_2 and link lengths a_1 and a_2. The hand is at the point (x, y) in standard coordinates.

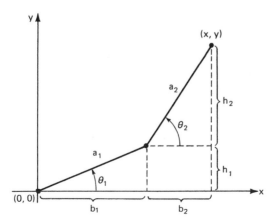

Figure 10-5 Manipulator from Fig. 10-4, with supplemental lines to generate trigonometric relations.

which yields the polynomial system

$$s_1 = \frac{h_1}{a_1} \qquad\qquad c_1 = \frac{b_1}{a_1}$$

$$s_2 = \frac{h_2}{a_2} \qquad\qquad c_2 = \frac{b_2}{a_2}$$

$$x = b_1 + b_2 \qquad y = h_1 + h_2$$

$$c_1^2 + s_1^2 = 1 \qquad c_2^2 + s_2^2 = 1.$$

This system has variables h_1, h_2, b_1, b_2, c_1, s_1, c_2, s_2 and constants a_1, a_2, x, y. It is a polynomial system of eight equations in eight unknowns with total degree 4. Its real solutions give the sines and cosines of the angles θ_1, θ_2, which we want, together with h_1, h_2, b_1, b_2, which we do not particularly want but which will be found as a side effect of the way we have set up the system. However, we see now how the indirect position problem (IPP) for this manipulator leads to a polynomial system.

I will not linger over this system, because I would like to reexamine the problem immediately, introducing a new formalism that leads to a new system. This new formalism is not especially natural for this problem, but it becomes useful for more complex planar and for three-dimensional cases. We will, in fact, spend the rest of this chapter studying the new formalism.

Going back to Fig. 10-4, let's change the way we measure the second axis angle, θ_2. Let's do it counterclockwise *from the first link*, as shown in Fig. 10-6. Further, let's introduce coordinate systems at the second joint and at the hand, with the x-axis of the new coordinate systems along the previous link (Fig. 10-7). If p is a geometric point in the plane, we can express p in any of the given three coordinate systems. Denote these coordinates for p in the first, second, and third coordinate systems by (x_1^0, y_1^0), (x_2^0, y_2^0), (x_3^0, y_3^0), respectively.

Let's spend a moment here recalling how to change coordinates from one of these coordinate systems to the next. If $\theta_1 = 0°$, then $(x_1, y_1) = (x_2 + a_1, y_2)$. That is, we change coordinates from the second to the first coordinate system merely by translating along the link. If $\theta_1 \neq 0$, we follow this translation by a θ_1-rotation, to obtain

$$(x_1, y_1) = (c_1(x_2 + a) - s_1 y_2, s_1(x_2 + a) + c_1 y_2).$$

This type of translation-rotation is sometimes called a "rigid motion," because it does not change the shape of objects, it just moves them. Another way of viewing the rotation is as a change of coordinates from the basis $\{(c_1, -s_1), (s_1, c_1)\}$ to $\{(1, 0), (0, 1)\}$. It is convenient to express this translation–rotation coordinate transformation as a single matrix multiplication, and we can do this as follows if we add a dummy coordinate "1":

$$\begin{bmatrix} x_1 \\ y_1 \\ 1 \end{bmatrix} = \begin{bmatrix} c_1 & -s_1 & 0 \\ s_1 & c_1 & 0 \\ 0 & 0 & 1 \end{bmatrix} \begin{bmatrix} 1 & 0 & a_1 \\ 0 & 1 & 0 \\ 0 & 0 & 1 \end{bmatrix} \begin{bmatrix} x_2 \\ y_2 \\ 1 \end{bmatrix}.$$

$\uparrow$ $\uparrow$

rotate   translate along link
by θ_1 so that second origin
                        becomes first origin

The product of the translation–rotation matrices is

$$A_1 = \begin{bmatrix} c_1 & -s_1 & c_1 a_1 \\ s_1 & c_1 & s_1 a_1 \\ 0 & 0 & 1 \end{bmatrix} \qquad (10\text{-}1)$$

and we have

$$\begin{bmatrix} x_1 \\ y_1 \\ 1 \end{bmatrix} = A_1 \begin{bmatrix} x_2 \\ y_2 \\ 1 \end{bmatrix}. \qquad (10\text{-}2)$$

Similarly, we have

$$\begin{bmatrix} x_2 \\ y_2 \\ 1 \end{bmatrix} = A_2 \begin{bmatrix} x_3 \\ y_3 \\ 1 \end{bmatrix}, \qquad (10\text{-}3)$$

where A_2 is defined like (10-1) but with 2's in place of 1's.

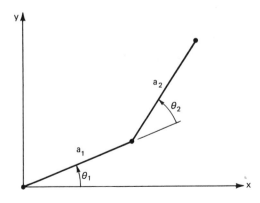

Figure 10-6 Manipulator from Fig. 10-4, showing second axis angle measured from the extended first link.

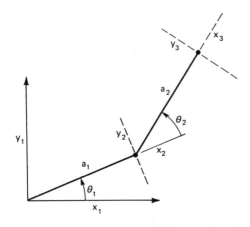

Figure 10-7 Manipulator from Fig. 10-6, showing coordinate systems generated by links. Each axis is its own origin. The standard coordinates are now labeled (x_1, y_1).

I realize that this matrix formalism may seem awkward and confusing if it is new to you. However, it is standard in the kinematic analysis of mechanisms ([Hartenberg and Denavit, 1964]). A good elementary matrix reference is [Shields, 1967], as is [Cullen, 1966]. If, at the moment, you feel this may be a stumbling block for you, try accepting (10-2) and (10-3) on faith and continuing.

Now let us derive a new polynomial system for our mechanism (refer to Fig. 10-7). Consider the matrix equation

$$\begin{bmatrix} x \\ y \\ 1 \end{bmatrix} = A_1 A_2 \begin{bmatrix} 0 \\ 0 \\ 1 \end{bmatrix}, \tag{10-4}$$

which expresses the fact that the hand position has coordinates $(0, 0)$ in the third coordinate system and (x, y) in the first. Since

$$A_1 A_2 = \begin{bmatrix} c_1 & -s_1 & c_1 a_1 \\ s_1 & c_1 & s_1 a_1 \\ 0 & 0 & 1 \end{bmatrix} \begin{bmatrix} c_2 & -s_2 & c_2 a_2 \\ s_2 & c_2 & s_2 a_2 \\ 0 & 0 & 1 \end{bmatrix}$$

$$= \begin{bmatrix} c_1 c_2 - s_1 s_2 & -c_1 s_2 - s_1 c_2 & c_1 c_2 a_2 - s_1 s_2 a_2 + c_1 a_1 \\ s_1 c_2 + c_1 s_2 & -s_1 s_2 + c_1 c_2 & s_1 c_2 a_2 + c_1 s_2 a_2 + s_1 a_1 \\ 0 & 0 & 1 \end{bmatrix},$$

we have

$$A_1 A_2 \begin{bmatrix} 0 \\ 0 \\ 1 \end{bmatrix} = \begin{bmatrix} c_1 c_2 a_2 - s_1 s_2 a_2 + c_1 a_1 \\ s_1 c_2 a_2 + c_1 s_2 a_2 + s_1 a_1 \\ 1 \end{bmatrix}.$$

Thus (10-4) yields the two equations

$$c_1 c_2 a_2 - s_1 s_2 a_2 + c_1 a_1 = x$$

$$s_1 c_2 a_2 + c_1 s_2 a_2 + s_1 a_1 = y,$$

which, together with

$$c_1^2 + s_1^2 = 1$$

$$c_2^2 + s_2^2 = 1,$$

form a system of four equations in the four unknowns c_1, s_1, c_2, s_2. The solutions of this system, whose total degree is 16, give the solutions to the IPP.

So we have derived a new polynomial system for the IPP. We will hold off considering reduction–scaling–solution for now. However, a glance at Fig. 10-7 shows for any reasonable choices of a_1 and a_2, there are only two or fewer solutions (perhaps none). Therefore, the new system that we have derived is a candidate for reduction.

Now let's consider some other mechanisms. First let's add a link to our two-link manipulator, as shown in Fig. 10-8. Convince yourself from the picture or a model (three strips of cardboard, three thumbtacks; go on, make one, it's fun) that you will (usually) have an infinite number of solutions (choices of axis angles) unless your hand position is beyond the reach of the manipulator. The associated polynomial system is

$$\begin{bmatrix} x \\ y \\ 1 \end{bmatrix} = A_1 A_2 A_3 \begin{bmatrix} 0 \\ 0 \\ 1 \end{bmatrix} \tag{10-5}$$

$$c_i^2 + s_i^2 = 1, \qquad i = 1, 2, 3,$$

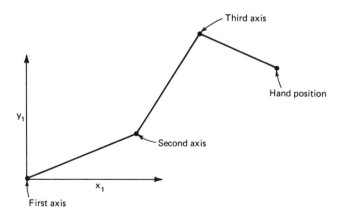

Figure 10-8 Three-axis manipulator.

which consists of five equations in six unknowns. If you submit this to
CONSOL (with, say, a dummy sixth equation), it will tend to find at least
one solution on each connected component of solutions. Thus it will find a
representative set of points out of the infinite number available. What more
could you ask of a method developed to solve n equations in n unknowns?

However, if we change the problem slightly, we will recover a system
with the same number of equations as unknowns. At the same time we will
redefine the IPP to ask a more sophisticated question. Suppose that we want
not only to specify the position of the hand but also the alignment of the
last joint; That is, we specify a point p and an arrow v (Fig. 10-9).

Changing v to v' changes the solution (see Fig. 10-10). Specifying both
the *position* and *orientation* of the hand is sometimes important in manipulator problems.

Since v gives the x-axis of the last coordinate system, we can express
this new condition as: "The x-axis in coordinate system 3 should be v in
coordinate system 1." Algebraically, with $v = (v_x, v_y)$, this is expressed by

$$\begin{bmatrix} v_x \\ v_y \\ 1 \end{bmatrix} = A_1 A_2 A_3 \begin{bmatrix} 1 \\ 0 \\ 1 \end{bmatrix}. \qquad (10\text{-}6)$$

To keep things concise, take (v_x, v_y) to be a point on the unit circle,
indicating "angle." Thus $v_x^2 + v_y^2 = 1$, and the two equations generated by
(10-6) are not independent. We can choose either one to augment (10-5),
yielding six equations in six unknowns. The solution sets of angles will put
the hand at (x, y) and the last link will be oriented along v.

There is another way to specify this last equation. Let's specify the
angle θ_0 that the x-axis of the last coordinate system is to make with the x-
axis of the first. Thus the sum of the three coordinate transformation angles

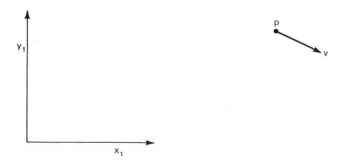

Figure 10-9 Desired hand position, p, and orientation vector, v, for the three-
axis manipulator.

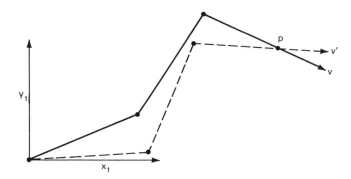

Figure 10-10   Changing the desired orientation of the hand, v (without changing the desired position, p), can change the axis angles.

is to be θ_0. Therefore, we want rotation through angle θ_0 to be the same as the successive rotations through the angles θ_3, θ_2, and θ_1:

$$\begin{bmatrix} c_1 & -s_1 \\ s_1 & c_1 \end{bmatrix} \begin{bmatrix} c_2 & -s_2 \\ s_2 & c_2 \end{bmatrix} \begin{bmatrix} c_3 & -s_3 \\ s_3 & c_3 \end{bmatrix} = \begin{bmatrix} c_0 & -s_0 \\ s_0 & c_0 \end{bmatrix},$$

where $c_0 = \sin \theta_0$, $s_0 = \sin \theta_0$. Now we can express the IPP as a single matrix equation

$$A_1 A_2 A_3 = A_0$$

where

$$A_0 \equiv \begin{bmatrix} c_0 & -s_0 & x \\ s_0 & c_0 & y \\ \hline 0 & 0 & 1 \end{bmatrix},$$

together with the auxiliary equations

$$c_i^2 + s_i^2 = 1$$

for $i = 1, 2, 3$. (Check this out. It works.)

The resulting system of nine equations in six unknowns is overspecified. It is as if we specified both the x-axis angle and the y-axis angle. Any three of the four equations given by the first and second columns of A_0 can be omitted.

Are you still there? Take a deep breath. The worst is over. All that follows uses the formalism above, admittedly later generalized to three dimensions.

What happens if we add another link? The inverse position problem (specifying position and orientation) then becomes: Solve $A_1 A_2 A_3 A_4 = A_0$. Generically, this always yields an infinite number of solutions (or none). But we can regain a system with a finite number of solutions by adding more restraints. For example, we could demand that link 3 have a particular ori-entation (Fig. 10-11). If this orientation is given by angle θ_*, then

$$A_1A_2A_3 \begin{bmatrix} 1 & 0 & 0 \\ 0 & 1 & 0 \\ \hline 0 & 0 & 0 \end{bmatrix} = \begin{bmatrix} c_* & -s_* & 0 \\ s_* & c_* & 0 \\ \hline 0 & 0 & 0 \end{bmatrix}$$

defines the new relation. As before, we choose only one of these four equa-tions, as they are all dependent. We can also consider *closed* mechanisms, in which the last joint axis is identical with the first (Fig. 10-12), with the associated relations

$$A_1A_2A_3A_4 = \begin{bmatrix} -1 & 0 & 0 \\ 0 & -1 & 0 \\ \hline 0 & 0 & 1 \end{bmatrix}.$$

The minus signs express the fact that the orientation is reversed at the end of the chain. Now we need an added relation which might simply be an orientation for one of the intermediate links. But more realistically, if we consider the link lengths a_1, a_2, a_3, and a_4 as unknowns and we specify points p_1, p_2, and p_3 that this *four-bar* should be able to pass through, then we end up with a system of equations in the unknowns a_1, a_2, a_3, and a_4 (and less important for this problem, the c_i and the s_i) whose solution gives us *designs* for the four-bar that synthesize a certain performance criteria (see Fig. 10-13). Naturally, the right number of points must be given to avoid over- or underspecification of the problem. We are not particularly focusing on n-bar design problems here, but the basics of their solution is available from the material in the chapter (see [Hartenberg and Denavit, 1964]).

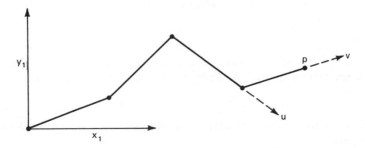

Figure 10-11 Four-axis manipulator, with a specified orientation for the third axis, as well as a specified hand position, p, and orientation, v. The resulting indirect position problem has only a finite number of solutions.

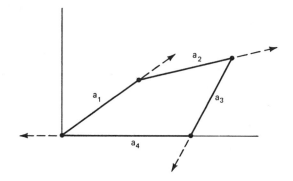

Figure 10-12   Closed mechanism.

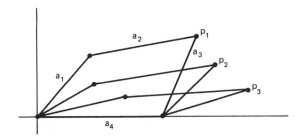

Figure 10-13   Specified precision points p_1, p_2, and p_3 generate restraints on the four-bar mechanism that allow us to solve for the design parameters a_1, a_2, a_3, and a_4.

10-3 REDUCTION BASICS FOR THE REVOLUTE JOINT PLANAR MANIPULATOR

We focus on the following "standard" IPP: Given constants

- a_1, a_2, a_3 (manipulator parameters)
- x_0, y_0, θ_0 (hand position and orientation parameters)

find θ_1, θ_2, θ_3 such that

$$A_1 A_2 A_3 = A_0, \tag{10-7}$$

where

$$A_j = \begin{bmatrix} c_j & -s_j & \vdots & c_j a_j \\ s_j & c_j & \vdots & s_j a_j \\ \hdashline 0 & 0 & \vdots & 1 \end{bmatrix}$$

for $j = 1, 2,$ and 3,

$$A_0 = \begin{bmatrix} c_0 & -s_0 & x_0 \\ s_0 & c_0 & y_0 \\ \hline 0 & 0 & 1 \end{bmatrix},$$

and

$$c_j = \cos \theta_j, \qquad s_j = \sin \theta_j$$

for $j = 0, 1, 2,$ and 3.

Since we want to solve (10-7) as a polynomial system, we take the c_j, s_j to be the variables and append the relations

$$c_j^2 + s_j^2 = 1, \qquad j = 1, 2, 3 \qquad (10\text{-}8)$$

to (10-7). Now (10-7) with (10-8) is a system of nine equations in six unknowns, but the four equations in the rotation part of (10-7) are equivalent and we need only choose (any) one of them to get a system of six equations in six unknowns. This system consists of three cubics and three quadrics, so it has total degree $3^3 \cdot 2^3 = 216$.

Now, taken literally, solving a system of three cubics and three quadrics is a formidable numerical problem. However, (10-7) with (10-8) is, in fact, reducible to an easy-to-solve system consisting of five linear and one quadric equation. Let's consider how to discover such a reduction systematically. Granted, in the two-dimensional case we do not really need a methodology. However, I am anticipating the next section.

To proceed systematically, let's consider "the degree properties of matrix relations." After collecting some useful facts, we will produce a reduction for (10-7).

Here is the approach to reduction that we will take. We can generate corollary equations to the basic matrix equation, (10-7), by performing matrix operations. From the basic and corollary equations, we will extract a new system more friendly to solution. For example,

$$A_2 A_3 = A_1^{-1} A_0 \qquad (10\text{-}9)$$

is a corollary to (10-7), and so is

$$A_1^T A_1 A_2 A_3 = A_1^T A_0 \qquad (10\text{-}10)$$

(A^T denotes the matrix transpose of A). We can generate an unlimited number of new equations. (Think about using other matrix operators, such as the trace.†) How shall we proceed to make systematic improvements?

---

† By definition, the trace of a matrix is the sum of its diagonal entries.

The guiding principle is: We want to produce a system of reduced total degree. We therefore want to investigate the degree properties of matrix relations. By collecting together degree-property facts, we gain insight into how to derive minimal-degree corollary equations. For example,

$$A_1A_2 = \left[\begin{array}{cc|c} c_1c_2 - s_1s_2 & -(c_1s_2 + s_1c_2) & a_2(c_1c_2 - s_1s_2) + a_1c_1 \\ c_1s_2 + s_1c_2 & c_1c_2 - s_1s_2 & a_2(c_1s_2 + s_1c_2) + a_1s_1 \\ \hline 0 & 0 & 1 \end{array} \right],$$

so we can display the degrees of the entries as a new matrix:

$$\begin{bmatrix} 2 & 2 & 2 \\ 2 & 2 & 2 \\ 0 & 0 & 0 \end{bmatrix}.$$

In this case, the degrees of the product are not surprising or particularly helpful in finding lower-degree expressions. But consider

$$(A_1A_2)^T(A_1A_2) = \left[\begin{array}{cc|c} 1 & 0 & a_1c_2 + a_2 \\ 0 & 1 & -a_1s_2 \\ \hline a_1c_2 + a_2 & -a_1s_2 & 2a_1a_2c_2 + a_1^2 + a_2^2 + 1 \end{array} \right].$$

In this case, the variables c_1 and s_1 have disappeared and the degree matrix is

$$\begin{bmatrix} 0 & 0 & 1 \\ 0 & 0 & 1 \\ 1 & 1 & 1 \end{bmatrix}.$$

This does suggest that $(A_1A_2)^T(A_1A_2)$ would be a useful form to develop as part of a corollary relation. Now look at

$$(A_1A_1^T)(A_2^TA_2) =$$

$$\left[\begin{array}{cc|c} a_1^2c_1^2 + a_1a_2c_1 + 1 & a_1^2c_1s_1 & a_1^2a_2c_1^2 + a_1(a_1^2 + 1)c_1 \\ a_1^2c_1s_1 + a_1a_2s_1 & -a_1^2c_1^2 + (a_1^2 + 1) & a_1^2a_2c_1s_1 + a_1(a_2^2 + 1)s_1 \\ \hline a_1c_1 + a_2 & -a_1s_1 & a_1a_2c_1 + (a_2^2 + 1) \end{array} \right].$$

The degree matrix is

$$\begin{bmatrix} 2 & 2 & 2 \\ 2 & 2 & 2 \\ 1 & 1 & 1 \end{bmatrix},$$

which is not particularly striking, but notice that $(A_1 A_1^T)(A_2^T A_2)$ has entries involving only c_1 and s_1. The variable c_2 and s_2 are missing. Also, the trace of $(A_1 A_1^T)(A_2^T A_2)$ is linear:

$$\text{trace} = 2a_1 a_2 c_1 + (a_1^2 + a_2^2 + 2)$$

and c_1 is the only variable that appears.

These examples illustrate that some matrix expressions can have lower degree or can omit variables that are present in the "parent" expressions. Our job is to generate corollaries to (10-7) that will include these simpler expressions. To illustrate how one might carry out this program systematically, I have listed in Table 10-1 a number of matrix expressions with their degree matrices and other relevant information. The significance of "cross terms only" is that if one variable is held constant, we can decrease the degrees by one from those shown (for example, if the constant matrix A_0 appears in the matrix expression).

Now let us reduce (10-7). For simplicity, let's assume that $a_1 \neq 0$, $a_2 \neq 0$, $a_3 \neq 0$. From Table 10-1, part 5 (denoted Table 10-1-5), we might be led to try

$$(A_1 A_2 A_3)^T (A_1 A_2 A_3) = A_0^T A_0, \tag{10-11}$$

yielding second-degree equations, which is better than the cubics of (10-7). Although this is an improvement, we can do better. From (10-7) we get

$$A_1 A_2 = A_0 A_3^{-1}, \tag{10-12}$$

from which we obtain

$$(A_1 A_2)^T (A_1 A_2) = (A_0 A_3^{-1})^T (A_0 A_3^{-1}). \tag{10-13}$$

Now by Table 10-1-3, the left-hand side of (10-13) consists of first-degree equations in c_2 and s_2 only, and by Table 10-1-7, the right-hand side consists of first-degree equations in c_3 and s_3 only. Thus (10-13) consists of nothing but linear equations in c_1, s_1, c_2, s_2. Similarly, we derive

$$A_2 A_3 = A_1^{-1} A_0 \tag{10-14}$$

from (10-7), and thus

$$(A_2 A_3)^T (A_2 A_3) = (A_1^{-1} A_0)^T (A_1^{-1} A_0). \tag{10-15}$$

Using Table 10-1-3 and 10-1-9 we conclude (as above) that (10-15) consists of linear equations in c_3, s_3, c_1, s_1. (*Note:* Our conclusions depend on the "cross terms only" entry in the table.) Since both (10-13) and (10-15) are symmetric matrix equations, each can yield no more than three independent equations; thus we have reduced (10-7) to a system of six linear equations in six unknowns.

At this point we have gone as far as general principles can take us. We must generate (10-13) and (10-15) and examine the new system consisting

TABLE 10–1   Degree matrices and related information for the two-dimensional IPP

		Matrix				Trace	
	Matrix	Degs of Entries	Missing Variables	Cross Terms Only?	Deg	Missing Variables	Cross Terms Only?
1.	$M = A_1 A_1^T$	2 2 1 2 2 1 1 1 0	No	No	0	No	No
	M^{-1}	0 0 1 0 0 1 1 1 0	No	No	0	No	No
2.	$M = A_1 A_2$	2 2 2 2 2 2 0 0 0	No	No	2	No	No
	M^{-1}	2 2 1 2 2 1 0 0 0	No	No	2	No	No
3.	$M = N^T N$, where $N = A_1 A_2$	0 0 1 0 0 1 1 1 1	c_1, s_1	Yes	1	c_1, s_1, s_2	Yes
	M^{-1}	2 2 1 2 2 1 1 1 0	c_1, s_1	Yes	1	c_1, s_1, s_2	Yes
4.	$M = A_1 A_2 A_3$	3 3 3 3 3 3 0 0 0	No	No	3	No	No
	M^{-1}	3 3 2 3 3 2 0 0 0	No	No	3	No	No
5.	$M = N^T N$, where $N = A_1 A_2 A_3$	0 0 2 0 0 2 2 2 2	c_1, s_1	Yes	2	c_1, s_1	Yes
	M^{-1}	4 4 2 4 4 2 2 2 0	c_1, s_1	No	2	c_1, s_1	Yes
6.	$M = A_1 A_2^{-1}$	2 2 1 2 2 1 0 0 0	No	Yes	2	No	Yes
	M^{-1}	2 2 1 2 2 1 0 0 0	No	Yes	2	No	Yes
7.	$M = N^T N$, where $N = A_1 A_2^{-1}$	0 0 1 0 0 1 1 1 0	c_1, s_1	Yes	0	No	No
	M^{-1}	2 2 1 2 2 1 1 1 0	c_1, s_1	No	0	No	No

TABLE 10–1   (continued)

		Matrix			Trace		
	Matrix	Degs of Entries	Missing Variables	Cross Terms Only?	Deg	Missing Variables	Cross Terms Only?
8.	$M = A_1^{-1}A_2$	2 2 2 2 2 2 0 0 0	No	Yes	2	No	Yes
	M^{-1}	2 2 2 2 2 2 0 0 0	No	Yes	2	No	Yes
9.	$M = N^T N$, where $N = A_1^{-1}A_2$	0 0 2 0 0 2 2 2 2	No	Yes	2	No	Yes
	M^{-1}	4 4 2 4 4 2 2 2 0	No	Yes	2	No	Yes
10.	$M = A_1 A_2^T$	2 2 1 2 2 1 1 1 0	No	Yes	2	No	Yes
	M^{-1}	2 2 1 2 2 1 1 1 0	No	Yes	2	No	Yes
11.	$M = A_1^T A_2$	2 2 2 2 2 2 2 2 2	No	Yes	2	No	Yes
	M^{-1}	2 2 0 2 2 0 0 0 0	No	Yes	2	No	Yes
12.	$M = N_1 N_2$, where $N_1 = A_1 A_1^T$ $N_2 = A_2^T A_2$	2 2 2 2 2 2 1 1 1	c_2, s_2	No	1	s_1, c_2, s_2	Yes
	M^{-1}	1 1 1 0 0 1 1 1 1	c_2, s_2	Yes	1	s_1, c_2, s_2	Yes
13.	$M = N_1 N_2$, where $N_1 = A_1^T A_1$ $N_2 = A_2 A_2^T$	2 2 1 2 2 1 2 2 1	c_1, s_1	No	1	c_1, s_1, s_2	Yes
	M^{-1}	1 0 1 1 0 1 1 1 1	c_1, s_1	Yes	1	c_1, s_1	Yes

of (10-13), (10-15), and (10-8). This new system consists of six linear and three quadric equations in six unknowns. We will want to pick out the lowest-degree independent subsystem of six equations. Because (10-13) and (10-15) were generated from (10-7), we might (in general) have to consider including some equations from (10-7) to obtain an independent system. However, we will see this is unnecessary here. The three distinct equations from (10-13) are

$$a_2 + c_2 a_1 = (x_0 c_0 + y_0 s_0 - a_3)c_3 + (x_0 s_0 - y_0 c_0)s_3 \quad (10\text{-}13\text{-}1)$$

$$- a_1 s_2 = (y_0 c_0 - x_0 s_0)c_3 + (x_0 c_0 + y_0 s_0 - a_3)s_3 \quad (10\text{-}13\text{-}2)$$

$$2a_1 a_2 c_2 + (a_1^2 + a_2^2 + 1) = x_0^2 + y_0^2$$
$$+ a_3^2 - 2a_3(x_0 c_0 + y_0 s_0) + 1 \quad (10\text{-}13\text{-}3)$$

and from (10-15)

$$a_3 + c_3 a_2 = (x_0 c_0 + y_0 s_0) - s_0 a_1 s_1 - c_0 a_1 c_1 \quad (10\text{-}15\text{-}1)$$

$$- a_2 s_3 = (-x_0 s_0 + y_0 c_0) + s_0 a_1 c_1 - c_0 a_1 s_1 \quad (10\text{-}15\text{-}2)$$

$$2a_2 a_3 c_3 + (a_2^2 + a_3^2 + 1) = -2x_0 a_1 c_1 - 2y_0 a_1 s_1$$
$$+ a_1^2 + x_0^2 + y_0^2 + 1 - 2a_1(x_0 c_1 + y_0 s_1). \quad (10\text{-}15\text{-}3)$$

Written as a linear system, (10-13-1) through (10-13-3) and (10-15-1) through (10-15-3) look like

$$
\begin{bmatrix}
0 & 0 & a_1 & 0 & -x_0 c_0 - y_0 s_0 + a_3 & -x_0 s_0 + y_0 c_0 \\
0 & 0 & 0 & -a_1 & x_0 s_0 - y_0 c_0 & -x_0 c_0 - y_0 s_0 + a_3 \\
0 & 0 & 2a_1 a_2 & 0 & 0 & 0 \\
\hdashline
c_0 a_1 & s_0 a_1 & 0 & 0 & a_2 & 0 \\
-s_0 a_1 & c_0 a_1 & 0 & 0 & 0 & -a_2 \\
2x_0 a_1 & 2y_0 a_1 & 0 & 0 & 2a_2 a_3 & 0
\end{bmatrix}
\begin{bmatrix}
c_1 \\ s_1 \\ c_2 \\ s_2 \\ c_3 \\ s_3
\end{bmatrix}
$$

$$
= \begin{bmatrix}
-a_2 \\
0 \\
x^2 + y^2 - a_1^2 - a_2^2 + a_3^2 - 2a_3(x_0 c_0 + y_0 s_0) \\
\hdashline
x_0 c_0 + y_0 s_0 - a_3 \\
-x_0 s_0 + y_0 c_0 \\
x_0^2 + y_0^2 + a_1^2 - a_2^2 - a_3^2
\end{bmatrix}. \quad (10\text{-}16)
$$

I have row reduced this linear system (via the GEL algorithm, as given in Appendix 5) to

$$
\begin{bmatrix}
0 & 0 & 0 & 0 & 2a_2(-s_0y_0 - c_0x_0 + a_3) & 2a_2(-s_0x_0 + c_0y_0) \\
0 & 0 & 0 & -a_1 & s_0x_0 - c_0y_0 & -s_0y_0 - c_0x_0 + a_3 \\
0 & 0 & -2a_1a_2 & 0 & 0 & 0 \\
\hline
a_1 & 0 & 0 & 0 & c_0a_2 & s_0a_2 \\
0 & a_1 & 0 & 0 & s_0a_2 & -c_0a_2 \\
0 & 0 & 0 & 0 & 2a_2(-s_0y_0 - c_0x_0 + a_3) & 2a_2(-s_0x_0 + c_0y_0)
\end{bmatrix}
\begin{bmatrix}
c_1 \\ s_1 \\ c_2 \\ s_2 \\ c_3 \\ s_3
\end{bmatrix}
$$

$$
=
\begin{bmatrix}
2s_0a_3y_0 + 2c_0a_3x_0 + a_1^2 - a_2^2 - a_3^2 - x_0^2 - y_0^2 \\
0 \\
2s_0a_3y_0 + 2c_0a_3x_0 + a_1^2 + a_2^2 - a_3^2 - x_0^2 - y_0^2 \\
\hline
-c_0a_3 + x_0 \\
-s_0a_3 + y_0 \\
2s_0a_3y_0 + 2c_0a_3x_0 + a_1^2 - a_2^2 - a_3^2 - x_0^2 - y_0^2
\end{bmatrix}.
\quad (10\text{-}17)
$$

Thus equation 1 in the reduced system (10-17) is the same as equation 6. The first five equations are independent, unless both $s_0y_0 + c_0x_0 = a_3$ and $s_0x_0 = c_0y_0$. Assuming that this is not the case, the first five equations, together with $c_3^2 + s_3^2 = 1$, yield a system of six equations in six unknowns which is a valid reduction of (10-7) and (10-8) to an independent system of six equations in six unknowns. If both $s_0y_0 + c_0x_0 = a_3$ and $s_0x_0 = c_0y_0$, then only the middle four of the linear equations are independent, and even including all of (10-7), these do not yield a system with a finite number of solutions. It is easy to confirm in this case that (10-7) with (10-8) has, in fact, an infinite number of solutions.

To summarize, we have reduced the original system, (10-7) with (10-8), of total degree 216, to (10-17) with $c_3^2 + s_3^2 = 1$, a system of total degree 2. The new system has an infinite number of solutions only when the original does. Our reduction strategy was guided by reference to basic matrix-degree relations as illustrated by Table 10-1.

In the next section we encounter the IPP for three-dimensional manipulators, where we will proceed in strict analogy with the two-dimensional case.

EXERCISES

10-1. Give a geometric interpretation of the condition $s_0x_0 = c_0y_0$; of the pair of conditions $s_0x_0 = c_0y_0$, $s_0y_0 + c_0x_0 = a_3$. Draw a manipulator diagram for the latter case and confirm that it has an infinite number of positions.

10-2. Carry out the row reduction of (10-16) to (10-17).

10-3. Verify that (10-17) with $c_3^2 + s_3^2 - 1 = 0$ is a system with a finite number of solutions if $s_0 y_0 + c_0 x_0 \neq a_3$ or $s_0 x_0 \neq c_0 y_0$.

10-4. Show how to solve the system (10-17) with $c_3^2 + s_3^2 - 1 = 0$ by elementary means. What if we use $c_1^2 + s_1^2 = 1$ or $c_2^2 + s_2^2 = 1$ instead of $c_3^2 + s_3^2 - 1 = 0$?

10-5. Draw pictures of the two-dimensional three-joint manipulator (like Fig. 10-8) to show the cases: two solutions, one solution, no solutions, an infinite number of solutions. Discuss the algebraic situation for each picture. For example, if there are no solutions, what will CONSOL find when it is asked to solve the system (10-17) with $c_3^2 + s_3^2 - 1 = 0$?

10-4 REVOLUTE JOINT SPATIAL MANIPULATORS

In previous sections we have considered planar manipulators, which are chains of links connected by joints in the plane. Here we consider spatial manipulators, chains of links in space. As before, we restrict ourselves to revolute joints. The basic ideas for defining spatial manipulators, for generating polynomial systems to solve the inverse position problem, and for reducing these systems are analogous to that for planar manipulators. The resulting models, however, are more complex, harder to visualize, and yield less trivial polynomial systems. The payoff for dealing with this added complexity is that we will now solve a significant problem: the inverse position problem for revolute joint spatial manipulators.

We extend the planar manipulator model to three dimensions in two steps. Looking at our two-link planar model in Fig. 10-14, we may visualize the joints pivoting on axes that stick straight out of the page. Imagine long

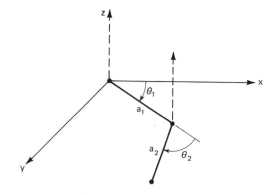

Figure 10-14 Two axis planar manipulator with joint axes parallel to the z-axis. Note link lengths a_1 and a_2 for the first and second links, respectively.

nails instead of thumbtacks. (Compare with Fig. 10-7.) These axes are parallel, we can say, to the z-axis. For the first step, extend the links along these axes by distances d_1 and d_2, as indicated in Fig. 10-15. Looking straight down, our model seems the same as before, but from an oblique view, we see that the links are now in different but parallel planes. For the second step, we "twist" the second axis by an angle α_1 in the plane normal to the first link (see Fig. 10-16). We establish coordinate systems at each joint, as before, with the joint being the z-axis, the link being the x-axis, and the y-axis chosen perpendicular to these. Then the coordinate transformation matrices from one joint coordinate system to the next provide us with our fundamental algebraic relations, just as in the planar case. The basic schematic is provided in Fig. 10-17 with associated coordinate transformation matrix

$$
A_i = \left[
\begin{array}{ccc:c}
c_i & -s_i\lambda_i & s_i\mu_i & a_ic_i \\
s_i & c_i\lambda_i & -c_i\mu_i & a_is_i \\
0 & \mu_i & \lambda_i & d_i \\
\hdashline
0 & 0 & 0 & 1
\end{array}
\right], \tag{10-18}
$$

where $c_i = \cos\theta_i$, $s_i = \sin\theta_i$, $\lambda_i = \cos\alpha_i$, $\mu_i = \sin\alpha_i$, and we have the basic relation

$$
p_i = A_ip_{i+1},
$$

where "p_i" denotes "point p's coordinates in ith coordinate system" and "p_{i+1}" denotes "point p's coordinates in $(i + 1)$st coordinate system." [Note from Fig. 10-17 that the first coordinate system has no link to define its x-axis. We chose this axis arbitrarily. Also, the last coordinate system (that of the hand) has no actual joint associated with it. Therefore, for the

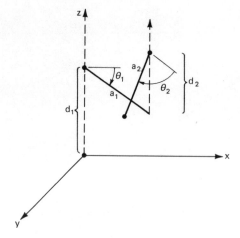

Figure 10-15 Spatial manipulator created from the planar manipulator in Fig. 10-14 by translating the links along the joint axes. The distances d_1 and d_2 are given for the first and second joints, respectively.

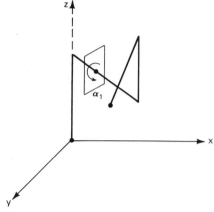

Figure 10-16 A new spatial manipulator created from the spatial manipulator in Fig. 10-15 by rotating the second joint axis. Thus the two joint axes for the two links are no longer parallel. The twist angle, α_1, represents a rotation in the plane normal to the first link. The manipulator still turns on θ_1 and θ_2, but it has a fixed twist given by α_1.

last coordinate system, we may choose $\alpha = 0$ and $d = 0$. The z-axis is arbitrary also.]

Physically, links may not look much like line segments in the normal planes of the joint axes (as we have here). However, we lose no generality with this model (see [Hartenberg and Devanit, 1964]). Thus we are not distinguishing between links and the common normals between joint axes.

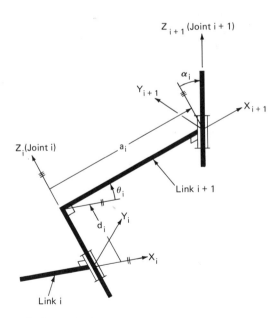

Figure 10-17 Basic manipulator notation. Parallel lines are indicated by tick marks. (Adapted from [Tsai and Morgan, 1985].)

Naturally, we may have as many (or few) joints as we like. However, we will study six-joint manipulators because a robot generally needs six joints to reach any point on its workspace with any orientation. Thus six-joint robots are a standard "general-purpose" industrial manipulator. (Five-joint robots are also common. See Exercise 10-6.) The formalism defined via (10-18) with Fig. 10-17 is developed in [Hartenberg and Denavit, 1964].

Here is a formal statement of the three-dimensional IPP for six revolute joint manipulators: Given the constants

$d_1, \ldots, d_6$ (manipulator parameters)

$a_1, \ldots, a_6$ (manipulator parameters)

$\alpha_1, \ldots, \alpha_6$ (manipulator parameters)

(p_x, p_y, p_z) (desired hand position)

$\left. \begin{array}{l} (l_x, l_y, l_z) \\ (m_x, m_y, m_z) \\ (n_x, n_y, n_z) \end{array} \right\}$ (desired hand orientation)

where the three vectors defining the desired hand orientation are mutually perpendicular unit vectors, find $\theta_1, \theta_2, \ldots, \theta_6$ such that

$$A_1 \cdot A_2 \cdots A_6 = A_0, \tag{10-19}$$

where A_i, $i = 1, \ldots, 6$, are given by (10-18) and

$$A_0 = \begin{bmatrix} l_x & m_x & n_x & \vdots & p_x \\ l_y & m_y & n_y & \vdots & p_y \\ l_z & m_z & n_z & \vdots & p_z \\ \hline 0 & 0 & 0 & \vdots & 1 \end{bmatrix}. \tag{10-20}$$

As before, we append the equations

$$c_i^2 + s_i^2 - 1 = 0 \tag{10-21}$$

for $i = 1, \ldots, 6$. Matrix equation (10-19) is a collection of 12 (noninde-pendent) polynomial equations in the 12 unknowns $c_1, s_1, c_2, s_2, \ldots, c_6, s_6$, which, together with the six equations (10-21), defines our basic system. Thus in analogy with the two-dimensional IPP, we have:

- The three-dimensional IPP is equivalent to solving a system of poly-nomial equations.
- If the manipulator cannot reach the position specified by A_0, then (10-19) with (10-21) will have no (real) solutions. Otherwise, it will usually have several real solutions. However, the nongeneric cases

"exactly one real solution" and "infinite number of real solutions" can occur.

- System (10-19) can be reduced by exploiting matrix relations.

The three-dimensional IPP differs from the two-dimensional IPP in that the system as it arises is virtually unsolvable. It has total degree 2,073,600, since (10-19) yields four sixth-degree and two fifth-degree equations and (10-21) contributes six quadrics. We approach reduction, as before, by exploiting matrix relations that tend to reduce degree. To be systematic, we should generate a catalog of basic relations, such as Table 10-1. This is especially true because although the IPP is an important problem, we are doing more than just solving the IPP. We are developing an approach to solving algebraic problems arising from models based on the rigid motion transformations (10-18). This includes many kinematic problems (see [Hartenberg and Denavit, 1964]). However, I will refrain from presenting a general catalog because:

- The catalog would be too large to fit conveniently here.
- The flavor of such a catalog is indicated by Table 10-1.
- I have so far discovered no organizational principles that would make such a catalog truly enlightening.

We will consider relations strictly derivative from (10-19) and that in the light of Table 10-1, seem likely to yield lower-degree equations. For illustration, let's consider one such relation:

$$A_3 A_4 A_5 = A_2^{-1} A_1^{-1} A_0 A_6^{-1}. \qquad (10\text{-}22)$$

Table 10-1 suggests this would yield third-degree equations, while the corollary relation

$$(A_3 A_4 A_5)^T (A_3 A_4 A_5) = (A_2^{-1} A_1^{-1} A_0 A_6^{-1})^T (A_2^{-1} A_1^{-1} A_0 A_6^{-1}) \qquad (10\text{-}23)$$

might generate second-degree equations. However, Table 10-1 does not apply directly, and to check we need to multiply out (10-22) and (10-23). Let

$$P = A_3 A_4 A_5,$$

and

$$Q = A_2^{-1} A_1^{-1} A_0 A_6^{-1}.$$

Then, letting $R = P - Q$ and $S = P^T P - Q^T Q$, we get the equations given in Table 10-2.

Although the listing of these equations looks formidable, we can study this listing for patterns the way any experimental scientist would. The key

TABLE 10–2　Equations from (10–22) and (10–23)[a]

$$R_{1,1} = -c_1c_2c_6l_x + c_1c_2s_6m_x\lambda_6 - c_1c_2s_6n_x\mu_6 - c_1s_2c_6l_y\lambda_1 + c_1s_2s_6m_y\lambda_6\lambda_1 - c_1s_2s_6n_y\mu_6\lambda_1$$
$$-s_1c_2c_6l_y + s_1c_2s_6m_y\lambda_6 - s_1c_2s_6n_y\mu_6 + s_1s_2c_6l_x\lambda_1 - s_1s_2s_6m_x\lambda_6\lambda_1$$
$$+ s_1s_2s_6n_x\mu_6\lambda_1 - s_2c_6l_z\mu_1 + s_2s_6m_z\lambda_6\mu_1 - s_2s_6n_z\mu_6\mu_1 + c_3c_4c_5$$
$$- c_3s_4s_5\lambda_4 - s_3c_4s_5\lambda_4\lambda_3 - s_3s_4c_5\lambda_3 + s_3s_5\mu_4\mu_3$$

$$R_{1,2} = -c_1c_2c_6m_x\lambda_6 + c_1c_2c_6n_x\mu_6 - c_1c_2s_6l_x - c_1s_2c_6m_y\lambda_6\lambda_1 + c_1s_2c_6n_y\mu_6\lambda_1$$
$$- c_1s_2s_6l_y\lambda_1 - s_1c_2c_6m_y\lambda_6 + s_1c_2c_6n_y\mu_6 - s_1c_2s_6l_y + s_1s_2c_6m_x\lambda_6\lambda_1$$
$$- s_1s_2c_6n_x\mu_6\lambda_1 + s_1s_2s_6l_x\lambda_1 - s_2c_6m_z\lambda_6\mu_1 + s_2c_6n_z\mu_6\mu_1 - s_2s_6l_z\mu_1 - c_3c_4s_5\lambda_5$$
$$- c_3s_4c_5\lambda_5\lambda_4 + c_3s_4\mu_5\mu_4 - s_3c_4c_5\lambda_5\lambda_4\lambda_3 + s_3c_4\mu_5\mu_4\lambda_3 + s_3s_4s_5\lambda_5\lambda_3$$
$$+ s_3c_5\lambda_5\mu_4\mu_3 \mp s_3\mu_5\mu_4\mu_3$$

$$R_{1,3} = -c_1c_2m_x\mu_6 - c_1c_2n_x\lambda_6 - c_1s_2m_y\mu_6\lambda_1 - c_1s_2n_y\lambda_6\lambda_1 - s_1c_2m_y\mu_6 - s_1c_2n_y\lambda_6$$
$$+ s_1s_2m_x\mu_6\lambda_1 + s_1s_2n_x\lambda_6\lambda_1 - s_2m_z\mu_6\mu_1 - s_2n_z\lambda_6\mu_1 + c_3c_4s_5\mu_5 + c_3s_4c_5\mu_5\lambda_4$$
$$+ c_3s_4\lambda_5\mu_4 + s_3c_4c_5\mu_5\lambda_4\lambda_3 + s_3c_4\lambda_5\mu_4\lambda_3 - s_3s_4s_5\mu_5\lambda_3 - s_3c_5\mu_5\mu_4\mu_3$$
$$+ s_3\lambda_5\mu_4\mu_3$$

$$R_{1,4} = d_1s_2\mu_1 + c_1c_2d_6m_x\mu_6 + c_1c_2d_6n_x\lambda_6 + c_1c_2l_xa_6 - c_1c_2p_x + c_1s_2d_6m_y\mu_6\lambda_1$$
$$+ c_1s_2d_6n_y\lambda_6\lambda_1 + c_1s_2l_ya_6\lambda_1 - c_1s_2p_y\lambda_1 + s_1c_2d_6m_y\mu_6 + s_1c_2d_6n_y\lambda_6$$
$$+ s_1c_2l_ya_6 - s_1c_2p_y - s_1s_2d_6m_x\mu_6\lambda_1 - s_1s_2d_6n_x\lambda_6\lambda_1 - s_1s_2l_xa_6\lambda_1 + s_1s_2p_x\lambda_1$$
$$+ c_2a_1 + s_2d_6m_z\mu_6\mu_1 + s_2d_6n_z\lambda_6\mu_1 + s_2l_za_6\mu_1 - s_2p_z\mu_1 + c_3c_4c_5a_5$$
$$+ c_3c_4a_4 + c_3s_4d_5\mu_4 - c_3s_4s_5a_5\lambda_4 + c_3a_3 + s_3d_4\mu_3 + s_3c_4d_5\mu_4\lambda_3$$
$$- s_3c_4s_5a_5\lambda_4\lambda_3 - s_3s_4c_5a_5\lambda_3 - s_3s_4a_4\lambda_3 + s_3d_5\lambda_4\mu_3 + s_3s_5a_5\mu_4\mu_3 + a_2$$

$$R_{2,1} = -c_1c_2c_6l_y\lambda_2\lambda_1 + c_1c_2s_6m_y\lambda_6\lambda_2\lambda_1 - c_1c_2s_6n_y\mu_6\lambda_2\lambda_1 + c_1s_2c_6l_x\lambda_2$$
$$- c_1s_2s_6m_x\lambda_6\lambda_2 + c_1s_2s_6n_x\mu_6\lambda_2 + c_1c_6l_y\mu_2\mu_1 - c_1s_6m_y\lambda_6\mu_2\mu_1 + c_1s_6n_y\mu_6\mu_2\mu_1$$
$$+ s_1c_2c_6l_x\lambda_2\lambda_1 - s_1c_2s_6m_x\lambda_6\lambda_2\lambda_1 + s_1c_2s_6n_x\mu_6\lambda_2\lambda_1 + s_1s_2c_6l_y\lambda_2$$
$$- s_1s_2s_6m_y\lambda_6\lambda_2 + s_1s_2s_6n_y\mu_6\lambda_2 - s_1c_6l_x\mu_2\mu_1 + s_1s_6m_x\lambda_6\mu_2\mu_1$$
$$- s_1s_6n_x\mu_6\mu_2\mu_1 - c_2c_6l_z\lambda_2\mu_1 + c_2s_6m_z\lambda_6\lambda_2\mu_1 - c_2s_6n_z\mu_6\lambda_2\mu_1 + c_3c_4s_5\lambda_4\lambda_3$$
$$+ c_3s_4c_5\lambda_3 - c_3s_5\mu_4\mu_3 + s_3c_4c_5 - s_3s_4s_5\lambda_4 - c_6l_z\mu_2\lambda_1 + s_6m_z\lambda_6\mu_2\lambda_1$$
$$- s_6n_z\mu_6\mu_2\lambda_1$$

$$R_{2,2} = -c_1c_2c_6m_y\lambda_6\lambda_2\lambda_1 + c_1c_2c_6n_y\mu_6\lambda_2\lambda_1 - c_1c_2s_6l_y\lambda_2\lambda_1 + c_1s_2c_6m_x\lambda_6\lambda_2$$
$$- c_1s_2c_6n_x\mu_6\lambda_2 + c_1s_2s_6l_x\lambda_2 + c_1c_6m_y\lambda_6\mu_2\mu_1 - c_1c_6n_y\mu_6\mu_2\mu_1 + c_1s_6l_y\mu_2\mu_1$$
$$+ s_1c_2c_6m_x\lambda_6\lambda_2\lambda_1 - s_1c_2c_6n_x\mu_6\lambda_2\lambda_1 + s_1c_2s_6l_x\lambda_2\lambda_1 + s_1s_2c_6m_y\lambda_6\lambda_2$$
$$- s_1s_2c_6n_y\mu_6\lambda_2 + s_1s_2s_6l_y\lambda_2 - s_1c_6m_x\lambda_6\mu_2\mu_1 + s_1c_6n_x\mu_6\mu_2\mu_1 - s_1s_6l_x\mu_2\mu_1$$
$$- c_2c_6m_z\lambda_6\lambda_2\mu_1 + c_2c_6n_z\mu_6\lambda_2\mu_1 - c_2s_6l_z\lambda_2\mu_1 + c_3c_4c_5\lambda_5\lambda_4\lambda_3 - c_3c_4\mu_5\mu_4\lambda_3$$
$$- c_3s_4s_5\lambda_5\lambda_3 - c_3c_5\lambda_5\mu_4\mu_3 - c_3\mu_5\mu_4\mu_3 - s_3c_4s_5\lambda_5 - s_3s_4c_5\lambda_5\lambda_4 + s_3s_4\mu_5\mu_4$$
$$- c_6m_z\lambda_6\mu_2\lambda_1 + c_6n_z\mu_6\mu_2\lambda_1 - s_6l_z\mu_2\lambda_1$$

$$R_{2,3} = -c_1c_2m_y\mu_6\lambda_2\lambda_1 - c_1c_2n_y\lambda_6\lambda_2\lambda_1 + c_1s_2m_x\mu_6\lambda_2 + c_1s_2n_x\lambda_6\lambda_2$$
$$+ c_1m_y\mu_6\mu_2\mu_1 + c_1n_y\lambda_6\mu_2\mu_1 + s_1c_2m_x\mu_6\lambda_2\lambda_1 + s_1c_2n_x\lambda_6\lambda_2\lambda_1$$
$$+ s_1s_2m_y\mu_6\lambda_2 + s_1s_2n_y\lambda_6\lambda_2 - s_1m_x\mu_6\mu_2\mu_1 - s_1n_x\lambda_6\mu_2\mu_1 - c_2m_z\mu_6\lambda_2\mu_1$$
$$- c_2n_z\lambda_6\lambda_2\mu_1 - c_3c_4c_5\mu_5\lambda_4\lambda_3 - c_3c_4\lambda_5\mu_4\lambda_3 + c_3s_4s_5\mu_5\lambda_3 + c_3c_5\mu_5\mu_4\mu_3$$
$$- c_3\lambda_5\mu_4\mu_3 + s_3c_4s_5\mu_5 + s_3s_4c_5\mu_5\lambda_4 + s_3s_4\lambda_5\mu_4 - m_z\mu_6\mu_2\lambda_1 - n_z\lambda_6\mu_2\lambda_1$$

TABLE 10–2   (continued)

$$R_{2,4} = d_1 c_2 \lambda_2 \mu_1 + d_1 \mu_2 \lambda_1 + c_1 c_2 d_6 m_y \mu_6 \lambda_2 \lambda_1 + c_1 c_2 d_6 n_y \lambda_6 \lambda_2 \lambda_1 + c_1 c_2 l_y a_6 \lambda_2 \lambda_1$$
$$- c_1 c_2 p_y \lambda_2 \lambda_1 - c_1 s_2 d_6 m_x \mu_6 \lambda_2 - c_1 s_2 d_6 n_x \lambda_6 \lambda_2 - c_1 s_2 l_x a_6 \lambda_2 + c_1 s_2 p_x \lambda_2$$
$$- c_1 d_6 m_y \mu_6 \mu_2 \mu_1 - c_1 d_6 n_y \lambda_6 \mu_2 \mu_1 - c_1 l_y a_6 \mu_2 \mu_1 + c_1 p_y \mu_2 \mu_1 - s_1 c_2 d_6 m_x \mu_6 \lambda_2 \lambda_1$$
$$- s_1 c_2 d_6 n_x \lambda_6 \lambda_2 \lambda_1 - s_1 c_2 l_x a_6 \lambda_2 \lambda_1 + s_1 c_2 p_x \lambda_2 \lambda_1 - s_1 s_2 d_6 m_y \mu_6 \lambda_2 - s_1 s_2 d_6 n_y \lambda_6 \lambda_2$$
$$- s_1 s_2 l_y a_6 \lambda_2 + s_1 s_2 p_y \lambda_2 + s_1 d_6 m_x \mu_6 \mu_2 \mu_1 + s_1 d_6 n_x \lambda_6 \mu_2 \mu_1 + s_1 l_x a_6 \mu_2 \mu_1$$
$$- s_1 p_x \mu_2 \mu_1 + d_2 \mu_2 + c_2 d_6 m_z \mu_6 \lambda_2 \mu_1 + c_2 d_6 n_z \lambda_6 \lambda_2 \mu_1 + c_2 l_z a_6 \lambda_2 \mu_1$$
$$- c_2 p_z \lambda_2 \mu_1 - s_2 \lambda_2 a_1 - c_3 d_4 \mu_3 - c_3 c_4 d_5 \mu_4 \lambda_3 + c_3 c_4 s_5 a_5 \lambda_4 \lambda_3$$
$$+ c_3 s_4 c_5 a_5 \lambda_3 + c_3 s_4 a_4 \lambda_3 - c_3 d_5 \lambda_4 \mu_3 - c_3 s_5 a_5 \mu_4 \lambda_3 + s_3 c_4 c_5 a_5 + s_3 c_4 a_4$$
$$+ s_3 s_4 d_5 \mu_4 - s_3 s_4 s_5 a_5 \lambda_4 + s_3 a_3 + d_6 m_z \mu_6 \mu_2 \mu_1 + d_6 n_z \lambda_6 \mu_2 \mu_1 + l_z a_6 \mu_2 \mu_1$$
$$- p_z \mu_2 \mu_1$$

$$R_{3,1} = c_1 c_2 c_6 l_y \mu_2 \mu_1 - c_1 c_2 s_6 m_y \lambda_6 \mu_2 \mu_1 + c_1 c_2 s_6 n_y \mu_6 \mu_2 \mu_1 - c_1 s_2 c_6 l_x \mu_2 + c_1 s_2 s_6 m_x \lambda_6 \mu_2$$
$$- c_1 s_2 s_6 n_x \mu_6 \mu_2 + c_1 c_6 l_y \lambda_2 \mu_1 - c_1 s_6 m_y \lambda_6 \lambda_2 \mu_1 + c_1 s_6 n_y \mu_6 \lambda_2 \mu_1$$
$$- s_1 c_2 c_6 l_x \mu_2 \mu_1 + s_1 c_2 s_6 m_x \lambda_6 \mu_2 \mu_1 - s_1 c_2 s_6 n_x \mu_6 \mu_2 \mu_1 - s_1 s_2 c_6 l_y \mu_2$$
$$+ s_1 s_2 s_6 m_y \lambda_6 \mu_2 - s_1 s_2 s_6 n_y \mu_6 \mu_2 - s_1 c_6 l_x \lambda_2 \mu_1 + s_1 s_6 m_x \lambda_6 \lambda_2 \mu_1 - s_1 s_6 n_x \mu_6 \lambda_2 \mu_1$$
$$+ c_2 c_6 l_z \mu_2 \mu_1 - c_2 s_6 m_z \lambda_6 \mu_2 \mu_1 + c_2 s_6 n_z \mu_6 \mu_2 \mu_1 + c_4 s_5 \lambda_4 \mu_3 + s_4 c_5 \mu_3$$
$$+ s_5 \mu_4 \lambda_3 - c_6 l_z \lambda_2 \mu_1 + s_6 m_z \lambda_6 \lambda_2 \mu_1 - s_6 n_z \mu_6 \lambda_2 \mu_1$$

$$R_{3,2} = c_1 c_2 c_6 m_y \lambda_6 \mu_2 \mu_1 - c_1 c_2 c_6 n_y \mu_6 \mu_2 \mu_1 + c_1 c_2 s_6 l_y \mu_2 \mu_1 - c_1 s_2 c_6 m_x \lambda_6 \mu_2$$
$$+ c_1 s_2 c_6 n_x \mu_6 \mu_2 - c_1 s_2 s_6 l_x \mu_2 + c_1 c_6 m_y \lambda_6 \lambda_2 \mu_1 - c_1 c_6 n_y \mu_6 \lambda_2 \mu_1$$
$$+ c_1 s_6 l_y \lambda_2 \mu_1 - s_1 c_2 c_6 m_x \lambda_6 \mu_2 \mu_1 + s_1 c_2 c_6 n_x \mu_6 \mu_2 \mu_1 - s_1 c_2 s_6 l_x \mu_2 \mu_1$$
$$- s_1 s_2 c_6 m_y \lambda_6 \mu_2 + s_1 s_2 c_6 n_y \mu_6 \mu_2 - s_1 s_2 s_6 l_y \mu_2 - s_1 c_6 m_x \lambda_6 \lambda_2 \mu_1 + s_1 c_6 n_x \mu_6 \lambda_2 \mu_1$$
$$- s_1 s_6 l_x \lambda_2 \mu_1 + c_2 c_6 m_z \lambda_6 \mu_2 \mu_1 - c_2 c_6 n_z \mu_6 \mu_2 \mu_1 + c_2 s_6 l_z \mu_2 \mu_1 + c_4 c_5 \lambda_5 \lambda_4 \mu_3$$
$$- c_4 \mu_5 \mu_4 \mu_3 - s_4 s_5 \lambda_5 \mu_3 + c_5 \lambda_5 \mu_4 \lambda_3 - c_6 m_z \lambda_6 \lambda_2 \mu_1 + c_6 n_z \mu_6 \lambda_2 \mu_1 - s_6 l_z \lambda_2 \mu_1$$
$$+ \mu_5 \lambda_4 \lambda_3$$

$$R_{3,3} = c_1 c_2 m_y \mu_6 \mu_2 \mu_1 + c_1 c_2 n_y \lambda_6 \mu_2 \mu_1 - c_1 s_2 m_x \mu_6 \mu_2 - c_1 s_2 n_x \lambda_6 \mu_2 + c_1 m_y \mu_6 \lambda_2 \mu_1$$
$$+ c_1 n_y \lambda_6 \lambda_2 \mu_1 - s_1 c_2 m_x \mu_6 \mu_2 \mu_1 - s_1 c_2 n_x \lambda_6 \mu_2 \mu_1$$
$$- s_1 s_2 m_y \mu_6 \mu_2 - s_1 s_2 n_y \lambda_6 \mu_2 - s_1 m_x \mu_6 \lambda_2 \mu_1$$
$$- s_1 n_x \lambda_6 \lambda_2 \mu_1 + c_2 m_z \mu_6 \mu_2 \mu_1$$
$$+ c_2 n_z \lambda_6 \mu_2 \mu_1 - c_4 c_5 \mu_5 \lambda_4 \mu_3 - c_4 \lambda_5 \mu_4 \mu_3 + s_4 s_5 \mu_5 \mu_3 - c_5 \mu_5 \lambda_4 \mu_3$$
$$- m_z \mu_6 \lambda_2 \mu_1 - n_z \lambda_6 \lambda_2 \mu_1 + \lambda_5 \lambda_4 \lambda_3$$

$$R_{3,4} = -d_1 c_2 \mu_2 \mu_1 + d_1 \lambda_2 \lambda_1 - c_1 c_2 d_6 m_y \mu_6 \mu_2 \mu_1 - c_1 c_2 d_6 n_y \lambda_6 \mu_2 \mu_1 - c_1 c_2 l_y a_6 \mu_2 \mu_1$$
$$+ c_1 c_2 p_y \mu_2 \mu_1 + c_1 s_2 d_6 m_x \mu_6 \mu_2 + c_1 s_2 d_6 n_x \lambda_6 \mu_2$$
$$+ c_1 s_2 l_x a_6 \mu_2 - c_1 s_2 p_x \mu_2 - c_1 d_6 m_y \mu_6 \lambda_2 \mu_1 - c_1 d_6 n_y \lambda_6 \lambda_2 \mu_1 - c_1 l_y a_6 \lambda_2 \mu_1$$
$$+ c_1 p_y \lambda_2 \mu_1 + s_1 c_2 d_6 m_x \mu_6 \mu_2 \mu_1 + s_1 c_2 d_6 n_x \lambda_6 \mu_2 \mu_1 + s_1 c_2 l_x a_6 \mu_2 \mu_1 - s_1 c_2 p_x \mu_2 \mu_1$$
$$+ s_1 s_2 d_6 m_y \mu_6 \mu_2 + s_1 s_2 d_6 n_y \lambda_6 \mu_2 + s_1 s_2 l_y a_6 \mu_2$$
$$- s_1 s_2 p_y \mu_2 + s_1 d_6 m_x \mu_6 \lambda_2 \mu_1 + s_1 d_6 n_x \lambda_6 \lambda_2 \mu_1 + s_1 l_x a_6 \lambda_2 \mu_1 - s_1 p_x \lambda_2 \mu_1 + d_2 \lambda_2$$
$$- c_2 d_6 m_z \mu_6 \mu_2 \mu_1 - c_2 d_6 n_z \lambda_6 \mu_2 \mu_1 - c_2 l_z a_6 \mu_2 \mu_1 + c_2 p_z \mu_2 \mu_1 + s_2 \mu_2 a_1 + d_3$$
$$+ d_4 \lambda_3 - c_4 d_5 \mu_4 \mu_3 + c_4 s_5 a_5 \lambda_4 \mu_3 + s_4 c_5 a_5 \mu_3 + s_4 a_4 \mu_3 + d_5 \lambda_4 \lambda_3 + s_5 a_5 \mu_4 \lambda_3$$
$$+ d_6 m_z \mu_6 \lambda_2 \mu_1 + d_6 n_z \lambda_6 \lambda_2 \mu_1 + l_z a_6 \lambda_2 \mu_1 - p_z \lambda_2 \mu_1$$

$$R_{4,j} = 0 \quad \text{for } j = 1, 4$$

$$S_{i,j} = 0 \quad \text{for } 1 \le i \le 3, \ 1 \le j \le 3$$

$$S_{1,4} = S_{4,1}$$

$$S_{2,4} = S_{4,2}$$

TABLE 10-2 (continued)

$S_{3,4} = S_{4,3}$

$S_{4,1} = d_1c_6l_z - d_1s_6m_z\lambda_6 + d_1s_6n_z\mu_6 - c_1d_2c_6l_y\mu_1 + c_1d_2s_6m_y\lambda_6\mu_1 - c_1d_2s_6n_y\mu_6\mu_1$
$\quad + c_1c_2c_6l_xa_2 - c_1c_2s_6m_x\lambda_6a_2 + c_1c_2s_6n_x\mu_6a_2 + c_1s_2c_6l_ya_2\lambda_1 - c_1s_2s_6m_y\lambda_6a_2\lambda_1$
$\quad + c_1s_2s_6n_y\mu_6a_2\lambda_1 + c_1c_6l_xa_1 - c_1s_6m_x\lambda_6a_1 + c_1s_6n_x\mu_6a_1 + s_1d_2c_6l_x\mu_1$
$\quad - s_1d_2s_6m_x\lambda_6\mu_1 + s_1d_2s_6n_x\mu_6\mu_1 + s_1c_2c_6l_ya_2 - s_1c_2s_6m_y\lambda_6a_2 + s_1c_2s_6n_y\mu_6a_2$
$\quad - s_1s_2c_6l_xa_2\lambda_1 + s_1s_2s_6m_x\lambda_6a_2\lambda_1 - s_1s_2s_6n_x\mu_6a_2\lambda_1 + s_1c_6l_ya_1 - s_1s_6m_y\lambda_6a_1$
$\quad + s_1s_6n_y\mu_6a_1 + d_2c_6l_z\lambda_1 - d_2s_6m_z\lambda_6\lambda_1 + d_2s_6n_z\mu_6\lambda_1 + s_2c_6l_za_2\mu_1$
$\quad - s_2s_6m_z\lambda_6a_2\mu_1 + s_2s_6n_z\mu_6a_2\mu_1 + d_3c_4s_5\lambda_4\mu_3$
$\quad + d_3s_4c_5\mu_3 + d_3s_5\mu_4\lambda_3 + d_4s_5\mu_4 + c_4c_5a_3$
$\quad - s_4s_5\lambda_4a_3 + c_5a_4 - c_6l_xp_x - c_6l_yp_y - c_6l_zp_z + c_6a_6 + s_6m_xp_x\lambda_6 - s_6n_xp_x\mu_6$
$\quad + s_6m_yp_y\lambda_6 - s_6n_yp_y\mu_6 + s_6m_zp_z\lambda_6 - s_6n_zp_z\mu_6 + a_5$

$S_{4,2} = d_1c_6m_z\lambda_6 - d_1c_6n_z\mu_6 + d_1s_6l_z - c_1d_2c_6m_y\lambda_6\mu_1 + c_1d_2c_6n_y\mu_6\mu_1 - c_1d_2s_6l_y\mu_1$
$\quad + c_1c_2c_6m_x\lambda_6a_2 - c_1c_2c_6n_x\mu_6a_2 + c_1c_2s_6l_xa_2 + c_1s_2c_6m_y\lambda_6a_2\lambda_1 - c_1s_2c_6n_y\mu_6a_2\lambda_1$
$\quad + c_1s_2s_6l_ya_2\lambda_1 + c_1c_6m_x\lambda_6a_1 - c_1c_6n_x\mu_6a_1 + c_1s_6l_xa_1 + s_1d_2c_6m_x\lambda_6\mu_1$
$\quad - s_1d_2c_6n_x\mu_6\mu_1 + s_1d_2s_6l_x\mu_1 + s_1c_2c_6m_y\lambda_6a_2 - s_1c_2c_6n_y\mu_6a_2 + s_1c_2s_6l_ya_2$
$\quad - s_1s_2c_6m_x\lambda_6a_2\lambda_1 + s_1s_2c_6n_x\mu_6a_2\lambda_1 - s_1s_2s_6l_xa_2\lambda_1 + s_1c_6m_y\lambda_6a_1 - s_1c_6n_y\mu_6a_1$
$\quad + s_1s_6l_ya_1 + d_2c_6m_z\lambda_6\lambda_1 - d_2c_6n_z\mu_6\lambda_1 + d_2s_6l_z\lambda_1 + s_2c_6m_z\lambda_6a_2\mu_1$
$\quad - s_1c_6n_z\mu_6a_2\mu_1 + s_2s_6l_za_2\mu_1 + d_3c_4c_5\lambda_5\mu_4\mu_3 - d_3c_4\mu_5\mu_4\mu_3 - d_3s_4s_5\lambda_5\mu_3$
$\quad + d_3c_5\lambda_5\mu_4\lambda_3 + d_3\mu_5\lambda_4\lambda_3 + d_4c_5\lambda_5\mu_4 + d_4\mu_5\lambda_4 - c_4s_5\lambda_5a_3 - s_4c_5\lambda_5\lambda_4a_3$
$\quad + s_4\mu_5\mu_4a_3 + d_5\mu_5 - s_5\lambda_5a_4 - c_6m_xp_x\lambda_6 + c_6n_xp_x\mu_6 - c_6m_yp_y\lambda_6$
$\quad + c_6n_yp_y\mu_6 - c_6m_zp_z\lambda_6 + c_6n_zp_z\mu_6 - s_6l_xp_x - s_6l_yp_y - s_6l_zp_z + s_6a_6$

$S_{4,3} = d_1m_z\mu_6 + d_1n_z\lambda_6 - c_1d_2m_y\mu_6\mu_1 - c_1d_2n_y\lambda_6\mu_1 + c_1c_2m_x\mu_6a_2 + c_1c_2n_x\lambda_6a_2$
$\quad + c_1s_2m_y\mu_6a_2\lambda_1 + c_1s_2n_y\lambda_6a_2\lambda_1 + c_1m_x\mu_6a_1 + c_1n_x\lambda_6a_1 + s_1d_2m_x\mu_6\mu_1$
$\quad + s_1d_2n_x\lambda_6\mu_1 + s_1c_2m_y\mu_6a_2 + s_1c_2n_y\lambda_6a_2 - s_1s_2m_x\mu_6a_2\lambda_1 - s_1s_2n_x\lambda_6a_2\lambda_1$
$\quad + s_1m_y\mu_6a_1 + s_1n_y\lambda_6a_1 + d_2m_z\mu_6\lambda_1 + d_2n_z\lambda_6\lambda_1 + s_2m_z\mu_6a_2\mu_1 + s_2n_z\lambda_6a_2\mu_1$
$\quad - d_3c_4c_5\mu_5\lambda_4\mu_3 - d_3c_4\lambda_5\mu_4\mu_3 + d_3s_4s_5\mu_5\mu_3 - d_3c_5\mu_5\mu_4\lambda_3 + d_3\lambda_5\lambda_4\lambda_3$
$\quad - d_4c_5\mu_5\mu_4 + d_4\lambda_5\lambda_4 + c_4s_5\mu_5a_3 + s_4c_5\mu_5\lambda_4a_3 + s_4\lambda_5\mu_4a_3 + d_5\lambda_5$
$\quad + s_5\mu_5a_4 + d_6 - m_xp_x\mu_6 - n_xp_x\lambda_6 - m_yp_y\mu_6 - n_yp_y\lambda_6 - m_zp_z\mu_6 - n_zp_z\lambda_6$

$S_{4,4} = - d_1^2 - 2d_1d_2\lambda_1 - 2d_1s_2a_2\mu_1 - 2d_1d_6m_z\mu_6 - 2d_1d_6n_z\lambda_6 - 2d_1l_za_6 + 2d_1p_z$
$\quad + 2c_1d_2d_6m_y\mu_6\mu_1 + 2c_1d_2d_6n_y\lambda_6\mu_1 + 2c_1d_2l_ya_6\mu_1 - 2c_1d_2p_y\mu_1$
$\quad - 2c_1c_2d_6m_x\mu_6a_1 - 2c_1c_2d_6n_x\lambda_6a_2 - 2c_1c_2l_xa_6a_2 + 2c_1c_2p_xa_2$
$\quad - 2c_1s_2d_6m_y\mu_6a_2\lambda_1 - 2c_1s_2d_6n_y\lambda_6a_2\lambda_1 - 2c_1s_2l_ya_6a_2\lambda_1 + 2c_1s_2p_ya_2\lambda_1$
$\quad - 2c_1d_6m_x\mu_6a_1 - 2c_1d_6n_x\lambda_6a_1 - 2c_1l_xa_6a_1 + 2c_1p_xa_1 - 2s_1d_2d_6m_x\mu_6\mu_1$
$\quad - 2s_1d_2d_6n_x\lambda_6\mu_1 - 2s_1d_2l_xa_6\mu_1 + 2s_1d_2p_x\mu_1 - 2s_1c_2d_6m_y\mu_6a_2 - 2s_1c_2d_6n_y\lambda_6a_2$
$\quad - 2s_1c_2l_ya_6a_2 + 2s_1c_2p_ya_2 + 2s_1s_2d_6m_x\mu_6a_2\lambda_1 + 2s_1s_2d_6n_x\lambda_6a_2\lambda_1$
$\quad + 2s_1s_2l_xa_6a_2\lambda_1 - 2s_1s_2p_xa_2\lambda_1 - 2s_1d_6m_y\mu_6a_1 - 2s_1d_6n_y\lambda_6a_1 - 2s_1l_ya_6a_1$
$\quad + 2s_1p_ya_1 - d_2^2 - 2d_2d_6m_z\mu_6\lambda_1 - 2d_2d_6n_z\lambda_6\lambda_1 - 2d_2l_za_6\lambda_1 + 2d_2p_z\lambda_1$
$\quad - 2c_2a_2a_1 - 2s_2d_6m_z\mu_6a_2\mu_1 - 2s_2d_6n_z\lambda_6a_2\mu_1 - 2s_2l_za_6a_2\mu_1$
$\quad + 2s_2p_za_2\mu_1 + d_3^2 + 2d_3d_4\lambda_3 - 2d_3c_4d_5\mu_4\mu_3$
$\quad + 2d_3c_4s_5a_5\lambda_4\mu_3 + 2d_3s_4c_5a_5\mu_3 + 2d_3s_4a_4\mu_3 + 2d_3d_5\lambda_4\lambda_3 + 2d_3s_5a_5\mu_4\lambda_3 + d_4^2$
$\quad + 2d_4d_5\lambda_4 + 2d_4s_5a_5\mu_4 + 2c_4c_5a_5a_3 + 2c_4a_4a_3 + 2s_4d_5\mu_4a_3 - 2s_4s_5a_5\lambda_4a_3 + d_5^2$
$\quad + 2c_5a_5a_4 - d_6^2 + 2d_6m_xp_x\mu_6 + 2d_6n_xp_x\lambda_6 + 2d_6m_yp_y\mu_6 + 2d_6n_yp_y\lambda_6$
$\quad + 2d_6m_zp_z\mu_6 + 2d_6n_zp_z\lambda_6 + 2l_xp_xa_6 - p_x^2 + 2l_yp_ya_6 - p_y^2 + 2l_zp_za_6 - p_z^2 - a_6^2$
$\quad + a_5^2 + a_4^2 + a_3^2 - a_2^2 - a_1^2$

^{a} $R = P - Q$ and $S = P^TP - Q^TQ$. See Section 10-4 for details.

patterns for us involve "degree." Let me summarize some of my own observations on these equations by noting two important facts:

- *First fact:*

$$
\text{degree } (P - Q) = \begin{bmatrix} 3 & 3 & 3 & 3 \\ 3 & 3 & 3 & 2 \\ 3 & 3 & 3 & 2 \\ 0 & 0 & 0 & 0 \end{bmatrix}
\tag{10-24}
$$

$$
\text{degree } (P^T P - Q^T Q) = \begin{bmatrix} 0 & 0 & 0 & 3 \\ 0 & 0 & 0 & 3 \\ 0 & 0 & 0 & 2 \\ 3 & 3 & 2 & 2 \end{bmatrix}
\tag{10-25}
$$

- *Second fact:* Although all the variables $c_1, s_1, \ldots, c_6, s_6$ show up, they appear only in certain combinations.

Let me expand on the second fact. In the cubic entries of $P - Q$ we see only products of variables of the form

$$x_3 y_4 z_5 \quad \text{and} \quad x_1 y_2 z_6,$$

where x, y, and z may be "c" or "s" or "1." (Here a subscripted 1 denotes "1.") Thus $c_3 s_4 s_5$, $s_3 c_4$, $c_3 c_5$, c_3 and $c_1 c_2 c_6$, $c_1 s_2 s_6$, $s_1 s_2$, c_6 all occur but not $c_3 c_6$ or $s_1 c_5$. For the quadric entries we see only

$$x_4 y_5 \quad \text{and} \quad x_1 y_2.$$

Similarly, in the cubic entries of $P^T P - Q^T Q$ we get only the following products of variables:

$$x_4 y_5 \quad \text{and} \quad x_1 y_2 z_6$$

and in the quadric entries only

$$x_4 y_5 \quad \text{and} \quad x_1 y_2.$$

Thus we can pick out four quadric equations in the variables $c_1, s_1, c_2, s_2,$ c_4, s_4, c_5, s_5 (two from $P - Q = 0$ and two from $P^T P - Q^T Q = 0$). Further, these quadrics have products of variables in only the following forms: $c_1 c_2$, $c_1 s_2$, $s_1 c_2$, $s_1 s_2$, $c_4 c_5$, $c_4 s_5$, $s_4 c_5$, and $s_4 s_5$. These four quadrics together with (10-21) for $i = 1, 2, 4, 5$ make a system of eight quadrics in eight unknowns whose solutions will yield $\theta_1, \theta_2, \theta_4,$ and θ_5. We can generate θ_3 and θ_6 with a supplementary calculation. (Implicit here is that the resulting system consists of independent equations.)

Equation (10-22) is not the only one that will yield such a system of quadrics. In Table 10-3 I have listed six variants. The purpose of this table is to indicate which corollary equations will lead to reasonable reductions of (10-19). Further, this table suggests general patterns in the way expressions in the A_j behave and offers insights into exploring further alternatives.

All the Table 10-3 relations have the degree structure shown in (10-24) and (10-25). We have

$$P - Q = 0 \quad \text{and} \quad P^T P - Q^T Q = 0,$$

where P and Q are the products of A_j indicated by the sequences of indices listed under "P" and "Q." For example, line 1 lists P as "123" and Q as "0654." This means that $P = A_1 A_2 A_3$ and $Q = A_0 A_6^{-1} A_5^{-1} A_4^{-1}$. Equation (10-22) is listed in line 3 as "345" under P and "2106" under Q. Each equation "$P - Q$" is a corollary to (10-19) obtained by multiplying (10-19) by various A_j or A_j^{-1}. (Note that $P^T P - Q^T Q$ is symmetric, the nonzero column equaling the transpose of the nonzero row.)

The terms in the cubic and quadric parts of $P - Q$ and $P^T P - Q^T Q$ are always restricted to certain products of variables, indicated in the other columns of Table 10-3. For example, line 1 shows "123;654" for the cubic part of $P - Q$ and "23;65" for the quadric. This means that the only products of variable that may appear with nonzero coefficients in the entries of $P - Q$ are $x_1 y_2 z_3$ and $x_4 x_5 x_6$ and only $x_2 y_3$ and $x_5 y_6$ may appear with nonzero coefficients in the entries of $P - Q$ denoted as quadric, where x, y, z may be "c," "s," or "1." The "23;654" and "23;65" under $P^T P - Q^T Q$ have the same meaning.

Focusing on the quadric parts of $P - Q$ and $P^T P - Q^T Q$, we see that each line yields four quadrics in the sines and cosines of the four angles. Thus these quadrics with the appropriate four equations from (10-21) give a system of eight quadrics in eight unknowns, just as does (10-22). These systems are valid reductions of (10-19), and (10-21). They reduce a system

TABLE 10-3   Products of variables in various corollary relations to (10-19)[a]

	P	Q	$P - Q$ Cubic	$P - Q$ Quadric	$P^T P - Q^T Q$ Cubic	$P^T P - Q^T Q$ Quadric
1.	123	0654	123;654	23;65	23;654	23;65
2.	234	1065	234;165	34;16	34;165	34;16
3.	345	2106	345;216	45;21	45;216	45;21
4.	4560	321	456;321	56;32	56;321	56;32
5.	5601	432	561;432	61;43	61;432	61;43
6.	6012	543	612;543	12;54	12;543	12;54

[a] SYS_1, SYS_2, and SYS_3 are generated by the quadric parts of lines 1, 2, and 3, respectively. See the text for more details.

of total degree over 2 million, (10-19) with (10-21), to one of total degree 256. When we solve one of these systems, we obtain four of the six angles we are seeking. Then the other two angles follow easily. Unfortunately, lines 1 and 4, 2 and 5, and 3 and 6 of Table 10-3 are algebraically equivalent, so we gain nothing by combining them.

It will be convenient to name the quadric systems generated by the first three lines of Table 10-3 as SYS_1, SYS_2, and SYS_3, respectively. Thus the third line yields the system of eight second-degree equations

$$R_{3,3} = 0$$

$$R_{3,4} = 0$$

$$S_{4,3} = 0 \qquad\qquad (SYS_3)$$

$$S_{4,4} = 0$$

$$c_j^2 + s_j^2 - 1 = 0 \qquad \text{for } j = 1, 2, 4, 5$$

in the eight unknowns c_1, s_1, c_2, s_2, c_4, s_4, c_5, and s_5, where these are the $R_{i,j}$, $S_{i,j}$ listed in Table 10-2. SYS_1 and SYS_2 are analogously defined from lines 1 and 2 of Table 10-3, respectively.

Shortly, we will pick one of these quadric reduced systems and compute some robot configurations. I would like to pause here to present an apology (in the ancient sense) for not just showing you this single reduction and getting on with the computations. There are two reasons. First, you need some insight into how reductions are generated to solve related problems (e.g., the IPP for five-degree-of freedom manipulators). Second, the state of this reduction (for six-degree-of-freedom manipulators) is not entirely satisfactory. The total degree of the reduced system is still too high. As we will see below, there is evidence that a practical lower-degree reduction ought to exist. I have presented the general material here in the hope that someone will be inspired to find a better reduction.†

The reduced system we will now focus on is SYS_3. This system tends to be not too badly scaled in practice, but we will use SCLGEN routinely in the runs that follow.

Before considering explicit examples, let us discuss the structure of the solution set of SYS_3. (These comments also apply to SYS_1 and SYS_2.) The most obnoxious characteristic is that SYS_3 has an infinite number of solutions at infinity. Thus we cannot make use of the strong form of Bezout's theorem. The second most important fact is that SYS_3 generally has only a few finite solutions. It is shown in [Duffy and Crane, 1980] using a different method of analysis that (10-19) can have no more than 32 finite solutions. Our reduction might double this number, because the squaring implicit in $P^T P - Q^T Q = 0$ generates "extraneous" solutions that we must eliminate

---

† Some recent progress is given in [Morgan and Sommese, 1986, 1987].

later. However, in my computations, I have never encountered more than 32 finite solutions (including extraneous).

The significance for CONSOL is at least $256 - 32 = 224$ paths diverge to infinity. We can use this expectation of 32 solutions as a rule of thumb (i.e., be suspicious if 225 paths diverge), but with the caution that it is a purely empirical observation. Of course, degenerate cases will have fewer finite solutions (e.g., 16, with 8 of these extraneous) or even an infinite number of finite solutions (e.g., where the rotation of a joint in a particular configuration does not change the hand position). Some finite solutions may have nonzero imaginary parts but only the real solutions are physically meaningful.

We should also consider on-line reduction. The reduction of (10-19) and (10-21) to SYS_3 is a preliminary, "off-line" reduction. It is fully general, and therefore it subsumes much algebraic complexity. However, in practice, we will choose particular numbers for $\alpha_1, \ldots, \alpha_6, d_1, \ldots, d_6, a_1, \ldots, a_6$, and the entries of A_0. The resulting coefficients may not be so complicated, especially because in practice the twist angles, $\alpha_1, \ldots, \alpha_6$, are often integer multiples of 0° or 90°. Thus if we reduce the system after the coefficients have been generated, we can maintain generality while improving efficiency. We can arrange to do this "by hand" or "automatically." The idea of classifying systems by twist angles is especially suited to this problem.

The following three examples define manipulators of increasing complexity. They are taken from [Tsai and Morgan, 1985]. The results of solving them using CONSOL8 are summarized in the various tables cited. In the experiments section I will suggest that you solve them yourself to get a feeling for the computational characteristics of the problems. *Warning*: Be prepared for long run times. Each problem requires the tracking of 256 paths, most of which go to singular solutions at infinity. Thus, without the projective transformation, we have to contend with at least 224 diverging paths. On the other hand, with the projective transformation, we will find 224 singular solutions at infinity. The most efficient way to solve these systems is to limit MAXNS (the maximum number of steps), in essence truncating paths going to infinity, whether or not the projective transformation is used (see Experiments 10-1 and 10-2). Much CPU time will be saved if you avoid CONSOL8T and write a specialized FFUN subroutine.

The hand position and orientation for all the examples are given in Table 10-4. For each of the three examples, there are three associated tables. Thus for Example 1, Table 10-5 gives the linkage parameters that define the manipulator, Table 10-6 summarizes the CONSOL output in terms of how many solutions were found in each of the three categories: real-significant, complex-significant, and extraneous. Table 10-7 lists the real-significant solutions that were found. Solutions to SYS_3 [i.e., equations (10-22) and (10-23) with (10-21)] that are not solutions to (10-19) with (10-21) are called

TABLE 10-4   Hand position and orientation for manipulator examples[a]

Given Vectors	x-Component	y-Component	z-Component
p	0.22441776	0.71549788	0.79551628
l	−0.71511545	0.65150320	0.25328538
m	−0.69899036	−0.66895464	−0.25280857
n	0.00473084	−0.35783135	0.93377425

[a] The notation is from equation (10–20).

TABLE 10-5   Linkage parameters for Example 1[a]

i	a_i (units)	d_i (units)	α_i (degrees)
1	0.0	0.0	−90.0
2	1.0	0.345	0.0
3	−0.047	0.0	90.0
4	0.0	1.0	−90.0
5	0.0	0.0	90.0
6	0.0	0.130	0.0

[a] The notation is from equation (10–18).

TABLE 10-6   Total number of solutions
found for Example 1

Real significant	8
Complex significant	0
Extraneous	8
Total	16

TABLE 10-7   Real solutions found for Example 1 (Joint displacements in degrees)

No.	θ_1	θ_2	θ_3	θ_4	θ_5	θ_6
1	−80.62	−76.06	−28.47	176.23	−125.23	34.71
2	−80.62	−76.06	−28.47	−3.77	125.33	−145.29
3	−80.62	162.66	−146.15	−36.03	5.23	−107.20
4	−80.62	162.66	−146.15	143.97	−5.23	72.80
5	47.89	−103.94	−146.15	−162.84	124.38	−84.24
6	47.89	−103.94	−146.15	17.16	−124.38	95.76
7	47.89	17.34	−28.47	−107.55	14.80	−166.02
8	47.89	17.34	−28.47	72.45	−14.80	13.98

TABLE 10–8 Linkage parameters for Example 2[a]

i	a_i (units)	d_i (units)	α_i (degrees)
1	0.45	0.5	80.0
2	0.55	0.6	93.0
3	0.75	0.4	120.0
4	0.75	1.0	120.0
5	0.55	0.4	93.0
6	0.45	0.6	80.0

[a] The notation is from equation (10–18).

TABLE 10–9 Total number of solutions
found for Example 2

Real significant	6
Complex significant	10
Extraneous	16
Total	32

TABLE 10–10 Real solutions found for Example 2 (Joint displacements in degrees)

No.	θ_1	θ_2	θ_3	θ_4	θ_5	θ_6
1	− 146.88	170.87	− 11.22	− 25.99	− 108.51	60.82
2	− 162.72	− 173.52	128.00	− 179.64	− 3.12	179.99
3	21.50	135.15	− 104.31	64.39	− 89.40	77.38
4	63.74	− 47.27	− 172.43	− 114.49	− 50.04	− 11.94
5	17.31	19.31	42.89	− 164.02	29.10	− 17.23
6	26.20	6.88	− 62.10	− 45.96	− 130.25	− 129.34

TABLE 10–11 Linkage parameters for Example 3[a]

i	a_i (units)	d_i (units)	α_i (degrees)
1	0.50	0.1875	80.0
2	1.00	0.375	15.0
3	0.125	0.25	120.0
4	0.625	0.875	75.0
5	0.3125	0.5	100.0
6	0.25	0.125	60.0

[a] The notation is from equation (10–18).

TABLE 10–12   Total number of solutions
found for Example 3

Real significant	12
Complex significant	4
Extraneous	16
Total	32

TABLE 10–13   Real solutions found for Example 3 (Joint displacements in degrees)

No.	θ_1	θ_2	θ_3	θ_4	θ_5	θ_6
1	167.68	83.55	168.07	65.84	-88.67	-44.77
2	-143.00	100.07	131.85	18.46	-59.49	-71.52
3	115.86	-168.65	-66.22	157.17	-111.41	156.71
4	107.56	2.00	-111.47	166.77	-173.54	-105.56
5	-106.07	-140.86	22.07	-161.28	35.54	134.45
6	-65.37	142.24	56.06	-70.90	-51.63	-116.13
7	120.52	31.27	-143.03	114.15	-143.62	-64.39
8	7.75	103.87	-113.21	-21.37	-79.90	82.26
9	-16.69	97.90	-25.97	-80.98	-25.72	-3.44
10	47.26	163.44	-119.49	28.32	-41.13	81.08
11	20.93	58.74	-125.17	-27.07	-125.66	106.21
12	38.93	-56.45	-149.20	12.28	72.23	67.43

"extraneous." We expect extraneous solutions because of the multiplicative formula used to generate (10-23). The other solutions are labeled "significant." Note that θ_3 and θ_6 are found by back substitution after θ_1, θ_2, θ_4, and θ_5 are computed.

Example 1: Manipulator with the Last Three Joint Axes Intersecting. In this case the system partially decouples. We notice that there are four different configurations of the upper arm (θ_1, θ_2, and θ_3) and there are two different configurations for the wrist joints (θ_4, θ_5, and θ_6), corresponding to each configuration of the upper arm.

Example 2: Symmetrical Manipulator. This manipulator is taken from the first example of [Duffy and Crane, 1980] except for a scale factor of 2. It is symmetric about the fourth joint axis. See Tables 10-8 through 10-10.

Example 3: General Manipulator. The linkage parameters given in Table 10-11 represent a manipulator with arbitrary linkage proportions. See also Tables 10-12 and 10-13.

EXERCISES

10-6. Five-degree-of-freedom revolute joint manipulators (5-R manipulators).
   (a) If a manipulator has five joints, we can express the IPP by the equation

$$A_1 A_2 A_3 A_4 A_5 = A_0, \qquad (10\text{-}26)$$

   in analogy with (10-19). Show that (10-26) can be reduced to two quadrics
   in c_2, s_2, c_3, s_3. Thus this IPP can be reduced to solving four quadrics
   in four unknowns, a total degree 16 system.
   (b) A 5-R manipulator cannot be put in an arbitrary specified position and
   orientation. That is, it can attain only certain positions and orientations.
   In industrial applications, the needed positions and orientations will be
   known for particular jobs, and sometimes a 5-R manipulator is sufficient.
   The best that can be done in general with a 5-R manipulator is to specify
   the position and the orientation of one axis of the hand, as follows. Let's
   define the "weak IPP" to be: given a desired hand position, (x_0, y_0, z_0),
   and a unit direction for the z-axis of the hand, (u_x, u_y, u_z), find the θ_1,
   $\ldots$, θ_5 so that

$$A_1 A_2 A_3 A_4 A_5 P_0 = P_1 \qquad (10\text{-}27)$$

   and

$$A_1 A_2 A_3 A_4 A_5 U_0 = U_1, \qquad (10\text{-}28)$$

   where

$$P_1 = \begin{bmatrix} x_0 \\ y_0 \\ z_0 \\ 1 \end{bmatrix}, \qquad P_0 = \begin{bmatrix} 0 \\ 0 \\ 0 \\ 1 \end{bmatrix}$$

$$U_1 = \begin{bmatrix} u_x \\ u_y \\ u_z \\ 0 \end{bmatrix}, \qquad U_0 = \begin{bmatrix} 0 \\ 0 \\ 1 \\ 0 \end{bmatrix}.$$

   Now, a 5-R manipulator can generally satisfy the weak IPP, unless the
   position is out of reach. Further, the IPP for 6-R manipulators can be
   reduced mathematically to the weak IPP for 5-R manipulators. That is,
   given A_0 for the IPP, there is an appropriate choice of P_1 and U_1 so that
   the solutions to the weak IPP, θ_1, θ_2, θ_3, θ_4, θ_5, yield the solutions of
   (10-19). Explain by a heuristic argument the reasonableness of this fact.
   Establish it by an algebraic derivation (see [Tsai and Morgan, 1985]).

10-7. Describe a way of reducing (10-19) and (10-21) to a system of six equations
   in six unknowns, of degrees 3, 3, 2, 2, 2, and 6. (*Hint*: Consider

$$A_1 A_2 A_3 A_4 = A_0 A_6^{-1} A_5^{-1}.)$$

10-8. Show that Example 1 can be solved by elementary methods, that is, by decoupling the system and reducing it to low degree polynomials in one variable.

10-9. In Example 1, the twist angles are

$$\alpha_1 = -90°$$

$$\alpha_2 = 0°$$

$$\alpha_3 = 90°$$

$$\alpha_4 = -90°$$

$$\alpha_5 = 90°$$

$$\alpha_6 = 0°.$$

Show that any system with these twist angles can be reduced to four quadrics in four unknowns. Can any system with these twist angles be solved by elementary methods?

EXPERIMENTS

10-1. Solve Example 1 using CONSOL8 with a smaller and smaller MAXNS, "maximum number of steps for a path," until you lose a solution. How does the run time change? If you had chosen the MAXNS that lost you a solution in the first place, how would you have detected that you had missed a solution? Do this also for Examples 2 and 3.

10-2. Replace SYS_3 by the projective transformation of SYS_3 (described in Chapter 3) and do Experiment 10-1.

PROJECTS

10-1. We say that a manipulator is "solvable in closed form" if its IPP can be reduced (for any A_0) to solving low degree polynomials in one variable. Characterize those systems that can be solved in closed form in terms of the twist angles, $\alpha_1, \ldots, \alpha_6$. (Exercises 10-8 and 10-9 are relevant. Part of the project is to define "low degree.") See [Peiper and Roth, 1969], [Duffy and Rooney, 1975], and [Duffy, 1980].

10-2. Explore using CONSOL8 how the number of solutions in the categories "real-significant," "complex-significant," and "extraneous" change as you vary the twist angles $\alpha_1, \ldots, \alpha_6$. Can you get 16 real-significant solutions? More than 16? (See [Roth et al., 1973] and [Duffy and Crane, 1980].)

Appendices

APPENDIX 1

Newton's Method

A1-1 INTRODUCTION

Newton's method is a powerful local solution method for smooth nonlinear systems, and it is the basic building block for all the CONSOL codes in this book. It functions in CONSOL as the "corrector" in the predictor–corrector path-tracking technique (see Section 4-5-1).

In this appendix I describe Newton's method for systems, discuss implementation considerations, and summarize the typical modes of local and global behavior that one observes in applying Newton's method to polynomial systems. (Newton's method for a single equation in one unknown is a special case.)

For linguistic simplicity I refer to "Newton's method" as "Newton" when convenient. If z^* is a solution to a system, $f(z) = 0$, then "z^* is nonsingular" if det $(df(z^*)) \neq 0$, where df denotes the Jacobian of f. Otherwise, z^* is singular (see Chapters 2 and 3).

A1-2 LOCAL BEHAVIOR

Newton works like this. Suppose that $f(z) = 0$ is a nonlinear system. Given a start point, z^0, Newton generates a sequence $z^1, z^2, z^3, \ldots$ of points. If z^0 started out close to a nonsingular solution, z^*, of $f(z) = 0$, then the z^1, $z^2, z^3, \ldots$ converge to z^*. In this sense, Newton is a "local method" because it improves solution approximations "locally." If z^0 is not close to a solution, the Newton iterates may converge to some solution, but they may not. The "global" behavior of Newton's method is difficult to predict. Two practical issues are:

(a) How can start points be chosen to guarantee convergence? (There is no precise practical answer to this question.)
(b) What happens when z^* is a singular solution? (Rule of thumb: Newton converges but more slowly than if z^* is nonsingular, and it finds fewer significant digits. However, "exceptional" bad behavior is possible.)

We will consider these issues further below.

Here is the formula for the Newton's method sequence:

$$z^{k+1} = z^k - \text{resid}^k, \qquad (A1\text{-}1)$$

where

$$\text{resid}^k = [df(z^k)]^{-1}f(z^k). \qquad (A1\text{-}2)$$

The formulas above tell us how to update z^k to z^{k+1} when we know

1. z^k, the previous Newton iterate
2. $f(z^k)$, the value of f at z^k, viewed as a column vector
3. $df(z^k)$, the Jacobian matrix of f evaluated at z^k

The vector "resid^k" is called the "residual" at the kth step. For polynomial systems, we can implement Newton in real or complex arithmetic. The defining formulas all make sense either way.

For example, if f is two equations in the two unknowns x and y, then $z = (x, y)$ and f is the column vector:

$$f(z) = f(x, y) = \begin{bmatrix} f_1(x, y) \\ f_2(x, y) \end{bmatrix}$$

and the Jacobian of f is

$$df(z) = df(x, y) = \begin{bmatrix} \dfrac{\partial f_1}{\partial x}(x, y) & \dfrac{\partial f_1}{\partial y}(x, y) \\ \dfrac{\partial f_2}{\partial x}(x, y) & \dfrac{\partial f_2}{\partial y}(x, y) \end{bmatrix}.$$

Now $[df(z^k)]^{-1}$ denotes the matrix inverse of $df(z^k)$ and $[df(z^k)]^{-1}f(z^k)$ denotes the matrix-vector product. Thus, taking

$$df = \begin{bmatrix} f_{1,1} & f_{1,2} \\ f_{2,1} & f_{2,2} \end{bmatrix},$$

we have

$$df^{-1} = \frac{1}{\det} \begin{bmatrix} f_{2,2} & -f_{1,2} \\ -f_{2,1} & f_{1,1} \end{bmatrix},$$

where $\det = f_{1,1}f_{2,2} - f_{1,2}f_{2,1}$ and

$$df^{-1}f = \frac{1}{\det} \begin{bmatrix} f_{2,2}f_1 - f_{1,2}f_2 \\ -f_{2,1}f_1 + f_{1,1}f_2 \end{bmatrix}.$$

Then the Newton update formula (A1-1) is

$$\begin{bmatrix} x^{k+1} \\ y^{k+1} \end{bmatrix} = \begin{bmatrix} x^k - \dfrac{1}{\det}(f_{2,2}f_1 - f_{1,2}f_2) \\ y^k - \dfrac{1}{\det}(-f_{2,1}f_1 + f_{1,1}f_2) \end{bmatrix},$$

where I have omitted the dependence of det, $f_{i,j}$ and f_i on z^k for notational simplicity.

For example, if $f_1 = x^2 + y^2 - 25$ and $f_2 = x^2 - y - 5$ [system (2-3)], the Newton update formula becomes

$$\begin{bmatrix} x^{k+1} \\ y^{k+1} \end{bmatrix} = \begin{bmatrix} \dfrac{x^k y^k + 0.5x^k + 0.5y^k y^k/x^k + 12.5/x^k + 5y^k/x^k}{1 + 2y^k} \\ \dfrac{y^k y^k + 20}{1 + 2y^k} \end{bmatrix}.$$

Using this formula, we evoke the Newton process as follows. We start with some $z^0 = (x^0, y^0)$ that we believe is close to a solution to $f(z) = 0$ and substitute x^0 and y^0 in the right-hand side, yielding x^1 and y^1, and then repeat the process, using x^1 and y^1 to get x^2 and y^2, and so on.

In one variable, the Newton update formula becomes

$$x^{k+1} = x^k - \frac{f(x^k)}{df(x^k)},$$

which is satisfactory as it stands. However, for systems of three or more variables, it is generally more efficient to use a linear solver to compute the residual in (A1-2) rather than generate the inverse. For systems of two variables, it is essentially a matter of convenience which approach we use. Thus (A1-2) can be rewritten as

$$df(z^k) \text{ resid}^k = f(z^k), \tag{A1-3}$$

and we can find resid^k by solving the linear system (A1-3) using an appropriate computer code without computing the matrix inverse of $df(z^k)$ (see Section 4-5-3).

The theory of Newton's method tells us that if z^0 is close enough to a solution, z^*, and df is not singular at z^*, then z^1, z^2, z^3, . . . will converge to z^*. (See [Kantorovich and Akilov, 1964], [Brown and Dennis, 1968], [Rall, 1974], [Kul'Chitskii and Shimelevich, 1974], [Rheinboldt, 1974], and [Ortega and Rheinboldt, 1970, Chap. 10].) In fact, convergence is proven to be "quadratic." This means that

$$| \text{resid}^{k+1} | \leq \gamma | \text{resid}^k | | \text{resid}^k | \qquad (A1\text{-}4)$$

for some constant $\gamma > 0$. Thus the residuals get small like the sequence

$$10^{-2}, 10^{-4}, 10^{-8}, 10^{-16}, \ldots$$

In other words, fast! In practice, z^0 is often not close enough to generate quadratic convergence at once, but (A1-4) will hold after some number of iterations have occurred. (See the computational example at the end of this Section.) However, proofs of convergence that estimate how close z^0 should be to z^* are conservative and of limited usefulness in practice. (See the references cited above.) In practice, Newton's method can work well even if z^0 is not "close" to z^*.

The Newton residual gives us an important approximation formula. Let z^* denote the actual solution that Newton is converging to (and that we do not know). Then we have

$$| z^k - z^* | \approx | \text{resid}^k | \qquad (A1\text{-}5)$$

when z^k is close enough to z^* and $df(z^*)$ is not singular. Equation (A1-5) comes from the Taylor's formula approximation

$$f(z^*) = f(z^k) + df(z^k) (z^k - z^*) + o(| z^k - z^* |^2).$$

We have $f(z^*) = 0$, since z^* is a solution, and we assume that $| z^k - z^* |$ is small enough so that $o(| z^k - z^* |^2)$ is negligible. By (A1-5) the Newton residual tells us how close we are to the solution. However, the practical catch is that we have to be "close enough" and "nonsingular." My experience is that when I see quadratic convergence [as in (A1-4)], then (A1-5) is valid. Further, if x^* is nonsingular, then z^k will eventually be accurate to the number of available digits. Thus, we can generally expect

$$| \text{resid}^k | \approx \epsilon_*$$

for sufficiently large k, where ϵ_* is defined by $1 + \epsilon_* \neq 1$ and $1 + \epsilon = 1$ for any positive ϵ smaller than ϵ_*.

If $df(z^*)$ is singular, then the local theory of Newton's method is much less complete and straightforward. Let's classify singular solutions according to two basic modes of Newton's method behavior. Thus we have: *nice singular solutions* and *nasty singular solutions*.

The behavior of Newton near a nice singular solution is linear convergence that attains half the available digits of accuracy; that is, if Newton is started at most points close enough† to a nice singular solution, then

$$| \text{resid}^{k+1} | \le \gamma | \text{resid}^k | \qquad \text{(A1-6)}$$

for some positive constant γ smaller than 1 (compare with (A1-4)) and eventually

$$| z^k - z^* | \approx \sqrt{\epsilon_*}$$

where ϵ_* is defined in the paragraph above. This sort of linear convergence is slow in comparison with quadratic convergence, and the loss of digits in the solution estimate can sometimes be important. However, often we can make do with this diminished performance of Newton near a nice singular solution, and sometimes we can even improve the performance by modifying the method (see [Grewank, 1985]).

The behavior of Newton near a nasty singular solution is more varied and less acceptable. What you might see is "arbitrarily slow convergence," where the Newton process seems to get stuck, or "cyclic behavior," where the iterates follow a repeating pattern. Other, less regular modes of behavior are possible. (However, even starting near a nasty singular solution, I have never observed "divergence to infinity," which is sometimes cited.) Problem set 4 in Section 4-7 includes a system with a nasty singular solution. I tend to label nasty solutions "exceptional," because I don't encounter them in practice.

The excellent survey [Grewank, 1985] gives general conditions under which it can be proven that singular solutions will be nice (see Theorem 2.1 and equation (3.8)). In [Grewank and Osborne, 1983] a collection of systems with nice and nasty singular solutions is analyzed, and a variety of possible types of Newton behavior is cataloged. (In the terminology of these papers, regular singular solutions are nice, and some irregular singular solutions are nice, but some are nasty.) See also [Rall, 1966], [Reddien, 1978], [Decker and Kelley, 1980-1, 1980-2, 1985], and the references cited in [Grewank, 1985]. These papers discuss special solution techniques for singular solutions and include arguments suggesting that some of the numerical difficulties of Newton at singular solutions will be shared generally by other numerical methods.

To summarize, Newton can be used to estimate nice singular solutions, but not nasty ones. However, its performance will be diminished from the nonsingular case. For now, I proceed with the assumption that the singular solutions I am going to encounter will be nice. If a nasty singular solution ever did come up, my first approach would be to try to revise the model (as in Chapter 7) rather than to modify the numerical method. This is consistent with the conclusions of [Grewank and Osborne, 1983].

---

† The good start points form a "starlike domain with density 1" rather than a ball (see [Grewank, 1985] Theorem 2.1).

TABLE A1–1 Newton iterates converging to the nonsingular solution $(X, Y) = (3, 4)$ of system (2–3) using a real start point[a]

IT	REAL(X)	IMAG(X)	REAL(Y)	IMAG(Y)	EQ	RESID	ERROR	DET
1	0.25D+01	0.0	0.35D+01	0.0	D+02	D+01	D+01	D+02
2	0.31D+01	0.0	0.40D+01	0.0	D+01	D+00	D+00	D+03
3	0.30D+01	0.0	0.40D+01	0.0	D−01	D−02	D−02	D+03
4	0.30D+01	0.0	0.40D+01	0.0	D−06	D−07	D−07	D+03
5	0.30D+01	0.0	0.40D+01	0.0	D−14	D−15	D−15	D+03
6	0.30D+01	0.0	0.40D+01	0.0	.	.	.	D+03
7	0.30D+01	0.0	0.40D+01	0.0	.	.	.	D+03
8	0.30D+01	0.0	0.40D+01	0.0	.	.	.	D+03
9	0.30D+01	0.0	0.40D+01	0.0	.	.	.	D+03
10	0.30D+01	0.0	0.40D+01	0.0	.	.	.	D+03

[a] The "D" indicates powers of 10. "IT" is the iteration number. "REAL(X)" and "IMAG(X)" are the real and imaginary parts of X, respectively, and similarly for "REAL(Y)" and "IMAG(Y)." "EQ" is an upper bound for the Euclidean norm of the equation values; thus "EQ = D+02" means "$\|(EQ(X, Y), EQ'(X, Y))\| \leq 10^2$" in the notation of Chapter 2. Similarly, "RESID," "ERROR," and "DET" give upper bounds for the norm of the Newton residual, for the norm of $(X, Y) − (3, 4)$, and for the absolute value of the determinant of the Jacobian at (X, Y), respectively. The dots alone in a column indicate that the number is zero.

TABLE A1-2 Newton iterates converging to nonsingular solution $(X, Y) = (3, 4)$ of system (2–3) using a complex start point[a]

IT	REAL(X)	IMAG(X)	REAL(Y)	IMAG(Y)	EQ	RESID	ERROR	DET
1	0.25D+01	0.50D+00	0.35D+01	0.50D+00	D+02	D+01	D+01	D+02
2	0.30D+01	-0.11D+00	0.40D+01	-0.62D-01	D+01	D+00	D+00	D+03
3	0.30D+01	0.78D-03	0.40D+01	0.10D-03	D-01	D-02	D-02	D+03
4	0.30D+01	-0.51D-06	0.40D+01	-0.92D-08	D-04	D-05	D-05	D+03
5	0.30D+01	-0.92D-13	0.40D+01	-0.37D-16	D-11	D-12	D-12	D+03
6	0.30D+01	-0.16D-27	0.40D+01	0.14D-29	D-15	D-16	D-15	D+03
7	0.30D+01	0.0	0.40D+01	0.12D-43	.	.	.	D+03
8	0.30D+01	0.0	0.40D+01	0.0	.	.	.	D+03
9	0.30D+01	0.0	0.40D+01	0.0	.	.	.	D+03
10	0.30D+01	0.0	0.40D+01	0.0	.	.	.	D+03

[a] See the footnote to Table A1-1.

TABLE A1–3 Newton iterates converging to singular solution $(X, Y) = (0, -5)$ of system (2–3) using a real start point[a]

IT	REAL(X)	IMAG(X)	REAL(Y)	IMAG(Y)	EQ	RESID	ERROR	DET
1	0.50D+00	0.0	-0.45D+01	0.0	D+01	D+01	D+01	D+02
2	0.22D+00	0.0	-0.50D+01	0.0	D+00	D+00	D+00	D+01
3	0.11D+00	0.0	-0.50D+01	0.0	D-01	D+00	D+00	D+01
4	0.55D-01	0.0	-0.50D+01	0.0	D-02	D-01	D+00	D+01
5	0.27D-01	0.0	-0.50D+01	0.0	D-02	D-01	D-01	D+00
6	0.14D-01	0.0	-0.50D+01	0.0	D-03	D-02	D-01	D+00
7	0.68D-02	0.0	-0.50D+01	0.0	D-03	D-02	D-02	D+00
8	0.34D-02	0.0	-0.50D+01	0.0	D-04	D-02	D-02	D+00
9	0.17D-02	0.0	-0.50D+01	0.0	D-05	D-03	D-02	D-01
10	0.85D-03	0.0	-0.50D+01	0.0	D-05	D-03	D-03	D-01
11	0.43D-03	0.0	-0.50D+01	0.0	D-06	D-03	D-03	D-01
12	0.21D-03	0.0	-0.50D+01	0.0	D-06	D-04	D-03	D-02
13	0.11D-03	0.0	-0.50D+01	0.0	D-07	D-04	D-03	D-02
14	0.53D-04	0.0	-0.50D+01	0.0	D-08	D-04	D-04	D-02
15	0.27D-04	0.0	-0.50D+01	0.0	D-08	D-04	D-04	D-03
16	0.13D-04	0.0	-0.50D+01	0.0	D-09	D-05	D-04	D-03
17	0.67D-05	0.0	-0.50D+01	0.0	D-09	D-05	D-05	D-03
18	0.33D-05	0.0	-0.50D+01	0.0	D-10	D-05	D-05	D-03
19	0.17D-05	0.0	-0.50D+01	0.0	D-11	D-06	D-05	D-04
20	0.83D-06	0.0	-0.50D+01	0.0	D-11	D-06	D-06	D-04
21	0.42D-06	0.0	-0.50D+01	0.0	D-12	D-06	D-06	D-04
22	0.21D-06	0.0	-0.50D+01	0.0	D-12	D-06	D-06	D-05
23	0.10D-06	0.0	-0.50D+01	0.0	D-13	D-06	D-06	D-05
24	0.52D-07	0.0	-0.50D+01	0.0	D-14	D-07	D-07	D-05
25	0.24D-07	0.0	-0.50D+01	0.0	D-14	D-07	D-07	D-06
26	0.81D-08	0.0	-0.50D-01	0.0	D-15	D-07	D-07	D-06
27	-0.70D-08	0.0	-0.50D+01	0.0	D-15	D-07	D-07	D-06
28	0.11D-07	0.0	-0.50D+01	0.0	D-15	D-07	D-07	D-06
29	-0.10D-08	0.0	-0.50D+01	0.0	D-15	D-06	D-08	D-07

30	0.12D−06	0.0	−0.50D+01	0.0	D−13	D−06	D−06	D−05
31	0.60D−07	0.0	−0.50D+01	0.0	D−13	D−07	D−06	D−05
32	0.28D−07	0.0	−0.50D+01	0.0	D−14	D−07	D−07	D−05
33	0.10D−07	0.0	−0.50D+01	0.0	D−15	D−07	D−07	D−06
34	−0.16D−08	0.0	−0.50D+01	0.0	D−15	D−06	D−08	D−07
35	0.76D−07	0.0	−0.50D+01	0.0	D−13	D−07	D−06	D−05
36	0.36D−07	0.0	−0.50D+01	0.0	D−14	D−07	D−07	D−05
37	0.17D−07	0.0	−0.50D+01	0.0	D−15	D−07	D−07	D−06
38	0.30D−08	0.0	−0.50D+01	0.0	D−15	D−07	D−08	D−06
39	−0.37D−07	0.0	−0.50D+01	0.0	D−14	D−07	D−07	D−05
40	−0.14D−07	0.0	−0.50D+01	0.0	D−15	D−07	D−07	D−06
41	−0.59D−08	0.0	−0.50D+01	0.0	D−15	D−07	D−07	D−06
42	0.15D−07	0.0	−0.50D+01	0.0	D−15	D−07	D−07	D−06
43	−0.13D−08	0.0	−0.50D+01	0.0	D−15	D−06	D−08	D−07
44	0.93D−07	0.0	−0.50D+01	0.0	D−13	D−07	D−06	D−05
45	0.45D−07	0.0	−0.50D+01	0.0	D−14	D−07	D−07	D−05
46	0.20D−07	0.0	−0.50D+01	0.0	D−15	D−07	D−07	D−06
47	0.74D−08	0.0	−0.50D+01	0.0	D−15	D−07	D−07	D−06
48	−0.92D−08	0.0	−0.50D+01	0.0	D−15	D−07	D−07	D−06
49	0.42D−08	0.0	−0.50D+01	0.0	D−15	D−07	D−08	D−06
50	−0.25D−07	0.0	−0.50D+01	0.0	D−14	D−07	D−07	D−06
51	−0.10D−07	0.0	−0.50D+01	0.0	D−15	D−07	D−07	D−06
52	0.20D−08	0.0	−0.50D+01	0.0	D−15	D−06	D−08	D−07
53	−0.59D−07	0.0	−0.50D+01	0.0	D−14	D−07	D−06	D−05
54	−0.26D−07	0.0	−0.50D+01	0.0	D−14	D−07	D−07	D−06
55	−0.98D−08	0.0	−0.50D+01	0.0	.	.	D−07	D−06
56	−0.98D−08	0.0	−0.50D+01	0.0	.	.	D−07	D−06
57	−0.98D−08	0.0	−0.50D+01	0.0	.	.	D−07	D−06
58	−0.98D−08	0.0	−0.50D+01	0.0	.	.	D−07	D−06
59	−0.98D−08	0.0	−0.50D+01	0.0	.	.	D−07	D−06
60	−0.98D−08	0.0	−0.50D+01	0.0	.	.	D−07	D−06

ᵃ See the footnote to Table A1–1. Here "ERROR" denotes an upper bound on $|(X, Y) - (0, -5)|$.

TABLE A1-4 Newton iterates converging to singular solution $(X, Y) = (0, -5)$ of system (2-3) using a complex start point[a]

IT	REAL(X)	IMAG(X)	REAL(Y)	IMAG(Y)	EQ	RESID	ERROR	DET
1	0.50D+00	0.50D+00	-0.45D+01	0.50D+00	D+02	D+01	D+01	D+02
2	0.22D+00	0.22D+00	-0.50D+01	-0.62D-01	D+01	D+00	D+00	D+02
3	0.11D+00	0.11D+00	-0.50D+01	0.10D-03	D-01	D+00	D+00	D+01
4	0.56D-01	0.54D-01	-0.50D+01	-0.92D-08	D-01	D-01	D+00	D+01
5	0.28D-01	0.27D-01	-0.50D+01	-0.37D-16	D-02	D-01	D-01	D+01
6	0.14D-01	0.13D-01	-0.50D+01	0.55D-20	D-02	D-01	D-01	D+00
7	0.70D-02	0.67D-02	-0.50D+01	0.55D-20	D-03	D-02	D-01	D+00
8	0.35D-02	0.34D-02	-0.50D+01	-0.54D-21	D-04	D-02	D-02	D+00
9	0.18D-02	0.17D-02	-0.50D+01	-0.19D-21	D-04	D-02	D-02	D-01
10	0.88D-03	0.84D-03	-0.50D+01	0.11D-23	D-05	D-02	D-02	D-01
11	0.44D-03	0.42D-03	-0.50D+01	0.22D-23	D-05	D-03	D-02	D-01
12	0.22D-03	0.21D-03	-0.50D+01	-0.39D-23	D-06	D-03	D-03	D-01
13	0.11D-03	0.10D-03	-0.50D+01	-0.46D-24	D-07	D-03	D-03	D-02
14	0.55D-04	0.52D-04	-0.50D+01	-0.11D-25	D-07	D-04	D-03	D-02
15	0.27D-04	0.26D-04	-0.50D+01	-0.17D-25	D-08	D-04	D-04	D-02
16	0.14D-04	0.13D-04	-0.50D+01	-0.97D-26	D-08	D-04	D-04	D-03
17	0.68D-05	0.66D-05	-0.50D+01	-0.26D-26	D-09	D-05	D-04	D-03
18	0.34D-05	0.33D-05	-0.50D+01	0.11D-27	D-10	D-05	D-05	D-03
19	0.17D-05	0.16D-05	-0.50D+01	-0.23D-27	D-10	D-05	D-05	D-04

20	0.86D−06	0.82D−06	−0.50D+01	−0.32D−28	D−11	D−05	D−05	D−04
21	0.43D−06	0.41D−06	−0.50D+01	−0.11D−28	D−12	D−06	D−05	D−04
22	0.21D−06	0.20D−06	−0.50D+01	−0.16D−30	D−12	D−06	D−06	D−04
23	0.11D−06	0.10D−06	−0.50D+01	−0.20D−30	D−13	D−06	D−06	D−05
24	0.53D−07	0.52D−07	−0.50D+01	0.50D−31	D−13	D−07	D−06	D−05
25	0.26D−07	0.26D−07	−0.50D+01	0.98D−31	D−14	D−07	D−07	D−05
26	0.11D−07	0.16D−07	−0.50D+01	0.18D−32	D−15	D−08	D−07	D−06
27	0.33D−08	0.11D−07	−0.50D+01	−0.33D−32	D−15	D−09	D−07	D−06
28	0.30D−09	0.97D−08	−0.50D+01	−0.39D−33	D−16	D−12	D−07	D−06
29	0.29D−12	0.97D−08	−0.50D+01	0.43D−35	D−19	D−21	D−07	D−06
30	0.26D−21	0.97D−08	−0.50D+01	−0.57D−37	D−28	D−37	D−07	D−06
31	0.47D−37	0.97D−08	−0.50D+01	0.47D−47	.	.	D−07	D−06
32	0.0	0.97D−08	−0.50D+01	0.11D−62	.	.	D−07	D−06
33	0.0	0.97D−08	−0.50D+01	0.0	.	.	D−07	D−06
34	0.0	0.97D−08	−0.50D+01	0.0	.	.	D−07	D−06
35	0.0	0.97D−08	−0.50D+01	0.0	.	.	D−07	D−06
36	0.0	0.97D−08	−0.50D+01	0.0	.	.	D−07	D−06
37	0.0	0.97D−08	−0.50D+01	0.0	.	.	D−07	D−06
38	0.0	0.97D−08	−0.50D+01	0.0	.	.	D−07	D−06
39	0.0	0.97D−08	−0.50D+01	0.0	.	.	D−07	D−06
40	0.0	0.97D−08	−0.50D+01	0.0	.	.	D−07	D−06

[a] See the footnote to Table A1-3.

Consider (2-3), whose Newton update formula was developed above. This system has three solutions, (± 3, 4) and (0, -5), with (± 3, 4) being nonsingular and (0, -5) being singular. In Tables A1-1 through A1-4, I have listed the Newton iterates beginning at a start point differing from the actual solution by 0.5 in each variable, for real and complex cases. These tables include the exponents of $| f(z^k) |$, $| \text{resid}^k |$, $| z^k - z^* |$, and $| \det (df(z^k)) |$, from which you can observe the speed of convergence and the singularity. [The *condition* of $df(z^k)$ is bounded above by $400/| \det (df(z^k)) |$, using (4-6); see Section 4-5-4.] Note especially the quadratic convergence in the residual for the nonsingular solution and the much slower convergence for the singular solution. Also note the way the residual (which we can generally compute) is a good estimate of the error (which we generally cannot compute) for the nonsingular solution. The relationship between error and the residual is less exact for the singular solution, especially as the iterates get close to the solution. Compare with $| f(z^k) |$. Note also the imaginary parts. The calculation was carried out in double precision on an IBM 3033, so about 14 decimal digits were available.

A1-3 A ROBUST IMPLEMENTATION

In addition to the basic theory, as described above, we need four tests. First, we need a test for possible singularity. We will, in essence, use two different methods, a singular and a nonsingular Newton's method, depending on whether or not the solution is judged to be "possibly singular" (see Section 4-5-4). Next we need a stopping rule. The Newton algorithm will successively generate iterates z^1, z^2, . . . , until the stopping rule is satisfied. Third, we need a way of evaluating the success of the method, and fourth we need a test to judge whether the final solution estimate is nonsingular, marginally singular, or singular.

Let z^k denote the current estimate of solution z^*. The test for possible singularity is given by a zero-epsilon, ϵ_{ps}. The solution is possibly singular if $C(df(z^k)) \geq 1/\epsilon_{ps}$, where $C(df(z^k))$ is the condition of $df(z^k)$, as described in Section 4-5-4. The stopping rule is given by a positive integer, MAXIT, and two zero-epsilons, ϵ and ϵ_0, with the rule being: stop if either MAXIT iterations have been completed or $| \text{resid}^k | \leq \epsilon$ or solving the linear system (A1-3) would divide by a number smaller than ϵ_0. The idea of the ϵ_0 part of the stopping rule is to prevent overflows when $df(z^k)$ is nearly singular. A variety of more sophisticated tests for avoiding overflow can be concocted. See, for example, the way subroutine D1MACH is used in subroutine DIVP in the package HOMPACK [Watson, et al., 1986]. The test for success of Newton is $| \text{resid}^k | \leq \epsilon_f$, where ϵ_f might be chosen several orders of magnitude larger than ϵ. The singular and nonsingular methods each have their own ϵ, ϵ_0, and ϵ_f. Finally, we have zero-epsilons ϵ_s and ϵ_{ns} to judge the singularity of the final estimate, z^{++}. If $C(df(z^{++})) \leq 1/\epsilon_{ns}$, then z^{++} is nonsingular. If $C(df(z^{++})) \geq 1/\epsilon_s$, it is singular. Otherwise, it is marginally singular. We proceed as follows.

Let z^+ be the initial solution estimate, the "guess" of the true solution, z^*. (Thus z^+ might be a random point in space or a continuation path's endpoint.) We are seeking z^{++}, a refinement of z^+. We evoke the Newton's method with start point $z^0 = z^+$ and the nonsingular MAXIT, ϵ, and ϵ_0. When it is done, we test the possible singularity of z^k, the last iterate. If z^k is judged nonsingular, we let $z^{++} = z^k$. Otherwise, we take MAXIT, ϵ, and ϵ_0 to be the singular versions, set $z^0 = z^k$, and evoke Newton again. (If possible, the singular Newton should be coded in extended-precision arithmetic.) Then we set z^{++} equal to the result of this second Newton's method. We test for success via ϵ_f and singularity via ϵ_{ns} and ϵ_s. (Regardless of the final singularity status of z^{++}, the singular ϵ_f is used if the singular Newton is evoked.) We return z^{++} together with two flags, one indicating "success" or "failure" and the other indicating "nonsingular," "marginally singular," or "singular."

Typical values for the parameters are: $\epsilon_{ps} = \epsilon_{ns} = 10^{-5}$, $\epsilon_s = 10^{-8}$, and

	Nonsingular	Singular
MAXIT	10	100
ϵ	10^{-12}	10^{-22}
ϵ_0	10^{-22}	10^{-22}
ϵ_f	10^{-8}	10^{-18}

The way I have presented the singular Newton, with MAXIT = 100 and coded in extended precision, it is computationally expensive. I have found it effective for nice singular solutions, even though expensive.

A1-4 NEWTON'S METHOD FOR CONTINUATION

There are basically two very different uses for Newton's method in the context of continuation: as a corrector during path tracking and to refine the path endpoints. Refining endpoints is really separate from continuation. It is a form of *postprocessing*. An implementation as described in Section A1-3 might be used for endpoint refinement, especially to make sure that singular solutions are computed as accurately as possible.

Using the continuations given in this book, singular solutions can be encountered only at the end of continuation paths. (CONSOL1 and CONSOL2 are exceptions.) In other words, singular solutions can occur only when $t = 1$ and the original system has a singular solution to which the continuation is converging. In this case, if the maximum number of iterations of the final Newton correction is increased, the accuracy of the endpoint will generally be improved. Compare the behavior shown in Tables A1-1 through A1-4 for singular and nonsingular solutions. It follows from the discussion in Section A1-2 that we cannot expect CONSOL to satisfactorily compute nasty singular solutions. I do not know of any general approach that does.

I set the Newton corrector parameters (for path tracking) as follows in CONSOL8 (see Section A1-3 for parameter definitions and Chapter 4 for a description of CONSOL8). MAXIT is an input parameter but is set equal to 10 at the end of the path; ϵ is set to the input EPSBIG until $t > 0.95$, when it is set to EPSSML = EPSBIG$*10^{-3}$ until $t = 1$, when it is set to ϵ_0; $\epsilon_0 = 10^{-22}$. The postprocessing Newton noted above is not included in the CONSOL8 code.

An implication of the reduced accuracy attainable for singular solutions is that extra care must be taken if we are looking for real-number solutions using CONSOL. Suppose, for example, that $\epsilon_* = 10^{-14}$ and z^* is a real singular solution, where ϵ_* is defined in Section A1-2. Then, beginning with a complex start point, the magnitudes of the imaginary parts of the z^k might stabilize at about 10^{-7} as the z^k converge to z^*. It is important to realize this in evaluating the results of computations when you are looking for real solutions. Otherwise, physically meaningful solutions may be missed. Note the computational example in Section A1-2.

A1-5 GLOBAL BEHAVIOR

I have been impressed with how frequently Newton's method for a polynomial system converges from a randomly chosen start point. Newton's method seems to be especially robust in complex arithmetic. However, for any system $f(z) = 0$, there are points from which the Newton iterates will not converge. Let

$$S_0 = \{z \in C^n \mid \det (df(z)) = 0\}$$

denote the singularity set of df. Then take

$$S_{m+1} = \{z \in C^n \mid w = z - [df(z)]^{-1}f(z) \text{ for some } w \in S_m\}.$$

In other words, S_{m+1} consists of the points that Newton sends to S_m. Then Newton cannot converge for start points in any S_m. Note that for $m > 0$, df is nonsingular at these "bad" start points. Such nonconvergent start points are often sparsely distributed in C^n and therefore avoidable by using random perturbations. That is, let z^0 be any point in C^n and add to z^0 a small random complex vector ϵ^0. Then $z^0 + \epsilon^0$ is a good candidate for a start point. Examples exist, however, where this trick will not work. See [Curry, et al., 1983]. And nothing will be of much use if $f(x) = 0$ has only nasty singular solutions and/or only solutions at infinity.

As suggested in the discussion of singular solutions in Section A1-2, we can make finer distinctions in the global modes of behavior for Newton than "convergence" and "nonconvergence." For example, Newton can "cycle," with the iterates forming a finite circuit that never settles down to a solution. (The example in [Petek, 1983] is simple and enlightening. See also [Curry, et al., 1983].) The global behavior of Newton's method is complicated and interesting but not especially relevant to CONSOL, which uses Newton strictly as a local solution technique. If you would like to read more about global issues, you might consult [Devaney, 1986] and the survey [Blanchard, 1984]. (The mathematics in the latter reference is advanced.)

APPENDIX 2

Emulating
Complex Operations
in Real Arithmetic

The arithmetic of complex numbers is based on a particular set of operations on pairs of real numbers, the familiar "$a + bi$" operations. The CONSOL code uses addition, subtraction, multiplication, division, powers, and absolute value. If our computer language and compiler handle complex operations satisfactorily, we can develop a CONSOL code without much concern for the complex nature of the arithmetic. We merely structure the program with the usual algebraic expressions, declaring the appropriate variables "complex." However, if any of the following conditions hold, we may need to write our own complex operations using only real declarations:

1. No complex declaration is available.
2. Complex variables must be single precision.
3. Complex operations are slow.

Case 1 holds for most BASIC compilers, and cases 2 and 3 hold for some FORTRAN compilers.

Here is the way I emulate complex operations in real arithmetic. I use a leading array index to distinguish the real and complex parts of complex variables, and do complex multiplication, division, and powers by subroutine calls. Addition, subtraction, and absolute values I do "in line," as explained below.

Thus "DIMENSION X(2)" defines a complex number, "DIMENSION A(2,10)" an array of 10 complex numbers, and "DIMENSION B(2,10,10)" a 10 × 10 array of complex numbers. For example,

$X(1)$ = real part of x
$X(2)$ = imaginary part of x

$A(1, 3)$ = real part of a_3
$B(2, 7, 9)$ = imaginary part of $b_{7,9}$.

The subroutines MUL, POW, and DIV multiply, take powers, and divide as follows:

"CALL MUL(X, Y, Z)" computes $z = x * y$

"CALL POW(N, X, Y)" computes $y = x^n$ (n must be a nonnegative integer)

"CALL DIV(X, Y, Z)" computes $z = x/y$.

Thus "MUL(A(1, 1), A(1, 2), A(1, 3))" generates $a_3 = a_1 * a_2$ and "POW(3, A(1, 7), X)" gives $x = a_7^3$.

For a more complicated example, consider computing the absolute value of the polynomial expression

$$\text{absf} = |\, 6x^5 - xy + 8\,|,$$

where x and y are complex numbers. The following code accomplishes this:

```
DIMENSION X(2), Y(2), Z(2), W(2), F(2)

CALL POW(5, X, Z)

CALL MUL(X, Y, W)

F(1) = 6*Z(1) - W(1) + 8

F(2) = 6*Z(2) - W(2)

ABSF = SQRT(F(1)**2 + F(2)**2)
```

It is "easy" to convert a program with declared complex variables into one without, where each multiplication, power, and division generates one subroutine call, and addition, subtraction, and absolute value are done in line as above. But it is tedious.

The complex-operation subroutines are listed in Appendix 6, Part B, Section 6. The various main codes in Appendix 6, Part B, illustrate their use. For example, see CONSOL2R.

APPENDIX 3

Some Real-Complex Calculus Formulas

In this appendix we consider continuation systems,

$$h: C^n \times I \to C^n, \tag{A3-1}$$

where $I = [0, 1]$. Our main concern is the continuation (3-25) for solving polynomial systems. We will see that the "complex polynomial calculus" way of computing the Jacobian of the continuation system is equivalent to the (more proper) real-calculus way. We will also see that the associated continuation paths are monotonically increasing in t.† [This will establish part (b) of the theorem in Section 3-4.] A brief discussion of parametrizations of paths and arc length is included. Although we focus on polynomials, the results extend naturally to nonpolynomial systems where $h_t(x) \equiv h(x, t)$ is complex analytic in x. In particular, continuation paths for such systems do not turn back in t (see Chapter 6).

The calculus of continuation is real-number calculus. Continuation paths are *real* one-dimensional submanifolds of space (see Appendix 4). Thus even though the continuation (A3-1) is defined over complex space, from the point of view of calculus (or topology or ordinary differential equations), we more properly have

$$h: R^{2n} \times I \to R^{2n},$$

where C^n is identified with R^{2n} under, say, the equivalence

$$(a_1 + a_2i, a_3 + a_4i, \ldots, a_{2n-1} + a_{2n}i) \to (a_1, a_2, a_3, \ldots, a_{2n-1}, a_{2n}).$$

---

† In [Morgan and Sommese, 1986] it is shown that paths are *strictly* increasing in t.

Viewing h as a smooth function on real space, $\partial h_j/\partial a_k$ and $\partial h_j/\partial t$ have their usual meaning from real multivariable calculus, and the Jacobian, dh, is a $2n \times (2n + 1)$ matrix whose null space gives the (one-real-dimensional) tangent line to the continuation curves.

However, we do not want to generate dh this way. Directly computing $\partial h_j/\partial a_k$ makes a mess of the simple polynomial formulas that define h. Happily, there is an easy equivalence between the proper $\partial h_j/\partial a_k$ and the less proper $\partial h_j/\partial x_k$. Let's clarify this equivalence.

Define $\theta: C^n \to R^{2n}$ by

$$\theta(x_1, x_2, \ldots, x_n) = (\text{Re}(x_1), \text{Im}(x_1), \ldots, \text{Re}(x_n), \text{Im}(x_n))$$

$$= (a_1, a_2, \ldots, a_{2n}),$$

where $x_k = a_{2k-1} + ia_{2k}$, $\text{Re}(x_k) = a_{2k-1}$, $\text{Im}(x_k) = a_{2k}$, a_{2k-1} and a_{2k} are real numbers for $k = 1$ to n, and $i^2 = -1$. Using θ, we can convert systems over C^n into systems over R^{2n}. Then (real) partial derivatives can be taken as $\partial/\partial a_j$ for $j = 1, 2, \ldots, 2n$.

Now, if u is an n vector of complex polynomials in m variables, we let $\theta(u)$ denote the corresponding $2n$ vector of real polynomials in $2m$ real variables, the "realification of u." Further, we let $d(\theta(u))$ be the $2n \times 2m$ real Jacobian matrix. On the other hand, $d(u) = du$ makes sense as the $n \times m$ complex Jacobian of u, derived from formal polynomial calculus, and a new function θ' can be defined so that $\theta'(d(u)) = d(\theta(u))$. Namely, if M denotes a complex $n \times m$ matrix with complex number $M_{k,l}$ as the (k, l)th entry, define $\theta'(M)$ as the $2n \times 2m$ real matrix with entries $a_{i,j}$ defined by

$$a_{2k-1,2l-1} = \text{Re}(M_{k,l})$$

$$a_{2k-1,2l} = -\text{Im}(M_{k,l})$$

$$a_{2k,2l-1} = \text{Im}(M_{k,l})$$

$$a_{2k,2l} = \text{Re}(M_{k,l})$$

for $k = 1$ to n and $l = 1$ to m. More informally, $\theta'(M)$ is derived from M by replacing the entry $M_{k,l}$ by the 2×2 block

$$\begin{bmatrix} \text{Re}(M_{k,l}) & -\text{Im}(M_{k,l}) \\ \text{Im}(M_{k,l}) & \text{Re}(M_{k,l}) \end{bmatrix}.$$

If we wish to exploit the $\theta'(d(u)) = d(\theta(u))$ relation for $u = h(z, t)$ with one real variable, t, we may proceed as if t were complex, forming $\theta'(d(u))$ and then stripping off the last column of $\theta'(d(u))$, the one associated with $\partial/\partial \text{Im}(t)$.

For example, consider the simple continuation

$$h_1(x, y, t) = x^2 + txy - 1 = 0$$
$$h_2(x, y, t) = y^2 + tx - 5 = 0,$$

$$(A3-2)$$

which is (6-18). Let's take $x = a + bi$ and $y = c + di$. Using polynomial calculus, we get

$$dh(x, y, t) = \begin{bmatrix} \dfrac{\partial h_1}{\partial x} & \dfrac{\partial h_1}{\partial y} & \dfrac{\partial h_1}{\partial t} \\[2mm] \dfrac{\partial h_2}{\partial x} & \dfrac{\partial h_2}{\partial y} & \dfrac{\partial h_2}{\partial t} \end{bmatrix}$$

$$= \begin{bmatrix} 2x + ty & tx & xy \\ t & 2y & x \end{bmatrix}$$

$$= \begin{bmatrix} 2(a + bi) + t(c + di) & t(a + bi) & (ac - bd) + (bc + ad)i \\ t & 2(c + di) & a + bi \end{bmatrix}.$$

Then, if we apply the transformation θ' to dh, we get the 4×6 matrix

$$\theta'(dh) = \begin{bmatrix} 2a + tc & -(2b + td) & ta & -tb & ac - bd & -(bc + ad) \\ 2b + td & 2a + tc & tb & ta & bc + ad & ac - bd \\ t & 0 & 2c & -2d & a & -b \\ 0 & t & 2d & 2c & b & a \end{bmatrix}.$$

Now, let us generate $d(\theta(h))$. We have

$$x^2 = (a + bi)^2 = (a^2 - b^2) + 2abi$$

$$y^2 = (c + di)^2 = (c^2 - d^2) + 2cdi$$

$$xy = (a + bi)(c + di) = (ac - bd) + (bc + ad)i$$

and

$$x^2 + txy - 1 = (a^2 - b^2) + t(ac - bd) - 1$$
$$+ [2ab + t(bc + ad)]i$$
$$y^2 + tx - 5 = (c^2 - d^2) + ta - 5$$
$$+ [2cd + tb]i.$$

So the system (A3-2) of two equations in two complex and one real unknown is equivalent to the system of four equations in five real unknowns

$$\theta(h(a, b, c, d, t)) = \begin{bmatrix} (a^2 - b^2) + t(ac - bd) - 1 \\ 2ab + t(bc + ad) \\ (c^2 - d^2) + ta - 5 \\ 2cd + tb \end{bmatrix} = \begin{bmatrix} 0 \\ 0 \\ 0 \\ 0 \end{bmatrix}. \qquad \text{(A3-3)}$$

Now

$$
d\theta(h(a,\,b,\,c,\,d,\,t)) = \begin{bmatrix}
\dfrac{\partial\theta h_1}{\partial a} & \dfrac{\partial\theta h_1}{\partial b} & \dfrac{\partial\theta h_1}{\partial c} & \dfrac{\partial\theta h_1}{\partial d} & \dfrac{\partial\theta h_1}{\partial t} \\[2ex]
\dfrac{\partial\theta h_2}{\partial a} & \dfrac{\partial\theta h_2}{\partial b} & \dfrac{\partial\theta h_2}{\partial c} & \dfrac{\partial\theta h_2}{\partial d} & \dfrac{\partial\theta h_2}{\partial t} \\[2ex]
\dfrac{\partial\theta h_3}{\partial a} & \dfrac{\partial\theta h_3}{\partial b} & \dfrac{\partial\theta h_3}{\partial c} & \dfrac{\partial\theta h_3}{\partial d} & \dfrac{\partial\theta h_3}{\partial t} \\[2ex]
\dfrac{\partial\theta h_4}{\partial a} & \dfrac{\partial\theta h_4}{\partial b} & \dfrac{\partial\theta h_4}{\partial c} & \dfrac{\partial\theta h_4}{\partial d} & \dfrac{\partial\theta h_4}{\partial t}
\end{bmatrix},
$$

where "θh_k" is the kth equation in the system (A3-3). Thus we see that

$$
d\theta(h(a,\,b,\,c,\,d,\,t)) = \begin{bmatrix}
2a + tc & -(2b + td) & ta & -tb & ac - bd \\
2b + td & 2a + tc & tb & ta & bc + ad \\
t & 0 & 2c & -2d & a \\
0 & t & 2d & 2c & b
\end{bmatrix}.
$$

Observe that $d\theta(h)$ is equal to $\theta'(dh)$ with the last column omitted.

We will complete this appendix with a proof that continuation paths for (A3-1) are monotonically increasing in t. However, first we need to consider a summary of some material on parametrizations and arc length.

A path is a purely geometrical object, a collection of points in space that looks locally like a (perhaps curved) piece of the real line. We like to *parametrize* paths, so that the points are (in essense) given by a formula. Thus "the unit semicircle in the upper half-plane" (Fig. A3-1) is a precise definition of a path, but

$$
c(s) = (\cos s,\ \sin s) \qquad \text{for } 0 \le s \le \pi \tag{A3-4}
$$

parametrizes the path using the *parameter s*. The parametrization can also be used to define the path, of course, but we think of the points in space as being the path. (Thus the path is the *image* of the parametrization.) We can

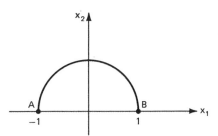

Figure A3-1 Unit semicircle in the upper half-plane.

parametrize the semicircle in different ways. For example,

$$c(s) = \left(\cos \frac{s}{2}, \sin \frac{s}{2} \right) \qquad \text{for } 0 \leq s \leq 2\pi$$

is another parametrization.

A particular way of parametrizing paths is by *arc length*. We would naturally expect to be able to assign to paths their length, but this is a surprisingly subtle concept. Pinning it down required the invention of a lot of the calculus. To bypass some of the subtleties, let's assume that between any two points, A and B, on a path, the length along the path from A to B is well defined. Thus for the semicircle above, if $A = (-1, 0)$ and $B = (1, 0)$, the distance from A to B is 2, but the arc length of the curve from A to B is π. It is a fact that we can always parametrize (smoothly embedded) paths by arc length just as (A3-4) does for the semicircle. That is, given a path in R^N with a single endpoint A, we can find a parametrization

$$c: [0, \infty) \to R^N$$

with $c(0) = A$ and so that the arc length from A to B is "s" when $c(s) = B$. If the path has two endpoints, the domain of c is $[0, L]$, where L is the path's length. See [Spivak, 1970] for a fuller discussion.

Given an ordinary differential equation

$$\frac{dc}{ds} = P(c), \tag{A3-5}$$

a solution to (A3-5) is a parametrization $c(s)$ obeying

$$\frac{dc}{ds}(s) = P(c(s)).$$

If we specify $c(0) = A$, then $c(s)$ parametrizes a path beginning at A. Now the parameter s turns out to be "arc length" exactly when

$$\left| \frac{dc}{ds}(s) \right| = 1$$

for all s, that is, when the (vector) length of dc/ds is always 1. For example, if we take $P(c) = P(c_1, c_2) = (-c_2, c_1)$, then $c(s) = (\cos s, \sin s)$ satisfies $(dc/ds)(s) = P(c(s))$ and $|(dc/ds)(s)| = (-\sin s)^2 + (\cos s)^2 = 1$ for all s.

From Chapter 3 we use equation (3-30) to get $(dx/dt, 1)$ as a tangent to the continuation curve in $R^{2n} \times I$. This yields a parametrization of the continuation curve in the variable t. We can, however, rescale this $(dx/dt, 1)$ to have length 1, after which it is still a point on the tangent line with the same orientation as before. But this rescaled vector is more properly $(dx/ds, dt/ds)$ because the associated continuation curve is parametrized by arc length. Of course, the curve is the same whether it is parametrized by

t or by s, but parametrizing by arc length has a regularizing effect that can be numerically beneficial. And if we are in a context in which continuation curves can turn back in t, we must parametrize by some variable other than t. In this case, arc length is the customary choice (see Chapter 6).

Now here is a proof that complex continuation paths cannot turn back in t. Let $A = d\theta(h)$, a $2n \times (2n + 1)$ real matrix. Denote the kth row of A by $A_k = (a_{k,1}, \ldots, a_{k,2n+1})$. Then

$$\det \begin{bmatrix} A_k \\ A \end{bmatrix} = 0$$

since row k is repeated. But then

$$0 = \det \begin{bmatrix} A_k \\ A \end{bmatrix} = \sum_{j=1}^{2n+1} (-1)^{1+j} a_{k,j} \det (A_{[j]})$$

for $k = 1$ to $2n$, by the Laplace expansion of the determinant by the first row, where $A_{[j]}$ denotes the $2n \times 2n$ matrix obtained from A by deleting column j (see [Cullen, 1966, Chap. 3]). Thus

$$w_j = (-1)^{1+j} \det (A_{[j]})$$

for $j = 1$ to $2n + 1$ defines the components of a solution, w, of $Aw = 0$. In other words, the vector w is a tangent vector to the continuation path.

Now, $w_{2n+1} = \det (A_{[2n+1]})$ and

$$\det (A_{[2n+1]}) \geq 0 \tag{A3-6}$$

because of the special 2×2 block structure of $A_{[2n+1]}$ (see the proposition, below). Formally, this structure is a consequence of the Cauchy–Riemann equations and holds whenever h is complex analytic in x (see [Garcia and Zangwill, 1981, Chap. 18]). But we may rescale w as $w/| w | = (dx/ds, dt/ds)$, where s denotes arc length. Thus (A3-6) tells us that we may consistently orient the continuation paths so that t is monotonically increasing as a function of arc length. In other words, paths do not "turn back in t."

We will complete the appendix with a proof of the proposition noted above. Let S denote the set of (not necessarily square) matrices made up of 2×2 blocks of the form

$$\begin{bmatrix} a & -b \\ b & a \end{bmatrix},$$

where a and b are real numbers. Let $A \in S$ and $B \in S$. Then $A + B \in S$ and $A \cdot B \in S$ if the addition and multiplication are defined. Also, if A is a 2×2 matrix in S, then $A^{-1} \in S$.

PROPOSITION.   If E is square and $E \in S$, then $\det (E) \geq 0$.

Proof. If every entry of E is zero, we are done. Otherwise, we may interchange rows and columns of E so that the determinant is unchanged and the first 2×2 block of E is nonzero. [Each column (row) interchange changes the sign of det (E), and we make an even number of them.] Now we set

$$E = \begin{bmatrix} A & B \\ C & D \end{bmatrix},$$

where A is the first 2×2 block of E and D is square. Then det (E) = det $(A) \cdot$ det $(D - C \cdot A^{-1} \cdot B)$ (see [Cullen, 1966, p. 72]). Thus det (E) has the same sign as det $(D - C \cdot A^{-1} \cdot B)$, and $D - C \cdot A^{-1} \cdot B \in S$. But $D - C \cdot A^{-1} \cdot B$ is a smaller matrix than E, and we may iterate the argument until it is 2×2. So, by induction, we are done.

APPENDIX 4

Proofs of Results
from Chapter 3

This Appendix contains proofs of parts (a) and (d) of the theorem in Section 3-4. However, we must first clarify some basic concepts from elementary differential topology. The Transversality Theorem and the Preimage Theorem (stated below) are the main results we need, together with the definitions of smooth manifold, tangent space, and the derivative of a smooth map between manifolds as a linear transformation of tangent spaces. Differential topology is not the only approach we could use. Concepts and methods from algebraic geometry are an alternative. See Note 2 at the end of this appendix.

In summarizing this basic material, I will draw on the excellent expository text [Guillemin and Pollack, 1974]. Take a look at the first chapter for an easy introduction to smooth manifolds. It includes most of what I will sketch through here. I will limit some of the statements of results to the special cases most relevant to proving the theorem from Chapter 3. Again, see [Guillemin and Pollack, 1974] for the more complete exposition.

We want to consider subsets of Euclidean space, in particular the complex Euclidean space, C^M, where M is a positive integer. Since we will be focusing mostly on topological ideas, it is natural to identify the complex numbers, C, with pairs of real numbers, R^2, via $a + bi \rightarrow (a, b)$. Similarly, we identify C^M with R^{2M} (see Appendix 3). Thus, if we establish basic results in R^N, where N is a positive integer, we will be able to apply them to the complex context.

Subsets of R^N can be "extremely bizarre" ([Guillemin and Pollack, 1974] p. 33). However, we want to deal with nice subsets: smooth paths or other "locally Euclidean" spaces. *Smooth manifolds* are the subsets of R^N that we want to consider. The key here is that we identify a subset, X, of

R^N as being nice if it is locally nice. That is, given a point $x \in X$ and a small ball about x, $N_\epsilon(x) = \{y \in R^N : |x - y| < \epsilon\}$, we want $N_\epsilon(x) \cap X \subset X$ to look nice. In fact, it should look like a ball in a smaller-dimensional Euclidean space. Think of a sphere in 3-space. Pick a point on the sphere and intersect the sphere with a small ball centered at the point. The resulting subset of R^3 looks like a (slightly curved) disk, that is, a "ball" in R^2 (see Fig. A4-1). We say that the sphere in R^3 is a two-dimensional submanifold of R^3. Generally, an m-dimensional manifold, X, looks locally like an m-dimensional ball. We sometimes think of covering X by "patches," each of which is smoothly identified with a ball in R^m, and we speak of "local coordinates" on X meaning the coordinates in R^m identified with points in a patch on X. The smoothness hypotheses implies that the manifold cannot look sharply bent (no corners) as a subset of R^N. Thus paths in space are 1-manifolds, because they look locally like intervals in R^1, and if a path has a sharp turn, it is not smooth. Since we depend on being able to track continuation paths numerically using tangent information, we need the paths to be smooth. Tangents do not exist at corners.

Let's look at these ideas a little more precisely. A subset U of R^m is "open" if for each $x \in U$, there is an $\epsilon > 0$ so that $N_\epsilon(x) \subset U$. For example, if U denotes the plane with the x-axis omitted, U is open as a subset of R^2. Similarly, $U \subset R^m$ is open if U is defined as R^m with a finite number of hyperplanes omitted. (The whole space and the empty set are open.)

If U and V are open subsets of Euclidean spaces, then $f: U \to V$ is "smooth" if for each $x \in U$, $df(x)$ (the Jacobian) exists. (Since we are focusing on polynomial systems, smoothness has not been an issue.) We say

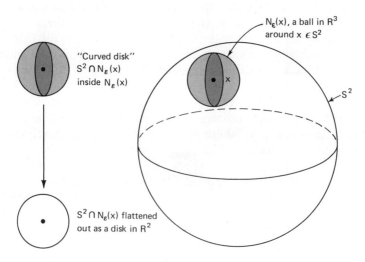

Figure A4-1 Sphere in R^3 as a two-dimensional manifold.

that f is "one-to-one" if $f(x^a) = f(x^b)$ implies that $x^a = x^b$; f is "onto" if for every $y \in V$, there is an $x \in U$ with $f(x) = y$; f has an inverse if there is a function $g: V \to U$ with $g(f(x)) = x$ and $f(g(y)) = y$ (notation: $g = f^{-1}$). If f is one-to-one and onto and has a smooth inverse, f is called a "diffeomorphism." If there exists such an f, we say that U and V are "diffeomorphic." Then $X \subset R^N$ is a *smooth m-dimensional manifold* if it is "locally diffeomorphic" to open subsets of R^m. In other words, if $x \in X$, there is an $\epsilon > 0$ such that $V = X \cap N_\epsilon(x)$ is diffeomorphic to some open subset U of R^m. We speak of the diffeomorphism $\phi: U \to V$ as a *parametrization* of V, and the inverse diffeomorphism $\phi^{-1}: V \to U$ is called a *coordinate system* on V. Using coordinates in U for points in V via ϕ^{-1}, we get the "local coordinates" mentioned above. Naturally, there is nothing unique about local coordinates; ϕ must be specified or understood.

If U and V are open subsets of R^m and R^k, respectively, and $f: U \to V$ is smooth, then, for each point $x \in R^m$, $df(x)$ is a matrix (the Jacobian), which can be viewed as a linear map $df(x): R^m \to R^k$. Using parametrizations, this way of viewing the derivative of a smooth map can be extended to smooth manifolds. First, we need to define tangent spaces. For each point x in a smooth m-dimensional manifold X, we have the parametrization $\phi: U \to V$, as above, with $x \in V$ and some $y \in U$ with $\phi(y) = x$. But $V \subset R^N$, the "ambient Euclidean space," so we may view ϕ as a map into R^N. Thus $d\phi(y): R^m \to R^N$ is well defined. Now we define "$T_x X$" to be the image of $d\phi(y)$ in R^N. This is the "tangent space to X at x." Actually, the parallel translate $x + T_x X$ intersects X at x and is a hyperplane in R^N approximating X near x, as one would expect of a tangent space. $T_x X$ passes through the origin in R^N and is parallel to this linear approximating hyperplane. Now, a smooth $f: X_1 \to X_2$ from smooth m-dimensional manifold X_1 into smooth k-dimensional manifold X_2 yields, for each point $x \in X_1$, a linear map $df(x): T_x X_1 \to T_{f(x)} X_2$, where $T_x X_1$ and $T_{f(x)} X_2$ denote the tangent spaces to X_1 at x and to X_2 at $f(x)$, respectively. Basically, we define the df map by passing to local coordinates. Given $f: X_1 \to X_2$, consider this square of maps and spaces:

$$
\begin{array}{ccc}
 & f & \\
X_1 & \to & X_2 \\
\phi_1 \uparrow & & \uparrow \phi_2, \\
U_1 & \to & U_2 \\
 & g &
\end{array}
$$

where g is defined by $g = \phi_2^{-1} \circ f \circ \phi_1$, and we will assume that ϕ_1, ϕ_2, U_1, U_2 have been chosen so that $f(\phi_1(U_1)) = \phi_2(U_2)$. We know how to differentiate ϕ_1, g, and ϕ_2, because they are maps of Euclidean spaces. This tells us how to define df. Let $x \in \phi_1(U_1)$ and let y be the point in U_1 such that $\phi_1(y) = x$. Then we have

$$\begin{array}{ccc} & df(x) & \\ T_x(X_1) & \rightarrow & T_{f(x)}(X_2) \\ d\phi_1(y) \uparrow & & \uparrow \quad d\phi_2(g(y)). \\ U_1 & \rightarrow & U_2 \\ & dg(y) & \end{array}$$

This defines $df(x)$ as $d\phi_2(g(y)) \circ dg(y) \circ [d\phi_1(y)]^{-1}$.

We need two basic theorems, the Preimage Theorem and the Transversality Theorem. These in turn are derived from two fundamental results, the Inverse Function Theorem and Sard's theorem. We will outline these results, and then prove parts (a) and (d) from the theorem in Chapter 3.

INVERSE FUNCTION THEOREM.   Suppose that U is an open subset of R^k, $f: U \to R^k$ is smooth, and $df(x^0)$ is nonsingular for some $x^0 \in U$. Then f is a local diffeomorphism in a neighborhood of x^0.

Another way of expressing the conclusion of this theorem is that there are open sets U^0 with $x^0 \in U^0 \subset U$ and $V^0 \subset R^k$ so that $f \mid U^0: U^0 \to V^0$ is a diffeomorphism from U^0 onto V^0. The theorem is not stated for manifolds, although a version of it holds for manifolds. Generally, any theorem about smooth maps of Euclidean spaces that is local in character has an immediate extension to manifolds, merely by restricting to local coordinates. Many advanced calculus books include a proof of the Inverse Function Theorem (see, e.g., [Spivak, 1965, p. 35]).

As an illustration of this theorem, consider the map $f: R^1 \to R^1$ defined by $f(x) = x^2$. Then $df(x) = 2x$, so $df(x^0) \neq 0$ except at $x^0 = 0$. We have two smooth inverses, $g(y) = \pm\sqrt{y}$, both of which are defined in a neighborhood of each point, y^0, in the image of f, except $y^0 = 0$ (see Fig. A4-2).

The following definition will be useful. Let U be an open subset of R^m. Suppose that $f: U \to R^k$ is smooth. We say that $y \in R^k$ is a *regular value* of f if for each $x \in U$ with $f(x) = y$, the rank of $df(x)$ is equal to k.

If you think of $df(x)$ as a $k \times m$ matrix (the Jacobian), then "$df(x)$ has rank k" means that the k rows of $df(x)$ are linearly independent. On the

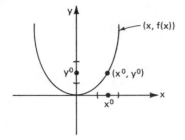

Figure A4-2   Graph of $y = f(x) = x^2$, to demonstrate the Inverse Function Theorem. Each point y^0 in the image of f is contained in an open interval on which an inverse is defined, except the point $y^0 = 0$.

other hand, if you think of $df(x)$ as a linear transformation, it means that $df(x)$ is onto R^k. Note that if $m < k$, this is impossible. However, we call y a regular value of f if there is no x with $f(x) = y$; that is, if y is not in the image of f. So if $m < k$, the only regular y are those not in the image of f. (If this does not seem logical to you, just consider it a convention.)

For example, if $f: R^2 \to R^1$ is defined by $f(x, y) = x^2 + y^2 - 1$, every point in R^1 is a regular value of f, except -1. On the other hand, if $f: R^1 \to R^2$ is defined by $f(x) = (x, x)$, no point in the image of f is a regular value, because $df(x)$ can never have rank 2. The regular values of this f are all the points $(x, y) \in R^2$ with $x \neq y$.

Now we can state the following.

PREIMAGE THEOREM.   Let W be an open subset of R^m. Suppose that $r: W \to R^k$ is smooth. If 0 is a regular value r, then either $r^{-1}(0)$ is empty or $r^{-1}(0)$ is a smooth $(m - k)$-dimensional submanifold of W.

Sketch of Proof. Let $x^0 \in r^{-1}(0)$. By the inverse function theorem, we can choose local coordinates about x^0 so that

$$r(x_1, x_2, \ldots, x_m) = (x_1, x_2, \ldots, x_k)$$

in local coordinates. Then, near x^0, $r^{-1}(0)$ looks like

$$\{(0, \ldots, 0, x_{k+1}, \ldots, x_m) : (x_{k+1}, \ldots, x_m) \in R^{m-k}\}.$$

See [Guillemin and Pollack, 1974, pp. 20–21] for a full proof.

Consider the examples given above. If $f(x, y) = x^2 + y^2 - 1$, then $f^{-1}(z)$ is a circle in R^2, unless $z = -1$, in which case it is the origin in R^2. Note that the circles are one-dimensional submanifolds, as predicted by the Preimage Theorem. If $f(x) = (x, x)$, then $f^{-1}(x, y)$ is empty unless $x = y$. In this case, $f^{-1}(x, x) = x$.

Recall that a set is *countable* if it can be indexed by the integers.

COROLLARY (WEAK VERSION OF BEZOUT'S THEOREM).   Let r be a system of m polynomial equations in m unknowns. If every solution to $r = 0$ is nonsingular, $r = 0$ has only a countable number of solutions.

Proof. By hypothesis, 0 is a regular value of $r: C^m \to C^m$. Thus either $r^{-1}(0)$ is empty or $r^{-1}(0)$ is a smooth zero-dimensional submanifold of C^m, by the Preimage Theorem. Therefore, each solution is geometrically isolated. But a subset of a Euclidean space consisting of geometrically isolated points is countable. (This is a consequence of the fact that any open cover of a Euclidean space has a countable subcover [Dugundji, 1967, Chap. 8, Sec. 6].)

Now we have the following amazing theorem.

SARD'S THEOREM.   Let Z be a smooth m-dimensional submanifold of R^N. If $f: Z \to R^k$ is any smooth map, then almost every point in R^k is a regular value of f.

Recall from Section 3-3 that "almost every point in R^k is a regular value" means that there is a subset, A, of R^k of measure zero and every point in $R^k - A$ is a regular value. In terms of the examples above, we see that $\{-1\}$ is a set of measure zero in R^1 and $\{(x, x): x \in R^1\}$ is a set of measure zero in R^2.

See [Guillemin and Pollack, 1974, App. A] for a proof. Sard's theorem yields the major technical result we need.

TRANSVERSALITY THEOREM.   Let W be an open subset of R^m. Suppose that $r: W \times R^k \to R^k$, and 0 is a regular value of r. Define $r_a: W \to R^k$ by $r_a(y) = r(y, a)$. Then 0 is a regular value of r_a for almost all $a \in R^k$.

Proof. Let $Z = r^{-1}(0)$, a smooth submanifold of $W \times R^k$ by the Pre-image Theorem. Define $\pi: W \times R^k \to R^k$ by $\pi(y, a) = a$. Consider $\pi_Z: Z \to R^k$, the restriction of π to Z. By Sard's theorem, a is a regular value of π_Z for almost all $a \in R^k$.

Let $(y, a) \in W \times R^k$ be chosen with $r(y, a) = 0$. Choose $\alpha \in R^k$. Since $dr(y, a)$ has rank k, there is a $(\beta, \gamma) \in R^m \times R^k$ such that $dr(y, a) \cdot (\beta, \gamma) = \alpha$. We want some $\delta \in R^m$ such that $dr_a(y) \cdot \delta = \alpha$. Note that $dr(y, a) \cdot (\delta, 0) = dr_a(y) \cdot \delta$ for $\delta \in R^m$.

By Sard's theorem, we may assume that $d\pi_Z(y, a)$ has rank k. Therefore, there is a $(\beta', \gamma) \in T_{(y, a)}Z$, the tangent space of Z at the point (y, a). But $Z = r^{-1}(0)$, so $dr(y, a) \cdot (\beta', \gamma) = 0$. Let $\delta = \beta - \beta'$. Then

$$dr_a(y) \cdot \delta = dr(y, a) \cdot ((\beta, \gamma) - (\beta', \gamma)) = dr(y, a) \cdot (\beta, \gamma) = \alpha.$$

This completes the proof.

See [Guillemin and Pollack, 1974, pp. 68–69].

Now let's prove parts (a) and (d) of the theorem from Chapter 3. We have two results to establish:

LEMMA A.   For almost all $q \in C^n$, 0 is a regular value of h restricted to $C^n \times (0, 1)$.

LEMMA B.   For almost all $p \in C^n$ and any $\epsilon > 0$, $h^{-1}(0)$ intersects $C^n \times (0, 1 - \epsilon)$ in a bounded subset of $C^n \times (0, 1)$.

Lemma A implies that $h^{-1}(0)$ consists of smooth paths in $C^n \times (0, 1)$. In fact, from the Preimage Theorem it follows that $h^{-1}(0)$ is a smooth one-dimensional submanifold of $C^n \times (0, 1)$. Part (a) of the theorem in Chapter

3 follows. Lemma B implies that these paths can become unbounded only as t approaches 1. Since we shall see that if a path diverges, $h = 0$ has a solution at infinity, part (d) of the theorem in Chapter 3 follows. (Strictly, the behavior of paths as t approaches 0 requires an additional argument. However, this is easy.)

We view $h: C^n \times (0, 1) \to C^n$ as a map $\bar{h}: R^{2n} \times (0, 1) \to R^{2n}$, where $h(x, t)$ is identified with $\bar{h}(y, t)$ and $y_{2i-1} = \text{Re}(x_i)$, $y_{2i} = \text{Im}(x_i)$, $\bar{h}_{2i-1}(y, t) = \text{Re}(h_i(x, t))$, and $\bar{h}_{2i}(y, t) = \text{Im}(h_i(x, t))$, where "Re" and "Im" denote the real and imaginary parts, respectively. We shall use these correspondences freely (see Appendix 3).

Proof of Lemma A. Observe that

$$\frac{\partial h_j}{\partial q_j}(x, t) = -(1 - t) \neq 0$$

for $j = 1, \ldots, n$ when $t \neq 1$. By the Transversality Theorem, this implies that the Jacobian of h with q fixed has real-rank $2n$ at solutions to $h = 0$, for almost all choices of $q \in C^n$. Thus 0 is a regular value of h.†

Proof of Lemma B. If Lemma B is false, there is a t^0 between 0 and 1 such that $h(x, t^0) = 0$ has a solution at infinity, x^0. To see this, let (x^s, t^s) denote a solution path diverging to infinity as $t^s \to t^0$. Then $x^s/|x^s|$ has a limit point x^0. (Note that $x^0 \neq 0$.) This (x^0, t^0) satisfies $h^0(x^0, t^0) = 0$, where h^0 is the homogeneous part of h, because

$$0 = h_j(x^s, t^s) = \hat{h}_j(x^s, 1, t^s) = \frac{1}{|x^s|^{d_j}} \hat{h}_j(x^s, 1, t^s)$$

$$= \hat{h}_j\left(\frac{x^s}{|x^s|}, \frac{1}{|x^s|}, t^s\right) \to \hat{h}_j(x^0, 0, t^0) = h_j^0(x^0, t^0)$$

as $t^s \to t^0$, for $j = 1$ to n. Here, $\hat{h}$ denotes the homogenization of h (with t fixed) (see Section 3-3).

The proof of Lemma B is organized into n cases. In case k, we prove that h^0 can have no solutions of the form $(x_1, \ldots, x_k, 0, \ldots, 0, t)$, with $x_j \neq 0$ for $j = 1$ to k and $0 < t < 1$, for almost all p. By the symmetry of the argument, this will imply that h^0 has no solutions with exactly k com-

---

† The same proof shows that for any t and for almost all q, 0 is a regular value of h with q and t fixed. In other words, for each $t \in (0, 1)$, there is a set of measure zero, A_t, in C^n such that if $q \notin A_t$, then $dh_x(x)$ is an invertible square matrix when $h(x) = 0$, where $dh_x(x)$ denotes the Jacobian of h with respect to x alone. But in implementations we will encounter only a finite number of t. This seems to suggest that if we choose q at random, we may assume that $dh_x(x)$ is always invertible. My experience confirms this suggestion, and the CONSOL codes assume it, except CONSOL9 (see Chapter 6). However, the issue of how badly conditioned $dh_x(x)$ will tend to become is somewhat controversial. I have not observed any difficulty. (See the footnote in Appendix 3.)

ponents nonzero, for almost all p. Since the proof is valid for $k = 1, \ldots, n$, the result follows.

Let

$$V_* = \{(x_1, \ldots, x_k) \in C^k : x_j \neq 0 \text{ for } j = 1, \ldots, k\}$$

$$V = V_* \times (0, 1),$$

and define

$$s_j(x_1, \ldots, x_k, t, p_1, \ldots, p_k)$$

$$= h_j^0(x_1, \ldots, x_k, 0, \ldots, 0, t, p_1, \ldots, p_k, p_{k+1}, \ldots, p_n)$$

for $j = 1, \ldots, k$, where $p_{k+1}, \ldots, p_n$ are arbitrary constants.

We will show that for almost all $(p_1, \ldots, p_k) \in C^k$, there is no $(x_1, \ldots, x_k, t) \in V$ satisfying $s(x_1, \ldots, x_k, t, p_1, \ldots, p_k) = 0$.

Since s is homogeneous in x, if $s(x, t, p) = 0$, then $s(c \cdot x, t, p) = 0$ for any $c \in C$. Thus we may assume that $x_j = 1$ for some j. Without loss of generality, let $x_k = 1$.

Now define

$$W = \{(y_1, \ldots, y_{2k-1}) \in R^{2k-1} : (y_{2j-1}, y_{2j}) \neq (0, 0) \text{ for }$$

$$j = 1 \text{ to } k - 1, y_{2k-1} \neq 0\}$$

and define $r: W \times R^{2k-1} \to R^{2k-1}$ by

$$r_{2j-1}(y, a) = \text{Re}[s_j(x, t, p)] \quad \text{for } j = 1 \text{ to } k$$

$$r_{2j}(y, a) = \text{Im}[s_j(x, t, p)] \quad \text{for } j = 1 \text{ to } k - 1,$$

where

$$y_1 = \text{Re}[x_1] \qquad\qquad a_1 = \text{Re}[p_1]$$

$$y_2 = \text{Im}[x_1] \qquad\qquad a_2 = \text{Im}[p_1]$$

$$\vdots \qquad\qquad\qquad\qquad \vdots$$

$$y_{2k-2} = \text{Im}[x_{k-1}] \qquad a_{2k-2} = \text{Im}[p_{k-1}]$$

$$y_{2k-1} = t \qquad\qquad\quad a_{2k-1} = \text{Re}[p_k].$$

Now,

$$\begin{bmatrix} \dfrac{\partial r_{2j-1}}{\partial a_{2j-1}} & \dfrac{\partial r_{2j}}{\partial a_{2j-1}} \\[2ex] \dfrac{\partial r_{2j-1}}{\partial a_{2j}} & \dfrac{\partial r_{2j}}{\partial a_{2j}} \end{bmatrix} = (1 - t) \begin{bmatrix} \text{Re}[z_j] & -\text{Im}[z_j] \\[1ex] \text{Im}[z_j] & \text{Re}[z_j] \end{bmatrix},$$

where $z_j = x_j^{d_j} \neq 0$ for $j = 1$ to $k - 1$ (because $x \in V_*$), and

$$\frac{\partial r_{2k-1}}{\partial a_{2k-1}} = (1 - t) \neq 0.$$

Thus the Jacobian of r at solutions has rank $2k - 1$.

Therefore, by the Transversality Theorem, r_a has 0 as a regular value, except perhaps when $a \in A_0$, where A_0 is some subset of R^{2k-1} of measure zero. By the corollary to the Preimage Theorem, $r_a = 0$ can have only a countable number of solutions. But we have one additional equation:

$$u = \text{Im}[s_k(x, t, p)] = 0 \tag{A4-1}$$

which must be satisfied if a solution (x, t) to $s = 0$ exists.

Now

$$u = \text{Im}[(1 - t)p_k x_k^{d_k} + t f_k(x)]$$

$$= \text{Im}[(1 - t)p_k \cdot 1 + t f_k(x)]$$

$$= (1 - t) \, \text{Im}[p_k] + t \, \text{Im}[f_k(x)],$$

so (A4-1) implies that

$$\text{Im}[p_k] = \left[\frac{-t}{1 - t}\right] \text{Im}[f_k(x)]. \tag{A4-2}$$

Since there are no more than a countable number of (x, t) that satisfy $r_a(x, t) = 0$ when $a \notin A_0$, there are no more than a countable number of $\text{Im}[p_k]$ that satisfy (A4-2) when $r_a(x, t) = 0$ and $a \notin A_0$. Thus the set of measure zero that p must avoid to keep $s = 0$ from having any solutions is the Cartesian product of A_0 and this countable set. (It is easy to show from the definition of "measure zero" in Section 3-3 that any countable subset of a Euclidean space has measure zero, and also that a Cartesian product of sets of measure zero has measure zero.) Thus Lemma B is proven.

NOTES

1. A number of papers on finding all solutions to polynomial systems by continuation have been published: for example, [Brunovský and Meravý, 1984], [Chow et al., 1979], [Drexler, 1977, 1978], [Garcia and Li, 1980], [Garcia and Zangwill, 1979–1, 1979–2, 1979–3, 1981], [Li, 1983], [Morgan, 1983, 1986–1, 1986–2], [Morgan and Sommese, 1986, 1987], and [Wright, 1985].

2. Differential topology does not yield the sharpest results on using continuation to solve polynomial systems. Algebraic geometry does. [To get a sense of this, compare Bezout's theorem (Chapter 3) to the corollary to the Preimage Theorem.] However, we do not have any elementary exposition, such as [Guillemin and

Pollack, 1974], which makes accessible the central results we need from algebraic geometry. Thus I have given the definition of multiplicity and Bezout's theorem not only without proof but without realistic expectation that the cited references will be readable by anyone this book is written for. The first two chapters of [Guillemin and Pollack, 1974] can be understood by an engineer or scientist who is willing to learn a little more advanced calculus. The algebraic geometry results we would like to use from [Fischer, 1977], [Fulton, 1984], and [Shafarevich, 1977] can be extracted by Ph.D.-level mathematicians and their advanced students. Few others will find the struggle worth it. Consequently, I have opted for the weaker statements of some results (e.g., the theorem in Chapter 3) that rest essentially on topology. However, you should be aware of this foundational weakness since you may one day want to use the stronger results. See [Morgan and Sommese, 1986, 1987] for an illustration.

APPENDIX 5

Gaussian Elimination
for System Reduction

Gaussian elimination is most often discussed as a technique for solving linear systems (as in [Forsythe et al., 1977, Chap. 3] or [Forsythe and Moler, 1967, Chap. 9]). However, in this appendix we consider elimination for the reduction of polynomial systems. The approach here is elementary, emphasizing row reduction of $n \times m$ matrices with $n < m$. (This is fully discussed in [Cullen, 1966] and [Hoffman and Kunze, 1961].) Chapter 7 describes how polynomial systems generate matrices and the relevance of matrix reduction to polynomial systems. Examples are given in Chapters 8 through 10.

Let A be an $n \times m$ matrix. The Gaussian elimination algorithm, GEL, will generate from A, by the process of row reduction, a matrix, A_r, with the following properties:

(a) Either all the rows of A_r are nonzero or the first n^* rows of A_r are nonzero and the rest zero.

(b) The ith nonzero row of A has a first nonzero element, $a_{i,j(i)}$, where $a_{i,j} = 0$ for $1 \leq j < j(i)$.

(c) $j(i) < j(i + 1)$ for $i = 1, n^*$.

(d) $a_{k,j(i)} = 0$ for $k \neq i$ for $i = 1, n^*$.

(e) $a_{i,j(i)} = 1$ for $i = 1, n^*$.

Typically, A_r might look like

$$\begin{bmatrix} 1 & * & 0 & 0 & * & * & 0 & * & * \\ 0 & 0 & 1 & 0 & * & * & 0 & * & * \\ 0 & 0 & 0 & 1 & * & * & 0 & * & * \\ 0 & 0 & 0 & 0 & 0 & 0 & 1 & * & * \end{bmatrix}$$

or

$$\begin{bmatrix} 0 & 0 & 1 & * & 0 \\ 0 & 0 & 0 & 0 & 1 \\ 0 & 0 & 0 & 0 & 0 \end{bmatrix},$$

where "$*$" indicates any value (including possibly zero), with zeros to the left of each "first nonzero element" of each row and zeros above and below each "first nonzero element."

There are three types of elementary row operations:

1. Add a scalar multiple of one row to another.
2. Multiply a row by a nonzero constant.
3. Interchange two rows.

Performing elementary row operations on A corresponds to taking algebraic combinations of the original polynomial equations that generated A. This does not change the (finite) solutions to the polynomial system. It can change the solutions at infinity. In fact, that is why we are doing it.

Here is the row reduction algorithm: We shall consider the n rows and m columns of A and "settle" them in sequence from $i = 1, 2, \ldots, n$, $j = 1, 2, \ldots, m$. Let i denote the "current row" and j denote the "current column." Begin with $i = 1$ and $j = 1$.

1. Look at the current row–current column entry. This is the "current entry."
   (a) If the current entry and all entries in the column below it are zero, go to the next column; that is, take $j = j + 1$.
   (b) If at least one entry at or below the current entry is nonzero, exchange with the current row the row with the largest (in absolute value) entry in the current column at or below the current entry. Now make every other entry in the current column zero by adding multiples of the (new) current row to every other row. Divide the current row through by the leading entry so that this entry becomes "1." Go to the next row ($i = i + 1$) and go to the next column ($j = j + 1$).
2. Continue in this way until $j > m$ or $i > n$.

See [Cullen, 1966] or [Hoffman and Kunze, 1961] for more details.

Often, one wishes to reduce matrices of symbols (rather than numbers) by hand or via a symbol-manipulating computer code (such as described in

[Hearn, 1983]), where the main issue is the exact cancellation of terms. This might be the case when we are looking for an underlying structure in the system. In this circumstance, I choose the pivots (the exchange rows) at each step by inspection, and I do not have GEL make the leading entries of the rows equal to "1." Sometimes, it is convenient to leave the rows in their original order, so that a permutation of rows would be required to put A_r in the triangular form described above. Sometimes, to avoid excessively complicated expressions, it is convenient to stop GEL before it completes the reduction, yielding a partially reduced system. See, for example, (10-17).

APPENDIX 6

Computer Programs

This appendix contains the computer programs referenced in the text. These codes are available in machine-readable form from the author. See the card enclosed with this book.

The programs are written in FORTRAN. Codes that use complex arithmetic are provided in two versions: one using the COMPLEX∗16 declaration and one using only real declarations. These "real" codes emulate complex operations using real arithmetic, as described in Appendix 2. Some codes have only a real version because they do not use complex arithmetic (e.g., CONSOL1, SCLGEN, SCLCEN). One complete set of the codes, including those with complex declarations, is given below in Part A. The real-arithmetic versions of the codes that have complex declarations are given in Part B. I use the following naming convention: PGM and PGMR for the complex FORTRAN program PGM and its real FORTRAN version, respectively. The codes are notated appropriately in comments at the top of each main routine.

I suggest that you use the codes in Part A unless your FORTRAN compiler does not allow a double-precision complex declaration or unless the complex arithmetic is limited in some other way (see Appendix 2). I would use the real versions rather than go to single precision. (All codes are written in double precision.) If you must change to single precision, then:

(a) The double-precision declarations that begin each subroutine must be changed to single precision.
(b) Calls to the double-precision routines DABS, DSQRT, DCOS, DSIN,

DFLOAT, and DLOG10 must be changed to their single-precision counterparts.

(c) Constants of the form "sxxx.Dsyy" must be changed to "sxxx.Esyy."

(d) Some long constants in subroutines STRPT and INPUTB may have to be shortened.

Aside from the issue of complex declarations, two or more versions of some subroutines are provided. For example, FFUNT is a version of FFUN for CONSOL8. I have given such versions separate names, but to use one of them, its name must be changed to the appropriate generic name. To use FFUNT with CONSOL8, its name must be changed to FFUN. Each such subroutine is identified with comments at the beginning that cite the associated generic name.

The programs are organized below by chapters, except that utility codes are grouped at the end of each part. A full table of contents is given, and the codes follow.

I have done my best to make the codes portable, but I also realize that some of these programs may not work for you "off the shelf." The best advice I can offer is to run the codes in sequence (while reading the book). The programs for Chapter 1 are relatively simple. By trying them first, it will be easier for you to get a sense of what your conversion issues are than with the programs in later chapters.

Table of Contents

PART A: COMPLETE FORTRAN PROGRAMS FOR THE BOOK, INCLUDING COMPLEX DECLARATIONS IF APPROPRIATE

Section 1: Programs for Chapter 1

```
C PROGRAM NAME:  CONSOL1
C
C PURPOSE:   SOLVES THE QUADRATIC EQUATION
C
C                 EQ = X**2 + A*X + B
C
C              USING THE CONTINUATION EQUATION
C
C              H   = X**2 + A*T*X + B.
C
C REFERENCE: CHAPTER 1 OF "SOLVING POLYNOMIAL SYSTEMS USING
C            CONTINUATION" BY ALEXANDER MORGAN
C
C NOTE:    1. CONSOL1 ASSUMES B IS NEGATIVE.
C          2. INPUT IS ON UNIT 5.  OUTPUT IS TO UNIT 6.
C             SAMPLE INPUT FOR DELT=.05, A=10., B=-.2:
C                   5.D-02
C                   1.D+01
C                   -2.D-01
C
C INPUT VARIABLES:
C
C     DELT        -- THE STEP SIZE FOR T
C     A           -- COEFFICIENT FOR EQ
C     B           -- COEFFICIENT FOR EQ
C
C
C OUTPUT VARIABLES:
C
C    FOR EACH STEP:
C
C     NUMSTP      -- CURRENT STEP NUMBER
C     T           -- CURRENT  VALUE OF T
C     X           -- CURRENT VALUE OF X
C     DH          -- CURRENT DERIVATIVE OF CONTINUATION EQUATION
C     RESID       -- NEWTON METHOD RESIDUAL (SHOULD BE NEAR 0)
C
C
C    AT THE END OF THE PATH:
C
C     NUMIT       -- NUMBER OF NEWTON ITERATIONS FOR THE PATH
C
C    AT THE END OF THE RUN:
C
C     ITOTIT      -- TOTAL NUMBER OF NEWTON ITERATIONS OVER ALL PATHS
C
C
C SUBROUTINES: DABS,MOD
C
C
C DECLARATIONS
      INTEGER ITER,ITOTIT,NS,NUMIT,NUMPAT,NUMSTP
      DOUBLE PRECISION A,B,DELT,DH,H,RESID,X,T
C
C
```

```
C INPUT STATEMENTS FOLLOW.
      READ(5,10) DELT
   10 FORMAT(D22.15)
      READ(5,10) A,B
C
      WRITE(6,20)
   20 FORMAT(/,'  CONSOL1',//)
      WRITE(6,25) DELT
   25 FORMAT(/,'  DELT =',D22.15,/)
      WRITE(6,30) A,B
   30 FORMAT('  COEFFICIENTS,    A=',D22.15,' B=',D22.15)
C
C COMPUTE NUMBER OF STEPS PER PATH
      NS = (1./DELT) + 1.
C
      WRITE(6,40)
   40 FORMAT(/)
C
      WRITE(6,50)  NS
   50 FORMAT(' NUMBER OF STEPS PER PATH =',I15//)
C
C INITIALIZE COUNTER FOR TOTAL NUMBER OF CORRECTOR ITERATIONS
      ITOTIT=0
C
C ****************************************************************
C *****                                                    *****
C *****      BEGINNING OF MAIN PART OF PROGRAM             *****
C *****                                                    *****
C ****************************************************************
C
C PATHS LOOP -- ITERATE THROUGH PATHS
C
      DO 60 NUMPAT = 1 ,  2
C
         WRITE(6,70) NUMPAT
   70    FORMAT(' PATH NUMBER =',I15)
C
C INITIALIZE COUNTERS AND CONSTANTS FOR THE PATH
         NUMIT=0
         T=0
C
C GENERATE A START POINT, S, FOR THE PATH
         X=DSQRT(-B)
         IF(NUMPAT.EQ.2)X=-X
C
         WRITE(6,80)X
   80    FORMAT(/,'  START X=',D22.15,//)
C
C STEPS LOOP -- ITERATE STEPS ALONG THE PATH
         DO 90 NUMSTP = 1 , NS
C
            T=T+DELT
            IF(T.GE.1.)T=1.
C
C
```

```
C NEWTON'S METHOD LOOP
                DO 100 ITER = 1,10
C
C EVALUATE THE CONTINUATION EQUATION AND DERIVATIVE
                H = X**2 + A*T*X + B
                DH = 2.*X + A*T
C
C GENERATE THE RESIDUAL AND UPDATE X
                RESID = H/DH
                X = X - RESID
C
C UPDATE NEWTON ITERATION COUNTER
                NUMIT = NUMIT + 1
C
C LEAVE NEWTON LOOP IF RESIDUAL IS SMALL ENOUGH
                IF(DABS(RESID) .LT. 1.D-12) GOTO 101
C
  100           CONTINUE
  101           CONTINUE
C
C
                IF(MOD(NUMSTP,NS/10).EQ.0 .OR. NUMSTP.GE.(NS-4)) THEN
                WRITE(6,110) NUMSTP
  110           FORMAT(/' STEP SUMMARY FOR STEP NUMBER',I15,' :')
                WRITE(6,115)
  115           FORMAT(' ----------------------------------------')
                WRITE(6,120) T
  120           FORMAT('  T  = ',D22.15)
                WRITE(6,130) X
  130           FORMAT('  X  =           ',D22.15)
                WRITE(6,135) H
  135           FORMAT('  H  =           ',D22.15)
                WRITE(6,140) RESID
  140           FORMAT(' RESIDUAL =     ',D22.15)
                WRITE(6,150) DH
  150           FORMAT('  DH =          ',D22.15)
                WRITE(6,115)
                WRITE(6,40)
                ENDIF
C
   90      CONTINUE
C             BOTTOM OF LOOP -- NUMSTP
C
           ITOTIT=ITOTIT + NUMIT
C
           WRITE(6,160) NUMIT
  160      FORMAT(' TOTAL NEWTON ITERATIONS ON PATH =',I15)
           WRITE(6,40)
C
   60  CONTINUE
C             BOTTOM OF LOOP -- NUMPAT
C
       WRITE(6,170) ITOTIT
  170  FORMAT(//' TOTAL NEWTON ITERATIONS =',I15//)
C
       STOP
       END
```

```
C PROGRAM NAME:  CONSOL2
C
C PURPOSE:    SOLVES THE QUADRATIC EQUATION
C
C                 EQ = X**2 + A*X + B
C
C               USING THE CONTINUATION EQUATION
C
C                 H  = X**2 + A*T*X + B.
C
C REFERENCE: CHAPTER 1 OF "SOLVING POLYNOMIAL SYSTEMS USING
C            CONTINUATION" BY ALEXANDER MORGAN
C
C NOTE: 1.    THIS PROGRAM USES THE FORTRAN "COMPLEX*16" DECLARATION.
C
C       2.    THIS PROGRAM GENERATES MANY UNDERFLOWS.  THIS IS
C             NOT AN ERROR CONDITION.  IF YOU NEED TO PREVENT
C             UNDERFLOWS FROM OCCURRING, SEE THE COMMENTS LABELED
C             "UNDERFLOWS" BELOW.
C
C       3.    INPUT IS ON UNIT 5.  OUTPUT IS TO UNIT 6.
C             SAMPLE INPUT FOR DELT=.05, A=10., B=.2:
C                 5.D-02
C                 1.D+01
C                 2.D-01
C
C INPUT VARIABLES:
C
C     DELT        -- THE STEP SIZE FOR T
C     A           -- COEFFICIENT FOR EQ
C     B           -- COEFFICIENT FOR EQ
C
C OUTPUT VARIABLES:
C
C   FOR EACH STEP:
C
C     NUMSTP      -- CURRENT STEP NUMBER
C     T           -- CURRENT VALUE OF T
C     X           -- CURRENT VALUE OF X
C     H           -- CURRENT VALUE OF THE CONTINUATION EQUATION
C     RESID       -- NEWTON METHOD RESIDUAL (SHOULD BE NEAR 0)
C     DH          -- CURRENT DERIVATIVE OF CONTINUATION EQUATION
C
C   AT THE END OF THE PATH:
C
C     NUMIT       -- NUMBER OF NEWTON ITERATIONS FOR THE PATH
C
C   AT THE END OF THE RUN:
C
C     ITOTIT      -- TOTAL NUMBER OF NEWTON ITERATIONS OVER ALL PATHS
C
C SUBROUTINES: DABS,DCMPLX,DIMAG,DREAL,DSQRT,MOD
C
C
```

```
C DECLARATIONS
      INTEGER ITER,ITOTIT,NS,NUMIT,NUMPAT,NUMSTP
      DOUBLE PRECISION A,B,DELT,EPSOO,T,XRESID
      COMPLEX*16 DH,H,RESID,X
C
C
C INPUT STATEMENTS FOLLOW.
      READ(5,10) DELT
  10  FORMAT(D22.15)
      READ(5,10) A,B
C
      WRITE(6,20)
  20  FORMAT(/,'  CONSOL2',//)
      WRITE(6,25) DELT
  25  FORMAT(/,'  DELT =',D22.15,/)
      WRITE(6,30) A,B
  30  FORMAT('  COEFFICIENTS,    A=',D22.15,'  B=',D22.15)
C
C
C COMPUTE NUMBER OF STEPS PER PATH
      NS = (1./DELT) + 1.
C
      WRITE(6,40)
  40  FORMAT(/)
C
      WRITE(6,50)  NS
  50  FORMAT(' NUMBER OF STEPS PER PATH =',I10//)
C
C INITIALIZE COUNTER FOR TOTAL NUMBER OF CORRECTOR ITERATIONS
      ITOTIT=0
C
C ***************************************************************
C *****                                                    *****
C *****        BEGINNING OF MAIN PART OF PROGRAM           *****
C *****                                                    *****
C ***************************************************************
C
C PATHS LOOP -- ITERATE THROUGH PATHS
      DO 60 NUMPAT = 1 , 2
C
          WRITE(6,70) NUMPAT
  70      FORMAT('  PATH NUMBER =',I10)
C
C INITIALIZE COUNTERS AND CONSTANTS FOR THE PATH
          NUMIT=0
          T=0.D0
C
C GENERATE A START POINT, X, FOR THE PATH
          IF(B.LE.0.D0) X=DCMPLX(DSQRT(DABS(B)),0.D0             )
          IF(B.GT.0.D0) X=DCMPLX(0.D0            ,DSQRT(DABS(B)))
          IF(NUMPAT.EQ.2)X=-X
C
          WRITE(6,80)X
  80      FORMAT(/,'   START X=',2D22.15,//)
C
```

```
C STEPS LOOP -- ITERATE STEPS ALONG THE PATH
         DO 90 NUMSTP = 1 , NS
            T=T+DELT
            IF(T.GE.1.DO)T=1.DO
C
C NEWTON'S METHOD LOOP
            DO 100 ITER = 1,100
C
C EVALUATE THE CONTINUATION EQUATION AND DERIVATIVE
               H = X**2 + A*T*X + B
               DH = 2.*X + A*T
C
C GENERATE THE RESIDUAL AND UPDATE X
               RESID = H/DH
               X = X - RESID
C
C$$$$$$$$$$$$$$$$$$$$$$$$$$$$$$$$$$$$$$$$$$$$$$$$$$$$$$$$$$$$$$$$$$$$$$$$$$$$
C UNDERFLOWS :  IF YOU MUST PREVENT UNDERFLOWS, I SUGGEST YOU
C              TAKE THE COMMENTS OFF THE FOLLOWING BLOCK OF
C              CODE.  THE VALUE FOR EPSOO MAY NEED TO BE ADJUSTED.
C              EPSOO=1.D-22
C              IF(DABS(DREAL(X)).LT.EPSOO) X=DCMPLX(0.DO,DIMAG(X))
C              IF(DABS(DIMAG(X)).LT.EPSOO) X=DCMPLX(DREAL(X),0.DO)
C$$$$$$$$$$$$$$$$$$$$$$$$$$$$$$$$$$$$$$$$$$$$$$$$$$$$$$$$$$$$$$$$$$$$$$$$$$$$
C
C UPDATE NEWTON ITERATION COUNTER
               NUMIT = NUMIT + 1
C
C LEAVE NEWTON LOOP IF RESIDUAL IS SMALL ENOUGH
               XRESID = DABS(DREAL(RESID))+DABS(DIMAG(RESID))
               IF(XRESID .LT. 1.D-12) GOTO 101
C
 100           CONTINUE
 101           CONTINUE
C
               IF(MOD(NUMSTP,NS/10).EQ.0.OR.NUMSTP.GE.(NS-4))THEN
               WRITE(6,110) NUMSTP
 110           FORMAT(/' STEP SUMMARY FOR STEP NUMBER',I10,' :')
               WRITE(6,115)
 115           FORMAT(' ---------------------------------------------')
               WRITE(6,120) T
 120           FORMAT(' T  = ',D22.15)
               WRITE(6,125)
 125           FORMAT('                    REAL PART        IMAGINARY PART')
               WRITE(6,130) X
 130           FORMAT(' X  =        ',D11.4,'        ',D11.4)
               WRITE(6,135)  H
 135           FORMAT(' H  =        ',D11.4,'        ',D11.4)
               WRITE(6,140) RESID
 140           FORMAT(' RESIDUAL =  ',D11.4,'        ',D11.4)
               WRITE(6,150) DH
 150           FORMAT(' DH  =       ',D11.4,'        ',D11.4)
               WRITE(6,115)
               WRITE(6,40)
               ENDIF
 90        CONTINUE
C              BOTTOM OF LOOP -- NUMSTP
```

```
C
            ITOTIT=ITOTIT + NUMIT
C
            WRITE(6,160) NUMIT
   160      FORMAT('  TOTAL NEWTON ITERATIONS ON PATH =',I5)
            WRITE(6,40)
C
    60   CONTINUE
C               BOTTOM OF LOOP -- NUMPAT
C
         WRITE(6,170) ITOTIT
   170   FORMAT(//'  TOTAL NEWTON ITERATIONS =',I10//)
C
         STOP
         END
```

```
C PROGRAM NAME:  CONSOL3
C
C PURPOSE:    SOLVES THE QUADRATIC EQUATION
C
C                   EQ = X**2 + A*X + B
C
C               USING THE CONTINUATION EQUATION
C
C                   H = X**2 + A*T*X + (T*B - (1.-T)*Q**2)
C
C REFERENCE: CHAPTER 1 OF "SOLVING POLYNOMIAL SYSTEMS USING
C            CONTINUATION" BY ALEXANDER MORGAN
C
C NOTE:  1.  THIS PROGRAM USES THE FORTRAN "COMPLEX*16" DECLARATION.
C
C        2.  THIS PROGRAM GENERATES MANY UNDERFLOWS.  THIS IS
C            NOT AN ERROR CONDITION.  IF YOU NEED TO PREVENT
C            UNDERFLOWS FROM OCCURRING, SEE THE COMMENTS LABELED
C            "UNDERFLOWS" BELOW.
C
C        3.  INPUT IS ON UNIT 5.  OUTPUT IS TO UNIT 6.
C            SAMPLE INPUT FOR DELT=.05, A=10., B=.2:
C                   5.D-02
C                   1.D+01
C                   2.D-01
C
C INPUT VARIABLES:
C
C     DELT          -- THE STEP SIZE FOR T
C     A             -- COEFFICIENT FOR EQ
C     B             -- COEFFICIENT FOR EQ
C
C OUTPUT VARIABLES:
C
C     FOR EACH STEP:
C
C     NUMSTP        -- CURRENT STEP NUMBER
C     T             -- CURRENT VALUE OF T
C     X             -- CURRENT VALUE OF X
C     H             -- CURRENT VALUE OF THE CONTINUATION EQUATION
C     RESID         -- NEWTON METHOD RESIDUAL (SHOULD BE NEAR 0)
C     DH            -- CURRENT DERIVATIVE OF CONTINUATION EQUATION
C
C
C     AT THE END OF THE PATH:
C
C     NUMIT                 NUMBER OF NEWTON ITERATIONS
C                                FOR THE PATH
C     AT THE END OF THE RUN:
C
C     ITOTIT                TOTAL NUMBER OF NEWTON ITERATIONS
C                                OVER ALL PATHS
C
C SUBROUTINES: DABS,DCMPLX,DIMAG,DREAL,MOD
C
```

```
C
C DECLARATIONS
      INTEGER ITER,ITOTIT,NS,NUMIT,NUMPAT,NUMSTP
      DOUBLE PRECISION A,B,DELT,EPSOO,T,XRESID
      COMPLEX*16 DH,H,Q,QSQ,RESID,X
C
C
C INPUT STATEMENTS FOLLOW.
      READ(5,10) DELT
   10 FORMAT(D22.15)
      READ(5,10) A,B
C
      WRITE(6,20)
   20 FORMAT(/,' CONSOL3',//)
      WRITE(6,25) DELT
   25 FORMAT(/,'  DELT =',D22.15,/)
      WRITE(6,30) A,B
   30 FORMAT('  COEFFICIENTS,    A=',D22.15,'  B=',D22.15)
C
C COMPUTE NUMBER OF STEPS PER PATH
      NS = (1./DELT) + 1.
C
      WRITE(6,40)
   40 FORMAT(/)
C
      WRITE(6,50)  NS
   50 FORMAT(' NUMBER OF STEPS PER PATH =',I5//)
C
C INITIALIZE COUNTER AND CONSTANTS FOR THE RUN
      ITOTIT=0
      Q = DCMPLX(.928746354D0, .265465386D0)
      QSQ=Q**2
C
C ****************************************************************
C *****                                                    *****
C *****       BEGINNING OF MAIN PART OF PROGRAM            *****
C *****                                                    *****
C ****************************************************************
C
C PATHS LOOP -- ITERATE THROUGH PATHS
      DO  60 NUMPAT = 1 ,  2
C
         WRITE(6,70) NUMPAT
   70    FORMAT('  PATH NUMBER =',I5)
C
C INITIALIZE COUNTERS AND CONSTANTS FOR THE PATH
         NUMIT=0
         T=0
C
C GENERATE A START POINT, S, FOR THE PATH
         X=Q
         IF(NUMPAT.EQ.2)X=-X
C
         WRITE(6,80)X
   80    FORMAT(/,'  START X=',2D22.15,//)
C
C
```

```
C STEPS LOOP -- ITERATE STEPS ALONG THE PATH
          DO 90 NUMSTP = 1 , NS
             T=T+DELT
             IF(T.GE.1.)T=1.
C
C NEWTON'S METHOD LOOP
             DO 100 ITER = 1,10
C
C EVALUATE THE CONTINUATION EQUATION AND DERIVATIVE
             H = X**2 + A*T*X + (T*B - (1.-T)*QSQ)
             DH = 2.*X + A*T
C
C GENERATE THE RESIDUAL AND UPDATE X
             RESID = H/DH
             X = X - RESID
C
C$$$$$$$$$$$$$$$$$$$$$$$$$$$$$$$$$$$$$$$$$$$$$$$$$$$$$$$$$$$$$$$$$$$$$$$$$$$
C UNDERFLOWS :  IF YOU MUST PREVENT UNDERFLOWS, I SUGGEST YOU
C              TAKE THE COMMENTS OFF THE FOLLOWING BLOCK OF
C              CODE.  THE VALUE FOR EPSOO MAY NEED TO BE ADJUSTED.
C              EPSOO=1.D-22
C              IF(DABS(DREAL(X)).LT.EPSOO) X=DCMPLX(O.DO,DIMAG(X))
C              IF(DABS(DIMAG(X)).LT.EPSOO) X=DCMPLX(DREAL(X),O.DO)
C$$$$$$$$$$$$$$$$$$$$$$$$$$$$$$$$$$$$$$$$$$$$$$$$$$$$$$$$$$$$$$$$$$$$$$$$$$$
C
C UPDATE NEWTON ITERATION COUNTER
             NUMIT = NUMIT + 1
C
C LEAVE NEWTON LOOP IF RESIDUAL IS SMALL ENOUGH
             XRESID = DABS(DREAL(RESID))+DABS(DIMAG(RESID))
             IF(XRESID .LT. 1.D-12) GOTO 101
C
 100         CONTINUE
 101         CONTINUE
C
             IF(MOD(NUMSTP,NS/10).EQ.O.OR.NUMSTP.GE.(NS-4))THEN
             WRITE(6,110) NUMSTP
 110         FORMAT(/'  STEP SUMMARY FOR STEP NUMBER',I5,' :')
             WRITE(6,115)
 115         FORMAT(' ---------------------------------------------')
             WRITE(6,120) T
 120         FORMAT('  T  = ',D22.15)
             WRITE(6,125)
 125         FORMAT('             REAL PART        IMAGINARY PART')
             WRITE(6,130) X
 130         FORMAT(' X  =        ',D11.4,'          ',D11.4)
             WRITE(6,135)  H
 135         FORMAT(' H  =        ',D11.4,'          ',D11.4)
             WRITE(6,140) RESID
 140         FORMAT(' RESIDUAL =  ',D11.4,'          ',D11.4)
             WRITE(6,150) DH
 150         FORMAT(' DH  =       ',D11.4,'          ',D11.4)
             WRITE(6,115)
             WRITE(6,40)
             ENDIF
 90       CONTINUE
C            BOTTOM OF LOOP -- NUMSTP
```

```
C
         ITOTIT=ITOTIT + NUMIT
C
         WRITE(6,160) NUMIT
 160     FORMAT(' TOTAL NEWTON ITERATIONS ON PATH =',I5)
         WRITE(6,40)
C
  60 CONTINUE
C              BOTTOM OF LOOP -- NUMPAT
C
      WRITE(6,170) ITOTIT
 170  FORMAT(//' TOTAL NEWTON ITERATIONS =',I5//)
C
      STOP
      END
```

```
C PROGRAM NAME:  CONSOL4
C
C PURPOSE:    SOLVES THE FIFTH-DEGREE EQUATION
C
C                EQ = X**5 + A*X + B
C
C            USING THE CONTINUATION EQUATION
C
C                H = X**5 + A*T*X + (T*B - (1.-T)*Q**5)
C
C REFERENCE: CHAPTER 1 OF "SOLVING POLYNOMIAL SYSTEMS USING
C               CONTINUATION" BY ALEXANDER MORGAN
C
C NOTE:   1.  THIS PROGRAM USES THE FORTRAN "COMPLEX*16" DECLARATION.
C
C         2.  THIS PROGRAM GENERATES MANY UNDERFLOWS.  THIS IS
C             NOT AN ERROR CONDITION.  IF YOU NEED TO PREVENT
C             UNDERFLOWS FROM OCCURRING, SEE THE COMMENTS LABELED
C             "UNDERFLOWS" BELOW.
C
C         3.  INPUT IS ON UNIT 5.  OUTPUT IS TO UNIT 6.
C             SAMPLE INPUT FOR DELT=.05, A=10., B=.2:
C                 5.D-02
C                 1.D+01
C                 2.D-01
C
C INPUT VARIABLES:
C
C     DELT          -- THE STEP SIZE FOR T
C     A             -- COEFFICIENT FOR EQ
C     B             -- COEFFICIENT FOR EQ
C
C OUTPUT VARIABLES:
C
C     FOR EACH STEP:
C
C     NUMSTP        -- CURRENT STEP NUMBER
C     T             -- CURRENT VALUE OF T
C     X             -- CURRENT VALUE OF X
C     H             -- CURRENT VALUE OF THE CONTINUATION EQUATION
C     RESID         -- NEWTON METHOD RESIDUAL (SHOULD BE NEAR 0)
C     DH            -- CURRENT DERIVATIVE OF CONTINUATION EQUATION
C
C
C     AT THE END OF THE PATH:
C
C     NUMIT         -- NUMBER OF NEWTON ITERATIONS FOR THE PATH
C
C     AT THE END OF THE RUN:
C
C     ITOTIT        -- TOTAL NUMBER OF NEWTON ITERATIONS OVER ALL PATHS
C
C
C SUBROUTINES: DABS,DCMPLX,DIMAG,DREAL,MOD
C
```

```
C
C DECLARATIONS
      INTEGER ITER,ITOTIT,NS,NUMIT,NUMPAT,NUMSTP
      DOUBLE PRECISION A,B,DELT,EPSOO,T,XRESID
      COMPLEX*16 DH,H,Q,QQ,RESID,X
C
C INPUT STATEMENTS FOLLOW.
      READ(5,10) DELT
  10  FORMAT(D22.15)
      READ(5,10) A,B
C
      WRITE(6,20)
  20  FORMAT(/,'  CONSOL4',//)
      WRITE(6,25) DELT
  25  FORMAT(/,'  DELT =',D22.15,/)
      WRITE(6,30) A,B
  30  FORMAT('  COEFFICIENTS,    A=',D22.15,'  B=',D22.15)
C
C COMPUTE NUMBER OF STEPS PER PATH
      NS = (1./DELT) + 1.
C
      WRITE(6,40)
  40  FORMAT(/)
C
      WRITE(6,50)  NS
  50  FORMAT(' NUMBER OF STEPS PER PATH =',I5//)
C
C INITIALIZE COUNTER AND CONSTANTS FOR THE RUN
      ITOTIT=0
      Q = DCMPLX(.928746354D0, .265465386D0)
      QQ=Q**5
C
C ******************************************************************
C *****                                                      *****
C *****        BEGINNING OF MAIN PART OF PROGRAM             *****
C *****                                                      *****
C ******************************************************************
C
C PATHS LOOP -- ITERATE THROUGH PATHS
      DO  60 NUMPAT = 1 ,  5
C
          WRITE(6,70) NUMPAT
  70      FORMAT('  PATH NUMBER =',I5)
C
C INITIALIZE COUNTERS AND CONSTANTS FOR THE PATH
          NUMIT=0
          T=0
C
C GENERATE A START POINT, X, FOR THE PATH
          IF(NUMPAT.EQ.1) X = DCMPLX( .034525822D0, .965323588D0)
          IF(NUMPAT.EQ.2) X = DCMPLX(-.907408223D0, .331137401D0)
          IF(NUMPAT.EQ.3) X = DCMPLX(-.595334945D0,-.760669419D0)
          IF(NUMPAT.EQ.4) X = DCMPLX( .539470992D0,-.801256956D0)
          IF(NUMPAT.EQ.5) X = DCMPLX( .928746354D0, .265465386D0)
C
          WRITE(6,80)X
  80      FORMAT(/,'  START X=',2D22.15,//)
```

320

```
C STEPS LOOP -- ITERATE STEPS ALONG THE PATH
          DO 90 NUMSTP = 1 , NS
          T=T+DELT
          IF(T.GE.1.)T=1.
C
C NEWTON'S METHOD LOOP
          DO 100 ITER = 1,10
C
C EVALUATE THE CONTINUATION EQUATION AND DERIVATIVE
          H = X**5 + A*T*X + (T*B - (1.-T)*QQ)
          DH = 5.*X**4 + A*T
C
C GENERATE THE RESIDUAL AND UPDATE X
          RESID = H/DH
          X = X - RESID
C$$$$$$$$$$$$$$$$$$$$$$$$$$$$$$$$$$$$$$$$$$$$$$$$$$$$$$$$$$$$$$$$$$$$$$$$$$
C UNDERFLOWS :  IF YOU MUST PREVENT UNDERFLOWS, I SUGGEST YOU
C              TAKE THE COMMENTS OFF THE FOLLOWING BLOCK OF
C              CODE.  THE VALUE FOR EPSOO MAY NEED TO BE ADJUSTED.
C              EPSOO=1.D-22
C              IF(DABS(DREAL(X)).LT.EPSOO) X=DCMPLX(0.DO,DIMAG(X))
C              IF(DABS(DIMAG(X)).LT.EPSOO) X=DCMPLX(DREAL(X),0.DO)
C$$$$$$$$$$$$$$$$$$$$$$$$$$$$$$$$$$$$$$$$$$$$$$$$$$$$$$$$$$$$$$$$$$$$$$$$$$
C
C UPDATE NEWTON ITERATION COUNTER
          NUMIT = NUMIT + 1
C
C LEAVE NEWTON LOOP IF RESIDUAL IS SMALL ENOUGH
          XRESID = DABS(DREAL(RESID))+DABS(DIMAG(RESID))
          IF(XRESID .LT. 1.D-12) GOTO 101
C
  100     CONTINUE
  101     CONTINUE
C
          IF(MOD(NUMSTP,NS/10).EQ.0.OR.NUMSTP.GE.(NS-4))THEN
          WRITE(6,110) NUMSTP
  110     FORMAT(/'  STEP SUMMARY FOR STEP NUMBER',I5,' :')
          WRITE(6,115)
  115     FORMAT(' ---------------------------------------------')
          WRITE(6,120) T
  120     FORMAT('  T  = ',D22.15)
          WRITE(6,125)
  125     FORMAT('               REAL PART        IMAGINARY PART')
          WRITE(6,130) X
  130     FORMAT('  X  =        ',D11.4,'        ',D11.4)
          WRITE(6,135) H
  135     FORMAT('  H  =        ',D11.4,'        ',D11.4)
          WRITE(6,140) RESID
  140     FORMAT(' RESIDUAL =   ',D11.4,'        ',D11.4)
          WRITE(6,150) DH
  150     FORMAT(' DH  =        ',D11.4,'        ',D11.4)
          WRITE(6,115)
          WRITE(6,40)
          ENDIF
C
  90      CONTINUE
C             BOTTOM OF LOOP -- NUMSTP
```

```
C
          ITOTIT=ITOTIT + NUMIT
C
          WRITE(6,160) NUMIT
  160     FORMAT(' TOTAL NEWTON ITERATIONS ON PATH =',I5)
          WRITE(6,40)
C
  60    CONTINUE
C            BOTTOM OF LOOP -- NUMPAT
C
        WRITE(6,170) ITOTIT
  170   FORMAT(//' TOTAL NEWTON ITERATIONS =',I5//)
C
        STOP
        END
```

```fortran
C PROGRAM NAME:   CONSOL5
C
C
C PURPOSE: SOLVES THE GENERAL POLYNOMIAL EQUATION
C
C    EQ = X**IDEG + A(IDEG)*X**(IDEG-1) + ... + A(2)*X + A(1)
C
C USING THE CONTINUATION EQUATION
C
C    H = X**IDEG + A(IDEG)*T*X**(IDEG-1) +  . . .
C
C                          + T*A(2)*X + (T*A(1) - (1.-T)*Q**IDEG)
C
C
C REFERENCE: CHAPTER 1 OF "SOLVING POLYNOMIAL SYSTEMS USING
C            CONTINUATION" BY ALEXANDER MORGAN
C
C
C NOTE:  1.  THIS PROGRAM USES THE FORTRAN "COMPLEX*16" DECLARATION.
C
C        2.  THIS PROGRAM GENERATES MANY UNDERFLOWS.  THIS IS
C            NOT AN ERROR CONDITION.  IF YOU NEED TO PREVENT
C            UNDERFLOWS FROM OCCURRING, SEE THE COMMENTS LABELED
C            "UNDERFLOWS" BELOW.
C
C        3.  INPUT IS ON UNIT 5.  OUTPUT IS TO UNIT 6.
C            SAMPLE INPUT FOR DELT=.08, IDEG=5, A(J)=J/10:
C                  8.D-02
C05
C                     1.D-01
C                     2.D-01
C                     3.D-01
C                     4.D-01
C                     5.D-01
C
C
C INPUT VARIABLES:
C
C    DELT            -- THE STEP SIZE FOR T
C    IDEG            -- DEGREE OF EQ
C    A               -- COEFFICIENTS FOR EQ, AN ARRAY OF LENGTH IDEG
C
C
C OUTPUT VARIABLES:
C
C    FOR EACH STEP:
C
C    NUMSTP          -- CURRENT STEP NUMBER
C    T               -- CURRENT VALUE OF T
C    X               -- CURRENT VALUE OF X
C    H               -- CURRENT VALUE OF THE CONTINUATION EQUATION
C    RESID           -- NEWTON METHOD RESIDUAL (SHOULD BE NEAR 0)
C    DH              -- CURRENT DERIVATIVE OF CONTINUATION EQUATION
C
```

```
C    AT THE END OF THE PATH:
C
C       NUMIT        -- NUMBER OF NEWTON ITERATIONS FOR THE PATH
C
C
C    AT THE END OF THE RUN:
C
C       ITOTIT       -- TOTAL NUMBER OF NEWTON ITERATIONS OVER ALL PATHS
C
C
C SUBROUTINES: DABS,DATAN,DCMPLX,DCOS,DIMAG,DREAL,DSIN,MOD
C
C DECLARATIONS
      INTEGER I,IDEG,IDEGM1,IDEGP1,ITER,ITOTIT,J,JP1,NS,NUMIT,NUMPAT,
     & NUMSTP
      DOUBLE PRECISION A,ANGLE,DELT,EPSOO,T,TWOPI,XRESID
      COMPLEX*16 DH,H,Q,QQ,RESID,X
C
C DIMENSIONS
      DIMENSION A(100)
C
C INPUT STATEMENTS FOLLOW.
      READ(5,10) DELT
   10 FORMAT(D22.15)
      READ(5,15) IDEG
   15 FORMAT(I2)
      READ(5,10) (A(I),I=1,IDEG)
C
      WRITE(6,20)
   20 FORMAT(/,'  CONSOL5',//)
      WRITE(6,25) DELT
   25 FORMAT(/,'  DELT =',D22.15,/)
      WRITE(6,28) IDEG
   28 FORMAT('  DEGREE OF THE EQUATION =',I5)
      WRITE(6,30)
   30 FORMAT(//,'  COEFFICIENTS:',/)
      WRITE(6,35)(I,A(I),I=1,IDEG)
   35 FORMAT('        A(',I2,')=',D22.15)
C
      WRITE(6,40)
   40 FORMAT(/)
C
C COMPUTE NUMBER OF STEPS PER PATH
      NS = (1./DELT) + 1.
C
      WRITE(6,40)
C
      WRITE(6,50)  NS
   50 FORMAT(' NUMBER OF STEPS PER PATH =',I5//)
C
C INITIALIZE COUNTER AND CONSTANTS FOR THE RUN
      ITOTIT=0
      IDEGM1=IDEG-1
      IDEGP1=IDEG+1
      Q = DCMPLX(.928746354D0, .265465386D0)
      QQ=Q**IDEG
C
```

```
C ****************************************************************
C *****                                                      *****
C *****        BEGINNING OF MAIN PART OF PROGRAM             *****
C *****                                                      *****
C ****************************************************************
C
C PATHS LOOP -- ITERATE THROUGH PATHS
       DO  60 NUMPAT = 1,IDEG
C
         WRITE(6,70) NUMPAT
  70     FORMAT(' PATH NUMBER =',I5)
C
C INITIALIZE COUNTERS AND CONSTANTS FOR THE PATH
         NUMIT=0
         T=0
C
C GENERATE A START POINT, S, FOR THE PATH
         TWOPI = 8.D0*DATAN(1.D0)
         ANGLE = TWOPI*NUMPAT/IDEG
         X = Q*DCMPLX(DCOS(ANGLE),DSIN(ANGLE))
C
         WRITE(6,80)X
  80     FORMAT(/,'  START X=',2D22.15,//)
C
C
C STEPS LOOP -- ITERATE STEPS ALONG THE PATH
         DO 90 NUMSTP = 1 , NS
C
           T=T+DELT
           IF(T.GE.1.)T=1.
C
C NEWTON'S METHOD LOOP
           DO 100 ITER = 1,10
C
C EVALUATE THE CONTINUATION EQUATION AND DERIVATIVE
           H = (T*A(1) - (1.-T)*QQ) + X**IDEG
           DH = IDEG*(X**IDEGM1)
           DO 105 J=1,IDEGM1
             JP1=J+1
             H = H + A(JP1)*T*(X**J)
             DH = DH + J*A(JP1)*T*(X**(J-1))
  105      CONTINUE
C
C  GENERATE THE RESIDUAL AND UPDATE X
           RESID = H/DH
           X = X - RESID
C
C$$$$$$$$$$$$$$$$$$$$$$$$$$$$$$$$$$$$$$$$$$$$$$$$$$$$$$$$$$$$$$$$$$$$$$$$$$$
C UNDERFLOWS :  IF YOU MUST PREVENT UNDERFLOWS, I SUGGEST YOU
C                TAKE THE COMMENTS OFF THE FOLLOWING BLOCK OF
C                CODE.  THE VALUE FOR EPSOO MAY NEED TO BE ADJUSTED.
C
C               EPSOO=1.D-22
C               IF(DABS(DREAL(X)).LT.EPSOO) X=DCMPLX(0.D0,DIMAG(X))
C               IF(DABS(DIMAG(X)).LT.EPSOO) X=DCMPLX(DREAL(X),0.D0)
C$$$$$$$$$$$$$$$$$$$$$$$$$$$$$$$$$$$$$$$$$$$$$$$$$$$$$$$$$$$$$$$$$$$$$$$$$$$
C
```

```
C
C UPDATE NEWTON ITERATION COUNTER
              NUMIT = NUMIT + 1
C
C LEAVE NEWTON LOOP IF RESIDUAL IS SMALL ENOUGH
              XRESID = DABS(DREAL(RESID))+DABS(DIMAG(RESID))
              IF(XRESID .LT. 1.D-12) GOTO 101
C
 100          CONTINUE
 101          CONTINUE
C
C
              IF(MOD(NUMSTP,NS/10).EQ.0.OR.NUMSTP.GE.(NS-4))THEN
              WRITE(6,110) NUMSTP
 110          FORMAT(/' STEP SUMMARY FOR STEP NUMBER',I5,' :')
              WRITE(6,115)
 115          FORMAT(' ------------------------------------------------')
              WRITE(6,120) T
 120          FORMAT('  T   = ',D22.15)
              WRITE(6,125)
 125          FORMAT('                 REAL PART        IMAGINARY PART')
              WRITE(6,130) X
 130          FORMAT(' X =          ',D11.4,'        ',D11.4)
              WRITE(6,135) H
 135          FORMAT(' H =          ',D11.4,'        ',D11.4)
              WRITE(6,140) RESID
 140          FORMAT(' RESIDUAL =   ',D11.4,'        ',D11.4)
              WRITE(6,150) DH
 150          FORMAT(' DH =         ',D11.4,'        ',D11.4)
              WRITE(6,115)
              WRITE(6,40)
              ENDIF
C
 90       CONTINUE
C            BOTTOM OF LOOP -- NUMSTP
C
          ITOTIT=ITOTIT + NUMIT
C
          WRITE(6,160) NUMIT
 160      FORMAT(' TOTAL NEWTON ITERATIONS ON PATH =',I5)
          WRITE(6,40)
C
 60   CONTINUE
C            BOTTOM OF LOOP -- NUMPAT
C
      WRITE(6,170) ITOTIT
 170  FORMAT(//' TOTAL NEWTON ITERATIONS =',I5//)
C
      STOP
      END
```

Section 2: Programs for Chapter 2

```
C PROGRAM NAME:  CONSOL6
C
C
C PURPOSE:   SOLVES A SYSTEM OF TWO SECOND-DEGREE
C            EQUATIONS IN TWO UNKNOWNS:
C
C   EQ(J) = A(J)*X**2 + B(J)*Y**2 + C(J)*X*Y + D(J)*X + E(J)*Y+ F(J)
C
C            FOR J=1,2 USING THE CONTINUATION EQUATION
C
C      H(J)  =  (1-T)*(P(J)**2 * X**2 - Q(J)**2) + T * EQ(J)
C
C
C REFERENCE: CHAPTER 2 OF "SOLVING POLYNOMIAL SYSTEMS USING
C            CONTINUATION" BY ALEXANDER MORGAN
C
C
C NOTE: 1.  THIS CODE IS WRITTEN USING FORTRAN'S "COMPLEX*16"
C           DECLARATION.
C
C       2.  THIS PROGRAM GENERATES MANY UNDERFLOWS.  THIS IS
C           NOT AN ERROR CONDITION.  IF YOU NEED TO PREVENT
C           UNDERFLOWS FROM OCCURRING, SEE THE COMMENTS LABELED
C           "UNDERFLOWS" BELOW.
C
C       3.  INPUT IS ON UNIT 5.  OUTPUT IS TO UNITS 6 AND 8.
C           (UNIT 6 IS THE FULL OUTPUT.  UNIT 8 CONTAINS THE SOLUTION LIST.)
C           SAMPLE INPUT FOLLOWS:
CTITLE:    SYSTEM OF TWO QUADRICS,   NO SOLUTIONS AT INFINITY
C               1.D-04      DELT
C             -.00098D 00   X(1)**2      *
C             978000.D+00   X(2)**2      *
C                 -9.8D+00  X(1)*X(2)    *     EQUATION 1
C             -235.0D+00    X(1)         *
C             88900.0D+00   X(2)         *
C               -1.000D+00  CONSTANT     *
C              -.0100D+00   X(1)**2      $
C              -.9840D+00   X(2)**2      $
C             -29.70D+00    X(1)*X(2)    $     EQUATION 2
C               .00987D+00  X(1)         $
C              -.1240D+00   X(2)         $
C              -.2500D+00   CONSTANT     $
C
C INPUT VARIABLES:
C
C    TITLE                            -- ANY 60 CHARACTERS
C    DELT                             -- THE STEP SIZE FOR T
C    A(J),B(J),C(J),D(J),E(J),F(J)    -- COEFFICIENTS FOR EQ(J), J=1 TO 2
C
C OUTPUT VARIABLES:
C
C   FOR EACH STEP:
C
C    NUMSTP     -- CURRENT STEP NUMBER
```

```
C      T                 -- CURRENT VALUE OF T
C      X,Y               -- CURRENT VALUES OF X AND Y
C      H(J)              -- CURRENT VALUES OF THE CONTINUATION EQUATIONS
C      XRESID            -- NORM OF NEWTON'S METHOD RESIDUAL (SHOULD BE NEAR 0)
C      XNRMDH            -- MATRIX NORM OF JACOBIAN DH
C      DET               -- DETERMINANT OF JACOBIAN
C
C
C   AT THE END OF THE PATH:
C
C      NUMIT          -- NUMBER OF NEWTON ITERATIONS FOR THE PATH
C
C   AT THE END OF THE RUN:
C
C      ITOTIT         -- TOTAL NUMBER OF NEWTON ITERATIONS OVER ALL PATHS
C
C
C SUBROUTINES: DABS,DCMPLX,DIMAG,DREAL,MOD
C
C
C DECLARATIONS
      INTEGER ITER,ITOTIT,J,NS,NUMIT,NUMPAT,NUMSTP,TITLE
      DOUBLE PRECISION A,B,C,D,DELT,E,EPSOO,F,T,TEST,TT,TWOPI,XNRMDH,
     & XRESID
      COMPLEX*16 DEQ,DET,DH,EQ,H,P,PP,Q,QQ,RESID,X,XX,XY,Y,YY
C
C
C DIMENSIONS
      DIMENSION A(2),B(2),C(2),D(2),DEQ(2,2),DH(2,2),E(2),EQ(2),F(2),
     & H(2),P(2),PP(2),Q(2),QQ(2),RESID(2),TITLE(60)
C
C
C INPUT STATEMENTS FOLLOW.
      READ(5,5) TITLE
    5 FORMAT(60A1)
      READ(5,10) DELT
   10 FORMAT(D22.15)
      READ(5,10) (  A(J),B(J),C(J),D(J),E(J),F(J),J=1,2)
C
      WRITE(6,20)
   20 FORMAT(/,'  CONSOL6',//)
      WRITE(6,5) TITLE
      WRITE(6,25) DELT
   25 FORMAT(/,'  DELT =',D22.15,/)
      WRITE(6,30)(J,A(J),B(J),C(J),D(J),E(J),F(J),J=1,2)
   30 FORMAT('  COEFFICIENTS FOR EQUATION ',I1,//,' A=',D22.15,/,
     &' B=',D22.15,/,' C=',D22.15,/,' D=',D22.15,/,' E=',D22.15,/,
     &' F=',D22.15,///)
C
C COMPUTE NUMBER OF STEPS PER PATH
      NS = (1.D0/DELT) + 1.D0
C
      WRITE(6,40)
   40 FORMAT(/)
C
      WRITE(6,50)  NS
   50 FORMAT(' NUMBER OF STEPS PER PATH =',I8//)
```

```
C
C
C
C INITIALIZE COUNTER AND CONSTANTS FOR THE RUN
      ITOTIT=0
      P(1)= DCMPLX( .12324754231D0, .76253746298D0)
      P(2)= DCMPLX( .93857838950D0,-.99375892810D0)
      Q(1)= DCMPLX( .58720452864D0, .01321964722D0)
      Q(2)= DCMPLX( .97884134700D0,-.14433009712D0)
      PP(1)=P(1)**2
      PP(2)=P(2)**2
      QQ(1)=Q(1)**2
      QQ(2)=Q(2)**2
C
C
C ****************************************************************
C *****                                                      *****
C *****        BEGINNING OF MAIN PART OF PROGRAM             *****
C *****                                                      *****
C ****************************************************************
C
C
C PATHS LOOP -- ITERATE THROUGH PATHS
C
      DO 60 NUMPAT = 1,4
C
          WRITE(6,70) NUMPAT
   70     FORMAT(' PATH NUMBER =',I8)
C
C
C INITIALIZE COUNTERS AND CONSTANTS FOR THE PATH
C
          NUMIT=0
          T=0.D0
C
C GENERATE A START POINT, (X,Y), FOR THE PATH
C
          X=Q(1)/P(1)
          Y=Q(2)/P(2)
          IF(NUMPAT.EQ.2 .OR. NUMPAT.EQ.4) X=-X
          IF(NUMPAT.EQ.3 .OR. NUMPAT.EQ.4) Y=-Y
C
C
          WRITE(6,80)X,Y
   80     FORMAT(/,' START X,Y=',2D11.4,'   ',2D11.4,//)
C
C STEPS LOOP -- ITERATE STEPS ALONG THE PATH
C
          DO 90 NUMSTP = 1 , NS
C
              T=T+DELT
              IF(T.GE.1.D0)T=1.D0
C
```

```
C NEWTON'S METHOD LOOP
               DO 100 ITER = 1,10
C
C EVALUATE THE ORIGINAL SYSTEM EQUATIONS AND DERIVATIVES.
C THE SECOND INDEX, 1 OR 2, INDICATES PARTIAL DERIVATIVE
C WITH RESPECT TO X OR Y RESPECTIVELY.
               XX=X**2
               YY=Y**2
               XY=X*Y
               DO 105 J=1,2
                  EQ(J) = A(J)*XX + B(J)*YY + C(J)*XY
     &                  + D(J)*X + E(J)*Y+ F(J)
                  DEQ(J,1) = 2.*A(J)*X + C(J)*Y + D(J)
                  DEQ(J,2) = 2.*B(J)*Y + C(J)*X + E(J)
  105          CONTINUE
C
C EVALUATE THE CONTINUATION EQUATIONS AND DERIVATIVES.
               TT=1.-T
               H(1)     = TT*(PP(1)*XX - QQ(1)) + T*EQ(1)
               DH(1,1) = TT*2.*PP(1)*X         + T*DEQ(1,1)
               DH(1,2) =                         T*DEQ(1,2)
C
               H(2)     = TT*(PP(2)*YY - QQ(2)) + T*EQ(2)
               DH(2,1) =                         T*DEQ(2,1)
               DH(2,2) = TT*2.*PP(2)*Y         + T*DEQ(2,2)
C
C GENERATE THE RESIDUAL AND UPDATE X,Y.
C SOLVE THE RESIDUAL EQUATION DH*RESID=H USING CRAMER'S RULE.
               DET=DH(1,1)*DH(2,2) - DH(1,2)*DH(2,1)
               RESID(1) = (H(1)*DH(2,2) - H(2)*DH(1,2))/DET
               RESID(2) = (H(2)*DH(1,1) - H(1)*DH(2,1))/DET
C
               X = X - RESID(1)
               Y = Y - RESID(2)
C
C$$$$$$$$$$$$$$$$$$$$$$$$$$$$$$$$$$$$$$$$$$$$$$$$$$$$$$$$$$$$$$$$$$$$$$$$$
C UNDERFLOWS:   IF YOU MUST PREVENT UNDERFLOWS, I SUGGEST YOU
C              TAKE THE COMMENTS OFF THE FOLLOWING BLOCK OF
C              CODE.  THE VALUE FOR EPSOO MAY NEED TO BE ADJUSTED.
C
C              EPSOO=1.D-22
C              IF(DABS(DREAL(X)).LT.EPSOO) X=DCMPLX(0.DO,DIMAG(X))
C              IF(DABS(DIMAG(X)).LT.EPSOO) X=DCMPLX(DREAL(X),0.DO)
C              IF(DABS(DREAL(Y)).LT.EPSOO) Y=DCMPLX(0.DO,DIMAG(Y))
C              IF(DABS(DIMAG(Y)).LT.EPSOO) Y=DCMPLX(DREAL(Y),0.DO)
C$$$$$$$$$$$$$$$$$$$$$$$$$$$$$$$$$$$$$$$$$$$$$$$$$$$$$$$$$$$$$$$$$$$$$$$$$
C
C UPDATE NEWTON ITERATION COUNTER
               NUMIT = NUMIT + 1
C
C LEAVE NEWTON LOOP IF RESIDUAL IS SMALL ENOUGH
               XRESID = DABS(DREAL(RESID(1)))+DABS(DIMAG(RESID(1)))
     &                + DABS(DREAL(RESID(2)))+DABS(DIMAG(RESID(2)))
               IF(XRESID .LT. 1.D-12) GOTO 101
C
  100          CONTINUE
  101          CONTINUE
```

```
C
                IF(MOD(NUMSTP,NS/10).EQ.0.OR.NUMSTP.GE.(NS-4))THEN
                   WRITE(6,110) NUMSTP
110                FORMAT(/' STEP SUMMARY FOR STEP NUMBER',I8,' :')
                   WRITE(6,115)
115                FORMAT(' -----------------------------------------------')
                   WRITE(6,120) T
120                FORMAT(' T   = ',D22.15)
                   WRITE(6,125)
125                FORMAT('              REAL PART          IMAGINARY PART')
                   WRITE(6,130) X
130                FORMAT(' X   =        ',D11.4,'            ',D11.4)
                   WRITE(6,131) Y
131                FORMAT(' Y   =        ',D11.4,'            ',D11.4)
                   WRITE(6,135)  (J,H(J),J=1,2)
135                FORMAT(' H(',I1,') =        ',D11.4,'            ',D11.4)
                   WRITE(6,140) XRESID
140                FORMAT(' NORM OF RESIDUAL = ',D11.4)
C
C COMPUTE THE MATRIX NORM OF DH, XNRMDH.
                XNRMDH = DABS(DREAL(DH(1,1)))+DABS(DIMAG(DH(1,1)))
     &                 + DABS(DREAL(DH(2,1)))+DABS(DIMAG(DH(2,1)))
                TEST   = DABS(DREAL(DH(1,2)))+DABS(DIMAG(DH(1,2)))
     &                 + DABS(DREAL(DH(2,2)))+DABS(DIMAG(DH(2,2)))
                IF(TEST.GT.XNRMDH) XNRMDH = TEST
                WRITE(6,145) XNRMDH
145             FORMAT(' NORM OF DH =',D11.4)
C
                WRITE(6,150) DET
150             FORMAT(' DET(DH) =     ',2D11.4)
                WRITE(6,115)
                WRITE(6,40)
             ENDIF
C
                IF(NUMSTP.EQ.NS) THEN
                   WRITE(8,70) NUMPAT
                   WRITE(8,130) X
                   WRITE(8,131) Y
                   WRITE(8,140) XRESID
                   WRITE(8,40)
                ENDIF
90       CONTINUE
C            BOTTOM OF LOOP -- NUMSTP
C
         ITOTIT=ITOTIT + NUMIT
C
         WRITE(6,160) NUMIT
160      FORMAT(' TOTAL NEWTON ITERATIONS ON PATH =',I8)
         WRITE(6,40)
60    CONTINUE
C            BOTTOM OF LOOP -- NUMPAT
C
      WRITE(6,170) ITOTIT
170   FORMAT(//' TOTAL NEWTON ITERATIONS =',I8//)
C
      STOP
      END
```

```
C PROGRAM NAME:  CONSOL7
C
C PURPOSE:    SOLVES A SYSTEM OF TWO POLYNOMIAL
C             EQUATIONS IN TWO UNKNOWNS:
C
C                    EQ(J)=0
C
C             FOR J=1,2, USING THE CONTINUATION EQUATIONS
C
C H(1) = (1-T)*(P(1)**IDEG(1) * X**IDEG(1)- Q(1)**IDEG(1))+ T*EQ(1)
C H(2) = (1-T)*(P(2)**IDEG(2) * Y**IDEG(2)- Q(2)**IDEG(2))+ T*EQ(2)
C
C             WHERE DEGREE(EQ(J))=IDEG(J).
C
C
C REFERENCE: CHAPTER 2 OF "SOLVING POLYNOMIAL SYSTEMS USING
C            CONTINUATION" BY ALEXANDER MORGAN
C
C
C NOTE: 1. THIS CODE REQUIRES THE USER TO PROVIDE SUBROUTINE "FFUN7"
C          TO RETURN EQ AND DEQ VALUES.  THREE SAMPLE SUBROUTINE
C          FFUN7'S ARE INCLUDED FOR PROBLEMS (2-19), (2-26), AND
C          (2-27), NAMED FFUN7A, FFUN7B, FFUN7C RESPECTIVELY.
C          TO USE ONE OF THESE SUBROUTINES, CHANGE THE NAME TO FFUN7.
C
C       2. THIS CODE IS WRITTEN USING FORTRAN'S "COMPLEX*16"
C          DECLARATION.
C
C       3. THIS PROGRAM GENERATES MANY UNDERFLOWS.  THIS IS
C          NOT AN ERROR CONDITION.  IF YOU NEED TO PREVENT
C          UNDERFLOWS FROM OCCURRING, SEE THE COMMENTS LABELED
C          "UNDERFLOWS" BELOW.
C
C       4. INPUT IS ON UNIT 5.  OUTPUT IS TO UNITS 6 AND 8.
C          (UNIT 6 IS THE FULL OUTPUT.  UNIT 8 CONTAINS THE SOUTION LIST.)
C          SAMPLE INPUT FOLLOWS:
CTHIS IS THE SAMPLE TITLE.  SIXTY COLUMNS ARE ALLOWED FOR IT.
C               1.1D-02        DELT
C05                            IDEG(1)
C07                            IDEG(2)
C               1.D-01         A(1)
C               2.D-01         B(1)
C               3.D-01         C(1)
C               4.D-01         D(1)
C               5.D-01         E(1)
C               6.D-01         F(1)
C               7.D-01         G(1)
C               1.D+01         A(2)
C               2.D-01         B(2)
C               3.D+01         C(2)
C               4.D-01         D(2)
C               5.D+01         E(2)
C               6.D-01         F(2)
C               7.D+01         G(2)
C
```

```
C INPUT VARIABLES:
C
C     DELT                                -- STEP SIZE FOR T
C     IDEG(J)                             -- DEGREE OF EQ(J)
C     A(J),B(J),C(J),D(J),E(J),F(J),G(J)  -- COEFFICIENTS FOR EQ(J)
C
C OUTPUT VARIABLES:
C
C     FOR EACH STEP:
C
C     NUMSTP       -- CURRENT STEP NUMBER
C     T            -- CURRENT VALUE OF T
C     X,Y          -- CURRENT VALUES OF X AND Y
C     H(J)         -- CURRENT VALUES OF THE CONTINUATION EQUATIONS
C     XRESID       -- NORM OF NEWTON'S METHOD RESIDUAL (SHOULD BE NEAR 0)
C     XNRMDH       -- MATRIX NORM OF JACOBIAN, DH
C     DET          -- DETERMINANT OF JACOBIAN, DH
C
C     AT THE END OF THE PATH:
C
C     NUMIT        -- NUMBER OF NEWTON ITERATIONS FOR THE PATH
C
C     AT THE END OF THE RUN:
C
C     ITOTIT       -- TOTAL NUMBER OF NEWTON ITERATIONS OVER ALL PATHS
C
C SUBROUTINES: DABS,DATAN,DCMPLX,DCOS,DIMAG,DREAL,DSIN,FFUN7,MOD
C
C
C
C DECLARATIONS
      INTEGER ICOUNT,IDEG,ITER,ITOTDG,ITEST,ITOTIT,J,NS,NUMIT,NUMPAT,
     & NUMSTP,TITLE
      DOUBLE PRECISION A,ANGLE,B,C,D,DELT,E,EPSOO,F,G,T,TEST,TT,TWOPI,
     & XNRMDH,XRESID
      COMPLEX*16 DEQ,DET,DH,EQ,H,P,PP,Q,QQ,RESID,X,XDG,XDGM1,XX,XY,
     & Y,YDG,YDGM1,YY
C
C DIMENSIONS
      DIMENSION DEQ(2,2),DH(2,2),EQ(2),H(2),ICOUNT(2),IDEG(2),
     & P(2),PP(2),Q(2),QQ(2),RESID(2),TITLE(60)
C
C COMMON
      COMMON/COEFS/ A(2),B(2),C(2),D(2),E(2),F(2),G(2)
C
C
C INPUT STATEMENTS FOLLOW.
      READ(5,5) TITLE
    5 FORMAT(60A1)
      READ(5,10) DELT
   10 FORMAT(D22.15)
      READ(5,15) IDEG
   15 FORMAT(I2)
      READ(5,10) ( A(J),B(J),C(J),D(J),E(J),F(J),G(J),J=1,2 )
C
      WRITE(6,20)
   20 FORMAT(/,'  CONSOL7',//)
```

```fortran
      WRITE(6,5) TITLE
      WRITE(6,25) DELT
  25  FORMAT(/,'   DELT =',D22.15,/)
      WRITE(6,30)  (J,IDEG(J),J=1,2)
  30  FORMAT(/,'   IDEG(',I1,')=',I2,'   IDEG(',I1,')=',I2,//)
      WRITE(6,35)(J,A(J),B(J),C(J),D(J),E(J),F(J),G(J),J=1,2)
  35  FORMAT('  COEFFICIENTS FOR EQUATION ',I1,//,' A=',D22.15,/,
     &' B=',D22.15,/,' C=',D22.15,/,' D=',D22.15,/,' E=',D22.15,/,
     &' F=',D22.15,/,' G=',D22.15,///)
C
C COMPUTE NUMBER OF STEPS PER PATH
      NS = (1.D0/DELT) + 1.D0
C
      WRITE(6,40)
  40  FORMAT(/)
C
      WRITE(6,50)  NS
  50  FORMAT(' NUMBER OF STEPS PER PATH =',I8//)
C
C INITIALIZE COUNTER AND CONSTANTS FOR THE RUN
C
C   COUNTER FOR TOTAL NUMBER OF CORRECTOR ITERATIONS
      ITOTIT=0
C
C   TOTAL DEGREE FOR THE SYSTEM
      ITOTDG=IDEG(1)*IDEG(2)
C
C   ICOUNT IS A COUNTER USED FOR GENERATING START POINTS.
      ICOUNT(1)=0
      ICOUNT(2)=1
C
      TWOPI=8.D0*DATAN(1.D0)
C
      P(1)= DCMPLX( .12324754231D0, .76253746298D0)
      P(2)= DCMPLX( .93857838950D0,-.99375892810D0)
C
      Q(1)= DCMPLX( .58720452864D0, .01321964722D0)
      Q(2)= DCMPLX( .97884134700D0,-.14433009712D0)
C
      PP(1)=P(1)**IDEG(1)
      PP(2)=P(2)**IDEG(2)
      QQ(1)=Q(1)**IDEG(1)
      QQ(2)=Q(2)**IDEG(2)
C
C *************************************************************
C *****                                                 *****
C *****       BEGINNING OF MAIN PART OF PROGRAM          *****
C *****                                                 *****
C *************************************************************
C
C PATHS LOOP -- ITERATE THROUGH PATHS
      DO  60 NUMPAT = 1,ITOTDG
C
          WRITE(6,70) NUMPAT
  70      FORMAT(' PATH NUMBER =',I8)
C
C
```

```
C INITIALIZE COUNTERS AND CONSTANTS FOR THE PATH
          NUMIT=0
          T=0.DO
C
C GENERATE A START POINT, (X,Y), FOR THE PATH
C
C ICOUNT IS A COUNTER USED TO INCREMENT EACH VARIABLE AROUND
C THE UNIT CIRCLE SO THAT EVERY COMBINATION OF START VALUES
C IS CHOSEN BY THE END OF THE RUN.  ICOUNT IS INITIALIZED ABOVE.
C
          ITEST = ICOUNT(1)
          IF( ITEST  .GE. IDEG(1) ) ICOUNT(1)=1
          IF( ITEST  .GE. IDEG(1) ) ICOUNT(2)=ICOUNT(2)+1
          IF( ITEST  .LT. IDEG(1) ) ICOUNT(1)=ICOUNT(1)+1
C
          ANGLE = (TWOPI*ICOUNT(1))/IDEG(1)
          X = DCMPLX(DCOS(ANGLE),DSIN(ANGLE))*Q(1)/P(1)
          ANGLE = (TWOPI*ICOUNT(2))/IDEG(2)
          Y = DCMPLX(DCOS(ANGLE),DSIN(ANGLE))*Q(2)/P(2)
C
C
C STEPS LOOP -- ITERATE STEPS ALONG THE PATH
          DO 90 NUMSTP = 1 , NS
              T=T+DELT
              IF(NUMSTP.EQ.NS)T=1.DO
C
C NEWTON'S METHOD LOOP
              DO 100 ITER = 1,10
C
C COMPUTE POWERS OF X AND Y TO BE USED IN H AND DH AND (OPTIONALLY)
C BY FFUN7
                  XDGM1 = X**(IDEG(1)-1)
                  YDGM1 = Y**(IDEG(2)-1)
                  XDG = X*XDGM1
                  YDG = Y*YDGM1
C
                  CALL FFUN7(X,Y,XDGM1,YDGM1,XDG,YDG, EQ,DEQ)
C
C EVALUATE THE CONTINUATION EQUATIONS AND DERIVATIVES.
                  TT=1.-T
                  H(1)    = TT*(PP(1)*XDG - QQ(1)) + T*EQ(1)
                  DH(1,1) = TT*IDEG(1)*PP(1)*XDGM1 + T*DEQ(1,1)
                  DH(1,2) =                          T*DEQ(1,2)
C
                  H(2)    = TT*(PP(2)*YDG - QQ(2)) + T*EQ(2)
                  DH(2,1) =                          T*DEQ(2,1)
                  DH(2,2) = TT*IDEG(2)*PP(2)*YDGM1 + T*DEQ(2,2)
C
C  GENERATE THE RESIDUAL AND UPDATE X,Y.
C  SOLVE THE RESIDUAL EQUATION DH*RESID=H USING CRAMER'S RULE.
                  DET=DH(1,1)*DH(2,2) - DH(1,2)*DH(2,1)
                  RESID(1) = (H(1)*DH(2,2) - H(2)*DH(1,2))/DET
                  RESID(2) = (H(2)*DH(1,1) - H(1)*DH(2,1))/DET
                  X = X - RESID(1)
                  Y = Y - RESID(2)
C
```

```fortran
C$$$$$$$$$$$$$$$$$$$$$$$$$$$$$$$$$$$$$$$$$$$$$$$$$$$$$$$$$$$$$$$$$$$$$$$$$$$
C UNDERFLOWS:     IF YOU MUST PREVENT UNDERFLOWS, I SUGGEST YOU
C                 TAKE THE COMMENTS OFF THE FOLLOWING BLOCK OF
C                 CODE.  THE VALUE FOR EPSOO MAY NEED TO BE ADJUSTED.
C
C                 EPSOO=1.D-22
C                 IF(DABS(DREAL(X)).LT.EPSOO) X=DCMPLX(0.DO,DIMAG(X))
C                 IF(DABS(DIMAG(X)).LT.EPSOO) X=DCMPLX(DREAL(X),0.DO)
C                 IF(DABS(DREAL(Y)).LT.EPSOO) Y=DCMPLX(0.DO,DIMAG(Y))
C                 IF(DABS(DIMAG(Y)).LT.EPSOO) Y=DCMPLX(DREAL(Y),0.DO)
C$$$$$$$$$$$$$$$$$$$$$$$$$$$$$$$$$$$$$$$$$$$$$$$$$$$$$$$$$$$$$$$$$$$$$$$$$$$
C
C UPDATE NEWTON ITERATION COUNTER
                  NUMIT = NUMIT + 1
C
C LEAVE NEWTON LOOP IF RESIDUAL IS SMALL ENOUGH
                  XRESID=DABS(DREAL(RESID(1)))+DABS(DIMAG(RESID(1)))
     &                  +DABS(DREAL(RESID(2)))+DABS(DIMAG(RESID(2)))
                  IF(XRESID .LT. 1.D-12) GOTO 101
C
  100             CONTINUE
  101             CONTINUE
C
C
                  IF(MOD(NUMSTP,NS/10).EQ.O.OR.NUMSTP.GE.(NS-4))THEN
                  WRITE(6,110) NUMSTP
  110             FORMAT(/' STEP SUMMARY FOR STEP NUMBER',I8,' :')
                  WRITE(6,115)
  115             FORMAT(' -----------------------------------------------')
                  WRITE(6,120) T
  120             FORMAT(' T   = ',D22.15)
                  WRITE(6,125)
  125             FORMAT('              REAL PART        IMAGINARY PART')
                  WRITE(6,130) X
  130             FORMAT(' X   =        ',D11.4,'        ',D11.4)
                  WRITE(6,131) Y
  131             FORMAT(' Y   =        ',D11.4,'        ',D11.4)
                  WRITE(6,135)  (J,H(J),J=1,2)
  135             FORMAT(' H(',I1,') =      ',D11.4,'        ',D11.4)
                  WRITE(6,140) XRESID
  140             FORMAT(' NORM OF RESIDUAL = ',D11.4)
C
C COMPUTE THE MATRIX NORM OF DH, XNRMDH.
                  XNRMDH = DABS(DREAL(DH(1,1)))+DABS(DIMAG(DH(1,1)))
     &                   + DABS(DREAL(DH(2,1)))+DABS(DIMAG(DH(2,1)))
                  TEST   = DABS(DREAL(DH(1,2)))+DABS(DIMAG(DH(1,2)))
     &                   + DABS(DREAL(DH(2,2)))+DABS(DIMAG(DH(2,2)))
                  IF(TEST.GT.XNRMDH) XNRMDH = TEST
                  WRITE(6,145) XNRMDH
  145             FORMAT(' NORM OF DH =',D11.4)
C
                  WRITE(6,150) DET
  150             FORMAT(' DET(DH) =     ',2D11.4)
                  WRITE(6,115)
                  WRITE(6,40)
                  ENDIF
C
```

```
                IF(NUMSTP.EQ.NS) THEN
                  WRITE(8,70) NUMPAT
                  WRITE(8,130) X
                  WRITE(8,131) Y
                  WRITE(8,140) XRESID
                  WRITE(8,40)
                ENDIF
C
 90        CONTINUE
C             BOTTOM OF LOOP -- NUMSTP
C
           ITOTIT=ITOTIT + NUMIT
C
           WRITE(6,160) NUMIT
 160       FORMAT(' TOTAL NEWTON ITERATIONS ON PATH =',I8)
           WRITE(6,40)
C
 60     CONTINUE
C             BOTTOM OF LOOP -- NUMPAT
C
        WRITE(6,170) ITOTIT
 170    FORMAT(//' TOTAL NEWTON ITERATIONS =',I8//)
C
        STOP
        END
```

```
              SUBROUTINE FFUN7A(X,Y,XDGM1,YDGM1,XDG,YDG, EQ,DEQ)
C
C PURPOSE:  EVALUATE THE SYSTEM EQUATIONS AND DERIVATIVES
C           FOR CONSOL7
C
C REFERENCE: CHAPTER 2 OF "SOLVING POLYNOMIAL SYSTEMS USING
C            CONTINUATION" BY ALEXANDER MORGAN.
C
C GENERIC NAME: FFUN7          FILE NAME: FFUN7A
C
C NOTE:   1. THE SECOND INDEX, 1 OR 2, INDICATES THE PARTIAL DERIVATIVE
C            WITH RESPECT TO X OR Y, RESPECTIVELY.
C
C         2. THIS SUBROUTINE IS SET UP FOR PROBLEM (2-19).
C
C INPUT:  X          -- INDEPENDENT VARIABLE
C         Y          -- INDEPENDENT VARIABLE
C         XDGM1      -- X**(IDEG(1)-1), WHERE IDEG(1) IS THE DEGREE OF EQ(1)
C         YDGM1      -- Y**(IDEG(2)-1), WHERE IDEG(2) IS THE DEGREE OF EQ(2)
C         XDG        -- X**IDEG(1)
C         YDG        -- Y**IDEG(2)
C
C OUTPUT: EQ         -- EQUATION VALUES
C         DEQ        -- DERIVATIVE VALUES
C
C DECLARATIONS
      DOUBLE PRECISION A,B,C,D,E,F,G
      COMPLEX*16 EQ,DEQ,X,X2,X3,X4,X5,X6,XDG,XDGM1,
     & Y,Y3,Y4,Y5,Y6,YDG,YDGM1
C
C DIMENSIONS
      DIMENSION EQ(2),DEQ(2,2)
C
C COMMON
      COMMON/COEFS/ A(2),B(2),C(2),D(2),E(2),F(2),G(2)
C
      X2=X**2
      X3=X*X2
      X4=XDGM1
      X5=XDG
      Y3=Y**3
      Y4=Y*Y3
      Y5=Y*Y4
      Y6=YDGM1
C
      EQ(1) = A(1)*X5+B(1)*X2*Y+C(1)*Y4+D(1)*X+E(1)*Y+F(1)
      DEQ(1,1) = 5.*A(1)*X4 + 2.*B(1)*X*Y + D(1)
      DEQ(1,2) = B(1)*X2 + 4.*C(1)*Y3 + E(1)
C
      EQ(2) = A(2)*X3+B(2)*X3*Y4+C(2)*X*Y6+D(2)*X+E(2)*Y+F(2)
      DEQ(2,1) = 3.*A(2)*X2+3.*B(2)*X2*Y4+C(2)*Y6+D(2)
      DEQ(2,2) = 4.*B(2)*X3*Y3 + 6.*C(2)*X*Y5 + E(2)
C
      RETURN
      END
```

```
      SUBROUTINE FFUN7B(X,Y,XDGM1,YDGM1,XDG,YDG, EQ,DEQ)
C
C PURPOSE:  EVALUATE THE SYSTEM EQUATIONS AND DERIVATIVES
C           FOR CONSOL7.
C
C REFERENCE: CHAPTER 2 OF "SOLVING POLYNOMIAL SYSTEMS USING
C            CONTINUATION" BY ALEXANDER MORGAN.
C
C GENERIC NAME: FFUN7          FILE NAME: FFUN7B
C
C NOTE:  1. THE SECOND INDEX, 1 OR 2, INDICATES THE PARTIAL DERIVATIVE
C           WITH RESPECT TO X OR Y, RESPECTIVELY.
C
C        2. THIS SUBROUTINE IS SET UP TO SOLVE PROBLEM (2-26).
C
C INPUT:  X         -- INDEPENDENT VARIABLE
C         Y         -- INDEPENDENT VARIABLE
C         XDGM1     -- X**(IDEG(1)-1), WHERE IDEG(1) IS THE DEGREE OF EQ(1)
C         YDGM1     -- Y**(IDEG(2)-1), WHERE IDEG(2) IS THE DEGREE OF EQ(2)
C         XDG       -- X**IDEG(1)
C         YDG       -- Y**IDEG(2)
C
C OUTPUT: EQ        -- EQUATION VALUES
C         DEQ       -- DERIVATIVE VALUES
C
C
C DECLARATIONS
      DOUBLE PRECISION A,B,C,D,E,F,G
      COMPLEX*16 EQ,DEQ,X,X2,XDG,XDGM1,Y,Y2,Y3,YDG,YDGM1
C
C DIMENSIONS
      DIMENSION EQ(2),DEQ(2,2)
C
C COMMON
      COMMON/COEFS/ A(2),B(2),C(2),D(2),E(2),F(2),G(2)
C
C
      X2=XDGM1
      Y2=YDGM1
      Y3=YDG
C
      EQ(1) = A(1)*X2*Y+B(1)*X*Y2+C(1)*X*Y+D(1)*Y3+E(1)*Y2+F(1)*Y+G(1)
      DEQ(1,1) = 2.*A(1)*X*Y + B(1)*Y2 + C(1)*Y
      DEQ(1,2) = A(1)*X2+2.*B(1)*X*Y+C(1)*X+3.*D(1)*Y2+2.*E(1)*Y+F(1)
C
      EQ(2) = A(2)*X2*Y + B(2)*X2 + C(2)*X*Y+ D(2)*X + E(2)*Y
      DEQ(2,1) = 2.*A(2)*X*Y + 2.*B(2)*X + C(2)*Y + D(2)
      DEQ(2,2) = A(2)*X2 + C(2)*X + E(2)
C
      RETURN
      END
```

```
      SUBROUTINE FFUN7C(X,Y,XDGM1,YDGM1,XDG,YDG, EQ,DEQ)
C
C PURPOSE:  EVALUATE THE SYSTEM EQUATIONS AND DERIVATIVES
C           FOR CONSOL7
C
C REFERENCE: CHAPTER 2 OF "SOLVING POLYNOMIAL SYSTEMS USING
C            CONTINUATION" BY ALEXANDER MORGAN.
C
C GENERIC NAME: FFUN7          FILE NAME: FFUN7C
C
C NOTE:   1. THE SECOND INDEX, 1 OR 2, INDICATES THE PARTIAL DERIVATIVE
C            WITH RESPECT TO X OR Y, RESPECTIVELY.
C
C         2. THIS SUBROUTINE IS SET UP FOR PROBLEM (2-27).
C
C INPUT:  X        -- INDEPENDENT VARIABLE
C         Y        -- INDEPENDENT VARIABLE
C         XDGM1    -- X**(IDEG(1)-1), WHERE IDEG(1) IS THE DEGREE OF EQ(1)
C         YDGM1    -- Y**(IDEG(2)-1), WHERE IDEG(2) IS THE DEGREE OF EQ(2)
C         XDG      -- X**IDEG(1)
C         YDG      -- Y**IDEG(2)
C
C OUTPUT: EQ       -- EQUATION VALUES
C         DEQ      -- DERIVATIVE VALUES
C
C
C DECLARATIONS
      DOUBLE PRECISION A,B,C,D,E,F,G
      COMPLEX*16 EQ,DEQ,X,X2,XDG,XDGM1,Y,Y2,Y3,YDG,YDGM1
C
C DIMENSIONS
      DIMENSION EQ(2),DEQ(2,2)
C
C COMMON
      COMMON/COEFS/ A(2),B(2),C(2),D(2),E(2),F(2),G(2)
C
C
      X2=X**2
      Y2=YDGM1
      Y3=YDG
C
      EQ(1) = A(1)*X2*Y2+B(1)*X*Y2+C(1)*X*Y+D(1)*Y3+E(1)*Y2+F(1)*Y+G(1)
      DEQ(1,1) = 2.*A(1)*X*Y2 + B(1)*Y2 + C(1)*Y
      DEQ(1,2)=2.*A(1)*X2*Y+2.*B(1)*X*Y+C(1)*X+3.*D(1)*Y2+2.*E(1)*Y+F(1)
C
      EQ(2) = A(2)*X2*Y + B(2)*X2 + C(2)*X*Y+ D(2)*X + E(2)*Y
      DEQ(2,1) = 2.*A(2)*X*Y + 2.*B(2)*X + C(2)*Y + D(2)
      DEQ(2,2) = A(2)*X2 + C(2)*X + E(2)
C
      RETURN
      END
```

```
C PROGRAM NAME:  CONSOL8
C
C
C PURPOSE:    SOLVES A SYSTEM OF N POLYNOMIAL EQUATIONS IN N UNKNOWNS
C             USING THE CONSOL CONTINUATION FOR POLYNOMIAL SYSTEMS.
C
C REFERENCE:  CHAPTER 4 OF "SOLVING POLYNOMIAL SYSTEMS USING
C             CONTINUATION" BY ALEXANDER MORGAN.
C
C NOTE:    1.  THIS PROGRAM USES THE FORTRAN "COMPLEX*16" DECLARATION.
C
C          2.  THIS PROGRAM GENERATES MANY UNDERFLOWS.  THIS IS
C              NOT AN ERROR CONDITION.  IF YOU NEED TO PREVENT
C              UNDERFLOWS FROM OCCURRING, SEE THE COMMENTS LABELED
C              "UNDERFLOWS" IN SUBROUTINE CORCT.
C
C          3.  THIS PROGRAM COMES WITH SUBROUTINES INPTA,INPTB,FFUN,AND
C              OTPUT SET UP TO SOLVE TWO SECOND-DEGREE EQUATIONS
C              IN TWO UNKNOWNS.  SEE INPTA.
C
C          4.  INPUT IS ON UNIT 5.  OUTPUT IS TO UNITS 6 AND 8.
C              SEE SUBROUTINE INPTA FOR SAMPLE INPUT.
C
C          5.  THE DIMENSIONS OF VARIABLES ARE SET TO ALLOW FOR A
C              MAXIMUM OF TEN EQUATIONS.  IF YOU WISH TO ALTER THIS
C              MAXIMUM, THEN ALL DIMENSIONS THAT ARE CURRENTLY SET
C              TO 10, 11, 20, AND 21 CAN BE CHANGED TO M, M+1,
C              2*M, AND 2*M+1, RESPECTIVELY, WHERE M IS AN UPPER BOUND
C              ON THE NUMBER OF EQUATIONS.
C
C
C USER SUPPLIED SUBROUTINES: INPTA, INPUTB, FFUN, OTPUT
C
C       "INPTA" SUPPLIES FREQUENTLY CHANGED CONSTANTS, INCLUDING
C       THE COEFFICIENTS A(I,J) FOR THE SYSTEM.
C
C       "INPTB" SUPPLIES LESS FREQUENTLY CHANGED CONSTANTS, AND
C       ORDINARILY NEED NOT BE CHANGED BY THE USER.
C
C       "FFUN" SUPPLIES THE EQUATION AND PARTIAL DERIVATIVE VALUES
C       FOR THE SYSTEM TO BE SOLVED.
C
C       "OTPUT" IS CALLED AT THE END OF EACH PATH.  IT CAN BE USED
C       TO REPROCESS THE X VALUES, AT THE USER'S DISCRETION.
C
C
C INPUT VARIABLES:  SEE SUBROUTINE INPTA FOR FORMAT.
C
C       TITLE        -- FIRST 60 CHARACTERS OF FIRST LINE OF INPUT
C       IFLGCR       -- IF =1, THEN CORRECTOR ITERATIONS ARE WRITTEN TO
C                       UNIT 8.  (SEE SUBROUTINE CORCT.)
C       IFLGST       -- IF =1, THEN STEP SUMMARY OUTPUT IS PRINTED AFTER
C                       EVERY STEP.
C       MAXNS        -- MAXIMUM NUMBER OF STEPS FOR A PATH
C       MAXIT        -- MAXIMUM NUMBER OF ITERATIONS TO CORRECT A STEP
```

```
C      EPSBIG       -- THE CORRECTOR RESIDUAL MUST BE LESS THAN EPS FOR
C                      A STEP TO "SUCCEED."  EPS=EPSBIG UNTIL T > .95
C      SSZBEG       -- BEGINNING STEP LENGTH
C      N            -- NUMBER OF EQUATIONS
C      IDEG         -- IDEG(I) IS THE DEGREE OF EQUATION I.
C      A            -- COEFFICIENTS FOR F (PASSED FROM SUBROUTINE INPTA
C                      TO SUBROUTINE FFUN VIA "COMMON/COEFS/")
C
C OUTPUT VARIABLES:
C
C    FOR EACH STEP (OPTIONAL OUTPUT: PRINTED IF IFLGST=1)
C
C      NUMSTP       -- CURRENT STEP NUMBER
C      STPSZE       -- CURRENT STEP SIZE
C      T            -- CURRENT VALUE OF T
C      X            -- X(J) IS THE J-TH INDEPENDENT VARIABLE
C      XRESID       -- NORM OF THE RESIDUAL OF THE LAST CORRECTOR ITERATION
C
C    FOR EACH PATH:
C
C      NUMPAT       -- PATH NUMBER
C      NUMSTP       -- TOTAL NUMBER OF STEPS
C      NSSUCC       -- NUMBER OF STEPS THAT SUCCEEDED
C      NSFAIL       -- NUMBER OF STEPS THAT FAILED
C      NUMIT        -- TOTAL CORRECTOR ITERATIONS
C      NUMLN        -- TOTAL NUMBER OF LINEAR SYSTEMS SOLVED
C      ARCLN2       -- ARC LENGTH OF PATH
C      T            -- VALUE AT END OF PATH (T=1, IF PATH CONVERGED)
C      X            -- VALUE AT END OF PATH (SEE ABOVE)
C      XRESID       -- VALUE AT END OF PATH (SEE ABOVE)
C      COND         -- THE CONDITION OF DF(X) (PRINTED IF DF(X) IS
C                      NONSINGULAR)
C      DET          -- DETERMINANT OF DF(X) (PRINTED IF DF(X) IS
C                      NONSINGULAR)
C
C    TOTALS OVER ALL PATHS:
C
C      ITOTIT       -- NUMBER OF CORRECTOR ITERATIONS
C      ITOTLN       -- NUMBER OF LINEAR SYSTEMS
C      TOTAR        -- ARC LENGTH
C
C
C SUBROUTINES (CALLED DIRECTLY OR INDIRECTLY):
C
C    USER SUPPLIED (OR MODIFIED)
C      FFUN
C      INPTA
C      INPTB
C      OTPUT
C
C    INITIALIZATION
C      INIT
C      STRPT
C
C    MAIN SUBROUTINES
C      CORCT
C      GFUN
```

```
C        HFUN
C        LINN
C        PREDC
C
C     UTILITY SUBROUTINES
C        DABS
C        DATAN
C        DCMPLX
C        DCOS
C        DIMAG
C        DREAL
C        DSIN
C        ICDC
C        MAXNC
C        LNFNC
C        LNSNC
C        PRNTS
C        PRNTV
C        XNORM
C        XNRM2
C
C
C DECLARATIONS
      INTEGER ICOUNT,IDEG,IFLGCR,IFLGST,ITOTDG,ITOTIT,ITOTLN,J,J2,J2M1,
     & MAXIT,MAXNS,MSG,N,N2,N2P1,NADV,NP1,NSFAIL,NSSUCC,NUMI,NUMIT,
     & NUMPAT,NUMSTP
      DOUBLE PRECISION ARCLN2,COND,DIMAG,DREAL,EPS,EPSO,EPSBIG,EPSSML,
     & FACTOR,OLDT,SSZBEG,SSZMIN,STPSZE,T,TOTAR,XNRM2,XRESID,Z
      COMPLEX*16 DET,OLDX,P,PDG,Q,QDG,R,TEMP,WORK,X
C
C DIMENSIONS
      DIMENSION ICOUNT(10),IDEG(10),OLDX(10),P(10),PDG(10),
     & Q(10),QDG(10),R(10),X(11),Z(21)
C
C
C CALL INPUT AND INITIALIZATION ROUTINES
      CALL INPTA(IFLGCR,IFLGST,MAXNS,MAXIT,EPSBIG,SSZBEG,N,IDEG)
      CALL INPTB(N,SSZBEG,EPSBIG,SSZMIN,EPSSML,EPSO,P,Q)
      CALL INIT(N,IDEG,P,Q, ITOTDG,PDG,QDG,R)
C
C
C INITIALIZE COUNTERS AND CONSTANTS FOR THE RUN
      ITOTIT=0
      ITOTLN=0
      TOTAR=0.DO
C
C
C ICOUNT IS A COUNTER USED BY SUBROUTINE STRPT.
      ICOUNT(1)=0
      DO 10 J=2,N
         ICOUNT(J)=1
  10  CONTINUE
C
      N2=2*N
      N2P1=N2+1
C
```

```
C
C     ****************************************************************
C     *****                                                      *****
C     *****          BEGINNING OF MAIN PART OF PROGRAM           *****
C     *****                                                      *****
C     ****************************************************************
C
C PATHS LOOP -- ITERATE THROUGH PATHS.
C
      DO 20 NUMPAT = 1,ITOTDG
C
          WRITE(6,30) NUMPAT
 30       FORMAT('  PATH NUMBER =',I5)
C
C INITIALIZE COUNTERS AND CONSTANTS FOR THE PATH.
          NSSUCC=0
          NSFAIL=0
          NUMIT=0
          STPSZE=SSZBEG
          T=0.D0
C
C THE COUNTER NADV CONTROLS THE "STEP INCREASE" LOGIC
          NADV=0
C
          ARCLN2=0.D0
C
C GET A START POINT, X, FOR THE PATH
          CALL STRPT(N,ICOUNT,IDEG,R ,X)
C
C THE FOLLOWING STATEMENT CAN BE USED TO SKIP SELECTED PATHS.
C THIS IS ESPECIALLY USEFUL IN SETTING UP RERUNS.
          IF(NUMPAT .EQ. 0 ) GOTO 51
C                                        JUST AFTER BOTTOM OF NUMSTP LOOP
C
C SAVE VALUES OF T AND X IN CASE INITIAL STEP MUST BE REDONE
C
          OLDT=T
C
          DO 40 J=1,N
              OLDX(J)=X(J)
 40       CONTINUE
C
C
C STEPS LOOP -- ITERATE STEPS ALONG THE PATH.
C
          DO 50 NUMSTP = 1,MAXNS
C
              CALL PREDC(N,IDEG,PDG,QDG,EPSO,STPSZE, T,X,MSG)
C
              EPS=EPSBIG
              IF( T  .GT.   .95D0 ) EPS=EPSSML
C
              CALL CORCT(IFLGCR,N,IDEG,MAXIT,EPS,EPSO,PDG,QDG,
     &                                   T,X,NUMI,MSG,XRESID)
C
              NUMIT = NUMIT + NUMI
C
```

```
              IF( MSG .EQ. 0 ) THEN
C STEP SUCCEEDED
C
                 NSSUCC=NSSUCC+1
C
                 DO 60 J=1,N
                    J2=2*J
                    J2M1=J2-1
                    TEMP = X(J)-OLDX(J)
                    Z(J2M1)= DREAL(TEMP)
                    Z(J2)  = DIMAG(TEMP)
   60            CONTINUE
                 Z(N2P1)=T-OLDT
                 ARCLN2=ARCLN2+XNRM2(N2P1,Z)
C
                 IF(IFLGST.EQ.1)  CALL PRNTS(N,NUMSTP,MSG,STPSZE,
     &                                      T,X,XRESID)
C
                 IF( T .GE. 1.D0 ) THEN
C MAKE A FINAL CORRECTION AND LEAVE LOOP.
                    WRITE(6,70)
   70            FORMAT(/' PATH CONVERGED')
C
                    FACTOR=(1.D0-OLDT)/(T-OLDT)
C
                    DO 80 J=1,N
                       X(J)=OLDX(J)+FACTOR*(X(J)-OLDX(J))
   80               CONTINUE
                    T=1.D0
C
                    CALL CORCT(IFLGCR,N,IDEG,10,EPSO,EPSO,PDG,QDG,
     &                                      T,X,NUMI,MSG,XRESID)
C
                    GOTO 51
C                     LEAVE LOOP
                 ENDIF
C
                 IF( NUMSTP .GE. MAXNS ) THEN
                    WRITE(6,90)
   90            FORMAT(/,' PATH FAILED, MAX NUM STPS EXCEEDED',/)
                    GOTO 51
C                     LEAVE LOOP
                 ENDIF
C
                 NADV=NADV+1
C NADV CONTROLS THE "STEP INCREASE" LOGIC.  AFTER 5 SUCCESSFUL STEPS,
C STPSZE IS DOUBLED.  WHEN A STEP FAILS, NADV IS SET TO 0.
                 IF( NADV .EQ. 5 ) STPSZE=2.D0*STPSZE
                 IF( NADV .EQ. 5 ) NADV=0
C
C SAVE VALUES OF T AND X IN CASE STEP MUST BE REDONE
C
                 OLDT=T
                 DO 100 J=1,N
                    OLDX(J)=X(J)
  100            CONTINUE
C
```

```
                     ELSE
C STEP FAILED
C
                     NSFAIL=NSFAIL+1
C
                     IF(IFLGST.EQ.1)   CALL PRNTS(N,NUMSTP,MSG,STPSZE,
      &                                      T,X,XRESID)
C
C RESET "STEP INCREASE COUNTER"
                     NADV=0
                     STPSZE=STPSZE/2.D0
C
                     IF( NUMSTP .GE. MAXNS ) THEN
                        WRITE(6,90)
                        GOTO 51
C                             LEAVE LOOP
                     ENDIF
C
                     IF( STPSZE .LT. SSZMIN ) THEN
                        WRITE(6,110)
  110                   FORMAT(/' PATH FAILED, SSZMIN EXCEEDED'/)
                        GOTO 51
C                             LEAVE LOOP
                     ENDIF
C
                     T=OLDT
                     DO 120 J=1,N
                        X(J)=OLDX(J)
  120                CONTINUE
C
                  ENDIF
C
   50       CONTINUE
C                 BOTTOM OF LOOP -- NUMSTP
C
   51       CONTINUE
C                 LEAVE LOOP -- NUMSTP
C
            ITOTIT=ITOTIT + NUMIT
            ITOTLN=ITOTLN + NUMIT +NUMSTP
            TOTAR= TOTAR + ARCLN2
C
            WRITE(6,190)
  190       FORMAT(/' FINAL VALUES FOR PATH'/)
            WRITE(6,195) NUMPAT
  195       FORMAT(' PATH NUMBER = ',I5)
            WRITE(6,200) NUMSTP
  200       FORMAT(' TOTAL NUMBER OF STEPS = ',I5)
            WRITE(6,205) NSSUCC
  205       FORMAT('      NUMBER OF STEPS THAT SUCCEEDED = ',I5)
            WRITE(6,210) NSFAIL
  210       FORMAT('      NUMBER OF STEPS THAT FAILED    = ',I5)
C
C
            WRITE(6,215) NUMIT
  215       FORMAT(' TOTAL CORRECTOR ITERATIONS ON PATH =',I5)
C
```

```fortran
          NUMIT=NUMIT+NUMSTP
          WRITE(6,216) NUMIT
  216     FORMAT(' TOTAL LINEAR SYSTEMS SOLVED ON PATH =',I5)
C
          WRITE(6,217) ARCLN2
  217     FORMAT(' PATH ARC LENGTH =',D11.4)
C
          CALL OTPUT(N,NUMPAT,IDEG,EPSO,X, COND,DET)
C
          CALL PRNTV(N,T,X,XRESID)
C
          IF (COND .GE. 0.D0) THEN
             WRITE(6,218) COND,DET
  218        FORMAT(
     &          /,' CONDITION AND DETERMINANT OF DF(X) AT END OF PATH',
     &          /,'    COND =', D11.4,
     &          /,'    DET  =',2D11.4)
          ELSE
             WRITE(6,219) EPSO
  219        FORMAT(' DF(X) IS SINGULAR WITH RESPECT TO EPSO =',D11.4)
          ENDIF
C
          WRITE(6,999)
  999     FORMAT(/)
C
   20 CONTINUE
C             BOTTOM OF LOOP -- NUMPAT
C
          WRITE(6,220) ITOTIT
  220     FORMAT(//' TOTAL CORRECTOR ITERATIONS =',I10)
C
          WRITE(6,225) ITOTLN
  225     FORMAT(//' TOTAL LINEAR SYSTEMS SOLVED =',I10)
C
          WRITE(6,230)  TOTAR
  230     FORMAT(//' TOTAL ARC LENGTH  =',D11.4//)
C
      STOP
      END
```

```
      SUBROUTINE PREDC(N,IDEG,PDG,QDG,EPSO,STPSZE,T,X, MSG)
C
C
C PURPOSE:   ON ENTRY TO PREDC, Z=(X,T) IS A POINT ON A
C            CONTINUATION CURVE.  PREDC FINDS A POINT,
C            DZ, ON THE TANGENT TO THE CURVE AT Z.  THEN
C            (X,T) = Z + DZ IS THE "PREDICTED POINT."
C
C
C INPUT:  N        -- NUMBER OF EQUATIONS
C         IDEG     -- IDEG(J) IS THE DEGREE OF THE J-TH EQUATION
C         PDG      -- CONSTANTS TO DEFINE G, THE START SYSTEM
C         QDG      -- CONSTANTS TO DEFINE G
C         EPSO     -- ZERO-EPSILON FOR SINGULARITY
C         STPSZE   -- CURRENT STEPSIZE
C         T        -- CURRENT VALUE OF T
C         X        -- CURRENT VALUE OF X
C
C OUTPUT: T        -- IF MSG=0, PREDICTED VALUE OF T
C         X        -- IF MSG=0, PREDICTED VALUE OF X
C         MSG      -- =1, IF SYSTEM JACOBIAN IS SINGULAR.
C                     =0, OTHERWISE.
C
C
C SUBROUTINES: DCMPLX,DREAL,HFUN,LINN,XNORM
C
C
C DECLARATIONS
      INTEGER IDEG,J,MSG,N,NP1,N2P1
      DOUBLE PRECISION DELT,DREAL,EPSO,FACTOR,STPSZE,T,XLNGTH,XNORM
      COMPLEX*16 DHT,DHX,DZ,H,PDG,QDG,RHS,X
C
C DIMENSIONS
      DIMENSION DHT(10),DHX(10,10),DZ(11),H(10),IDEG(1),
     & PDG(1),QDG(1),RHS(10),X(1)
C
C
      NP1=N+1
      N2P1=2*N+1
      MSG = 0
C
      CALL HFUN(N,IDEG,PDG,QDG,T,X, H,DHX,DHT)
C
      DO 10 J=1,N
         RHS(J)=-DHT(J)
   10 CONTINUE
C
C SOLVE DHX*DZ=RHS.  MSG=1 MEANS DHX IS SINGULAR WITH RESPECT TO EPSO.
      CALL LINN(N,EPSO,DHX,RHS, MSG,DZ)
      IF( MSG .EQ. 1 ) WRITE(6,15)
   15 FORMAT(//'  FROM PREDC, MSG=1, ERROR '//)
      IF( MSG .EQ. 1 ) RETURN
C
      DZ(NP1)=DCMPLX(1.D0,0.D0)
C
C
```

```
C DZ IS NOW THE TANGENT POINT TO BE ADDED TO Z=(X,T), EXCEPT
C THAT ITS LENGTH MAY NEED TO BE CHANGED.
C
C ADJUST THE LENGTH OF DZ BY THE STEP LENGTH, STPSZE.
C
      XLNGTH = XNORM(N2P1,DZ)
C
      FACTOR=STPSZE/XLNGTH
C
C IF THIS FACTOR WILL MAKE T > 1, THEN SHORTEN FACTOR SO THAT T WILL
C EQUAL 1.
      IF(T+FACTOR .GT. 1.D0) FACTOR=1.D0-T
C
      DO 20 J=1,NP1
         DZ(J)=DZ(J)*FACTOR
   20 CONTINUE
C
C THE LENGTH OF DZ IS NOW CORRECT.
C
C UPDATE Z=(X,T)
      DO 30 J=1,N
         X(J)=X(J)+DZ(J)
   30 CONTINUE
      T=T+DREAL(DZ(NP1))
C
      RETURN
      END
```

```
              SUBROUTINE CORCT(IFLGCR,N,IDEG,MAXIT,EPS,EPSO,PDG,QDG,
           &                                  T,X,NUMI,MSG,XRESID)
C
C PURPOSE:  MOVE THE POINT Z=(X,T) BACK TO THE CONTINUATION CURVE
C
C INPUT:   IFLGCR    -- =1, IF CORRECTOR ITERATIONS ARE TO BE WRITTEN
C                            TO UNIT 8.
C                       =0, OTHERWISE.
C          N         -- NUMBER OF EQUATIONS
C          IDEG      -- IDEG(J) IS THE DEGREE OF THE J-TH EQUATION
C          MAXIT     -- MAXIMUM NUMBER OF ITERATIONS ALLOWED
C          EPS       -- ZERO-EPSILON FOR CORRECTOR RESIDUAL
C          EPSO      -- ZERO-EPSILON FOR SINGULARITY
C          PDG       -- CONSTANTS TO DEFINE G, THE START SYSTEM
C          QDG       -- CONSTANTS TO DEFINE G
C          T         -- PREDICTED VALUE OF T
C          X         -- PREDICTED VALUE OF X
C
C OUTPUT:  T         -- IF MSG=0, CORRECTED VALUE OF T
C          X         -- IF MSG=0, CORRECTED VALUE OF X
C          NUMI      -- NUMBER OF ITERATIONS USED FOR CORRECTION
C          MSG       -- =1, IF SYSTEM JACOBIAN IS SINGULAR.
C                       =2, IF CORRECTOR FAILED TO REDUCE RESIDUAL
C                            TO BE LESS THAN EPS WITHIN MAXIT ITERATIONS.
C                       =0, OTHERWISE.
C          XRESID    -- NORM OF CORRECTOR RESIDUAL
C
C SUBROUTINES: DABS,DCMPLX,DIMAG,DREAL,HFUN,LINN,XNORM
C
C
C DECLARATIONS
        INTEGER IDEG,IFLGCR,J,MAXIT,MSG,N,N2,NUMI
        DOUBLE PRECISION DIMAG,DREAL,EPS,EPSO,EPSOO,T,XL,XNORM,XRESID
        COMPLEX*16 DHX,DHT,H,PDG,QDG,RESID,X
C
C DIMENSIONS
        DIMENSION  DHT(10),DHX(10,10),H(10),IDEG(1),
       & PDG(1),QDG(1),RESID(10),X(1)
C
C
        N2=2*N
        MSG = 0
C
        DO 10 NUMI = 1,MAXIT
C
            CALL HFUN(N,IDEG,PDG,QDG,T,X, H,DHX,DHT)
C
C SOLVE DHX*RESID=H.  MSG=1 MEANS DHX IS SINGULAR WITH RESPECT TO EPSO.
            CALL LINN(N,EPSO,DHX,H, MSG,RESID)
C
            IF( MSG .EQ. 1 ) WRITE(6,20)
   20       FORMAT(//' FROM CORCT, MSG=1, ERROR '//)
            IF( MSG .EQ. 1 ) RETURN
C
            DO 30 J=1,N
                X(J)= X(J)-RESID(J)
   30       CONTINUE
```

```
C
C$$$$$$$$$$$$$$$$$$$$$$$$$$$$$$$$$$$$$$$$$$$$$$$$$$$$$$$$$$$$$$$$$$$$$$$$$$$$$$
C UNDERFLOWS :  IF YOU MUST PREVENT UNDERFLOWS, I SUGGEST YOU
C                 TAKE THE COMMENTS OFF THE FOLLOWING BLOCK OF
C                 CODE.  THE VALUE FOR EPSOO MAY NEED TO BE ADJUSTED.
C
C          EPSOO=1.D-22
C
C          DO 35 J=1,N
C             IF(DABS(DREAL(X(J))).LE.EPSOO) X(J)=DCMPLX(0.D0,DIMAG(X(J)))
C             IF(DABS(DIMAG(X(J))).LE.EPSOO) X(J)=DCMPLX(DREAL(X(J)),0.D0)
C 35        CONTINUE
C$$$$$$$$$$$$$$$$$$$$$$$$$$$$$$$$$$$$$$$$$$$$$$$$$$$$$$$$$$$$$$$$$$$$$$$$$$$$$$
C
C COMPUTE THE LENGTH OF RESID
          XRESID=XNORM(N2,RESID)
C
          IF(IFLGCR.EQ.1) THEN
          XL =XNORM(N2,X)
          WRITE(8,40) T,XL,XRESID
 40        FORMAT(' T=',D11.4,' XL=',D11.4,' XRESID =',D11.4)
          IF( XRESID  .LT.  EPS ) WRITE(8,999)
 999       FORMAT(/)
          ENDIF
C
          IF( XRESID  .LT.  EPS ) RETURN
C
 10       CONTINUE
C
      MSG=2
C
      RETURN
C
      END
```

```
      SUBROUTINE HFUN(N,IDEG,PDG,QDG,T,X, H,DHX,DHT)
C
C PURPOSE: EVALUATES H, THE CONTINUATION EQUATION
C
C INPUT:   N         -- NUMBER OF EQUATIONS
C          IDEG      -- IDEG(J) IS THE DEGREE OF THE J-TH EQUATION
C          PDG       -- CONSTANTS TO DEFINE G, THE START SYSTEM
C          QDG       -- CONSTANTS TO DEFINE G
C          T         -- CURRENT VALUE OF T
C          X         -- CURRENT VALUE OF X
C
C OUTPUT:  H         -- H IS THE CONTINUATION SYSTEM, A SYSTEM OF N
C                       COMPLEX POLYNOMIALS IN THE N+1 UNKNOWNS X AND T.
C          DHX       -- DHX IS THE N X N (COMPLEX) MATRIX OF PARTIAL
C                       DERIVATIVES OF H WITH RESPECT TO X.  DHX(J,K) IS
C                       THE PARTIAL OF H(J) WITH RESPECT TO X(K).
C          DHT       -- DHT IS THE N X 1 MATRIX OF PARTIAL DERIVATIVES
C                       OF H WITH RESPECT TO T.
C
C SUBROUTINES:  DCMPLX,FFUN,GFUN
C
C DECLARATIONS
      INTEGER IDEG,J,K,N
      DOUBLE PRECISION ONEMT,T
      COMPLEX*16 DF,DG,DHT,DHX,F,G,H,PDG,QDG,X,XDG,XDGM1
C
C DIMENSIONS
      DIMENSION DF(10,11),DG(10),DHT(10),DHX(10,10),F(10),G(10),
     & H(10),IDEG(1),PDG(1),QDG(1),X(1),XDG(10),XDGM1(10)
C
C COMPUTE THE (IDEG-1)-TH AND IDEG-TH POWER OF X
C
      DO 10 J=1,N
        IF(IDEG(J).EQ.1) THEN
            XDGM1(J)=DCMPLX(1.D0,0.D0)
          ELSE
            XDGM1(J)=X(J)**(IDEG(J)-1)
        ENDIF
        XDG(J)   =X(J)*XDGM1(J)
  10  CONTINUE
C
      CALL GFUN(N,IDEG,PDG,QDG,  XDGM1,XDG, G,DG)
      CALL FFUN(N,                X,XDGM1,XDG, F,DF)
C
      ONEMT=1.D0 - T
      DO 20 J=1,N
        DO 30 K=1,N
            DHX(J,K)= T*DF(J,K)
  30      CONTINUE
        DHX(J,J)= DHX(J,J) + ONEMT*DG(J)
C
        DHT(J)=   F(J) -          G(J)
        H(J)  = T*F(J) + ONEMT*  G(J)
  20  CONTINUE
C
      RETURN
      END
```

```
      SUBROUTINE GFUN(N,IDEG,PDG,QDG,XDGM1,XDG, G,DG)
C
C PURPOSE: EVALUATES G, THE START SYSTEM
C
C INPUT:  N       -- NUMBER OF EQUATIONS
C         IDEG    -- IDEG(J) IS THE DEGREE OF THE J-TH EQUATION
C         PDG     -- CONSTANTS TO DEFINE G
C         QDG     -- CONSTANTS TO DEFINE G
C         XDGM1   -- XDGM1(I) = X(I)**(IDEG(I)-1)
C         XDG     -- XDG(I) = X(I)**IDEG(I)
C
C OUTPUT: G       -- G IS THE START SYSTEM.
C         DG      -- DG(J) IS THE PARTIAL DERIVATIVE OF G(J) WITH
C                    RESPECT TO X(J).
C
C DECLARATIONS
      INTEGER IDEG,J,N
      COMPLEX*16 DG,G,PDG,PXDGM1,PXDG,QDG,XDG,XDGM1
C
C DIMENSIONS
      DIMENSION  DG(1),G(1),IDEG(1),PDG(1),PXDG(10),PXDGM1(10),QDG(1),
     & XDG(1),XDGM1(1)
C
C
C COMPUTE THE PRODUCT OF PDG AND XDGM1
      DO 10 J=1,N
          PXDGM1(J) = PDG(J)*XDGM1(J)
   10 CONTINUE
C
C COMPUTE THE PRODUCT OF PDG AND XDG
      DO 20 J=1,N
          PXDG(J) = PDG(J)*XDG(J)
   20 CONTINUE
C
      DO 30 J=1,N
          G(J) = PXDG(J) - QDG(J)
          DG(J)= IDEG(J)*PXDGM1(J)
   30 CONTINUE
C
      RETURN
      END
```

```
      SUBROUTINE LINN(N,EPSO,DHX,RHS, MSG,RESID)
C
C PURPOSE: SOLVES THE COMPLEX LINEAR SYSTEM DHX*RESID=RHS
C
C
C INPUT:  N        -- NUMBER OF EQUATIONS
C         EPSO     -- ZERO-EPSILON FOR SINGULARITY
C         DHX      -- N X N COMPLEX JACOBIAN OF H WITH RESPECT TO X.
C         RHS      -- COMPLEX VECTOR OF LENGTH N
C
C OUTPUT: MSG      -- =1, IF DHX IS SINGUALAR WITH RESPECT TO EPSO.
C                     =0, OTHERWISE
C         RESID    -- IF MSG=0, SOLUTION TO THE LINEAR SYSTEM
C                     DHX*RESID=RHS
C
C SUBROUTINES: DCMPLX,LNFNC,LNSNC
C
C DECLARATIONS
      INTEGER IR,IROW,J,LDIM,MSG,N
      DOUBLE PRECISION EPSO
      COMPLEX*16 DHX,RESID,RHS
C
C DIMENSIONS
      DIMENSION DHX(10,1),IROW(10),RESID(1),RHS(1)
C
C
      LDIM=10
C
      CALL LNFNC(LDIM,N,EPSO,DHX, MSG,IROW)
      IF( MSG .EQ. 1)  RETURN
C
      CALL LNSNC(LDIM,N,IROW,DHX,RHS, RESID)
C
      RETURN
      END
```

```
          SUBROUTINE INIT(N,IDEG,P,Q, ITOTDG,PDG,QDG,R)
C
C PURPOSE: COMPUTES POWERS OF P AND Q,
C          COMPUTES R=Q/P, USED IN SUBROUTINE STRPT, AND
C          COMPUTES THE TOTAL DEGREE, ITOTDG.
C
C INPUT:  N        -- NUMBER OF EQUATIONS
C         IDEG     -- IDEG(J) IS THE DEGREE OF THE J-TH EQUATION
C         P        -- CONSTANTS TO DEFINE G, THE START SYSTEM
C         Q        -- CONSTANTS TO DEFINE G
C
C OUTPUT: ITOTDG   -- THE TOTAL DEGREE OF THE SYSTEM
C         PDG      -- CONSTANTS TO DEFINE G
C         QDG      -- CONSTANTS TO DEFINE G
C         R        -- R(J)=Q(J)/P(J), USED IN SUBROUTINE STRPT
C
C
C DECLARATIONS
      INTEGER IDEG,ITOTDG,J,N
      COMPLEX*16 P,PDG,Q,QDG,R
C
C DIMENSIONS
      DIMENSION  IDEG(1),P(1),PDG(1),Q(1),QDG(1),R(1)
C
C
C COMPUTE THE IDEG-TH POWER OF P
      DO 10 J=1,N
        PDG(J)=P(J)**IDEG(J)
   10 CONTINUE
C
C COMPUTE THE IDEG-TH POWER OF Q
      DO 20 J=1,N
        QDG(J)=Q(J)**IDEG(J)
   20 CONTINUE
C
C COMPUTE R=Q/P
      DO 30 J=1,N
        R(J)=Q(J)/P(J)
   30 CONTINUE
C
C COMPUTE ITOTDG
      ITOTDG=1
      DO 40 J=1,N
          ITOTDG=ITOTDG*IDEG(J)
   40 CONTINUE
C
      RETURN
      END
```

```
          SUBROUTINE STRPT(N,ICOUNT,IDEG,R ,X)
C
C PURPOSE: COMPUTES START POINT FOR EACH PATH.
C
C
C INPUT:  N        -- NUMBER OF EQUATIONS
C         ICOUNT   -- A COUNTER USED TO INCREMENT EACH (NORMALIZED)
C                     VARIABLE AROUND THE UNIT CIRCLE, SO THAT EVERY
C                     COMBINATION OF START VALUES IS CHOSEN.  ICOUNT
C                     IS INITIALIZED IN THE MAIN ROUTINE.
C         IDEG     -- IDEG(J) IS THE DEGREE OF THE J-TH EQUATION.
C         R        -- CONSTANTS TO DEFINE START POINTS
C
C OUTPUT: X        -- START POINT FOR A PATH
C
C
C SUBROUTINES: DATAN,DCMPLX,DCOS,DSIN
C
C
C DECLARATIONS
      INTEGER ICOUNT,IDEG,ITEST,J,N
      DOUBLE PRECISION ANGLE,TWOPI,XCOUNT,XDEG
      COMPLEX*16 R,X
C
C DIMENSIONS
      DIMENSION  ICOUNT(1),IDEG(1),R(1),X(1)
C
C
      DO 10 J=1,N
         ITEST = ICOUNT(J)
         IF( ITEST  .GE.   IDEG(J) ) ICOUNT(J)=1
         IF( ITEST  .LT.   IDEG(J) ) ICOUNT(J)=ICOUNT(J)+1
         IF( ITEST  .LT.   IDEG(J) ) GOTO 20
C                                             LEAVE LOOP
   10 CONTINUE
   20 CONTINUE
C
      TWOPI=8.D0*DATAN(1.D0)
C
      DO 30 J=1,N
         ANGLE = (TWOPI/IDEG(J))*ICOUNT(J)
         X(J) = R(J)*DCMPLX(DCOS(ANGLE),DSIN(ANGLE))
   30 CONTINUE
C
      RETURN
      END
```

```
      SUBROUTINE PRNTS(N,NUMSTP,MSG,STPSZE,T,X,XRESID)
C
C PURPOSE: PRINTS STEP SUMMARY.
C
C INPUT:   N        -- NUMBER OF EQUATIONS
C          NUMSTP   -- STEP NUMBER
C          MSG      -- ERROR MESSAGE
C          STPSZE   -- STEPSIZE
C          T        -- CURRENT VALUE OF T
C          X        -- CURRENT VALUE OF X
C          XRESID   -- NORM OF CORRECTOR RESIDUAL
C
C OUTPUT: TO UNIT 6
C
C SUBROUTINES: PRNTV
C
C
C DECLARATIONS
      INTEGER MSG,N,NUMSTP
      DOUBLE PRECISION STPSZE,T,XRESID
      COMPLEX*16 X
C
C DIMENSIONS
      DIMENSION  X(1)
C
C
      WRITE(6,100) NUMSTP
  100 FORMAT(/' STEP SUMMARY FOR STEP NUMBER',I5,' :')
C
      IF(MSG.EQ.0) WRITE(6,110) NUMSTP
  110 FORMAT(' STEP ',I5,' SUCCEEDED')
      IF(MSG.EQ.1) WRITE(6,120) NUMSTP
  120 FORMAT(' STEP ',I5,' FAILED DUE TO SINGULARITY')
      IF(MSG.EQ.2) WRITE(6,130) NUMSTP
  130 FORMAT(' STEP ',I5,' FAILED DUE TO NON-CONVERGENCE')
C
      WRITE(6,140) STPSZE
  140 FORMAT(' STPSZE =',D11.4)
C
      CALL PRNTV(N,T,X,XRESID)
C
      RETURN
      END
```

```fortran
      SUBROUTINE PRNTV(N,T,X,XRESID)
C
C PURPOSE: PRINTS SELECTED VARIABLE VALUES
C
C
C INPUT:   N         -- NUMBER OF EQUATIONS
C          T         -- CURRENT VALUE OF T
C          X         -- CURRENT VALUE OF X
C          XRESID    -- NORM OF CORRECTOR RESIDUAL
C
C OUTPUT:  TO UNIT 6
C
C
C DECLARATIONS
      INTEGER J,N
      DOUBLE PRECISION T,XRESID
      COMPLEX*16 X
C
C DIMENSIONS
      DIMENSION  X(1)
C
C
      WRITE(6,100)
  100 FORMAT(' --------------------------------------------------------')
C
      WRITE(6,110) T
  110 FORMAT('  T  = ',D22.15)
C
      WRITE(6,120)
  120 FORMAT('                  REAL PART           IMAGINARY PART' )
C
      DO 125 J=1,N
          WRITE(6,130) J,X(J)
  125 CONTINUE
  130 FORMAT(' X(',I5,') = ', D11.4,'           ',D11.4)
C
      WRITE(6,150) XRESID
  150 FORMAT(' NORM OF RESIDUAL = ',D11.4)
C
      WRITE(6,100)
C
      RETURN
      END
```

```fortran
      SUBROUTINE INPTA(IFLGCR,IFLGST,MAXNS,MAXIT,EPSBIG,SSZBEG,N,IDEG)
C
C PURPOSE:   INITIALIZES FREQUENTLY CHANGED CONSTANTS.
C            INITIALIZES COEFFICIENTS FOR SYSTEM.
C
C INPUT:   READ FROM UNIT 5.  SEE CODE BELOW FOR FORMAT.
C
C       TITLE       -- FIRST 60 CHARACTERS OF FIRST LINE OF INPUT
C       IFLGCR      -- =1, IF THE CORRECTOR ITERATIONS ARE TO BE WRITTEN
C                       TO UNIT 8.  (SEE SUBROUTINE CORCT.)
C       IFLGST      -- =1, IF A STEP SUMMARY OUTPUT IS PRINTED AFTER
C                       EVERY STEP.
C       MAXNS       -- MAXIMUM NUMBER OF STEPS FOR A PATH
C       MAXIT       -- MAXIMUM NUMBER OF ITERATIONS TO CORRECT A STEP
C       EPSBIG      -- THE CORRECTOR RESIDUAL MUST BE LESS THAN EPS FOR
C                       A STEP TO "SUCCEED."  EPS=EPSBIG UNTIL T > .95
C       SSZBEG      -- BEGINNING STEP LENGTH
C       N           -- NUMBER OF EQUATIONS
C       IDEG        -- IDEG(I) IS THE DEGREE OF EQUATION I.
C       A           -- COEFFICIENTS FOR F (PASSED FROM SUBROUTINE INPTA
C                       TO SUBROUTINE FFUN VIA "COMMON/COEFS/")
C
C SAMPLE INPUT FOLLOWS:
CTITLE:    SYSTEM OF TWO QUADRICS,    NO SOLUTIONS AT INFINITY
C     0        IFLGCR
C     0        IFLGST
C   200        MAXNS
C     3        MAXIT
C                   1.D-02      EPSBIG
C                   1.D-02      SSZBEG
C     2        N
C     2        IDEG(1)
C     2        IDEG(2)
C             -.00098D 00    X(1)**2       *
C             978000.D+00    X(2)**2       *
C                 -9.8D+00   X(1)*X(2)     *      EQUATION 1
C             -235.0D+00     X(1)          *
C             88900.0D+00    X(2)          *
C             -1.000D+00     CONSTANT      *
C             -.0100D+00     X(1)**2       $
C             -.9840D+00     X(2)**2       $
C             -29.70D+00     X(1)*X(2)     $      EQUATION 2
C              .00987D+00    X(1)          $
C             -.1240D+00     X(2)          $
C             -.2500D+00     CONSTANT      $
C
C OUTPUT:   IFLGCR,IFLGST,MAXNS,MAXIT,EPSBIG,SSZBEG,N,IDEG
C
C COMMON:   /COEFS/  IS USED TO PASS COEFFICIENT VALUES IN ARRAY A FROM
C                    SUBROUTINE INPTA TO SUBROUTINE FFUN.
C
C
C DECLARATIONS
      INTEGER I,IDEG,IFLGCR,IFLGST,J,MAXNS,MAXIT,N,TITLE
      DOUBLE PRECISION A,EPSBIG,SSZBEG
```

```
C
C DIMENSIONS
      DIMENSION IDEG(1),TITLE(60)
C
      COMMON/COEFS/A(2,6)
C          A(I,1),... A(I,6) ARE THE COEFFICIENTS FOR EQUATION I
C
C
C  INPUT STATEMENTS FOLLOW.
      READ(5,10) TITLE
 10   FORMAT(60A1)
      READ(5,20) IFLGCR
 20   FORMAT(I5)
      READ(5,20) IFLGST
      READ(5,20) MAXNS
      READ(5,20) MAXIT
      READ(5,30) EPSBIG
 30   FORMAT(D22.15)
      READ(5,30) SSZBEG
      READ(5,20) N
      DO 40 I=1,N
         READ(5,20) IDEG(I)
 40   CONTINUE
      DO 50 I=1,N
      DO 50 J=1,6
         READ(5,30) A(I,J)
 50   CONTINUE
C
C
      WRITE(6,60)
 60   FORMAT(' CONSOL8',//)
      WRITE(6,70) TITLE
 70   FORMAT(' ',60A1)
C
      WRITE(6,100) IFLGCR
 100  FORMAT(/' FLAG TO PRINT OUT CORRECTOR RESIDUALS =',I2)
      WRITE(6,101) IFLGST
 101  FORMAT(/' FLAG TO PRINT OUT STEP SUMMARIES =',I2)
      WRITE(6,105) MAXNS
 105  FORMAT(/' MAX NUM STEPS PER PATH =',I5)
      WRITE(6,110) MAXIT
 110  FORMAT(/' MAX NUM ITERATIONS PER STEP =',I5)
      WRITE(6,115) EPSBIG
 115  FORMAT(/' EPSBIG =',D11.4/)
      WRITE(6,120) SSZBEG
 120  FORMAT(/' BEGINNING STEP SIZE = ',D11.4/)
C
      WRITE(6,122) N
 122  FORMAT(/' NUMBER OF EQUATIONS = ',I5/)
C
      DO 124 I=1,N
         WRITE(6,125) I,IDEG(I)
 125  FORMAT(' DEG(',I5,')=',I5)
 124  CONTINUE
C
      WRITE(6,130)
 130  FORMAT(/' COEFFICIENTS :')
```

```
C
      DO 135 I=1,N
        WRITE(6,140) I
140     FORMAT(' EQUATION ',I5)
C
        DO 142 J=1,6
          WRITE(6,145) I,J,A(I,J)
145       FORMAT('   A(',I5,',',I5,') =',D22.15)
142     CONTINUE
C
        WRITE(6,999)
999     FORMAT(/)
135   CONTINUE
C
      RETURN
      END
```

```
      SUBROUTINE INPTB(N,SSZBEG,EPSBIG,SSZMIN,EPSSML,EPSO,P,Q)
C
C PURPOSE: INITIALIZES LESS FREQUENTLY CHANGED CONSTANTS
C
C NOTE:    IF YOU CHANGE THE DIMENSIONS OF VARIABLES IN CONSOL8
C          IN ORDER TO CHANGE THE MAXIMUM NUMBER OF EQUATIONS,
C          DO NOT CHANGE ANY OF THE DIMENSIONS IN THIS SUBROUTINE.
C
C INPUT:   N         -- NUMBER OF EQUATIONS
C          SSZBEG    -- BEGINNING STEPSIZE
C          EPSBIG    -- ZERO-EPSILON FOR PATH TRACKING RESIDUAL,
C                       UNTIL T > .95
C
C OUTPUT: SSZMIN     -- MINIMUM STEPSIZE ALLOWED
C          EPSSML    -- ZERO-EPSILON FOR PATH TRACKING RESIDUAL,
C                       AFTER T > .95
C          EPSO      -- ZERO-EPSILON FOR SINGULARITY
C          P         -- "RANDOM" CONSTANTS TO DEFINE G, THE START SYSTEM
C          Q         -- "RANDOM" CONSTANTS TO DEFINE G
C
C
C SUBROUTINES: DCMPLX,MOD
C
C
C DECLARATIONS
      INTEGER J,JJ,N
      DOUBLE PRECISION EPSO,EPSBIG,EPSSML,PP,QQ,SSZBEG,SSZMIN
      COMPLEX*16 P,Q
C
C DIMENSIONS
      DIMENSION  P(1),PP(2,10),Q(1),QQ(2,10)
C
C
      SSZMIN = 1.D-5*SSZBEG
      EPSSML = 1.D-3*EPSBIG
      EPSO   = 1.D-22
C
      PP(1, 1)= .12324754231D0
         PP(2, 1)= .76253746298D0
      PP(1, 2)= .93857838950D0
         PP(2, 2)=-.99375892810D0
      PP(1, 3)=-.23467908356D0
         PP(2, 3)= .39383930009D0
      PP(1, 4)= .83542556622D0
         PP(2, 4)=-.10192888288D0
      PP(1, 5)=-.55763522521D0
         PP(2, 5)=-.83729899911D0
      PP(1, 6)=-.78348738738D0
         PP(2, 6)=-.10578234903D0
      PP(1, 7)= .03938347346D0
         PP(2, 7)= .04825184716D0
      PP(1, 8)=-.43428734331D0
         PP(2, 8)= .93836289418D0
      PP(1, 9)=-.99383729993D0
         PP(2, 9)=-.40947822291D0
      PP(1,10)= .09383736736D0
         PP(2,10)= .26459172298D0
```

```
C
      QQ( 1, 1)= .58720452864D0
        QQ(2, 1)= .01321964722D0
      QQ( 1, 2)= .97884134700D0
        QQ(2, 2)=-.14433009712D0
      QQ( 1, 3)= .39383737289D0
        QQ(2, 3)= .41543223411D0
      QQ( 1, 4)=-.03938376373D0
        QQ(2, 4)=-.61253112318D0
      QQ( 1, 5)= .39383737388D0
        QQ(2, 5)=-.26454678861D0
      QQ( 1, 6)=-.00938376766D0
        QQ(2, 6)= .34447867861D0
      QQ( 1, 7)=-.04837366632D0
        QQ(2, 7)= .48252736790D0
      QQ( 1, 8)= .93725237347D0
        QQ(2, 8)=-.54356527623D0
      QQ( 1, 9)= .39373957747D0
        QQ(2, 9)= .65573434564D0
      QQ( 1,10)=-.39380038371D0
        QQ(2,10)= .98903450052D0
C
      DO 10 J=1,N
      JJ=MOD(J-1,10)+1
      P(J)=DCMPLX(PP(1,JJ),PP(2,JJ))
      Q(J)=DCMPLX(QQ(1,JJ),QQ(2,JJ))
  10  CONTINUE
C
      WRITE(6,100) SSZMIN
 100  FORMAT(/' MINIMUM STPSZE = ',D11.4)
      WRITE(6,105) EPSSML
 105  FORMAT(/' EPSSML =',D11.4)
      WRITE(6,110) EPSO
 110  FORMAT(/' EPSO   =',D11.4)
C
      WRITE(6,115) (P(J),J=1,N)
 115  FORMAT(/' P   =',2D11.4)
C
      WRITE(6,120) (Q(J),J=1,N)
 120  FORMAT(/' Q   =',2D11.4)
C
      WRITE(6,999)
 999  FORMAT(/)
C
      RETURN
      END
```

```
      SUBROUTINE FFUN(N,X,XDGM1,XDG, F,DF)
C
C PURPOSE: EVALUATES F AND DF.
C
C INPUT:   N        -- NUMBER OF EQUATIONS
C          X        -- CURRENT VALUE OF X
C          XDGM1    -- XDGM1(J) = X(J)**(IDEG(J)-), WHERE IDEG(J) IS
C                      THE DEGREE OF THE J-TH EQUATIONS, F(J).
C          XDG      -- XDG(J) = X(J)**IDEG(J)
C
C OUTPUT: F         -- F(J) IS THE J-TH EQUATION IN THE SYSTEM WE WANT
C                      TO SOLVE.
C         DF        -- DF(J,K) IS THE PARTIAL DERIVATIVE OF F(J) WITH
C                      RESPECT TO X(K).
C
C COMMON: /COEFS/ IS USED TO PASS COEFFICIENT VALUES FROM
C                 SUBROUTINE INPTA TO SUBROUTINE FFUN.
C
C *********************************************************
C THIS SUBROUTINE IS FOR TWO QUADRIC EQUATIONS
C
C     A(J,1)*XDG(1) +A(J,2)*XDG(2) +A(J,3)*X(1)*X(2)
C
C                  +A(J,4)*X(1) +A(J,5)*X(2) +A(J,6) = 0
C
C FOR J=1 TO 2.
C *********************************************************
C
C DECLARATIONS
      INTEGER J,N
      DOUBLE PRECISION A
      COMPLEX*16 DF,F,ONE,X,X12,XDG,XDGM1
C
C DIMENSIONS
      DIMENSION DF(10,1),F(1),X(1),XDG(1),XDGM1(1)
C
C COMMON
      COMMON/COEFS/A(2,6)
C
C
      ONE=DCMPLX(1.D0,0.D0)
      X12=X(1)*X(2)
C
      DO 10 J=1,N
        F(J)=    A(J,1)*XDG(1) + A(J,2)*XDG(2) +   A(J,3)*X12
     &         + A(J,4)*X(1)   + A(J,5)*X(2)   +   A(J,6)*ONE
        DF(J,1)=2*A(J,1)*X(1)+A(J,3)*X(2)+A(J,4)*ONE
        DF(J,2)=2*A(J,2)*X(2)+A(J,3)*X(1)+A(J,5)*ONE
   10 CONTINUE
C
      RETURN
      END
```

```
      SUBROUTINE OTPUT(N,NUMPAT,IDEG,EPSO,X, COND,DET)
C
C PURPOSE: OTPUT IS CALLED BY THE MAIN ROUTINE AFTER EACH PATH IS COMPLETED.
C          OUTPUT CAN BE USED TO TRANSFORM THE VARIABLES OR FOR ANY OTHER
C          PURPOSE.  THIS VERSION OF OTPUT COMPUTES THE CONDITION AND
C          DETERMINANT OF THE JACOBIAN OF F AT X.
C
C
C INPUT:  N       -- NUMBER OF EQUATIONS
C         NUMPAT  -- CURRENT PATH NUMBER
C         IDEG    -- IDEG(J) IS THE DEGREE OF THE J-TH EQUATION.
C         EPSO    -- ZERO-EPSILON FOR SINGULARITY.
C         X       -- CURRENT X VALUE
C
C OUTPUT: COND    -- CONDITION OF THE JACOBIAN, DF, AT X.  COND=-1
C                    MEANS THAT DF IS SINGULAR WITH RESPECT TO EPSO.
C         DET     -- IF COND >= 0, DETERMINANT OF DF.
C
C
C SUBROUTINES: DCMPLX,FFUN,ICDC
C
C
C DECLARATIONS
      INTEGER IDEG,IWORK,J,MSG,N,NUMPAT
      DOUBLE PRECISION COND,EPSO
      COMPLEX*16 DF,F,DET,ONE,WORK,WORK2,X,XDG,XDGM1
C
C DIMENSIONS
      DIMENSION DF(10,10),F(10),IDEG(1),IWORK(10),WORK(10),
     & WORK2(10,10),X(1),XDG(10),XDGM1(10)
C
C
      ONE=DCMPLX(1.D0,0.D0)
C
      DO 10 J=1,N
        IF(IDEG(J).EQ.1)THEN
           XDGM1(J)=ONE
         ELSE
           XDGM1(J)=X(J)**(IDEG(J)-1)
        ENDIF
        XDG(J)=X(J)*XDGM1(J)
  10  CONTINUE
C
      CALL FFUN(N,X,XDGM1,XDG, F,DF)
C
      CALL ICDC(10,N,EPSO,DF, WORK,IWORK,WORK2,COND,DET,MSG)
C
      IF(MSG.EQ.1) COND=-1
C
      RETURN
      END
```

```
      SUBROUTINE INPTAT(IFLGCR,IFLGST,MAXNS,MAXIT,EPSBIG,SSZBEG,N,IDEG)
C
C
C PURPOSE:   INITIALIZES FREQUENTLY CHANGED CONSTANTS.
C            INITIALIZES COEFFICIENTS FOR SYSTEM.
C
C
C GENERIC NAME: INPTA          FILE NAME: INPTAT
C
C
C REFERENCE:   CHAPTER 5 OF "SOLVING POLYNOMIAL SYSTEMS USING
C              CONTINUATION" BY ALEXANDER MORGAN.
C
C NOTE:   1. FOR SOLVING A GENERAL POLYNOMIAL SYSTEM USING TABLEAU INPUT.
C            CHANGE TO GENERIC NAME AND USE WITH CONSOL8.
C
C         2. IT IS ASSUMED THAT NO EQUATION HAS MORE THAN 30 TERMS.  THIS
C            CAN BE INCREASED BY CHANGING THE "30" IN THE COMMON BLOCKS
C            /EQUAT/ AND /COEFS/ TO THE APPROPRIATE VALUE IN THIS
C            SUBROUTINE AND IN SUBROUTINE FFUNT.
C
C
C INPUT:   READ FROM UNIT 5.  SEE CODE BELOW FOR FORMAT.
C
C       TITLE       -- FIRST 60 CHARACTERS OF FIRST LINE OF INPUT
C       IFLGPT      -- =1, IF THE PROJECTIVE TRANSFORMATION IS TO BE USED.
C       IFLGSC      -- =1, IF THE SCLGEN SCALING IS TO BE USED.
C       IFLGCR      -- =1, IF THE CORRECTOR ITERATIONS ARE TO BE WRITTEN
C                       TO UNIT 8.  (SEE SUBROUTINE CORCT.)
C       IFLGST      -- =1, IF A STEP SUMMARY OUTPUT IS TO BE PRINTED AFTER
C                       EVERY STEP.
C       MAXNS       -- MAXIMUM NUMBER OF STEPS FOR A PATH
C       MAXIT       -- MAXIMUM NUMBER OF ITERATIONS TO CORRECT A STEP
C       EPSBIG      -- THE CORRECTOR RESIDUAL MUST BE LESS THAN EPS FOR
C                       A STEP TO "SUCCEED."  EPS=EPSBIG UNTIL T > .95
C       SSZBEG      -- BEGINNING STEP LENGTH
C       N           -- NUMBER OF EQUATIONS
C
C THE FOLLOWING THREE SETS OF INPUT VARIABLES ARE GROUPED BY EQUATION
C AND TERM, AS IN THE SAMPLE BELOW:
C       NUMTRM(I)   -- THE NUMBER OF TERMS IN THE I-TH EQUATION,
C                       I=1 TO N.
C       DEG(I,K,J)  -- DEGREE OF X(K) IN THE J-TH TERM OF THE I-TH
C                       EQUATION, I=1 TO N, K=1 TO N, J=1 TO NUMTRM(I).
C       A(I,J)      -- COEFFICIENT OF THE J-TH TERM OF THE I-TH EQUATION,
C                       I=1 TO N, J=1 TO NUMTRM(I).
C
C SAMPLE INPUT FOLLOWS:
CTITLE:   SYSTEM OF TWO QUADRICS,     NO SOLUTIONS AT INFINITY
C   1                   IFLGPT
C   1                   IFLGSC
C   0                   IFLGCR
C   0                   IFLGST
C   200                 MAXNS
C   3                   MAXIT
```

```
C                    1.D-02     EPSBIG
C                    1.D-02     SSZBEG
C        2                      N
C        6                      NUMTRM(1)
C        2                      DEG(1,1,1)
C        0                      DEG(1,2,1)
C                  -.00098D 00   A(1,1)
C        0                      DEG(1,1,2)
C        2                      DEG(1,2,2)
C                978000.D 00     A(1,2)
C        1                      DEG(1,1,3)
C        1                      DEG(1,2,3)
C                  -9.8D 00      A(1,3)
C        1                      DEG(1,1,4)
C        0                      DEG(1,2,4)
C                -235.0D 00      A(1,4)
C        0                      DEG(1,1,5)
C        1                      DEG(1,2,5)
C               88900.0D 00     A(1,5)
C        0                      DEG(1,1,6)
C        0                      DEG(1,2,6)
C                 -1.000D 00     A(1,6)
C        6                      NUMTRM(2)
C        2                      DEG(2,1,1)
C        0                      DEG(2,2,1)
C                  -.0100D 00    A(2,1)
C        0                      DEG(2,1,2)
C        2                      DEG(2,2,2)
C                  -.9840D 00    A(2,2)
C        1                      DEG(2,1,3)
C        1                      DEG(2,2,3)
C                 -29.70D 00     A(2,3)
C        1                      DEG(2,1,4)
C        0                      DEG(2,2,4)
C                  .00987D 00    A(2,4)
C        0                      DEG(2,1,5)
C        1                      DEG(2,2,5)
C                  -.1240D 00    A(2,5)
C        0                      DEG(2,1,6)
C        0                      DEG(2,2,6)
C                  -.2500D 00    A(2,6)
C
C
C OUTPUT:  IFLGCR,IFLGST,MAXNS,MAXIT,EPSBIG,SSZBEG,N,IDEG
C
C
C COMMON:   /EQUAT/  IS USED TO PASS EQUATION DEFINITION PARAMETERS
C                    FROM SUBROUTINE INPTAT TO SUBROUTINE FFUNT.
C           /COEFS/  IS USED TO PASS COEFFICIENT VALUES FROM
C                    SUBROUTINE INPTAT TO SUBROUTINE FFUNT.
C           /OTPVT/  IS USED TO PASS THE VARIABLE SCALE FACTORS
C                    FROM SUBROUTINE INPTAT TO SUBROUTINE OTPUTT.
C           /FLGPT/  IS USED TO PASS PROJECTIVE TRANSFORMATION FLAG
C                    FROM SUBROUTINE INPTAT TO SUBROUTINE INPTBT.
C
C SUBROUTINE: SCLGEN
C
```

```
C DECLARATIONS
      INTEGER I,IDEG,IIDEG,IERR,IFLGPT,IFLGSC,IFLGCR,IFLGST,IPVT,
     & J,K,KDEG,KKDEG,L,MAXNS,MAXIT,MODE,N,NP1,NT,NUMT,NNUMT,TITLE
      DOUBLE PRECISION A,AA,ALPHA,ASCL,BETA,EPSBIG,FACE,FACV,GAMMA,
     & SSZBEG
C
C DIMENSIONS
      DIMENSION A(10,30),AA(10,30),ALPHA(20,20),BETA(20),FACE(10),
     & GAMMA(20),IDEG(1),IIDEG(20),IPVT(20),KKDEG(10,11,30),NNUMT(10),
     & TITLE(60)
C
C IN THE COMMON BLOCKS BELOW, "10" DENOTES THE NUMBER
C OF EQUATIONS, AND "30" THE MAXIMUM NUMBER OF TERMS
C OVER ALL EQUATIONS.  ("11" IS THE NUMBER OF VARIABLES, INCLUDING
C XNP1 (THE PROJECTIVE VARIABLE))
C
C COMMON
      COMMON/EQUAT/NUMT(10),KDEG(10,11,30)
      COMMON/COEFS/ASCL(10,30)
      COMMON/OTPVT/FACV(10)
      COMMON/FLGPT/IFLGPT
C
C
C INPUT STATEMENTS FOLLOW.
      READ(5,10) TITLE
   10 FORMAT(60A1)
      READ(5,20) IFLGPT
   20 FORMAT(I5)
      READ(5,20) IFLGSC
      READ(5,20) IFLGCR
      READ(5,20) IFLGST
      READ(5,20) MAXNS
      READ(5,20) MAXIT
      READ(5,30) EPSBIG
   30 FORMAT(D22.15)
      READ(5,30) SSZBEG
      READ(5,20) N
C
      NP1=N+1
      DO 40 J=1,N
         READ(5,20) NUMT(J)
         IDEG(J)=0
         NT=NUMT(J)
         DO 50 K=1,NT
            IIDEG(K)=0
            DO 60 L=1,N
               READ(5,20) KDEG(J,L,K)
               IIDEG(K)=IIDEG(K)+KDEG(J,L,K)
   60       CONTINUE
            IF(IIDEG(K).GT.IDEG(J))IDEG(J)=IIDEG(K)
            READ(5,30) A(J,K)
   50    CONTINUE
         DO 70 K=1,NT
            KDEG(J,NP1,K)=IDEG(J)-IIDEG(K)
   70    CONTINUE
   40 CONTINUE
C
```

```
C
      WRITE(6,80)
  80  FORMAT(' CONSOL8 WITH TABLEAU INPUT',//)
      WRITE(6,90) TITLE
  90  FORMAT(' ',60A1)
C
      WRITE(6,100) IFLGPT
 100  FORMAT(/,' IF IFLGPT=1, USE PROJ TRANS.  IFLGPT=',I2)
      WRITE(6,101) IFLGSC
 101  FORMAT(/,' IF IFLGSC=1, USE SCLGEN.  IFLGSC=',I2)
      WRITE(6,102) IFLGCR
 102  FORMAT(/
     & ' IF IFLGCR=1, PRINT OUT CORRECTOR RESIDUALS.  IFLGCR =',I2)
      WRITE(6,103) IFLGST
C
 103  FORMAT(/' IF IFLGST=1, PRINT OUT STEP SUMMARIES.  IFLGST =',I2)
      WRITE(6,105) MAXNS
 105  FORMAT(/' MAX NUM STEPS PER PATH =',I5)
      WRITE(6,110) MAXIT
 110  FORMAT(/' MAX NUM ITERATIONS PER STEP =',I5)
C
      WRITE(6,115) EPSBIG
 115  FORMAT(/' EPSBIG =',D11.4/)
      WRITE(6,120) SSZBEG
 120  FORMAT(/' BEGINNING STEP SIZE = ',D11.4/)
C
      WRITE(6,125) N
 125  FORMAT(/' NUMBER OF EQUATIONS = ',I5/)
C
C
      DO 130 J=1,N
         WRITE(6,99)
  99     FORMAT(/)
         WRITE(6,135) J,NUMT(J)
 135     FORMAT(' NUMT(',I2,')=',I5)
         IDEG(J)=0
         NT=NUMT(J)
         DO 140 K=1,NT
            IIDEG(K)=0
            DO 145 L=1,N
               WRITE(6,150) J,L,K,KDEG(J,L,K)
 150           FORMAT(' KDEG(',I2,',',I2,',',I2,')=',I5)
               IIDEG(K)=IIDEG(K)+KDEG(J,L,K)
 145        CONTINUE
            IF(IIDEG(K).GT.IDEG(J))IDEG(J)=IIDEG(K)
               WRITE(6,155) J,K,A(J,K)
 155           FORMAT(' A(',I2,',',I2,')=',D22.15)
 140     CONTINUE
         WRITE(6,160) J,IDEG(J)
 160     FORMAT(' IDEG(',I2,')=',I5)
         DO 165 K=1,NT
            KDEG(J,NP1,K)=IDEG(J)-IIDEG(K)
               WRITE(6,170) J,NP1,K,KDEG(J,NP1,K)
 170           FORMAT(' KDEG(',I2,',',I2,',',I2,')=',I5)
 165     CONTINUE
 130  CONTINUE
C
```

```
      IF(IFLGSC .EQ. 1) THEN
         CALL SCLGEN(10,11,20,N,NUMT,KDEG,0,0.D0,A,
     &                        NNUMT,KKDEG,AA,ALPHA,BETA,GAMMA,IPVT,
     &                                    FACV,FACE,ASCL,IERR)
         IF(IERR.EQ.1) THEN
            WRITE(6,172)
172         FORMAT(//,'  SCLGEN WARNING: MATRIX ALPHA SINGULAR',//)
         ENDIF
      ELSE
         IERR=0
         DO 175 I=1,N
            FACV(I)=0.D0
            FACE(I)=0.D0
            NT=NUMT(I)
            DO 175 J=1,NT
               ASCL(I,J)=A(I,J)
175      CONTINUE
      ENDIF
C
C
      WRITE(6,185)(FACE(I),I=1,N)
185   FORMAT(' EQUATION SCALE FACTORS ='/ 30(D22.15/))
C
      WRITE(6,190)(FACV(I),I=1,N)
190   FORMAT('  VARIABLE SCALE FACTORS ='/ 30(D22.15/))
C
      WRITE(6,99)
      WRITE(6,195)
195   FORMAT(' SCALED COEFFICIENTS  WITH DEGREES OF VARIABLES'/)
C
      DO 200 I=1,N
      NT=NUMT(I)
      DO 205 J=1,NT
C
         WRITE(6,210)  ASCL(I,J),(KDEG(I,K,J),K=1,N)
210      FORMAT(' ASCL =',D22.15,' KDEG=',30I4)
205   CONTINUE
      WRITE(6,99)
200   CONTINUE
C
      RETURN
      END
```

```
      SUBROUTINE INPTBT(N,SSZBEG,EPSBIG, SSZMIN,EPSSML,EPSO,P,Q)
C
C PURPOSE: INITIALIZES LESS FREQUENTLY CHANGED CONSTANTS
C
C GENERIC NAME: INPTB          FILE NAME: INPTBT
C
C REFERENCE:  CHAPTER 5 OF "SOLVING POLYNOMIAL SYSTEMS USING
C             CONTINUATION" BY ALEXANDER MORGAN.
C
C NOTE:  1. FOR SOLVING A GENERAL POLYNOMIAL SYSTEM, USING TABLEAU
C           INPUT.  CHANGE TO GENERIC NAME AND USE WITH CONSOL8.
C
C        2. IF YOU CHANGE THE DIMENSIONS OF VARIABLES IN CONSOL8
C           IN ORDER TO CHANGE THE MAXIMUM NUMBER OF EQUATIONS,
C           CHANGE THE DIMENSION OF  CL  IN COMMON BLOCK "/PROJT/"
C           FROM CL(11) TO CL(M+1) WHERE M IS THE NEW UPPER BOUND
C           ON THE NUMBER OF EQUATIONS.  DO NOT CHANGE ANY OF
C           THE OTHER DIMENSIONS IN THIS SUBROUTINE.
C
C
C INPUT:  N        -- NUMBER OF EQUATIONS
C         SSZBEG   -- BEGINNING STEPSIZE
C         EPSBIG   -- ZERO-EPSILON FOR PATH TRACKING RESIDUAL,
C                     UNTIL T > .95
C
C OUTPUT: SSZMIN   -- MINIMUM STEPSIZE ALLOWED
C         EPSSML   -- ZERO-EPSILON FOR PATH TRACKING RESIDUAL,
C                     AFTER T > .95
C         EPSO     -- ZERO-EPSILON FOR SINGULARITY
C         P        -- "RANDOM" CONSTANTS TO DEFINE G, THE START SYSTEM
C         Q        -- "RANDOM" CONSTANTS TO DEFINE G
C
C
C COMMON:  /FLGPT/  IS USED TO PASS THE PROJECTIVE TRANSFORMATION
C                   FLAG FROM SUBROUTINE INPTAT TO SUBROUTINE INPTBT.
C          /PROJT/  IS USED TO PASS PROJECTIVE TRANSFORMATION
C                   CONSTANTS FROM SUBROUTINE INPTBT TO SUBROUTINES
C                   FFUNT AND OTPUTT.
C
C SUBROUTINES: DCMPLX,MOD
C
C
C   DECLARATIONS
      INTEGER IFLGPT,J,JJ,N,NP1
      DOUBLE PRECISION CCL,EPSO,EPSBIG,EPSSML,PP,QQ,SSZBEG,SSZMIN
      COMPLEX*16 CL,P,Q
C
C   DIMENSIONS
      DIMENSION CCL(2,11),P(1),PP(2,10),Q(1),QQ(2,10)
C
C   COMMON
      COMMON/PROJT/CL(11)
      COMMON/FLGPT/IFLGPT
C
C
      NP1=N+1
      SSZMIN = 1.D-5*SSZBEG
```

```
                    EPSSML = 1.D-3*EPSBIG
                    EPSO   = 1.D-12
C
                    PP(1, 1)= .12324754231D0
                        PP(2, 1)= .76253746298D0
                    PP(1, 2)= .93857838950D0
                        PP(2, 2)=-.99375892810D0
                    PP(1, 3)=-.23467908356D0
                        PP(2, 3)= .39383930009D0
                    PP(1, 4)= .83542556622D0
                        PP(2, 4)=-.10192888288D0
                    PP(1, 5)=-.55763522521D0
                        PP(2, 5)=-.83729899911D0
                    PP(1, 6)=-.78348738738D0
                        PP(2, 6)=-.10578234903D0
                    PP(1, 7)= .03938347346D0
                        PP(2, 7)= .04825184716D0
                    PP(1, 8)=-.43428734331D0
                        PP(2, 8)= .93836289418D0
                    PP(1, 9)=-.99383729993D0
                        PP(2, 9)=-.40947822291D0
                    PP(1,10)= .09383736736D0
                        PP(2,10)= .26459172298D0
C
                    QQ(1, 1)= .58720452864D0
                        QQ(2, 1)= .01321964722D0
                    QQ(1, 2)= .97884134700D0
                        QQ(2, 2)=-.14433009712D0
                    QQ(1, 3)= .39383737289D0
                        QQ(2, 3)= .41543223411D0
                    QQ(1, 4)=-.03938376373D0
                        QQ(2, 4)=-.61253112318D0
                    QQ(1, 5)= .39383737388D0
                        QQ(2, 5)=-.26454678861D0
                    QQ(1, 6)=-.00938376766D0
                        QQ(2, 6)= .34447867861D0
                    QQ(1, 7)=-.04837366632D0
                        QQ(2, 7)= .48252736790D0
                    QQ(1, 8)= .93725237347D0
                        QQ(2, 8)=-.54356527623D0
                    QQ(1, 9)= .39373957747D0
                        QQ(2, 9)= .65573434564D0
                    QQ(1,10)=-.39380038371D0
                        QQ(2,10)= .98903450052D0
C
C
                    CCL(1, 1)=-.03485644332D0
                        CCL(2, 1)= .28554634336D0
                    CCL(1, 2)= .91453454766D0
                        CCL(2, 2)= .35354566613D0
                    CCL(1, 3)=-.36568737635D0
                        CCL(2, 3)= .45634642477D0
                    CCL(1, 4)=-.89089767544D0
                        CCL(2, 4)= .34524523544D0
                    CCL(1, 5)= .13523462465D0
                        CCL(2, 5)= .43534535555D0
                    CCL(1, 6)=-.34523544445D0
```

```
            CCL(2, 6)=  .00734522256D0
         CCL(1, 7)=-.80004678763D0
            CCL(2, 7)=-.009387123644D0
         CCL(1, 8)=-.875432124245D0
            CCL(2, 8)=  .00045687651D0
         CCL(1, 9)=  .65256352333D0
            CCL(2, 9)=-.12356777452D0
         CCL(1,10)=  .09986798321548D0
            CCL(2,10)=-.56753456577D0
         CCL(1,11)=  .29674947394739D0
            CCL(2,11)=  .93274302173D0
C
C
      DO 10 J=1,N
        JJ=MOD(J-1,10)+1
        P(J)=DCMPLX(PP(1,JJ),PP(2,JJ))
        Q(J)=DCMPLX(QQ(1,JJ),QQ(2,JJ))
   10 CONTINUE
C
      DO 20 J=1,NP1
        JJ=MOD(J-1,11)+1
        CL(J)=DCMPLX(CCL(1,JJ),CCL(2,JJ))
   20 CONTINUE
C
C
C IF WE WISH NOT TO USE THE PROJECTIVE TRANSFORMATION, THEN WE
C RESET CL TO THE NOMINAL VALUES BELOW.
      IF(IFLGPT .NE. 1) THEN
          DO 30 J=1,N
              CL(J)=DCMPLX(0.D0,0.D0)
   30     CONTINUE
          CL(NP1)=DCMPLX(1.D0,0.D0)
      END IF
C
      WRITE(6,100) SSZMIN
  100 FORMAT(/' MINIMUM STPSZE = ',D11.4)
      WRITE(6,105) EPSSML
  105 FORMAT(/' EPSSML =',D11.4)
      WRITE(6,110) EPSO
  110 FORMAT(/' EPSO   =',D11.4)
C
      WRITE(6,115)  (P(J),J=1,N)
  115 FORMAT(/' P    =',2D11.4)
C
      WRITE(6,120)  (Q(J),J=1,N)
  120 FORMAT(/' Q    =',2D11.4)
C
      WRITE(6,125)  (CL(J),J=1,NP1)
  125 FORMAT(/' CL   =',2D22.15)
C
C
      WRITE(6,999)
  999 FORMAT(/)
C
      RETURN
      END
```

```
      SUBROUTINE FFUNT(N,X,XDGM1,XDG, F,DF)
C
C PURPOSE: EVALUATES F AND DF.
C
C GENERIC NAME: FFUN             FILE NAME: FFUNT
C
C REFERENCE:   CHAPTER 5 OF "SOLVING POLYNOMIAL SYSTEMS USING
C              CONTINUATION" BY ALEXANDER MORGAN.
C
C NOTE:  1. FOR SOLVING A GENERAL POLYNOMIAL SYSTEM, USING TABLEAU
C           INPUT.  CHANGE TO GENERIC NAME AND USE WITH CONSOL8.
C
C        2. THIS SUBROUTINE DOESN'T USE XDGM1 OR XDG.
C
C        3. EQUATION DEFINITION PARAMETERS, EQUATION COEFFICIENTS,
C           AND THE PROJECTIVE TRANSFORMATION CONSTANTS ARE
C           PASSED TO FFUNT VIA COMMON, AS NOTED BELOW.
C
C
C INPUT:  N         -- NUMBER OF EQUATIONS
C         X         -- CURRENT VALUE OF X
C         XDGM1     -- XDGM1(J) = X(J)**(IDEG(J)-), WHERE IDEG(J) IS
C                      THE DEGREE OF THE J-TH EQUATIONS, F(J).
C         XDG       -- XDG(J) = X(J)**IDEG(J)
C OUTPUT: F         -- F(J) IS THE J-TH EQUATION IN THE SYSTEM WE WANT
C                      TO SOLVE.
C         DF        -- DF(J,K) IS THE PARTIAL DERIVATIVE OF F(J) WITH
C                      RESPECT TO X(K).
C
C
C  COMMON:  /EQUAT/  IS USED TO PASS EQUATION DEFINITION PARAMETERS
C                    FROM SUBROUTINE INPTAT TO SUBROUTINE FFUNT.
C           /COEFS/  IS USED TO PASS COEFFICIENT VALUES FROM
C                    SUBROUTINE INPTAT TO SUBROUTINE FFUNT.
C           /PROJT/  IS USED TO PASS PROJECTIVE TRANSFORMATION
C                    CONSTANTS FROM SUBROUTINE INPTBT TO SUBROUTINES
C                    FFUNT AND OTPUTT.
C
C
C SUBROUTINES: XNORM
C
C DECLARATIONS
      INTEGER IFLGPT,J,K,KDEG,L,M,N,NP1,NT,NUMT
      DOUBLE PRECISION ASCL,DABS,EPSDIV,XNORM
      COMPLEX*16 CL,DF,DSTRM,DTRM,DXNP1,F,FACV,ONE,STRM,TEMP,TRM,
     & XDG,XDGM1,X,XX,ZERO
C
C IN THE DIMENSIONED VARIABLES BELOW, "10" DENOTES THE NUMBER
C OF EQUATIONS, AND "30" THE MAXIMUM NUMBER OF TERMS
C OVER ALL EQUATIONS.  ("11" IS THE NUMBER OF VARIABLES, INCLUDING
C XNP1.)
C
C DIMENSIONS
      DIMENSION  DF(10,1),DSTRM(10,11,30),DTRM(10,11,30),
     & DXNP1(10),F(1),STRM(10,30),TRM(10,30),X(1),XDG(1),
     & XDGM1(1),XX(10,11,30)
```

```
C
C COMMON
      COMMON/EQUAT/NUMT(10),KDEG(10,11,30)
      COMMON/COEFS/ASCL(10,30)
      COMMON/PROJT/CL(11)
C
C
      ZERO = DCMPLX(0.D0,0.D0)
      ONE  = DCMPLX(1.D0,0.D0)
C
C EPSDIV IS USED TO DETERMINE IF PARTIAL DERIVATIVES OF F CAN
C BE SAFELY DERIVED FROM THE TERMS OF F BY DIVISION BY X.
C THE FOLLOWING AD HOC CRITERION IS USED: "A/B" IS DEFINED
C IF ABS(B) .GE. EPSDIV.  A MORE EXACT CRITERION CAN BE DEVELOPED
C AS FOLLOWS.  LET MACHMAX BE THE LARGEST FLOATING POINT NUMBER.
C THEN "A/B" IS DEFINED IF ABS(A)/MACHMAX .LT. ABS(B).
      EPSDIV=1.D-8
C
      NP1=N+1
C
      TEMP=CL(NP1)
      DO 3  J=1,N
          TEMP = TEMP + CL(J)*X(J)
          DXNP1(J)=CL(J)
   3  CONTINUE
      X(NP1)=TEMP
C
C
      DO 10 J=1,N
      DO 10 L=1,NP1
          NT=NUMT(J)
          DO 10 K=1,NT
              IF(KDEG(J,L,K).EQ.0) THEN
                  XX(J,L,K)=ONE
                ELSE
                  XX(J,L,K) = X(L)**KDEG(J,L,K)
              ENDIF
  10  CONTINUE
C
C
      DO 20 J=1,N
          NT=NUMT(J)
          DO 20 K=1,NT
              STRM(J,K)=ONE
              DO 22 L=1,NP1
                  STRM(J,K)=STRM(J,K)*XX(J,L,K)
  22          CONTINUE
          TRM(J,K)=ASCL(J,K)*STRM(J,K)
  20  CONTINUE
C
      DO 30 J=1,N
          F(J)=ZERO
          NT=NUMT(J)
          DO 30 K=1,NT
              F(J)= F(J) + TRM(J,K)
  30  CONTINUE
C
```

```
          DO 40 J=1,N
          DO 40 M=1,NP1
             NT=NUMT(J)
             DO 40 K=1,NT
                IF(KDEG(J,M,K).EQ.0) THEN
                   DSTRM(J,M,K)=ZERO
                   GOTO 40
                ENDIF
                IF(XNORM(2,X(M)).GE.EPSDIV) THEN
                   DSTRM(J,M,K) = KDEG(J,M,K)*STRM(J,K)/X(M)
                   GOTO 40
                ENDIF
C

                DSTRM(J,M,K)=ONE
                DO 44 L=1,NP1
                   IF(L.NE.M)DSTRM(J,M,K)=XX(J,L,K)*DSTRM(J,M,K)
   44           CONTINUE
                IF(KDEG(J,M,K).EQ.1) THEN
                   TEMP=ONE
                 ELSE
                   TEMP = X(M)**(KDEG(J,M,K)-1)
                ENDIF
                DSTRM(J,M,K)=KDEG(J,M,K)*DSTRM(J,M,K)*TEMP
   40     CONTINUE
C
      DO 50 J=1,N
      DO 50 M=1,NP1
         NT=NUMT(J)
         DO 50 K=1,NT
            DTRM(J,M,K)=ASCL(J,K)*DSTRM(J,M,K)
   50 CONTINUE
C
      DO 60 J=1,N
      DO 60 M=1,NP1
         DF(J,M)=ZERO
         NT=NUMT(J)
         DO 60 K=1,NT
            DF(J,M)= DF(J,M) + DTRM(J,M,K)
   60 CONTINUE
C
      DO 70 J=1,N
      DO 70 K=1,N
         DF(J,K)=DF(J,K)+DF(J,NP1)*DXNP1(K)
   70 CONTINUE
C
      RETURN
      END
```

```
          SUBROUTINE OTPUTT(N,NUMPAT,IDEG,EPSO,X, COND,DET)
C
C PURPOSE: OTPUTT IS CALLED BY CONSOL8 AFTER EACH PATH IS COMPLETED.
C          OTPUTT COMPUTES THE CONDITION AND DETERMINANT OF THE
C          JACOBIAN OF F AT THE SOLUTION X, AND THEN UNTRANSFORMS
C          AND UNSCALES X.  FINALLY, IT  IDENTIFIES SOLUTIONS AS
C          REAL OR COMPLEX, AND  WRITES A BRIEF SOLUTION SUMMARY
C          TO UNIT 8.
C
C
C GENERIC NAME: OTPUT          FILE NAME: OTPUTT
C
C
C REFERENCE:  CHAPTER 5 OF "SOLVING POLYNOMIAL SYSTEMS USING
C             CONTINUATION" BY ALEXANDER MORGAN.
C
C NOTE:  1. FOR SOLVING A GENERAL POLYNOMIAL SYSTEM, USING TABLEAU
C           INPUT.  CHANGE TO GENERIC NAME AND USE WITH CONSOL8.
C
C        2. TWO CONSTANTS MAY NEED TO BE CHANGED: THE
C           ZERO-EPSILON TO DETERMINE IF A SOLUTION IS AT
C           INFINITY (EPSFIN) AND THE ZERO-EPSILON TO DETERMINE
C           IF A SOLUTION IS REAL (EPSREL).  SEE BELOW.
C
C INPUT:  N        -- NUMBER OF EQUATIONS
C         NUMPAT   -- CURRENT PATH NUMBER
C         IDEG     -- IDEG(J) IS THE DEGREE OF THE J-TH EQUATION.
C         EPSO     -- ZERO-EPSILON FOR SINGULARITY.
C         X        -- COMPUTED PATH ENDPOINT
C
C OUTPUT: X        -- IF THE PROJECTIVE COORDINATE, XNP1, HAS NORM
C                     GREATER THAN EPSFIN, THEN X IS THE PATH ENDPOINT
C                     UNTRANSFORMED AND UNSCALED.  OTHERWISE, X IS
C                     UNCHANGED.
C         COND     -- CONDITION OF THE JACOBIAN, DF, AT THE COMPUTED
C                     PATH ENDPOINT (THAT IS, TRANSFORMED AND SCALED).
C                     COND=-1 MEANS THAT DF IS SINGULAR WITH RESPECT
C                     TO EPSO.
C         DET      -- IF COND >= 0, DETERMINANT OF DF.
C
C COMMON:   /OTPVT/  IS USED TO PASS VARIABLE SCALE FACTORS FROM
C                    SUBROUTINE INPTAT TO SUBROUTINE OTPUTT.
C           /PROJT/  IS USED TO PASS PROJECTIVE TRANSFORMATION CONSTANTS
C                    FROM SUBROUTINE INPTBT TO SUBROUTINES FFUNT AND
C                    OTPUTT.
C
C SUBROUTINES: DABS,DCMPLX,DIMAG,FFUNT(CALLED AS FFUN),ICDC,XNORM
C
C DECLARATIONS
      INTEGER IDEG,IFLGPT,ITEST,IWORK,J,MSG,N,NP1,NUMPAT
      DOUBLE PRECISION COND,DABS,DIMAG,EPSO,EPSFIN,EPSREL,FAC,FACV,T,
     & XNORM
      COMPLEX*16 CL,DET,DF,DUMMY1,DUMMY2,F,ONE,TEMP,WORK,WORK2,X
C DIMENSIONS
      DIMENSION DF(10,11),DUMMY1(10),DUMMY2(10),F(10),IDEG(1),
     & IWORK(10),WORK(10),WORK2(10,10),X(1)
```

```fortran
C
C COMMON
      COMMON/OTPVT/FACV(10)
      COMMON/PROJT/CL(11)
C
      NP1=N+1
      EPSREL=1.D-4
      EPSFIN=1.D-22
      ONE=DCMPLX(1.D0,0.D0)
C
      CALL FFUN(N,X,DUMMY1,DUMMY2, F,DF)
      CALL ICDC(10,N,EPSO,DF, WORK,IWORK,WORK2,COND,DET,MSG)
      IF(MSG.EQ.1) COND=-1
C
      TEMP=CL(NP1)
      DO 20 J=1,N
          TEMP = TEMP + CL(J)*X(J)
  20  CONTINUE
      X(NP1)=TEMP
      WRITE(6,30) X(NP1)
  30  FORMAT(/,'  XNP1=',2D20.10,/)
      IF(XNORM(2,X(NP1)) .LE. EPSFIN) RETURN
C
C UNTRANSFORM VARIABLES
      DO 40 J=1,N
          X(J)=X(J)/X(NP1)
  40  CONTINUE
C
C UNSCALE VARIABLES
      DO 50 J=1,N
          FAC=10.D0**FACV(J)
          X(J)=FAC*X(J)
  50  CONTINUE
C
      WRITE(8,60) NUMPAT
  60  FORMAT('  NUMPAT=',I5)
C
C DESIGNATE SOLUTIONS "REAL" OR "COMPLEX"
      ITEST=0
      DO 70 J=1,N
          IF(DABS(DIMAG(X(J))) .GE. EPSREL) ITEST=1
  70  CONTINUE
      IF(ITEST.EQ.1) THEN
          WRITE(6,80)
          WRITE(8,80)
  80      FORMAT('  COMPLEX SOLUTION  ')
      ELSE
          WRITE(6,90)
          WRITE(8,90)
  90      FORMAT('  REAL SOLUTION  ')
      ENDIF
C
      WRITE(8,100) (X(J),J=1,N)
 100  FORMAT(2D22.15)
C
      RETURN
      END
```

```
      SUBROUTINE SCLGEN(DIM,DIMP1,DIM2,N,NUMT,DEG,MODE,EPSO,COEF,
     &                  NNUMT,DDEG,CCOEF,ALPHA,BETA,GAMMA,IPVT,
     &                  FACV,FACE,COESCL,IERR)
```

C
C PURPOSE: SCALE THE COEFFICIENTS OF A POLYNOMIAL SYSTEM OF N EQUATIONS
C   IN N UNKNOWNS.
C
C REFERENCE: CHAPTER 5 OF "SOLVING POLYNOMIAL SYSTEMS USING
C   CONTINUATION" BY ALEXANDER MORGAN.
C
C DESCRIPTION: THE J-TH TERM OF THE I-TH EQUATION LOOKS LIKE:
C
C   COEF(I,J) * X(1)**DEG(I,1,J) ... X(N)**DEG(I,N,J)
C
C THE I-TH EQUATION IS SCALED BY 10**FACE(I). THE K-TH VARIABLE
C IS SCALED BY 10**FACV(K). IN OTHER WORDS, X(K)=(10**FACV(K)) * Y(K),
C WHERE Y SOLVES THE SCALED EQUATION. THE SCALED EQUATION HAS THE
C SAME FORM AS THE ORIGINAL EQUATION, EXCEPT THAT COESCL(I,J) REPLACES
C COEF(I,J), WHERE
C
C   COESCL(I,J)=COEF(I,J)* 10**(FACE(I) + FACV(1)*DEG(I,1,J)+ ...
C
C   +FACV(N)*DEG(I,N,J))
C
C THE CRITERION FOR GENERATING FACE AND FACV IS THAT OF MINIMIZING THE SUM
C OF SQUARES OF THE EXPONENTS OF THE SCALED COEFFICIENTS. IT TURNS OUT
C THAT THIS CRITERION REDUCES TO SOLVING A SINGLE LINEAR SYSTEM,
C ALPHA*X = BETA, AS DEFINED IN THE CODE BELOW. FURTHER, THE FORM OF THE
C POLYNOMIAL SYSTEM ALONE DETERMINES MATRIX ALPHA. THUS, IF A SEQUENCE
C OF SYSTEMS WITH THE SAME FORM, BUT PERHAPS WITH DIFFERENT COEFFICIENTS,
C IS BEING SCALED, WE CALL SCLGEN WITH MODE=1 FOR ALL CALLS AFTER THE
C FIRST ONE. THEN SCLGEN DOES NOT RECOMPUTE OR REFACTOR THE MATRIX ALPHA
C AFTER THE FIRST CALL.
C
C NOTE: THE COEFFICIENTS ARE ASSUMED TO BE REAL NUMBERS. THERE
C   IS NO COMPLEX VERSION OF THIS PROGRAM.
C
C
C INPUT: DIM   -- KEY DIMENSION PASSED TO SCLGEN. MUST MATCH
C   LEADING DIMENSIONS OF DEG, DDEG, COEF, CCOEF,
C   AND COESCL.
C
C   DIMP1   -- =DIM+1. KEY DIMENSION PASSED TO SCLGEN. MUST
C   MATCH DIMENSION OF SECOND INDEX OF DEG AND DDEG.
C
C   DIM2   -- =2*DIM. MUST MATCH THE LEADING DIMENSION OF
C   ALPHA.
C
C   N   -- NUMBER OF EQUATIONS AND NUMBER OF VARIABLES.
C
C   NUMT   -- NUMT(I) IS THE NUMBER OF TERMS IN THE I-TH
C   EQUATION.
C
C   DEG   -- DEG(I,K,J) IS THE DEGREE OF THE K-TH VARIABLE
C   IN THE J-TH TERM OF THE I-TH EQUATION.

```
C
C              MODE            -- IF MODE=1, THEN THIS IS NOT THE FIRST CALL TO
C                                 SCLGEN AND THE MATRIX ALPHA IS ASSUMED TO BE
C                                 ALREADY FACTORED.  IF MODE=1, THEN IERR SHOULD
C                                 EQUAL O.
C
C              EPSO            -- ZERO-EPSILON FOR COEFFICIENTS.
C
C              COEF            -- COEF(I,J) IS THE COEFFICIENT OF THE J-TH TERM
C                                 OF THE I-TH EQUATION.  IT MAY BE ZERO.
C
C
C OUTPUT(WORKSPACE):
C              NNUMT           -- NNUMT(I) IS THE NUMBER OF TERMS IN THE I-TH
C                                 EQUATION WITH COEFFICIENTS WHOSE ABSOLUTE
C                                 VALUES ARE GREATER THAN EPSO.
C
C              DDEG            -- DDEG(I,K,J) IS THE DEGREE OF THE K-TH VARIABLE
C                                 IN THE J-TH TERM OF THE I-TH EQUATION, WHERE
C                                 THE TERMS WITH COEFFICIENTS WITH ABSOLUTE VALUE
C                                 LESS THAN EPSO HAVE BEEN DELETED.
C
C              CCOEF           -- CCOEF(I,J) IS THE COEFFICIENT OF THE J-TH TERM
C                                 OF THE I-TH EQUATION, WHERE THE TERMS WITH
C                                 COEFFICIENTS WITH ABSOLUTE VALUE LESS THAN EPSO
C                                 HAVE BEEN DELETED.
C
C              ALPHA           -- THIS MATRIX IS DEFINED BELOW, FROM DDEG.  IF
C                                 IERR=0, THEN ALPHA IS RETURNED IN L-U FACTORED
C                                 FORM BY LINEAR EQUATION SOLVER.
C
C              BETA            -- THIS COLUMN ARRAY IS DEFINED BELOW, FROM COEF.
C
C              GAMMA           -- WORKSPACE.
C
C              IPVT            -- THE COLUMN IPVT IS USED BY THE LINEAR SOLVER.
C
C OUTPUT:      FACV            -- FACV(I) IS THE SCALE FACTOR FOR THE I-TH VARIABLE.
C
C              FACE            -- FACE(I) IS THE SCALE FACTOR FOR THE I-TH EQUATION.
C
C              COESCL          -- COESCL(I,J) IS THE SCALED VERSION OF COEF(I,J).
C
C              IERR            -- =1, IF A SINGULAR MATRIX ALPHA WAS GENERATED;
C                                 =0, OTHERWISE.  IN EITHER CASE, A SCALING OF THE
C                                 COEFFICIENTS IN COEF IS RETURNED IN COESCL.
C
C SUBROUTINES: DABS,DLOG10,LNFN,LNSN
C
C DECLARATIONS
      INTEGER DEG,DDEG,DIM,DIMP1,DIM2,I,IERR,IPVT,J,JJ,K,
     & MODE,N,N2,NT,NUMPI,NUMPK,NUMPS,NUMT,NNUMT,S
      DOUBLE PRECISION ALPHA,BETA,COEF,CCOEF,
     & COESCL,EPSO,FACE,FACV,GAMMA,SUM,TEMP
C
C
```

```fortran
C DIMENSIONS
      DIMENSION ALPHA(DIM2,1),BETA(1),COEF(DIM,1),CCOEF(DIM,1),
     & COESCL(DIM,1),DEG(DIM,DIMP1,1),DDEG(DIM,DIMP1,1),FACE(1),
     & FACV(1),GAMMA(1),IPVT(1),NUMT(1),NNUMT(1)
C
C
      N2=2*N
C
C DELETE NEAR ZERO TERMS
      DO  10 I=1,N
         JJ=0
         NT=NUMT(I)
         DO 20 J=1,NT
            IF(DABS(COEF(I,J)) .GT. EPSO) THEN
               JJ=JJ+1
               CCOEF(I,JJ)=COEF(I,J)
               DO 30 K=1,N
                  DDEG(I,K,JJ)=DEG(I,K,J)
 30            CONTINUE
            ENDIF
 20      CONTINUE
         NNUMT(I)=JJ
 10   CONTINUE
C
C GENERATE LOGS OF COEFFICIENTS
      DO 40 I=1,N
         NT=NNUMT(I)
         DO 40 J=1,NT
            COESCL(I,J)=DLOG10(DABS(CCOEF(I,J)))
 40   CONTINUE
C
C GENERATE EQUATION AND VARIABLE SCALE FACTORS
C
C SKIP OVER THE GENERATION AND DECOMPOSITON OF MATRIX ALPHA IF MODE=1.
      IF(MODE.NE.1) THEN
C
C GENERATE THE MATRIX ALPHA
         IERR=0
         DO 50 S=1,N
         DO 50 K=1,N
            ALPHA(S,K)=0
 50      CONTINUE
C
         DO 60 S=1,N
           ALPHA(S,S)=NNUMT(S)
 60      CONTINUE
C
         DO 70 S=1,N
         DO 70 I=1,N
            SUM=0
            NT=NNUMT(I)
            DO 80 J=1,NT
              SUM=SUM+DDEG(I,S,J)
 80         CONTINUE
            NUMPS=N+S
            ALPHA(NUMPS,I)=SUM
 70      CONTINUE
```

```
C
            DO 90 S=1,N
             DO 90 K=1,N
              SUM=0
              DO 100 I=1,N
                NT=NNUMT(I)
                DO 100 J=1,NT
                 SUM=SUM+DDEG(I,S,J)*DDEG(I,K,J)
  100         CONTINUE
              NUMPS=N+S
              NUMPK=N+K
              ALPHA(NUMPS,NUMPK)=SUM
  90        CONTINUE
C
            DO 110 S=1,N
            DO 110 K=1,N
              SUM=0
              NT=NNUMT(S)
              DO 115 J=1,NT
                SUM=SUM+DDEG(S,K,J)
  115         CONTINUE
              NUMPK=N+K
              ALPHA(S,NUMPK)=SUM
  110       CONTINUE
C
C COMPUTE THE L-U DECOMPOSITION OF ALPHA
            CALL LNFN(DIM2,N2,EPSO,ALPHA,IERR,IPVT)
C
            IF(IERR .EQ. 1) THEN
                DO 118 I=1,N2
                IF(DABS(ALPHA(IPVT(I),I)) .LE. EPSO)ALPHA(IPVT(I),I)=1.DO
  118           CONTINUE
            ENDIF
C
      ENDIF
C CONTROL PASSES TO HERE IF MODE=1.
C
C GENERATE THE COLUMN BETA
      DO 120 S=1,N
        SUM=0
        NT=NNUMT(S)
        DO 130 J=1,NT
          SUM=SUM+COESCL(S,J)
  130   CONTINUE
        BETA(S)=-SUM
  120 CONTINUE
C
      DO 140 S=1,N
        SUM=0
        DO 150 I=1,N
          NT=NNUMT(I)
          DO 150 J=1,NT
            SUM=SUM+COESCL(I,J)*DDEG(I,S,J)
  150   CONTINUE
        NUMPS=N+S
        BETA(NUMPS)=-SUM
  140 CONTINUE
```

```
C
C
C SOLVE THE LINEAR SYSTEM ALPHA*X = BETA
      CALL LNSN(DIM2,N2,IPVT,ALPHA,BETA, GAMMA)
C
C GENERATE FACE, FACV,AND THE MATRIX COESCL
      DO 160 I=1,N
        FACE(I)=GAMMA(I)
        NUMPI=N+I
        FACV(I)=GAMMA(NUMPI)
  160   CONTINUE
C
C GENERATE THE SCALED COEFFICIENTS
      DO 170 I=1,N
        NT= NUMT(I)
        DO 170 J=1,NT
          TEMP=DABS(COEF(I,J))
          IF(TEMP.EQ.0.DO) THEN
            COESCL(I,J)=0.DO
          ELSE
            SUM=FACE(I)+DLOG10( TEMP )
            DO 180 K=1,N
              SUM=SUM+FACV(K)*DEG(I,K,J)
  180       CONTINUE
            COESCL(I,J)=10.DO**SUM
            IF(COEF(I,J) .LT. 0.DO) COESCL(I,J)=-COESCL(I,J)
          ENDIF
  170   CONTINUE
C
      RETURN
      END
```

```
      SUBROUTINE SCLCEN(DIM,DIMP1,N,NUMT,DEG,EPSO,COEF,
     &                          NNUMT,DDEG,CCOEF,FACV,FACE,COESCL,IERR)
```

C
C
C
C PURPOSE: SCALE THE COEFFICIENTS OF A POLYNOMIAL SYSTEM OF N EQUATIONS
C   IN N UNKNOWNS.
C
C REFERENCE: CHAPTER 5 OF "SOLVING POLYNOMIAL SYSTEMS USING
C   CONTINUATION" BY ALEXANDER MORGAN.
C
C DESCRIPTION: THE JTH TERM OF THE I-TH EQUATION LOOKS LIKE:
C
C   COEF(I,J) * X(1)**DEG(I,J,1) ... X(N)**DEG(I,J,N)
C
C THE I-TH EQUATION IS SCALED BY 10**FACE(I). THE KTH VARIABLE IS SCALED
C BY 10**FACV(K). IN OTHER WORDS, X(K) = 10**FACV(K) * Y(K), WHERE Y
C SOLVES THE SCALED EQUATION. THE SCALED EQUATION HAS THE SAME FORM AS
C THE ORIGINAL EQUATION, EXCEPT THAT COESCL(I,J) REPLACES COEF(I,J), WHERE
C
C   COESCL(I,J)=COEF(I,J)* 10**(FACE(I) + FACV(1)*DEG(I,J,1)+ ...
C
C   +FACV(N)*DEG(I,J,N))
C
C AND FACV IS INPUT BY THE USER. THE CRITERION FOR GENERATING FACE IS THAT
C OF MINIMIZING THE SUM OF SQUARES OF THE EXPONENTS OF THE SCALED
C COEFFICIENTS. IT TURNS OUT THAT THIS CRITERION REDUCES TO A SIMPLE FORMULA,
C GIVEN BELOW.
C
C NOTE: THE COEFFICIENTS ARE ASSUMED TO BE REAL NUMBERS. THERE IS NO COMPLEX
C   VERSION OF THIS PROGRAM.
C
C INPUT: DIM   -- KEY DIMENSION PASSED TO SCLCEN. MUST MATCH
C   LEADING DIMENSIONS OF DEG, COEF, AND COESCL
C   IN THE CALLING ROUTINE.
C
C   DIMP1   -- KEY DIMENSION PASSED TO SCLCEN. MUST MATCH
C   DIMENSION OF INDEX 2 OF DEG AND DDEG.
C
C   N   -- NUMBER OF EQUATIONS AND NUMBER OF VARIABLES.
C
C   NUMT   -- NUMT(I) IS THE NUMBER OF TERMS IN THE I-TH
C   EQUATION.
C
C   DEG   -- DEG(I,K,J) IS THE DEGREE OF THE KTH VARIABLE
C   IN THE JTH TERM OF THE I-TH EQUATION.
C
C   COEF   -- COEF(I,J) IS THE COEFFICIENT OF THE JTH TERM
C   OF THE I-TH EQUATION. IT MAY BE ZERO.
C
C   EPSO   -- ZERO-EPSILON FOR COEFFICIENTS.
C
C   FACV   -- FACV(I) IS THE SCALE FACTOR FOR THE I-TH VARIABLE.
C
C

384

```
C OUTPUT (WORKSPACE):
C             NNUMT          -- NNUMT(I) IS THE NUMBER OF TERMS IN THE I-TH
C                               EQUATION BIGGER THAN EPSO.
C
C             DDEG           -- DDEG(I,K,J) IS THE DEGREE OF THE KTH VARIABLE
C                               IN THE JTH TERM OF THE I-TH EQUATION, WHERE THE
C                               TERMS LESS THAN EPSO HAVE BEEN DELETED.
C
C             CCOEF          -- CCOEF(I,J) IS THE COEFFICIENT OF THE JTH TERM OF
C                               THE I-TH EQUATION GREATER THAN EPSO.
C
C
C OUTPUT:  FACE              -- FACE(I) ARE THE EQUATION SCALE FACTORS.
C
C          COESCL            -- COESCL(I,J) IS THE SCALED VERSION OF COEF(I,J).
C
C          IERR              -- =1, IF SCALING FAILED BECAUSE NNUMT(J)=0 FOR
C                               SOME J.  =0, OTHERWISE.
C
C
C SUBROUTINES: DABS,DLOG10
C
C DECLARATIONS
      INTEGER DEG,DDEG,DIM,DIMP1,I,IERR,J,JJ,K,N,NT,NUMT,NNUMT,S
      DOUBLE PRECISION COEF,CCOEF,COESCL,EPSO,FACE,FACV,SUM,TEMP
C
C DIMENSIONS
      DIMENSION CCOEF(DIM,1),COEF(DIM,1),COESCL(DIM,1),
     & DDEG(DIM,DIMP1,1),DEG(DIM,DIMP1,1),FACE(1),FACV(1),NNUMT(1),
     & NUMT(1)
C
C
      IERR=0
C
C DELETE NEAR ZERO TERMS
      DO   10 I=1,N
        JJ=0
        NT=NUMT(I)
        DO   20 J=1,NT
          IF(DABS(COEF(I,J)) .GT. EPSO) THEN
            JJ=JJ+1
            CCOEF(I,JJ)=COEF(I,J)
            DO   30 K=1,N
              DDEG(I,K,JJ)=DEG(I,K,J)
  30        CONTINUE
          ENDIF
  20    CONTINUE
        NNUMT(I)=JJ
  10  CONTINUE
C
      DO 40 I=1,N
        IF (NNUMT(I).EQ.0) THEN
          IERR=1
          RETURN
        ENDIF
  40  CONTINUE
C
```

```fortran
C
      IERR=0
C
C GENERATE LOGS OF COEFFICIENTS
      DO 50 I=1,N
          NT=NNUMT(I)
          DO 50 J=1,NT
              COESCL(I,J)=DLOG10(DABS(CCOEF(I,J)))
  50  CONTINUE
C
C
C GENERATE THE EQUATION SCALE FACTORS
      DO 60 I=1,N
          SUM=0
          NT =NNUMT(I)
          DO 70 J=1,NT
            SUM=SUM+COESCL(I,J)
             DO 70 K=1,N
               SUM=SUM+FACV(K)*DDEG(I,K,J)
  70      CONTINUE
          FACE(I)=-SUM/NNUMT(I)
  60  CONTINUE
C
C GENERATE THE SCALED COEFFICIENTS
      DO 80 I=1,N
          NT=NUMT(I)
          DO 80 J=1,NT
            TEMP=DABS(COEF(I,J))
            IF(TEMP.EQ.0.D0) THEN
                COESCL(I,J)=0.D0
            ELSE
                SUM=FACE(I)+DLOG10( TEMP )
                DO 90 K=1,N
                  SUM=SUM+FACV(K)*DEG(I,K,J)
  90            CONTINUE
                COESCL(I,J)=10.D0**SUM
                IF(COEF(I,J) .LT. 0.D0) COESCL(I,J)=-COESCL(I,J)
            ENDIF
  80  CONTINUE
C
      RETURN
      END
```

Section 5: Programs for Chapter 6

```
C PROGRAM NAME:  CONSOL9
C
C
C PURPOSE:    FIND A SOLUTION TO A SYSTEM OF M POLYNOMIAL EQUATIONS
C             IN M UNKNOWNS, F(Y)=0, USING A USER-SUPPLIED
C             CONTINUATION SYSTEM PROVIDED IN SUBROUTINE HFNN.
C             (ALONG WITH HFNN, THE USER PROVIDES INPA, INPB, STPT,
C             AND FUNN, AS DESCRIBED BELOW.)
C
C
C REFERENCE:  CHAPTER 6 OF "SOLVING SYSTEMS OF POLYNOMIALS USING
C             CONTINUATION" BY ALEXANDER MORGAN.
C
C
C NOTE:     1. UNLIKE CONSOL8, THIS CODE IS WRITTEN TO IMPLEMENT
C             CONTINUATIONS IN REAL ARITHMETIC.  TO EMPHASIZE
C             THE DIFFERENCE WITH THE OTHER CONSOL CODES,
C             THE INDEPENDENT VARIABLE IS DENOTED "Y" INSTEAD OF
C             "X" AND THE NUMBER OF EQUATIONS IS "M" INSTEAD OF "N."
C
C             HOWEVER, THIS CODE CAN BE USED TO SOLVE SYSTEMS THAT
C             ARE DEFINED IN TERMS OF COMPLEX ARITHMETIC, WITH THE
C             CONVERSION FROM COMPLEX TO REAL OCCURRING IN SUBROUTINE
C             HFNN.  FOR AN EXAMPLE OF HOW THIS IS DONE, SEE THE SAMPLE
C             SUBROUTINE HFNN3, DESCRIBED IN NOTE 3, BELOW.
C
C           2. THIS CODE FEATURES A VARIABLE BALANCE COLUMN TO ALLOW
C             PATHS TO "TURN BACK IN T."  (SEE CHAPTER 6.)
C
C           3. THREE SAMPLE SETS OF "USER SUPPLIED" SUBROUTINES -- INPA, INPB,
C             STPT, HFNN, AND FFNN -- ARE INCLUDED, INDEXED BY "1," "2," AND
C             "3."  HFNN1  IS SET UP TO SOLVE THE FIRST TEN EXPERIMENTS IN
C             CHAPTER6.  HFNN2  IMPLEMENTS THE CONVEX-LINEAR CONTINUATION IN
C             REAL ARITHMETIC, AND HFNN3  IMPLEMENTS THE SAME CONTINUATION IN
C             COMPLEX ARITHMETIC.  TO USE ONE OF THESE SAMPLE SUBROUTINE
C             SETS, YOU MUST CHANGE THE NAMES OF THE ROUTINES TO THEIR GIVEN
C             GENERIC NAMES.  THE SETS AND THEIR ROUTINES ARE LISTED HERE:
C
C             SAMPLE SET 1:  INPA1, INPB1, STPT1, HFNN1, FFNN1.
C             SAMPLE SET 2:  INPA2, INPB2, STPT2, HFNN2, FFNNQ2, FFNNB2.
C             SAMPLE SET 3:  INPA3, INPB3, STPT3, HFNN3, FFNNQ3, FFNNB3.
C
C             SETS 2 AND 3 HAVE TWO "FFNN" ROUTINES, WHICH DEFINE TWO
C             DIFFERENT SYSTEMS TO SOLVE.
C
C           4. IT IS FREQUENTLY CONVENIENT TO MODIFY THE SUBROUTINES
C             OTPT OR PRTS TO SUIT THE NEEDS OF A PARTICULAR EXPERIMENT.
C             FOR EXAMPLE, I USE PRTS TO WRITE A DATE FILE TO UNIT 8,
C             WHEN I WANT TO GRAPH THE SCHEMATIC OF A CONTINUATION PATH.
C
C           5. DIMENSIONS ARE SET FOR A MAXIMUM OF M=20 EQUATIONS.
C             ALL DIMENSIONS CURRENTLY SET TO 20 AND 21 CAN BE CHANGED
C             TO L AND L+1 RESPECTIVELY, WHERE L IS A NEW UPPER BOUND
C             ON THE NUMBER OF EQUATIONS.
C
C
```

```
C USER SUPPLIED SUBROUTINES:  INPA, INPB, STPT, HFNN, FFNN
C
C           "INPA" SUPPLIES FREQUENTLY CHANGED CONSTANTS, INCLUDING
C           THE COEFFICIENTS FOR THE SYSTEM.
C
C           "INPB" SUPPLIES LESS FREQUENTLY CHANGED CONSTANTS.
C
C           "STPT" SUPPLIES A START POINT, (Y,T), FOR THE PATH.
C
C           "HFNN" SUPPLIES THE EQUATION AND PARTIAL
C           DERIVATIVE VALUES FOR THE CONTINUATION SYSTEM.
C
C           "FFNN" SUPPLIES THE EQUATION AND PARTIAL DERIVATIVE
C           VALUES FOR THE SYSTEM TO BE SOLVED.
C
C INPUT VARIABLES: SEE SUBROUTINE INPA FOR FORMAT.
C
C        TITLE      -- FIRST 60 CHARACTERS OF FIRST LINE OF INPUT
C        IFLGCR     -- IF =1, THEN CORRECTOR ITERATIONS ARE WRITTEN TO
C                      UNIT 8.  (SEE SUBROUTINE CORCT.)
C        IFLGST     -- IF =1, THEN STEP SUMMARY OUTPUT IS PRINTED AFTER
C                      EVERY STEP.
C        MAXNS      -- MAXIMUM NUMBER OF STEPS FOR A PATH
C        MAXIT      -- MAXIMUM NUMBER OF ITERATIONS TO CORRECT A STEP
C        EPSBIG     -- THE CORRECTOR RESIDUAL MUST BE LESS THAN EPS FOR
C                      A STEP TO "SUCCEED."  EPS=EPSBIG UNTIL T > .95
C        SSZBEG     -- BEGINNING STEP LENGTH
C        M          -- NUMBER OF EQUATIONS
C
C OUTPUT VARIABLES:
C
C    FOR EACH STEP (OPTIONAL OUTPUT: PRINTED IF IFLGST=1.  SEE
C    SUBROUTINE PRTS)
C
C        NUMSTP     -- CURRENT STEP NUMBER
C        STPSZE     -- CURRENT STEP SIZE
C        T          -- CURRENT VALUE OF T
C        Y          -- Y(J)  IS THE J-TH INDEPENDENT VARIABLE
C        XRESID     -- NORM OF THE RESIDUAL OF THE LAST CORRECTOR ITERATION
C
C    FOR THE PATH:
C
C        NUMPAT     -- PATH NUMBER
C        NUMSTP     -- TOTAL NUMBER OF STEPS
C        NSSUCC     -- NUMBER OF STEPS THAT SUCCEEDED
C        NSFAIL     -- NUMBER OF STEPS THAT FAILED
C        NUMIT      -- TOTAL CORRECTOR ITERATIONS
C        NUMLN      -- TOTAL NUMBER OF LINEAR SYSTEMS SOLVED
C        ARCLN2     -- ARC LENGTH OF PATH
C        T          -- VALUE AT END OF PATH (T=1, IF PATH CONVERGED)
C        Y          -- VALUE AT END OF PATH (SEE ABOVE)
C        XRESID     -- VALUE AT END OF PATH (SEE ABOVE)
C        COND       -- THE CONDITION OF DF(Y) (PRINTED IF DF(Y) IS
C                      NONSINGULAR)
C        DET        -- DETERMINANT OF DF(Y) (PRINTED IF DF(Y) IS
C                      NONSINGULAR)
C
```

```
C SUBROUTINES:
C
C   USER SUPPLIED (OR MODIFIED)
C     INPA
C     INPB
C     STPT
C     HFNN
C     FFNN
C
C   MAIN SUBROUTINES
C     PRED
C     CORC
C     LIN
C
C   UTILITY SUBROUTINES
C     IRMAX
C     ICD
C     LNFP
C     LNSP
C     OTPT
C     PRTS
C     PRTV
C     XNORM
C     XNRM2
C
C DECLARATIONS
      INTEGER  IFLGCR,IFLGST,ITOTIT,J,KSNEW,KSTAR,M,MAXIT,MAXNS,MP1,
     & MSG,NADV,NSFAIL,NSSUCC,NUMI,NUMIT,NUMPAT,NUMSTP
      DOUBLE PRECISION  ARCLN2,COND,DET,DZOLD,EPS,EPSO,EPSBIG,EPSSML,
     & FACTOR,SSZBEG,SSZMIN,STPSZE,T,TOLD,XNRM2,XRESID,Y,YOLD,Z
C
C DIMENSIONS
      DIMENSION DZOLD(21),Y(20),YOLD(20),Z(21)
C
C INPUT ROUTINES.
      CALL INPA(IFLGCR,IFLGST,MAXNS,MAXIT,EPSBIG,SSZBEG,M)
      CALL INPB(M,SSZBEG,EPSBIG,SSZMIN,EPSSML,EPSO)
C
C INITIALIZE COUNTERS AND CONSTANTS.
C
      MP1=M+1
      ITOTIT=0
      ARCLN2=0.D0
      NSSUCC=0
      NSFAIL=0
      NUMIT=0
      STPSZE=SSZBEG
      T=0.D0
C THE COUNTER NADV CONTROLS THE "STEP INCREASE" LOGIC.
      NADV=0
C KSNEW IS THE NEW BALANCE COLUMN, INITIALIZED HERE.
      KSNEW=MP1
C
      DO 10 J=1,MP1
          DZOLD(J)=0.D0
 10   CONTINUE
C
```

```
C SUBROUTINE STPT RETURNS A START POINT, Y, FOR THE PATH.
      CALL STPT(M,Y)
C
C SAVE VALUES OF T AND Y IN CASE INITIAL STEP MUST BE REDONE.
      TOLD=T
      DO 20 J=1,M
         YOLD(J)=Y(J)
 20   CONTINUE
C
C *****************************************************************
C *****                                                     *****
C *****        BEGINNING OF MAIN PART OF PROGRAM            *****
C *****                                                     *****
C *****************************************************************
C
C STEPS LOOP -- ITERATE STEPS ALONG THE PATH.
      DO 30 NUMSTP = 1,MAXNS
C KSTAR IS THE BALANCE COLUMN FOR THIS STEP.
         KSTAR=KSNEW
         CALL PRED(M,KSTAR,EPSO,STPSZE,T,Y,DZOLD,KSNEW,MSG)
C
         EPS=EPSBIG
         IF( T .GT. .95DO ) EPS=EPSSML
C
         CALL CORC(IFLGCR,M,KSTAR,MAXIT,EPS,EPSO,T,Y,NUMI,MSG,XRESID)
         NUMIT = NUMIT + NUMI
C
         IF( MSG .EQ. 0 ) THEN
C STEP SUCCEEDED.
C
            NSSUCC=NSSUCC+1
C
            DO 40 J=1,M
               Z(J)=Y(J)-YOLD(J)
 40         CONTINUE
            Z(MP1)=T-TOLD
            ARCLN2=ARCLN2+XNRM2(MP1,Z)
C
            IF(IFLGST.EQ.1) CALL PRTS(M,NUMSTP,MSG,STPSZE,T,Y,XRESID)
C
            IF( T .GE. 1.DO ) THEN
C MAKE A FINAL CORRECTION AND LEAVE LOOP.
C
               WRITE(6,50)
 50            FORMAT(/,' PATH CONVERGED')
C
               KSTAR=MP1
               FACTOR=(1.DO-TOLD)/(T-TOLD)
               DO 60 J=1,M
                  Y(J)=YOLD(J)+FACTOR*(Y(J)-YOLD(J))
 60            CONTINUE
               T=1.DO
               CALL CORC(IFLGCR,M,KSTAR,10,EPSO,EPSO,T,Y,NUMI,MSG,
     &                                                        XRESID)
               GOTO 31
C                     LEAVE LOOP
            ENDIF
```

```
C
              IF( NUMSTP .GE. MAXNS ) THEN
                  WRITE(6,70)
 70               FORMAT(/,' PATH FAILED, MAX NUM STEPS EXCEEDED')
                  GOTO 31
C                     LEAVE LOOP
              ENDIF
C
C NADV CONTROLS THE "STEP INCREASE" LOGIC.  AFTER 5 SUCCESSFUL
C STEPS, STPSZE IS DOUBLED.  WHEN A STEP FAILS, NADV IS SET TO 0.
              NADV=NADV+1
              IF( NADV .EQ. 5 ) THEN
                  STPSZE=2.DO*STPSZE
                  NADV=0
              ENDIF
C
C SAVE VALUES OF T AND Y IN CASE STEP MUST BE REDONE
              TOLD=T
              DO 80 J=1,M
                  YOLD(J)=Y(J)
 80           CONTINUE
C
          ELSE
C STEP FAILED
C
              NSFAIL=NSFAIL+1
              IF(IFLGST.EQ.1) CALL PRTS(M,NUMSTP,MSG,STPSZE,T,Y,XRESID)
C
C RESET "STEP INCREASE COUNTER"
              NADV=0
              STPSZE=STPSZE/2.DO
C
              IF( NUMSTP .GE. MAXNS ) THEN
                  WRITE(6,70)
                  GOTO 31
C                     LEAVE LOOP
              ENDIF
C
              IF( STPSZE .LT. SSZMIN ) THEN
                  WRITE(6,90)
 90               FORMAT(/,' PATH FAILED, SSZMIN EXCEEDED'/)
                  GOTO 31
C                     LEAVE LOOP
              ENDIF
C
              T=TOLD
              DO 100 J=1,M
                  Y(J)=YOLD(J)
 100          CONTINUE
C
          ENDIF
C
 30   CONTINUE
C         BOTTOM OF LOOP -- NUMSTP
C
 31   CONTINUE
C         LEAVE LOOP -- NUMSTP
```

```
C
C
      ITOTIT=ITOTIT + NUMIT
C
      WRITE(6,190)
  190 FORMAT(/' FINAL VALUES FOR PATH'/)
      WRITE(6,200) NUMSTP
  200 FORMAT('  TOTAL NUMBER OF STEPS = ',I5)
      WRITE(6,205) NSSUCC
  205 FORMAT('        NUMBER OF STEPS THAT SUCCEEDED = ',I5)
      WRITE(6,210) NSFAIL
  210 FORMAT('        NUMBER OF STEPS THAT FAILED    = ',I5)
C
      WRITE(6,215) NUMIT
  215 FORMAT(' NUMBER OF CORRECTOR ITERATIONS =',I10)
C
      NUMIT=NUMIT+NUMSTP
      WRITE(6,217) NUMIT
  217 FORMAT(' NUMBER OF LINEAR SYSTEMS SOLVED =',I10)
C
      WRITE(6,218) ARCLN2
  218 FORMAT(' ARC LENGTH OF PATH =',D11.4,//)
C
        CALL OTPT(M,NUMPAT,EPSO,Y, COND,DET)
C
        CALL PRTV(M,T,Y,XRESID)
C
        IF (COND .GE. 0.D0) THEN
           WRITE(6,219) COND,DET
  219      FORMAT(
     &       /,' CONDITION AND DETERMINANT OF DF(Y) AT END OF PATH',
     &       /,'     COND =', D11.4,
     &       /,'     DET  =',2D11.4)
        ELSE
           WRITE(6,220) EPSO
  220      FORMAT(' DF(Y) IS SINGULAR WITH RESPECT TO EPSO =',D11.4)
        ENDIF
C
      STOP
      END
```

```
          SUBROUTINE PRED(M,KSTAR,EPSO,STPSZE,T,Y,DZOLD, KSNEW,MSG)
C
C PURPOSE: ON ENTRY TO PRED, Z=(Y,T) IS A POINT ON THE CONTINUATION
C          CURVE.  PRED FINDS A POINT, DZ, ON THE TANGENT TO THE CURVE
C          AT Z.  THEN Z + DZ WILL BE THE "PREDICTED POINT."
C
C INPUT:  M          -- NUMBER OF EQUATIONS
C         KSTAR      -- THE INDEX OF THE BALANCE COLUMN
C         EPSO       -- ZERO-EPSILON FOR SINGULARITY
C         STPSZE     -- CURRENT STEPSIZE
C         T          -- CURRENT VALUE OF T
C         Y          -- CURRENT VALUE OF Y
C         DZOLD      -- CURRENT TANGENT TO THE CURVE AT Z=(Y,T)
C
C OUTPUT: T          -- IF MSG=0, PREDICTED VALUE OF T
C         Y          -- IF MSG=0, PREDICTED VALUE OF Y
C         DZOLD      -- DZ, THE TANGENT TO THE INPUT Z=(Y,T)
C         MSG        -- =1, IF SYSTEM JACOBIAN IS SINGULAR.
C                       =0, OTHERWISE.
C         KSNEW      -- NEW INDEX OF THE BALANCE COLUMN
C
C
C SUBROUTINES:  HFNN,IRMAX,LIN,XNORM
C
C
C DECLARATIONS
      INTEGER IRMAX,J,JJ,KSNEW,KSTAR,M,MP1,MSG
      DOUBLE PRECISION DELT,DH,DOT,DZ,DZOLD,EPSO,FACTOR,H,RESID,
     & STPSZE,T,XLNGTH,XNORM,Y
C
C DIMENSIONS
      DIMENSION DH(20,21),DZ(21),DZOLD(21),H(20),RESID(20),Y(20)
C
      MSG = 0
      MP1=M+1
C
C H IS A SYSTEM OF M EQUATIONS IN THE M+1 UNKNOWNS Y AND T.  DH IS
C THE ASSOCIATED M X (M+1) JACOBIAN MATRIX.
      CALL HFNN(M,T,Y, H,DH)
C
C THIS CALL TO SUBROUTINE LIN SOLVES DH<KSTAR>*RESID=-DH(KSTAR), WHERE
C DH(KSTAR) IS THE KSTAR-TH COLUMN OF DH AND DH<KSTAR> IS THE SQUARE
C MATRIX FORMED FROM DH BY OMITTING THE KSTAR-TH COLUMN OF DH.
C MSG=1 MEANS DH<KSTAR> IS SINGULAR WITH RESPECT TO EPSO.
      CALL LIN(M,KSTAR,EPSO,DH, MSG,RESID)
C
      IF( MSG .EQ. 1 ) WRITE(6,10)
  10  FORMAT(//' FROM PRED, MSG=1, ERROR '//)
      IF( MSG .EQ. 1 ) RETURN
C
      DO 20 J=1,M
         JJ=J
         IF( J .GE. KSTAR ) JJ=J+1
         DZ(JJ) = RESID(J)
  20  CONTINUE
      DZ(KSTAR)=1.D0
C
```

```
C DZ IS NOW THE TANGENT POINT TO BE ADDED TO Z=(Y,T), EXCEPT
C THAT ITS LENGTH MAY NEED TO BE CHANGED.
C
C ADJUST THE LENGTH OF DZ BY THE STEP LENGTH, STPSZE.
C
      XLNGTH = XNORM(MP1,DZ)
C
      FACTOR=STPSZE/XLNGTH
C
C IF THIS FACTOR MAKES T > 1, THEN SHORTEN FACTOR SO THAT T=1.
      DELT = DZ(MP1)*FACTOR
      IF(T+DELT .GT. 1.D0) FACTOR=(1.D0-T)/DZ(MP1)
C
      DO 30 J=1,MP1
          DZ(J)=DZ(J)*FACTOR
   30 CONTINUE
C
C THE LENGTH OF DZ IS NOW CORRECT.
C
C FIND THE BIGGEST ENTRY OF DZ AND SET KSNEW.
C
      KSNEW=IRMAX(MP1,DZ)
C
C APPLY DOT PRODUCT TEST TO SEE IF THE DIRECTION OF DZ MUST
C BE REVERSED.  IF SO, REVERSE IT.
C
      DOT=0.D0
      DO 40 J=1,MP1
          DOT=DOT+DZOLD(J)*DZ(J)
   40 CONTINUE
C
      IF( DOT .LT. 0.D0) THEN
          DO 50 J=1,MP1
              DZ(J)=-DZ(J)
   50     CONTINUE
      ENDIF
C
C
C SAVE CURRENT PREDICTION IN DZOLD AND UPDATE Z=(Y,T).
C
      DO 60 J=1,MP1
          DZOLD(J)=DZ(J)
   60 CONTINUE
C
      DO 70 J=1,M
          Y(J) = Y(J) + DZ(J)
   70 CONTINUE
      T = T + DZ(MP1)
C
      RETURN
      END
```

```
      SUBROUTINE CORC(IFLGCR,M,KSTAR,MAXIT,EPS,EPSO,T,Y,NUMI,MSG,XRESID)
C
C
C PURPOSE: MOVE THE POINT Z=(Y,T) BACK TO THE CONTINUATION CURVE
C
C
C INPUT:   IFLGCR   -- =1, IF CORRECTOR ITERATIONS ARE TO BE WRITTEN
C                         TO UNIT 8.
C                      =0, OTHERWISE.
C          M        -- NUMBER OF EQUATIONS
C          KSTAR    -- INDEX OF THE BALANCE COLUMN
C          MAXIT    -- MAXIMUM NUMBER OF ITERATIONS ALLOWED
C          EPS      -- ZERO-EPSILON FOR CORRECTOR RESIDUAL
C          EPSO     -- ZERO-EPSILON FOR SINGULARITY
C          T        -- PREDICTED VALUE OF T
C          Y        -- PREDICTED VALUE OF Y
C
C OUTPUT:  T        -- CORRECTED VALUE OF T
C          Y        -- CORRECTED VALUE OF Y
C          NUMI     -- NUMBER OF ITERATIONS USED FOR CORRECTION
C          MSG      -- =1, IF SYSTEM JACOBIAN IS SINGULAR.
C                      =2, IF CORRECTOR FAILED TO REDUCE RESIDUAL
C                         TO BE LESS THAN EPS WITHIN MAXIT ITERATIONS.
C                      =0, OTHERWISE.
C          XRESID   -- NORM OF CORRECTOR RESIDUAL
C
C
C SUBROUTINES: HFNN,LIN,XNORM
C
C
C DECLARATIONS
      INTEGER IFLGCR,J,K,KK,KSTAR,M,MAXIT,MP1,MSG,NUMI
      DOUBLE PRECISION DH,EPS,EPSO,H,RESID,T,XNORM,XRESID,Y,YL
C
C DIMENSIONS
      DIMENSION DH(20,21),H(20),RESID(20),Y(20)
C
      MP1=M+1
      MSG = 0
C
      DO 10 NUMI = 1,MAXIT
C
          CALL HFNN(M,T,Y, H,DH)
C
C REPLACE THE KSTAR-TH COLUMN OF DH BY -H.
          DO 20 J=1,M
              DH(J,KSTAR)=-H(J)
 20       CONTINUE
C
          CALL LIN(M,KSTAR,EPSO,DH, MSG,RESID)
C
          IF( MSG .EQ. 1 ) THEN
              WRITE(6,30)
 30           FORMAT(//'  FROM CORC, MSG=1, ERROR '//)
              RETURN
          ENDIF
C
```

```
          DO 40 K=1,M
              IF( K .LT. KSTAR ) KK=K
              IF( K .GE. KSTAR ) KK=K+1
              IF(KK.LE.M) Y(KK)= Y(KK)-RESID(K)
  40      CONTINUE
          IF( KSTAR .LT. MP1 ) T=T-RESID(M)
C
C COMPUTE THE LENGTH OF RESID.
C
          XRESID=XNORM(M,RESID)
          YL    =XNORM(M,Y)
C
          IF( IFLGCR .EQ. 1 ) THEN
              WRITE(8,50) KSTAR,T,YL,XRESID
  50          FORMAT(' KSTAR=',I2,' T=',D11.4,' YL=',D11.4,
     &                                          ' XRESID=',D11.4)
              IF( XRESID  .LT.  EPS )WRITE(8,999)
 999          FORMAT(/)
          ENDIF
C
          IF( XRESID  .LT.  EPS ) RETURN
C
  10      CONTINUE
C
      MSG=2
C
      RETURN
      END
```

```
      SUBROUTINE LIN(M,KSTAR,EPSO,DH, MSG,RESID)
C
C PURPOSE:   SOLVES THE LINEAR SYSTEM DH<KSTAR>*RESID=-DH(KSTAR), WHERE
C            DH(KSTAR) IS THE KSTAR-TH COLUMN OF DH AND DH<KSTAR> IS THE
C            SQUARE MATRIX FORMED FROM DH BY OMITTING THE KSTAR-TH COLUMN
C            OF DH.
C INPUT:   M,KSTAR,EPSO,DH
C
C OUTPUT:  MSG,RESID
C
C
C INPUT:   M         -- NUMBER OF EQUATIONS
C          KSTAR     -- INDEX OF THE BALANCE COLUMN
C          EPSO      -- ZERO-EPSILON FOR SINGULARITY
C          DH        -- M X (M+1) COMPLEX JACOBIAN OF H WITH RESPECT TO Y
C                       AND T
C
C OUTPUT: MSG        -- =1, IF DH<KSTAR> IS SINGULAR WITH RESPECT TO EPSO
C                       =0, OTHERWISE
C         RESID      -- IF MSG=0, SOLUTION TO THE LINEAR SYSTEM
C                       DH<KSTAR>*RESID=DH(KSTAR)
C
C SUBROUTINES: LNFP,LNSP
C
C
C DECLARATIONS
      INTEGER ICOL,IROW,J,KSTAR,LDIM,M,MP1,MSG
      DOUBLE PRECISION DH,EPSO,RESID
C
C DIMENSIONS
      DIMENSION DH(20,21),ICOL(20),IROW(20),RESID(20)
C
C
      LDIM=20
      MP1=M+1
C
      CALL LNFP(LDIM,M,KSTAR,EPSO,DH, MSG,IROW,ICOL)
      IF( MSG .EQ. 1)  RETURN
C
      CALL LNSP(LDIM,M,KSTAR,IROW,ICOL,DH, RESID)
C
      RETURN
      END
```

```
      SUBROUTINE PRTS(M,NUMSTP,MSG,STPSZE,T,Y,XRESID)
C
C
C PURPOSE: PRINTS STEP SUMMARY.
C
C
C NOTE: TWO VERSIONS OF THIS SUBROUTINE, A "FULL" AND A "BRIEF" VERSION
C       ARE GIVEN BELOW, WITH THE BRIEF BEING COMMENTED OUT.  THE BRIEF
C       VERSION MIGHT BE USED TO GENERATE A TABLE OF VALUES TO GRAPH THE
C       PATH SCHEMATIC.
C
C
C INPUT:  M        -- NUMBER OF EQUATIONS
C         NUMSTP   -- STEP NUMBER
C         MSG      -- ERROR MESSAGE
C         STPSZE   -- STEPSIZE
C         T        -- CURRENT VALUE OF T
C         Y        -- CURRENT VALUE OF Y
C         XRESID   -- NORM OF CORRECTOR RESIDUAL
C
C OUTPUT: TO UNIT 6 AND/OR TO UNIT 8
C
C
C SUBROUTINES:  PRTV,XNRM2
C
C
C DECLARATIONS
      INTEGER MSG,M,NUMSTP
      DOUBLE PRECISION STPSZE,T,XNRM2,XRESID,Y,YL
C
C DIMENSIONS
      DIMENSION  Y(1)
C
C
C------------------------------------------------------------------------
C FULL VERSION
C--------------
C
      WRITE(6,100) NUMSTP
  100 FORMAT(/'  STEP SUMMARY FOR STEP NUMBER',I5,' :')
C
      IF(MSG.EQ.0) WRITE(6,110) NUMSTP
  110 FORMAT(' STEP ',I5,' SUCCEEDED')
      IF(MSG.EQ.1) WRITE(6,120) NUMSTP
  120 FORMAT('  STEP ',I5,' FAILED DUE TO SINGULARITY')
      IF(MSG.EQ.2) WRITE(6,130) NUMSTP
  130 FORMAT('  STEP ',I5,' FAILED DUE TO NON-CONVERGENCE')
C
      WRITE(6,140) STPSZE
  140 FORMAT('  STPSZE =',D11.4)
C
      CALL PRTV(M,T,Y,XRESID)
C
C----------------------
C END OF FULL VERSION
C------------------------------------------------------------------------
C
```

```
C------------------------------------------------------------------
C BRIEF VERSION
C---------------
C
C RETURN IF STEP FAILED
CC    IF(MSG.NE.0) RETURN
C
C GENERATE THE EUCLIDEAN NORM OF Y
CC    YL= XNRM2(M,Y)
C
CC    WRITE(6,10) T,YL
CC10   FORMAT(' T=',D22.15,'   Y=',D22.15)
CC    WRITE(8,20) T,YL
CC20   FORMAT(2D22.15)
C
C---------------------
C END OF BRIEF VERSION
C------------------------------------------------------------------
C
      RETURN
      END
```

```
      SUBROUTINE PRTV(M,T,Y,XRESID)
C
C PURPOSE: PRINTS SELECTED VARIABLE VALUES
C
C
C INPUT:  M         -- NUMBER OF EQUATIONS
C         T         -- CURRENT VALUE OF T
C         Y         -- CURRENT VALUE OF Y
C         XRESID    -- NORM OF CORRECTOR RESIDUAL
C
C OUTPUT:  TO UNIT 6
C
C
C DECLARATIONS
      INTEGER J,M
      DOUBLE PRECISION T,XRESID,Y
C
C DIMENSIONS
      DIMENSION  Y(1)
C
C
      WRITE(6,100)
  100 FORMAT(' ------------------------------------------------------')
C
      WRITE(6,110) T
  110 FORMAT('  T  = ',D22.15)
C
C
      DO 125 J=1,M
          WRITE(6,130) J,Y(J)
  125 CONTINUE
  130 FORMAT('  Y(',I5,')  = ', D22.15)
C
      WRITE(6,150) XRESID
  150 FORMAT(' NORM OF RESIDUAL = ',D22.15)
C
      WRITE(6,100)
C
      RETURN
      END
```

```
      SUBROUTINE OTPT(M,NUMPAT,EPSO,Y, COND,DET)
C
C PURPOSE: OTPUT IS CALLED AT THE END OF THE PATH.  OTPUT CAN BE USED
C          TO TRANSFORM THE VARIABLES OR FOR ANY OTHER PURPOSE.
C          THIS VERSION OF OTPT COMPUTES THE CONDITION AND
C          DETERMINANT OF THE JACOBIAN OF F AT Y.  COMMENTED OUT IS
C          CODE TO "UNPROJECTIVIZE" SOLUTIONS, AS MIGHT BE WANTED
C          IN RUNNING EXPERIMENT (6-6) AND OTHER PROJECTIVIZED
C          SYSTEMS.
C
C INPUT:  M        -- NUMBER OF EQUATIONS
C         NUMPAT   -- CURRENT PATH NUMBER
C         EPSO     -- ZERO-EPSILON FOR SINGULARITY.
C         Y        -- CURRENT Y VALUE
C
C OUTPUT: COND     -- CONDITION OF THE JACOBIAN, DF, AT Y.  COND=-1
C                     MEANS THAT DF IS SINGULAR WITH RESPECT TO EPSO.
C         DET      -- IF COND >= 0, DETERMINANT OF DF.
C
C
C SUBROUTINES: FFNN,ICD
C
C
C DECLARATIONS
      INTEGER IDEG,IWORK,J,M,MM1,MSG,NUMPAT
      DOUBLE PRECISION COND,DET,DF,EPSO,F,TEMP,WORK,WORK2,Y
C
C DIMENSIONS
      DIMENSION DF(20,20),F(20),IDEG(1),IWORK(20),WORK(20),
     & WORK2(20,20),Y(1)
C
C
      CALL FFNN(M,Y, F,DF)
      CALL ICD(20,M,EPSO,DF, WORK,IWORK,WORK2,COND,DET,MSG)
      IF(MSG.EQ.1) COND=-1
C
CC    IF(DABS(Y(M)) .GE. EPSO) THEN
CC       TEMP=Y(M)
CC       MM1=M-1
CC       DO 10 J=1,MM1
CC          Y(J)=Y(J)/TEMP
CC 10    CONTINUE
CC    ENDIF
C
      RETURN
      END
```

```
      SUBROUTINE INPA1(IFLGCR,IFLGST,MAXNS,MAXIT,EPSBIG,SSZBEG,M)
C
C
C PURPOSE:  FOR EXPERIMENTS 6-1 THROUGH 6-10.
C           INITIALIZES FREQUENTLY CHANGED CONSTANTS.
C
C GENERIC NAME: INPA          FILE NAME: INPA1
C
C
C REFERENCE:  CHAPTER 6 OF "SOLVING POLYNOMIAL SYSTEMS USING
C             CONTINUATION" BY ALEXANDER MORGAN.
C
C
C INPUT:  READ FROM UNIT 5.  SEE CODE BELOW FOR FORMATS.
C
C      TITLE       -- FIRST 60 CHARACTERS OF FIRST LINE OF INPUT
C      IFLGCR      -- =1, IF THE CORRECTOR ITERATIONS ARE TO BE WRITTEN
C                     TO UNIT 8.  (SEE SUBROUTINE CORC.)
C      IFLGST      -- =1, IF A STEP SUMMARY OUTPUT IS PRINTED AFTER
C                     EVERY STEP.
C      MAXNS       -- MAXIMUM NUMBER OF STEPS FOR A PATH
C      MAXIT       -- MAXIMUM NUMBER OF ITERATIONS TO CORRECT A STEP
C      EPSBIG      -- THE CORRECTOR RESIDUAL MUST BE LESS THAN EPS FOR
C                     A STEP TO "SUCCEED." EPS=EPSBIG UNTIL T > .95
C      SSZBEG      -- BEGINNING STEP LENGTH
C      M           -- NUMBER OF EQUATIONS
C
C SAMPLE INPUT FOLLOWS:
CTITLE:    NOTES TO IDENTIFY EXPERIMENT
C   0                      IFLGCR
C   0                      IFLGST
C 200                      MAXNS
C 3                        MAXIT
C               1.D-02     EPSBIG
C               1.D-02     SSZBEG
C   1                      M
C
C
C OUTPUT:  IFLGCR,IFLGST,MAXNS,MAXIT,EPSBIG,SSZBEG
C
C
C DECLARATIONS
      INTEGER IFLGCR,IFLGST,M,MAXIT,MAXNS,TITLE
      DOUBLE PRECISION EPSBIG,SSZBEG
C
C DIMENSIONS
      DIMENSION TITLE(60)
C
C
C INPUT STATEMENTS FOLLOW.
      READ(5,10) TITLE
  10  FORMAT(60A1)
      READ(5,20) IFLGCR
  20  FORMAT(I5)
      READ(5,20) IFLGST
```

```
      READ(5,20) MAXNS
      READ(5,20) MAXIT
      READ(5,30) EPSBIG
30    FORMAT(D22.15)
      READ(5,30) SSZBEG
      READ(5,20) M
C
C
      WRITE(6,40)
40    FORMAT('  CONSOL9, FOR EXPERIMENTS 6-1 THROUGH 6-10',//)
      WRITE(6,50) TITLE
50    FORMAT(' ',60A1)
      WRITE(6,60) IFLGCR
60    FORMAT(/' FLAG TO PRINT OUT CORRECTOR RESIDUALS =',I2)
      WRITE(6,70) IFLGST
70    FORMAT(/' FLAG TO PRINT OUT STEP SUMMARIES =',I2)
      WRITE(6,80) MAXNS
80    FORMAT(/' MAX NUM STEPS PER PATH =',I5)
      WRITE(6,90) MAXIT
90    FORMAT(/' MAX NUM ITERATIONS PER STEP =',I5)
      WRITE(6,100) EPSBIG
100   FORMAT(/' EPSBIG =',D22.15/)
      WRITE(6,110) SSZBEG
110   FORMAT(/' BEGINNING STEP SIZE = ',D22.15/)
      WRITE(6,120) M
120   FORMAT(/' NUMBER OF EQUATIONS = ',I5/)
C
      RETURN
      END
```

```
      SUBROUTINE INPB1(M,SSZBEG,EPSBIG,SSZMIN,EPSSML,EPSO)
C
C PURPOSE: FOR EXPERIMENTS 6-1 THROUGH 6-10.
C          INITIALIZES LESS FREQUENTLY CHANGED CONSTANTS
C
C
C GENERIC NAME: INPB              FILE NAME: INPB1
C
C
C INPUT:  M         -- NUMBER OF EQUATIONS
C         SSZBEG    -- BEGINNING STEPSIZE
C         EPSBIG    -- ZERO-EPSILON FOR PATH TRACKING RESIDUAL,
C                      UNTIL T > .95
C
C OUTPUT: SSZMIN    -- MINIMUM STEPSIZE ALLOWED
C         EPSSML    -- ZERO-EPSILON FOR PATH TRACKING RESIDUAL,
C                      AFTER T > .95
C         EPSO      -- ZERO-EPSILON FOR SINGULARITY
C
C
C DECLARATIONS
      INTEGER M
      DOUBLE PRECISION EPSO,EPSBIG,EPSSML,SSZBEG,SSZMIN
C
C
      SSZMIN = 1.D-5*SSZBEG
      EPSSML = 1.D-3*EPSBIG
      EPSO   = 1.D-22
C
      WRITE(6,10) SSZMIN
  10  FORMAT(/' MINIMUM STPSZE = ',D22.15)
      WRITE(6,20) EPSSML
  20  FORMAT(/' EPSSML =',D22.15)
      WRITE(6,30) EPSO
  30  FORMAT(/' EPSO   =',D22.15)
C
      RETURN
      END
```

```
      SUBROUTINE STPT1(M,Y)
C
C PURPOSE: FOR EXPERIMENTS 6-1 THROUGH 6-10.
C          GENERATES A START POINT FOR THE PATH.
C
C
C GENERIC NAME: STPT          FILE NAME: STPT1
C
C
C NOTE: THIS CODE IS DIVIDED INTO PARTS, LABELED FOR THE VARIOUS
C       EXPERIMENTS. MAKE SURE THE PARTS YOU DON'T WANT ARE
C       "COMMENTED OUT."
C
C
C INPUT:  M          -- NUMBER OF EQUATIONS
C
C OUTPUT: Y          -- START POINT FOR PATH
C
C COMMON:  /RCOM/    IS USED FOR EXPERIMENT 6-6 TO PASS THE R ARRAY
C                    FROM SUBROUTINE STPT TO SUBROUTINE FFNN.
C          /QCOM/    IS USED FOR EXPERIMENTS 6-7 THROUGH 6-10 TO PASS
C                    THE Q ARRAY FROM SUBROUTINE STPT TO SUBROUTINE HFNN.
C
C DECLARATIONS
      INTEGER J,M
      DOUBLE PRECISION Q,R,Y
C
C DIMENSIONS
      DIMENSION Y(1)
C
C COMMON
      COMMON/RCOM/R(3)
      COMMON/QCOM/Q(20)
C
C
C THE ARRAY "R" IS USED FOR EXPERIMENT 6-6
      R(1) = 1.762 537 462 980 D0
      R(2) =-0.993 758 928 100 D0
      R(3) = 0.393 839 300 090 D0
C
C
C START POINT FOR EXPERIMENT 6-1
      Y(1)=-.1759D0
C
C
C START POINT FOR EXPERIMENTS 6-2 AND 6-3
      Y(1)=-.3247D0
C
C
C START POINT FOR EXPERIMENT 6-4
      Y(1) = 1.D0
      Y(2) = DSQRT(5.D0)
C
C
C START POINT FOR EXPERIMENT 6-5
      Y(1) =  5.D0
      Y(2) = -9.D0/5.D0
```

```
C
C
C START POINTS FOR EXPERIMENT 6-6.  THREE CHOICES OF START POINT ARE
C GIVEN. COMMENT OUT THE TWO YOU DON'T WANT.  (ONE START POINT IS A BAD
C CHOICE BECAUSE OF SINGULARITY.)
C
C FIRST START POINT FOR EXPERIMENT 6-6
      Y(3) = 1.D0/(5.D0*R(1) - 9.D0*R(2)/5.D0 +R(3))
      Y(1) = 5.D0*Y(3)
      Y(2) = -9.D0*Y(3)/5.D0
C
C SECOND START POINT FOR EXPERIMENT 6-6
        Y(1)=0.D0
        Y(2)=1.D0/R(2)
        Y(3)=0
C
C THIRD START POINT FOR EXPERIMENT 6-6
        Y(1)=1.D0/R(1)
        Y(2)=0.D0
        Y(3)=0
C
C
C START POINTS FOR EXPERIMENTS 6-7 THROUGH 6-10 ARE READ FROM
C UNIT 5 AND SAVED IN Q.
      READ(5,10)(Y(J),J=1,M)
   10 FORMAT(D22.15)
      DO 20 J=1,M
        Q(J) = Y(J)
   20 CONTINUE
C
C
C THE FOLLOWING IS FOR ALL EXPERIMENTS.
      WRITE(6,30)
   30 FORMAT(/,'  START POINT:')
      WRITE(6,40) ( J,Y(J),J=1,M )
   40 FORMAT(/,'    Y(',I2,') = ',D22.15)
C
      RETURN
      END
```

```
      SUBROUTINE HFNN1(M,T,Y, H,DH)
C
C PURPOSE:   FOR EXPERIMENTS 6-1 THROUGH 6-10.
C            EVALUATES THE CONTINUATION EQUATION,H, AND ITS PARTIAL
C            DERIVATIVES, DH.
C
C
C GENERIC NAME: HFNN          FILE NAME: HFNN1
C
C
C NOTE: 1. THE BLOCKS OF CODE ARE MARKED FOR EACH EXPERIMENT.  MAKE SURE
C          THE PARTS YOU DON'T WANT ARE "COMMENTED OUT."
C
C       2. THE CONTINUATION FOR THE FIRST THREE EXPERIMENTS IS OF THE FORM
C          H(Y,T) =  FA(Y) + FB(T).
C          THE CONTINUATION FOR THE SECOND THREE EXPERIMENTS IS OF THE FORM
C          H(Y,T) = F(Y) + G(Y,T).
C          THE LAST FOUR EXPERIMENTS ARE EACH IN TWO PARTS.  ONE PART USES
C          CONTINUATION (6-15): H(Y,T)=F(Y)-(1-T)*F(Q).  THE OTHER PART USES
C          CONTINUATION (6-16): H(Y,T)=(1-T)*(Y-Q)+T*F(Y).
C
C
C INPUT:  M          -- NUMBER OF EQUATIONS
C         T          -- CURRENT VALUE OF T
C         Y          -- CURRENT VALUE OF Y
C
C OUTPUT: H          -- H IS THE CONTINUATION SYSTEM, A SYSTEM OF M
C                       REAL POLYNOMIALS IN THE M+1 UNKNOWNS Y AND T.
C         DH         -- DH IS THE M X (M+1) MATRIX OF PARTIAL
C                       DERIVATIVES OF H WITH RESPECT TO Y AND T.
C                       DH(J,K) IS FOR J=1 TO M AND K=1 TO M IS
C                       THE PARTIAL OF H(J) WITH RESPECT TO Y(K).
C                       DH(J,M+1) FOR J=1 TO M IS THE PARTIAL DERIVATIVE
C                       OF H(J) WITH RESPECT TO T.
C
C COMMON:  /QCOM/     IS USED FOR EXPERIMENTS 6-7 THROUGH 6-10 TO PASS
C                     THE Q ARRAY FROM SUBROUTINE STPT TO SUBROUTINE HFNN.
C
C
C DECLARATIONS
      INTEGER IDEG,J,K,M,MP1
      DOUBLE PRECISION DF,DFA,DFB,DG,DH,DUMMY,F,FA,FB,FQ,G,H,ONEMT,Q,T,Y
C
C DIMENSIONS
      DIMENSION DF(20,20),DG(3,4),DH(20,1),DUMMY(20,20),F(20),FQ(20),
     & G(3),H(1),Y(1)
C
C
C COMMON
      COMMON/QCOM/Q(20)
C
C
      MP1=M+1
      ONEMT=1.D0-T
C
      CALL FFNN(M,Y, F,DF)
C
```

```
C $$$$$  BEGINNING OF BLOCK FOR EXPERIMENTS 6-1 THROUGH 6-3  $$$$$
C
      FA=F(1)
      DFA=DF(1,1)
C
C DEFINE FB AND DFB
C
C BEGINNING OF PART FOR EXPERIMENT 6-1.
      FB  = .45D0*(1.D0-2.D0*T)
      DFB=-.9D0
C END OF PART FOR EXPERIMENT 6-1.
C
C BEGINNING OF PART FOR EXPERIMENT 6-2.
        FB=(1.D0-2.D0*T)**3
        DFB=-6.D0*(1.D0-2.D0*T)**2
C END OF PART FOR EXPERIMENT 6-2.
C
C BEGINNING OF PART FOR EXPERIMENT 6-3.
        IDEG=2
        FB=(1.D0-2.D0*T)**IDEG
        DFB=-2.D0*IDEG*(1.D0-2.D0*T)**(IDEG-1)
C END OF PART FOR EXPERIMENT 6-3.
C
C END OF DEFINITION OF FB AND DFB
C
C
C DEFINE H AND DH
        H(1)   = FA+FB
        DH(1,1) = DFA
        DH(1,2) = DFB
C $$$$$      END OF BLOCK FOR EXPERIMENTS 6-1 THROUGH 6-3      $$$$$
C
C
C $$$$$  BEGINNING OF BLOCK FOR EXPERIMENTS 6-4 THROUGH 6-6  $$$$$
C
C BEGINNING OF PART FOR EXPERIMENT 6-4.
      G(1) = -ONEMT*Y(1)*Y(2)
      G(2) = -ONEMT*Y(1)
      DG(1,1)=-ONEMT*Y(2)
      DG(1,2)=-ONEMT*Y(1)
      DG(1,3)= Y(1)*Y(2)
      DG(2,1)=-ONEMT
      DG(2,2)= 0.D0
      DG(2,3)= Y(1)
C END OF PART FOR EXPERIMENT 6-4.
C
C BEGINNING OF PART FOR EXPERIMENT 6-5.
        G(1) =  -ONEMT*Y(1)**2 + ONEMT*10.D0
        G(2) =  -ONEMT*Y(2)**2
        DG(1,1)=-2.D0*ONEMT*Y(1)
        DG(1,2)= 0.D0
        DG(1,3)= Y(1)**2 - 10.D0
        DG(2,1)= 0.D0
        DG(2,2)=-2.D0*ONEMT*Y(2)
        DG(2,3)= Y(2)**2
C END OF PART FOR EXPERIMENT 6-5.
C
```

```
C BEGINNING OF PART FOR EXPERIMENT 6-6.
      G(1) =  -ONEMT*Y(1)**2 + 10.DO*ONEMT * Y(3)**2
      DG(1,1)=-2.DO*ONEMT*Y(1)
      DG(1,2)= 0.DO
      DG(1,3)= 20.DO*ONEMT * Y(3)
      DG(1,4)= Y(1)**2 - 10.DO*Y(3)**2
      G(2) = -ONEMT*Y(2)**2
      DG(2,1)= 0.DO
      DG(2,2)=-2.DO*ONEMT*Y(2)
      DG(2,3)= 0.DO
      DG(2,4)= Y(2)**2
      G(3) = 0.DO
      DG(3,1)= 0.DO
      DG(3,2)= 0.DO
      DG(3,3)= 0.DO
      DG(3,4)= 0.DO
C END OF PART FOR EXPERIMENT 6-6.
C
C THE FOLLOWING PART IS FOR EACH OF EXPERIMENTS 6-4 THROUGH 6-6
      DO 10 J=1,M
         DO 20 K=1,M
            DH(J,K)=DF(J,K)+DG(J,K)
 20      CONTINUE
         DH(J,MP1)=DG(J,MP1)
         H(J) = F(J) + G(J)
 10   CONTINUE
C
C $$$$$     END OF BLOCK FOR EXPERIMENTS 6-4 THROUGH 6-6     $$$$$
C
C
C $$$$$  BEGINNING OF BLOCK FOR EXPERIMENTS 6-7 THROUGH 6-10 $$$$$
C
C THIS PART IS FOR CONTINUATION (6-15): H(Y,T)=F(Y)-(1-T)*F(Q).
      CALL FFNN(M,Q,  FQ,DUMMY)
      DO  30 J=1,M
         H(J) = F(J)-ONEMT*FQ(J)
         DH(J,MP1)=FQ(J)
         DO 30 K=1,M
            DH(J,K)= DF(J,K)
 30      CONTINUE
C END OF PART FOR CONTINUATION (6-15).
C
C THIS PART IS FOR CONTINUATION (6-16): H(Y,T)=(1-T)*(Y-Q)+T*F(Y).
      DO 40 J=1,M
         H(J) =   ONEMT*(Y(J)-Q(J))+T*F(J)
         DH(J,MP1)= -(Y(J)-Q(J))+F(J)
            DO 50 K=1,M
               DH(J,K)= T*DF(J,K)
 50         CONTINUE
         DH(J,J)= DH(J,J)+ONEMT
 40   CONTINUE
C END OF PART FOR CONTINUATION (6-16).
C
C $$$$$     END OF BLOCK FOR EXPERIMENTS 6-7 THROUGH 6-10     $$$$$
C
      RETURN
      END
```

```
      SUBROUTINE FFNN1(M,Y, F,DF)
C
C PURPOSE: FOR EXPERIMENTS 6-1 THROUGH 6-10.
C          RETURNS SYSTEM AND PARTIAL DERIVATIVE VALUES FOR THE SYSTEM
C          TO BE SOLVED.
C
C GENERIC NAME: FFNN           FILE NAME: FFNN1
C
C NOTE:  THE PARTS OF THE CODE FOR EACH EXPERIMENT ARE MARKED.  MAKE SURE
C        THE PARTS YOU DON'T WANT ARE "COMMENTED OUT."
C
C
C INPUT:  M       -- NUMBER OF EQUATIONS
C         Y       -- CURRENT VALUE OF Y
C
C OUTPUT: F       -- F(J) IS THE VALUE OF THE J-TH EQUATION AT Y
C         DF      -- DF(J,K) IS THE PARTIAL DERIVATIVE OF F(J) WITH
C                    RESPECT TO Y(K)
C
C COMMON:  /RCOM/   IS USED FOR EXPERIMENT 6-6 TO PASS THE R ARRAY
C                   FROM SUBROUTINE STPT TO SUBROUTINE FFNN.
C
C
C SUBROUTINES: DATAN,DEXP
C
C
C DECLARATIONS
      INTEGER I,J,M
      DOUBLE PRECISION A,B,C,CC,DATAN,DF,DFA,EE,F,FA,PI,R,Y
C
C DIMENSIONS
      DIMENSION A(5,5),B(5,5),C(5),DF(20,20),F(1),Y(1)
C
C COMMON
      COMMON/RCOM/R(3)
C
C
C "PI" IS USED IN EXPERIMENTS 6-7 THROUGH 6-9.
      PI = 4.DO*DATAN(1.DO)
C
C
C BEGINNING OF PART FOR EXPERIMENTS 6-1 THROUGH 6-3.
C NOTE: "F(1)" IS REFERRED TO IN CHAPTER 6 AS "FA".
      F(1)   =   Y(1)**3 - 3.DO*Y(1)**2  + 2.DO*Y(1)
      DF(1,1) = 3.DO*Y(1)**2 - 6.DO*Y(1) + 2.DO
C END OF PART FOR EXPERIMENTS 6-1 THROUGH 6-3.
C
C
C BEGINNING OF PART FOR EXPERIMENTS 6-4 AND 6-5.
      F(1) = Y(1)**2 + Y(1)*Y(2) -1.DO
      DF(1,1) = 2.DO*Y(1) + Y(2)
      DF(1,2) = Y(1)
      F(2) = Y(2)**2 + Y(1)      - 5.DO
      DF(2,1) = 1.DO
      DF(2,2) = 2.DO*Y(2)
C END OF PART FOR EXPERIMENTS 6-4 AND 6-5.
C
```

410

```
C BEGINNING OF PART FOR EXPERIMENT 6-6.
      F(1) = Y(1)**2 + Y(1)*Y(2) - Y(3)**2
      DF(1,1) = 2.D0*Y(1) + Y(2)
      DF(1,2) = Y(1)
      DF(1,3) = -2.D0*Y(3)
      F(2) = Y(2)**2 + Y(1)*Y(3) - 5.D0*Y(3)**2
      DF(2,1) = Y(3)
      DF(2,2) = 2.D0*Y(2)
      DF(2,3) = Y(1) - 10.D0*Y(3)
      F(3) = R(1)*Y(1) + R(2)*Y(2) + R(3)*Y(3) -1.D0
      DF(3,1) = R(1)
      DF(3,2) = R(2)
      DF(3,3) = R(3)
C END OF PART FOR EXPERIMENT 6-6.
C
C
C BEGINNING OF PART FOR EXPERIMENT 6-7.
      F(1) = Y(1)**2 - Y(2) + 1.D0
      DF(1,1) = 2.D0*Y(1)
      DF(1,2) = -1.D0
      F(2) = Y(1) + DCOS(.5D0*PI*Y(2))
      DF(2,1) = 1.D0
      DF(2,2) =-.5D0*PI*DSIN(.5D0*PI*Y(2))
C END OF PART FOR EXPERIMENTS 6-7.
C
C BEGINNING OF PART FOR EXPERIMENT 6-8.
      F(1) = 1.D0 - 2.D0*Y(2) + .13D0*DSIN(4.D0*PI*Y(2)) - Y(1)
      DF(1,1) =  -1.D0
      DF(1,2) =  -2.D0 + 4.D0*PI * .13D0*DCOS(4.D0*PI*Y(2))
      F(2) = Y(2) - .5D0*DSIN(2.D0*PI*Y(1))
      DF(2,1) = -2.D0*PI * .5D0*DCOS(2.D0*PI*Y(1))
      DF(2,2) = 1.D0
C END OF PART FOR EXPERIMENTS 6-8.
C
C
C BEGINNING OF PART FOR EXPERIMENT 6-9.
      F(1) = .5D0*DSIN(Y(1)*Y(2)) - .25D0*(Y(2)/PI) - .5D0*Y(1)
      DF(1,1) = Y(2) * .5D0*DCOS(Y(1)*Y(2)) - .5D0
      DF(1,2) = Y(1) * .5D0*DCOS(Y(1)*Y(2)) - .25D0/PI
      EE=DEXP(1.D0)
      CC=1.D0 - 1.D0/(4.D0*PI)
      F(2)=CC*(DEXP(2.D0*Y(1))-EE) + EE*Y(2)/PI - 2.D0*EE*Y(1)
      DF(2,1) = 2.D0*CC*DEXP(2.D0*Y(1)) - 2.D0*EE
      DF(2,2) = EE/PI
C END OF PART FOR EXPERIMENTS 6-9.
C
C
C BEGINNING OF PART FOR EXPERIMENT 6-10.
      A( 1, 1)= 0.52D+02
      A( 1, 2)= 0.33D+02
      A( 1, 3)= 0.39D+02
      A( 1, 4)= 0.68D+02
      A( 1, 5)= 0.41D+02
      A( 2, 1)= 0.15D+02
      A( 2, 2)= 0.92D+02
      A( 2, 3)= 0.72D+02
      A( 2, 4)= 0.52D+02
```

```
            A( 2, 5)= 0.31D+02
            A( 3, 1)= 0.50D+02
            A( 3, 2)= 0.63D+02
            A( 3, 3)= 0.26D+02
            A( 3, 4)= 0.41D+02
            A( 3, 5)= 0.64D+02
            A( 4, 1)= 0.95D+02
            A( 4, 2)= 0.14D+02
            A( 4, 3)= 0.61D+02
            A( 4, 4)= 0.39D+02
            A( 4, 5)= 0.72D+02
            A( 5, 1)= 0.32D+02
            A( 5, 2)= 0.30D+01
            A( 5, 3)= 0.98D+02
            A( 5, 4)= 0.69D+02
            A( 5, 5)= 0.82D+02
            B( 1, 1)= 0.45D+02
            B( 1, 2)= 0.14D+02
            B( 1, 3)= 0.74D+02
            B( 1, 4)= 0.10D+02
            B( 1, 5)= 0.54D+02
            B( 2, 1)= 0.90D+01
            B( 2, 2)= 0.80D+01
            B( 2, 3)= 0.46D+02
            B( 2, 4)= 0.67D+02
            B( 2, 5)= 0.40D+02
            B( 3, 1)= 0.69D+02
            B( 3, 2)= 0.76D+02
            B( 3, 3)= 0.73D+02
            B( 3, 4)= 0.97D+02
            B( 3, 5)= 0.44D+02
            B( 4, 1)= 0.23D+02
            B( 4, 2)= 0.54D+02
            B( 4, 3)= 0.63D+02
            B( 4, 4)= 0.99D+02
            B( 4, 5)= 0.72D+02
            B( 5, 1)= 0.29D+02
            B( 5, 2)= 0.70D+02
            B( 5, 3)= 0.42D+02
            B( 5, 4)= 0.30D+02
            B( 5, 5)= 0.87D+02
            DO 10 I=1,M
               C(I)=I
      10    CONTINUE
            DO 20 I=1,M
               F(I) = - C(I)
               DO 20 J=1,M
                  F(I)=F(I) + A(I,J)*DSIN(Y(J)) - B(I,J)*DCOS(Y(J))
      20    CONTINUE
            DO 30 I=1,M
            DO 30 J=1,M
               DF(I,J) = A(I,J)*DCOS(Y(J)) + B(I,J)*DSIN(Y(J))
      30    CONTINUE
C END OF PART FOR EXPERIMENTS 6-10.
C
            RETURN
            END
```

```
            SUBROUTINE INPA2(IFLGCR,IFLGST,MAXNS,MAXIT,EPSBIG,SSZBEG,M)
C
C PURPOSE:  SAMPLE INPUT ROUTINE FOR THE REAL CONVEX-LINEAR CONTINUATION,
C              FOR USE WITH CONSOL9.  INITIALIZES FREQUENTLY CHANGED CONSTANTS.
C
C GENERIC NAME: INPA          FILE NAME: INPA2
C
C REFERENCE:  CHAPTER 6 OF "SOLVING POLYNOMIAL SYSTEMS USING
C              CONTINUATION" BY ALEXANDER MORGAN.
C
C NOTE: INPA MUST BE CUSTOMIZED FOR EACH EXPERIMENT.
C        IT IS CURRENTLY WRITTEN TO READ 12 COEFFICIENTS FROM UNIT 5.
C
C INPUT:  READ FROM UNIT 5.  SEE CODE BELOW FOR FORMATS.
C         TITLE      -- FIRST 60 CHARACTERS OF FIRST LINE OF INPUT
C         IFLGCR     -- =1, IF THE CORRECTOR ITERATIONS ARE TO BE WRITTEN
C                        TO UNIT 8.  (SEE SUBROUTINE CORC.)
C         IFLGST     -- =1, IF A STEP SUMMARY OUTPUT IS PRINTED AFTER
C                        EVERY STEP.
C         MAXNS      -- MAXIMUM NUMBER OF STEPS FOR A PATH
C         MAXIT      -- MAXIMUM NUMBER OF ITERATIONS TO CORRECT A STEP
C         EPSBIG     -- THE CORRECTOR RESIDUAL MUST BE LESS THAN EPS FOR
C                        A STEP TO "SUCCEED."  EPS=EPSBIG UNTIL T > .95
C         SSZBEG     -- BEGINNING STEP LENGTH
C         M          -- THE NUMBER OF EQUATIONS
C         A          -- COEFFICIENT ARRAY
C
C SAMPLE INPUT FOLLOWS:
C TITLE: TWO SECOND-DEGREE EQUATIONS, SOLUTIONS X=+-3,Y=+-4
C    1          IFLGCR
C    1          IFLGST
C   200         MAXNS
C    3          MAXIT
C                   1.D-02      EPSBIG
C                   1.D-02      SSZBEG
C    2      M
C                   1.D+00      Y(1)**2     *
C                   1.D+00      Y(2)**2     *
C                   0.D+00      Y(1)*Y(2)   *   EQUATION 1
C                   0.D+00      Y(1)        *
C                   0.D+00      Y(2)        *
C                 -25.D+00      CONSTANT    *
C                   1.D+00      Y(1)**2     $
C                   0.D+00      Y(2)**2     $
C                   0.D+00      Y(1)*Y(2)   $   EQUATION 2
C                   0.D+00      Y(1)        $
C                   0.D+00      Y(2)        $
C                  -9.D+00      CONSTANT    $
C
C OUTPUT:  IFLGCR,IFLGST,MAXNS,MAXIT,EPSBIG,SSZBEG,M
C
C COMMON:  /COEFS/  IS USED TO PASS COEFFICIENT VALUES FROM SUBROUTINE
C                   INPA TO SUBROUTINE FFNN.
C
C DECLARATIONS
      INTEGER IFLGCR,IFLGST,J,K,M,MAXIT,MAXNS,TITLE
      DOUBLE PRECISION A,EPSBIG,SSZBEG
```

```
C
C DIMENSIONS
      DIMENSION TITLE(60)
C
C COMMON
      COMMON/COEFS/A(2,6)
C
C INPUT STATEMENTS FOLLOW.
      READ(5,10) TITLE
 10   FORMAT(60A1)
      READ(5,20) IFLGCR
 20   FORMAT(I5)
      READ(5,20) IFLGST
      READ(5,20) MAXNS
      READ(5,20) MAXIT
      READ(5,30) EPSBIG
 30   FORMAT(D22.15)
      READ(5,30) SSZBEG
      READ(5,20) M
      DO 35 J=1,2
      DO 35 K=1,6
         READ(5,30) A(J,K)
 35   CONTINUE
C
      WRITE(6,40)
 40   FORMAT(' CONSOL9, FOR REAL CONVEX-LINEAR CONTINUATION',//)
      WRITE(6,50) TITLE
 50   FORMAT(' ',60A1)
      WRITE(6,60) IFLGCR
 60   FORMAT(/,' FLAG TO PRINT OUT CORRECTOR RESIDUALS =',I2)
      WRITE(6,70) IFLGST
 70   FORMAT(/,' FLAG TO PRINT OUT STEP SUMMARIES =',I2)
      WRITE(6,80) MAXNS
 80   FORMAT(/,' MAX NUM STEPS PER PATH =',I5)
      WRITE(6,90) MAXIT
 90   FORMAT(/,' MAX NUM ITERATIONS PER STEP =',I5)
      WRITE(6,100) EPSBIG
 100  FORMAT(/,' EPSBIG =',D22.15/)
      WRITE(6,110) SSZBEG
 110  FORMAT(/,' BEGINNING STEP SIZE = ',D22.15/)
      WRITE(6,120) M
 120  FORMAT(/,' NUMBER OF EQUATIONS = ',I5/)
      WRITE(6,130)
 130  FORMAT(/,' COEFFICIENTS :')
      DO 140 J=1,2
         WRITE(6,150) J
 150     FORMAT(' EQUATION ',I5)
         DO 160 K=1,6
            WRITE(6,170) J,K,A(J,K)
 170        FORMAT('  A(',I5,',',I5,') =',D22.15)
 160     CONTINUE
         WRITE(6,999)
 999     FORMAT(/)
 140  CONTINUE
C
      RETURN
      END
```

```
      SUBROUTINE INPB2(M,SSZBEG,EPSBIG,SSZMIN,EPSSML,EPSO)
C
C PURPOSE:  SAMPLE INPUT ROUTINE FOR THE REAL CONVEX-LINEAR
C           CONTINUATION, FOR USE WITH CONSOL9.  INITIALIZES
C           LESS FREQUENTLY CHANGED CONSTANTS.
C
C GENERIC NAME: INPB             FILE NAME: INPB2
C
C
C NOTE: WHILE INPB NEED NOT BE CUSTOMIZED FOR EACH EXPERIMENT,
C       THE CHOICE OF THE PARAMETERS P AND Q CAN EFFECT THE
C       CONVERGENCE OF THE METHOD.  NOTE THAT IN THIS VERSION OF
C       THE SUBROUTINE, I HAVE CHOSEN P TO HAVE OFF-DIAGONAL ELEMENTS
C       EQUAL TO ZERO.  ALSO NOTE THAT THE DETERMINANT OF P MUST
C       EQUAL THE DETERMINANT OF THE JACOBIAN OF F AT ANY SOLUTION
C       TO BE FOUND.  THUS, THE SIGNS OF THE ELEMENTS OF P ARE IMPORTANT.
C       CODE IS IN PLACE, BUT COMMENTED OUT, TO READ Q AND THE DIAGONAL OF
C       P FROM UNIT 5.
C
C INPUT:  M       -- NUMBER OF EQUATIONS
C         SSZBEG  -- BEGINNING STEPSIZE
C         EPSBIG  -- ZERO-EPSILON FOR PATH TRACKING RESIDUAL,
C                    UNTIL T > .95
C
C OUTPUT: SSZMIN  -- MINIMUM STEPSIZE ALLOWED
C         EPSSML  -- ZERO-EPSILON FOR PATH TRACKING RESIDUAL,
C                    AFTER T > .95
C         EPSO    -- ZERO-EPSILON FOR SINGULARITY
C
C
C COMMON:  /RANPAR/  IS USED TO PASS RANDOM PARAMETERS FROM SUBROUTINE
C                    INPB TO SUBROUTINES STPT AND HFNN.
C
C DECLARATIONS
      INTEGER I,J,M
      DOUBLE PRECISION EPSO,EPSBIG,EPSSML,P,Q,SSZBEG,SSZMIN
C
C COMMON
      COMMON/RANPAR/ P(20,20),Q(20)
C
C
      SSZMIN = 1.D-5*SSZBEG
      EPSSML = 1.D-3*EPSBIG
      EPSO   = 1.D-22
C
      P( 1, 1)= .12324754231D0
      P( 2, 2)= .76253746298D0
      P( 3, 3)= .93857838950D0
      P( 4, 4)= .99375892810D0
      P( 5, 5)= .23467908356D0
      P( 6, 6)= .39383930009D0
      P( 7, 7)= .83542556622D0
      P( 8, 8)= .10192888288D0
      P( 9, 9)= .55763522521D0
      P(10,10)= .83729899911D0
      P(11,11)= .78348738738D0
      P(12,12)= .10578234903D0
```

```
      P(13,13)= .03938347346D0
      P(14,14)= .04825184716D0
      P(15,15)= .43428734331D0
      P(16,16)= .93836289418D0
      P(17,17)= .99383729993D0
      P(18,18)= .40947822291D0
      P(19,19)= .09383736736D0
      P(20,20)= .26459172298D0
C
      Q( 1)= .58720452864D0
      Q( 2)= .01321964722D0
      Q( 3)= .97884134700D0
      Q( 4)=-.14433009712D0
      Q( 5)= .39383737289D0
      Q( 6)= .41543223411D0
      Q( 7)=-.03938376373D0
      Q( 8)=-.61253112318D0
      Q( 9)= .39383737388D0
      Q(10)=-.26454678861D0
      Q(11)=-.00938376766D0
      Q(12)= .34447867861D0
      Q(13)=-.04837366632D0
      Q(14)= .48252736790D0
      Q(15)= .93725237347D0
      Q(16)=-.54356527623D0
      Q(17)= .39373957747D0
      Q(18)= .65573434564D0
      Q(19)=-.39380038371D0
      Q(20)= .98903450052D0
C
C
      DO 4 I=1,M
      DO 4 J=1,M
        IF(I.NE.J)P(I,J)=0.D0
    4   CONTINUE
C
C
CC      READ(5,8)(P(J,J),J=1,M)
CC  8   FORMAT(D22.15)
CC      READ(5,8)(Q(J),J=1,M)
C
C
      WRITE(6,10) SSZMIN
   10 FORMAT(/' MINIMUM STPSZE = ',D22.15)
      WRITE(6,20) EPSSML
   20 FORMAT(/' EPSSML =',D22.15)
      WRITE(6,30) EPSO
   30 FORMAT(/' EPSO   =',D22.15)
      WRITE(6,40) ((J,K,P(J,K),K=1,M),J=1,M)
   40 FORMAT(/' P(',I2,',',I2,') =',D22.15)
      WRITE(6,50) (J,Q(J),J=1,M)
   50 FORMAT(/' Q(',I2,') =',D22.15)
      WRITE(6,999)
  999 FORMAT(/)
C
      RETURN
      END
```

```
      SUBROUTINE STPT2(M,Y)
C
C PURPOSE:  GENERATE A START POINT FOR THE PATH FOR THE REAL
C           CONVEX-LINEAR CONTINUATION,FOR USE WITH CONSOL9.
C
C GENERIC NAME: STPT          FILE NAME: STPT2
C
C INPUT:  M        -- NUMBER OF EQUATIONS
C
C OUTPUT: Y        -- START POINT FOR PATH
C
C COMMON:  /RANPAR/  IS USED TO PASS RANDOM PARAMETERS FROM SUBROUTINE
C                    INPB TO SUBROUTINES STPT AND HFNN.
C
C SUBROUTINES: LNFN,LNSN
C
C
C DECLARATIONS
      INTEGER IROW,J,M,MSG
      DOUBLE PRECISION P,PCOPY,Q,Y
C
C DIMENSIONS
      DIMENSION IROW(20),PCOPY(20,20),Y(1)
C
C COMMON
      COMMON/RANPAR/P(20,20),Q(20)
C
C
      DO 10 I=1,M
      DO 10 J=1,M
        PCOPY(I,J)=P(I,J)
  10  CONTINUE
      CALL LNFN(20,M,0.D0,PCOPY, MSG,IROW)
      CALL LNSN(20,M,IROW,PCOPY,Q, Y)
C
C
      WRITE(6,20)
  20  FORMAT(/,' START POINT:')
      WRITE(6,30) ( J,Y(J),J=1,M )
  30  FORMAT('    Y(',I2,') = ',D22.15)
C
      RETURN
      END
```

```
      SUBROUTINE HFNN2(M,T,Y, H,DH)
C
C PURPOSE: EVALUATES THE CONTINUATION EQUATION,H, AND ITS PARTIAL
C          DERIVATIVES, DH, FOR THE REAL CONVEX-LINEAR CONTINUATION,
C          FOR USE WITH CONSOL9.
C
C GENERIC NAME: HFNN              FILE NAME: HFNN2
C
C NOTE: THE CONVEX-LINEAR CONTINUATION IS DEFINED BY
C     H(Y,T) = (1-T)*(P*Y-Q) + T*F(Y).
C
C INPUT:  M          -- NUMBER OF EQUATIONS
C         T          -- CURRENT VALUE OF T
C         Y          -- CURRENT VALUE OF Y
C
C OUTPUT: H          -- H IS THE CONTINUATION SYSTEM, A SYSTEM OF M
C                       REAL POLYNOMIALS IN THE M+1 UNKNOWNS Y AND T.
C         DH         -- DH IS THE M X (M+1) MATRIX OF PARTIAL
C                       DERIVATIVES OF H WITH RESPECT TO Y AND T.
C                       DH(J,K) IS FOR J=1 TO M AND K=1 TO M IS
C                       THE PARTIAL OF H(J) WITH RESPECT TO Y(K).
C                       DH(J,M+1) FOR J=1 TO M IS THE PARTIAL DERIVATIVE
C                       OF H(J) WITH RESPECT TO T.
C
C COMMON:  /RANPAR/  IS USED TO PASS RANDOM PARAMETERS FROM SUBROUTINE
C                    INPB TO SUBROUTINES STPT AND HFNN.
C
C SUBROUTINES: FFNN
C
C DECLARATIONS
      INTEGER J,K,M,MP1
      DOUBLE PRECISION DF,DH,F,H,ONEMT,P,PL,Q,T,Y
C
C DIMENSIONS
      DIMENSION DF(20,20),DH(20,1),DHT(20),DHY(20,20),F(20),H(1),
     & PL(20),Y(1)
C
C COMMON
      COMMON/RANPAR/P(20,20),Q(20)
C
      CALL FFNN(M,Y, F,DF)
C
      MP1=M+1
      ONEMT=1.D0-T
      DO 10 J=1,M
          PL(J)=0.D0
          DO 20 K=1,M
              DH(J,K)=ONEMT*P(J,K) + T*DF(J,K)
              PL(J)=PL(J) + P(J,K)*Y(K)
  20      CONTINUE
          DH(J,MP1) =       - ( PL(J) - Q(J) ) +      F(J)
          H(J)      = ONEMT*( PL(J) - Q(J) ) + T*F(J)
  10  CONTINUE
C
      RETURN
      END
```

```
      SUBROUTINE FFNNQ2(M,Y, F,DF)
C
C PURPOSE:  SAMPLE SYSTEM-EVALUATION ROUTINE FOR THE REAL
C           CONVEX-LINEAR CONTINUATION, FOR USE WITH CONSOL9.
C           SUBROUTINE FFNN RETURNS F AND DF VALUES.
C
C GENERIC NAME: FFNN          FILE NAME: FFNNQ2
C
C NOTE: THIS SUBROUTINE IS SET UP FOR THE CASE THAT F IS TWO SECOND
C       DEGREE POLYNOMIALS IN TWO UNKNOWNS. WITH REAL COEFFICIENTS
C       AND REAL Y:
C
C         F(J) = A(J,1)*Y(1)**2 +A(J,2)*Y(2)**2 +A(J,3)*Y(1)*Y(2)
C
C                                  +A(J,4)*Y(1) +A(J,5)*Y(2) +A(J,6)
C
C         FOR  J=1 TO 2.
C
C
C
C INPUT:  M        -- NUMBER OF EQUATIONS
C         Y        -- CURRENT VALUE OF Y
C
C OUTPUT: F        -- F(J) IS THE VALUE OF THE J-TH EQUATION AT Y
C         DF       -- DF(J,K) IS THE PARTIAL DERIVATIVE OF F(J) WITH
C                     RESPECT TO Y(K)
C
C COMMON: /COEFS/  IS USED TO PASS COEFFICIENT VALUES FROM SUBROUTINE
C                  INPA  TO SUBROUTINE  FFNN.
C
C
C DECLARATIONS
      INTEGER J,M
      DOUBLE PRECISION A,DF,F,Y,Y12,YSQ
C
C DIMENSIONS
      DIMENSION DF(20,1),F(1),Y(1),YSQ(2)
C
C COMMON
      COMMON/COEFS/A(2,6)
C
C
      YSQ(1)=Y(1)**2
      YSQ(2)=Y(2)**2
      Y12=Y(1)*Y(2)
C
      DO 10 J=1,2
         F(J)=A(J,1)*YSQ(1)+A(J,2)*YSQ(2) +A(J,3)*Y12
     &       + A(J,4)*Y(1) +A(J,5)*Y(2) +A(J,6)
         DF(J,1)=2*A(J,1)*Y(1)+A(J,3)*Y(2)+A(J,4)
         DF(J,2)=2*A(J,2)*Y(2)+A(J,3)*Y(1)+A(J,5)
   10 CONTINUE
C
      RETURN
      END
```

```
      SUBROUTINE FFNNB2(M,Y, F,DF)
C
C PURPOSE:   SAMPLE SYSTEM-EVALUATION ROUTINE FOR THE REAL
C            CONVEX-LINEAR CONTINUATION, FOR USE WITH CONSOL9.
C            SUBROUTINE FFNN RETURNS F AND DF VALUES.
C
C GENERIC NAME: FFNN              FILE NAME: FFNNB2
C
C NOTE: THIS SUBROUTINE IS SET UP FOR BROWN'S ALMOST LINEAR SYSTEM.
C       (SEE BELOW, AND EXPERIMENT 6-17.)
C
C INPUT:  M        -- NUMBER OF EQUATIONS
C         Y        -- CURRENT VALUE OF Y
C
C OUTPUT: F        -- F(J) IS THE VALUE OF THE J-TH EQUATION AT Y
C         DF       -- DF(J,K) IS THE PARTIAL DERIVATIVE OF F(J) WITH
C                     RESPECT TO Y(K)
C
C DECLARATIONS
      INTEGER J,M,MM1,MP1
      DOUBLE PRECISION A,DF,F,Y,YY,YP
C
C DIMENSIONS
      DIMENSION DF(20,1),F(1),Y(1)
C
      MP1=M+1
      MM1=M-1
C
      YY=0.D0
      YP=1.D0
      DO 10 J=1,M
         YY=YY+Y(J)
         YP=YP*Y(J)
  10  CONTINUE
C
      DO 20 J=1,MM1
         F(J)=Y(J)+YY-MP1
  20  CONTINUE
      F(M)=YP-1.D0
C
      DO 30 I=1,MM1
      DO 30 J=1,M
        DF(I,J)=1.D0
        IF(I.EQ.J)DF(I,J)=2.D0
  30  CONTINUE
C
      DO 40 J=1,M
      DF(M,J)=1.D0
      DO 40 K=1,M
        IF(J.NE.K)DF(M,J)=DF(M,J)*Y(K)
  40  CONTINUE
C
      RETURN
      END
```

```
      SUBROUTINE INPA3(IFLGCR,IFLGST,MAXNS,MAXIT,EPSBIG,SSZBEG,M)
C
C PURPOSE:  SAMPLE INPUT ROUTINE FOR THE COMPLEX CONVEX-LINEAR CONTINUATION,
C           FOR USE WITH CONSOL9.  INITIALIZES FREQUENTLY CHANGED CONSTANTS.
C
C GENERIC NAME: INPA          FILE NAME: INPA3
C
C REFERENCE:  CHAPTER 6 OF "SOLVING POLYNOMIAL SYSTEMS USING
C             CONTINUATION" BY ALEXANDER MORGAN.
C
C NOTE: INPA MUST BE CUSTOMIZED FOR EACH EXPERIMENT.
C       IT IS CURRENTLY WRITTEN TO READ 12 COEFFICIENTS FROM UNIT 5.
C
C INPUT:  READ FROM UNIT 5.  SEE CODE BELOW FOR FORMATS.
C     TITLE        -- FIRST 60 CHARACTERS OF FIRST LINE OF INPUT
C     IFLGCR       -- =1, IF THE CORRECTOR ITERATIONS ARE TO BE WRITTEN
C                        TO UNIT 8.  (SEE SUBROUTINE CORC.)
C     IFLGST       -- =1, IF A STEP SUMMARY OUTPUT IS PRINTED AFTER
C                        EVERY STEP.
C     MAXNS        -- MAXIMUM NUMBER OF STEPS FOR A PATH
C     MAXIT        -- MAXIMUM NUMBER OF ITERATIONS TO CORRECT A STEP
C     EPSBIG       -- THE CORRECTOR RESIDUAL MUST BE LESS THAN EPS FOR
C                        A STEP TO "SUCCEED." EPS=EPSBIG UNTIL T > .95
C     SSZBEG       -- BEGINNING STEP LENGTH
C     M            -- TWICE THE NUMBER OF EQUATIONS
C     A            -- COEFFICIENT ARRAY
C
C SAMPLE INPUT FOLLOWS:
C TITLE: TWO SECOND-DEGREE EQUATIONS, SOLUTIONS X=+-3,Y=+-4
C   1        IFLGCR
C   1        IFLGST
C 200        MAXNS
C   3        MAXIT
C            1.D-02       EPSBIG
C            1.D-02       SSZBEG
C   4    M
C            1.D+00       Y(1)**2       *
C            1.D+00       Y(2)**2       *
C            0.D+00       Y(1)*Y(2)     *   EQUATION 1
C            0.D+00       Y(1)          *
C            0.D+00       Y(2)          *
C          -25.D+00       CONSTANT      *
C            1.D+00       Y(1)**2       $
C            0.D+00       Y(2)**2       $
C            0.D+00       Y(1)*Y(2)     $   EQUATION 2
C            0.D+00       Y(1)          $
C            0.D+00       Y(2)          $
C           -9.D+00       CONSTANT      $
C
C OUTPUT:  IFLGCR,IFLGST,MAXNS,MAXIT,EPSBIG,SSZBEG,M
C
C COMMON:  /COEFS/  IS USED TO PASS COEFFICIENT VALUES FROM SUBROUTINE
C                   INPA  TO SUBROUTINE  FFNN.
C
C DECLARATIONS
      INTEGER IFLGCR,IFLGST,J,K,M,MAXIT,MAXNS,TITLE
      DOUBLE PRECISION A,EPSBIG,SSZBEG
```

```
C
C DIMENSIONS
      DIMENSION TITLE(60)
C
C COMMON
      COMMON/COEFS/A(2,6)
C
C INPUT STATEMENTS FOLLOW.
      READ(5,10) TITLE
  10  FORMAT(60A1)
      READ(5,20) IFLGCR
  20  FORMAT(I5)
      READ(5,20) IFLGST
      READ(5,20) MAXNS
      READ(5,20) MAXIT
      READ(5,30) EPSBIG
  30  FORMAT(D22.15)
      READ(5,30) SSZBEG
      READ(5,20) M
      DO 35 J=1,2
      DO 35 K=1,6
          READ(5,30) A(J,K)
  35  CONTINUE
C
      WRITE(6,40)
  40  FORMAT(' CONSOL9, FOR COMPLEX CONVEX-LINEAR CONTINUATION',//)
      WRITE(6,50) TITLE
  50  FORMAT(' ',60A1)
      WRITE(6,60) IFLGCR
  60  FORMAT(/,' FLAG TO PRINT OUT CORRECTOR RESIDUALS =',I2)
      WRITE(6,70) IFLGST
  70  FORMAT(/,' FLAG TO PRINT OUT STEP SUMMARIES =',I2)
      WRITE(6,80) MAXNS
  80  FORMAT(/,' MAX NUM STEPS PER PATH =',I5)
      WRITE(6,90) MAXIT
  90  FORMAT(/,' MAX NUM ITERATIONS PER STEP =',I5)
      WRITE(6,100) EPSBIG
 100  FORMAT(/,' EPSBIG =',D22.15/)
      WRITE(6,110) SSZBEG
 110  FORMAT(/,' BEGINNING STEP SIZE = ',D22.15/)
      WRITE(6,120) M
 120  FORMAT(/,' NUMBER OF EQUATIONS = ',I5/)
      WRITE(6,130)
 130  FORMAT(/,' COEFFICIENTS :')
      DO 140 J=1,2
          WRITE(6,150) J
 150      FORMAT(' EQUATION ',I5)
          DO 160 K=1,6
              WRITE(6,170) J,K,A(J,K)
 170          FORMAT('    A(',I5,',',I5,') =',D22.15)
 160      CONTINUE
          WRITE(6,999)
 999      FORMAT(/)
 140  CONTINUE
C
      RETURN
      END
```

```
      SUBROUTINE INPB3(M,SSZBEG,EPSBIG,SSZMIN,EPSSML,EPSO)
C
C PURPOSE:  SAMPLE INPUT ROUTINE FOR THE COMPLEX CONVEX-LINEAR
C           CONTINUATION, FOR USE WITH CONSOL9.  INITIALIZES
C           LESS FREQUENTLY CHANGED CONSTANTS.
C
C GENERIC NAME: INPB          FILE NAME: INPB3
C
C NOTE: WHILE INPB NEED NOT BE CUSTOMIZED FOR EACH EXPERIMENT,
C       THE CHOICE OF THE PARAMETERS P AND QQ CAN EFFECT THE
C       CONVERGENCE OF THE METHOD.  NOTE THAT IN THIS VERSION OF
C       THE SUBROUTINE, I HAVE CHOSEN P TO HAVE OFF-DIAGONAL ELEMENTS
C       EQUAL TO ZERO.  CODE IS IN PLACE, BUT COMMENTED OUT, TO READ
C       QQ AND THE DIAGONAL OF P FROM UNIT 5.
C
C INPUT:  M        -- TWICE THE NUMBER OF EQUATIONS
C         SSZBEG   -- BEGINNING STEPSIZE
C         EPSBIG   -- ZERO-EPSILON FOR PATH TRACKING RESIDUAL, UNTIL T > .95
C
C OUTPUT: SSZMIN   -- MINIMUM STEPSIZE ALLOWED
C         EPSSML   -- ZERO-EPSILON FOR PATH TRACKING RESIDUAL, AFTER T > .95
C         EPSO     -- ZERO-EPSILON FOR SINGULARITY
C
C COMMON:  /RANPAR/  IS USED TO PASS RANDOM PARAMETERS FROM SUBROUTINE
C                    INPB TO SUBROUTINES STPT AND HFNN.
C
C SUBROUTINES: DCMPLX
C
C DECLARATIONS
      INTEGER I,J,K,M,MD2
      DOUBLE PRECISION EPSO,EPSBIG,EPSSML,SSZBEG,SSZMIN
      COMPLEX*16 P,Q,ZERO
C
C COMMON
      COMMON/RANPAR/ P(10,10),Q(10)
C
      MD2=M/2
      ZERO=DCMPLX(0.D0,0.D0)
      SSZMIN = 1.D-5*SSZBEG
      EPSSML = 1.D-3*EPSBIG
      EPSO   = 1.D-22
C
      P( 1, 1)= DCMPLX( .12324754231D0, .76253746298D0)
      P( 2, 2)= DCMPLX( .93857838950D0, .99375892810D0)
      P( 3, 3)= DCMPLX( .23467908356D0, .39383930009D0)
      P( 4, 4)= DCMPLX( .83542556622D0, .10192888288D0)
      P( 5, 5)= DCMPLX( .55763522521D0, .83729899911D0)
      P( 6, 6)= DCMPLX( .78348738738D0, .10578234903D0)
      P( 7, 7)= DCMPLX( .03938347346D0, .04825184716D0)
      P( 8, 8)= DCMPLX( .43428734331D0, .93836289418D0)
      P( 9, 9)= DCMPLX( .99383729993D0, .40947822291D0)
      P(10,10)= DCMPLX( .09383736736D0, .26459172298D0)
C
      Q( 1)= DCMPLX( .58720452864D0, .01321964722D0)
      Q( 2)= DCMPLX( .97884134700D0,-.14433009712D0)
      Q( 3)= DCMPLX( .39383737289D0, .41543223411D0)
      Q( 4)= DCMPLX(-.03938376373D0,-.61253112318D0)
```

```
              Q( 5)= DCMPLX( .39383737388D0,-.26454678861D0)
              Q( 6)= DCMPLX(-.00938376766D0, .34447867861D0)
              Q( 7)= DCMPLX(-.04837366632D0, .48252736790D0)
              Q( 8)= DCMPLX( .93725237347D0,-.54356527623D0)
              Q( 9)= DCMPLX( .39373957747D0, .65573434564D0)
              Q(10)= DCMPLX(-.39380038371D0, .98903450052D0)
       C
              DO 4 J=1,MD2
              DO 4 K=1,MD2
                IF(J.NE.K)P(J,K)=ZERO
         4    CONTINUE
       C
       CC     READ(5,8)(P(J,J),J=1,MD2)
       CC  8  FORMAT(D22.15)
       CC     READ(5,8)(Q(J),J=1,MD2)
       C
              WRITE(6,10) SSZMIN
        10    FORMAT(/' MINIMUM STPSZE = ',D22.15)
              WRITE(6,20) EPSSML
        20    FORMAT(/' EPSSML =',D22.15)
              WRITE(6,30) EPSO
        30    FORMAT(/' EPSO   =',D22.15)
              WRITE(6,40)  ((J,K,P(J,K),K=1,MD2),J=1,MD2)
        40    FORMAT(/' P(',I2,',',I2,')  =',2D22.15)
              WRITE(6,50)  (J,Q(J),J=1,MD2)
        50    FORMAT(/' Q(',I2,')  =',2D22.15)
              WRITE(6,999)
       999    FORMAT(/)
              RETURN
              END
```

```
      SUBROUTINE STPT3(M,Y)
C
C PURPOSE:  GENERATE A START POINT FOR THE PATH FOR THE COMPLEX
C           CONVEX-LINEAR CONTINUATION,FOR USE WITH CONSOL9.
C
C GENERIC NAME: STPT          FILE NAME: STPT3
C
C INPUT: M          -- TWICE THE NUMBER OF EQUATIONS
C
C OUTPUT: Y          -- START POINT FOR PATH
C
C COMMON:  /RANPAR/  IS USED TO PASS RANDOM PARAMETERS FROM SUBROUTINE
C                    INPB TO SUBROUTINES STPT AND HFNN.
C
C SUBROUTINES: LNFNC,LNSNC
C
C
C DECLARATIONS
      INTEGER I,IROW,J,M,MD2,MSG
      DOUBLE PRECISION Y
      COMPLEX*16 P,PCOPY,Q
C
C DIMENSIONS
      DIMENSION IROW(10),PCOPY(10,10),Y(1)
C
C COMMON
      COMMON/RANPAR/P(10,10),Q(10)
C
C
      MD2=M/2
C
      DO 10 J=1,MD2
      DO 10 K=1,MD2
        PCOPY(J,K)=P(J,K)
  10  CONTINUE
      CALL LNFNC(10,MD2,0.D0,PCOPY, MSG,IROW)
      CALL LNSNC(10,MD2,IROW,PCOPY,Q, Y)
C
C
      WRITE(6,20)
  20  FORMAT(/,'  START POINT:')
      WRITE(6,30) ( J,Y(J),J=1,M )
  30  FORMAT('   Y(',I2,') = ',D22.15)
C
      RETURN
      END
```

```
      SUBROUTINE HFNN3(M,T,Y, H,DH)
C
C PURPOSE: EVALUATES THE CONTINUATION EQUATION,H, AND ITS PARTIAL
C          DERIVATIVES, DH, FOR THE COMPLEX CONVEX-LINEAR CONTINUATION,
C          FOR USE WITH CONSOL9.
C
C GENERIC NAME: HFNN            FILE NAME: HFNN3
C
C NOTE:   THIS SUBROUTINE EVALUATES THE COMPLEX CONTINUATION AND
C         THEN "REALIFIES" IT FOR USE BY CONSOL9.  THUS, FIRST
C         IT EVALUATES THE COMPLEX CONVEX-LINEAR CONTINUATION
C         DEFINED BY CH(X,T) =  (1-T)*(P*X - Q) + T*F(X), WHERE
C         X IS A COMPLEX VECTOR OF LENGTH M/2 IDENTIFIED WITH Y.
C         THEN IT "REALIFIES" CH AND DCH, TO OBTAIN H AND DH,
C         RESPECTIVELY.
C
C
C INPUT:  M          -- TWICE THE NUMBER OF EQUATIONS
C         T          -- CURRENT VALUE OF T
C         Y          -- CURRENT VALUE OF Y
C
C OUTPUT: H          -- H IS THE REALIFICATION OF THE COMPLEX CONTINUATION, CH.
C                       THUS, H IS A SYSTEM OF M REAL POLYNOMIALS IN THE M+1
C                       UNKNOWNS Y AND T.
C         DH         -- DH IS AN M X (M+1) MATRIX, THE REALIFICATION OF THE
C                       JACOBIAN, DCH, OF CH.  DCH(J,K) IS THE PARTIAL OF
C                       CH(J) WITH RESPECT TO X(K), FOR J=1 TO M/2 AND K=1
C                       TO M/2.  DCH(J,M+1) IS THE PARTIAL DERIVATIVE
C                       OF CH(J) WITH RESPECT TO T, FOR J=1 TO M/2.
C
C
C COMMON:   /RANPAR/  IS USED TO PASS RANDOM PARAMETERS FROM SUBROUTINE
C                     INPB TO SUBROUTINES STPT AND HFNN.
C
C
C SUBROUTINES:  DCMPLX,DIMAG,DREAL,FFNN
C
C
C DECLARATIONS
      INTEGER J,J2,J2M1,K,K2,K2M1,M,MD2,MP1
      DOUBLE PRECISION DH,H,ONEMT,T,Y
      COMPLEX*16 CH,DCH,DF,F,ONE,P,PL,Q,X,ZERO
C
C DIMENSIONS
      DIMENSION CH(10),DCH(10,11),DF(10,10),DH(20,1),F(10),H(1),
     & PL(10),X(10),Y(1)
C
C COMMON
      COMMON/RANPAR/P(10,10),Q(10)
C
      MD2 =M/2
      ONE =DCMPLX(1.D0,0.D0)
      ZERO=DCMPLX(0.D0,0.D0)
C
      MP1=M+1
      ONEMT=1.D0-T
C
```

426

```
      DO 5 J=1,MD2
         J2=2*J
         J2M1=J2-1
         X(J)=DCMPLX(Y(J2M1),Y(J2))
   5  CONTINUE
C

      CALL FFNN(MD2,X, F,DF)
C

      DO 10 J=1,MD2
         PL(J)=ZERO
         DO 20 K=1,MD2
            DCH(J,K)=ONEMT*P(J,K) + T*DF(J,K)
            PL(J)=PL(J) + P(J,K)*X(K)
  20     CONTINUE
         DCH(J,MP1) =   - ( PL(J) - Q(J) ) +   F(J)
         CH(J)     = ONEMT*( PL(J) - Q(J) ) + T*F(J)
  10  CONTINUE
C
C

      DO 30 J=1,MD2
         J2=2*J
         J2M1=J2-1
         DO 40 K=1,MD2
            K2=2*K
            K2M1=K2-1
            DH(J2M1,K2M1)= DREAL(DCH(J,K))
            DH(J2 ,K2 )= DH(J2M1,K2M1)
            DH(J2 ,K2M1)= DIMAG(DCH(J,K))
            DH(J2M1,K2 )=-DH(J2 ,K2M1)
  40     CONTINUE
C

         DH(J2M1,MP1)= DREAL(DCH(J,MP1))
         DH(J2 ,MP1)= DIMAG(DCH(J,MP1))
         H(J2M1)    = DREAL( CH (J))
         H(J2 )     = DIMAG( CH (J))
  30 CONTINUE
C

      RETURN
      END
```

```
      SUBROUTINE FFNNQ3(N,X, F,DF)
C
C
C PURPOSE:   SAMPLE SYSTEM-EVALUATION ROUTINE FOR THE COMPLEX
C            CONVEX-LINEAR CONTINUATION, FOR USE WITH CONSOL9.
C            SUBROUTINE FFNN RETURNS F AND DF VALUES.
C
C GENERIC NAME: FFNN            FILE NAME: FFNNQ3
C
C NOTE: THIS SUBROUTINE IS SET UP FOR THE CASE THAT F IS TWO SECOND
C       DEGREE POLYNOMIALS IN TWO UNKNOWNS WITH REAL COEFFICIENTS
C       AND COMPLEX X:
C
C          F(J) = A(J,1)*X(1)**2 +A(J,2)*X(2)**2 +A(J,3)*X(1)*X(2)
C
C                                  +A(J,4)*X(1) +A(J,5)*X(2) +A(J,6)
C
C       FOR  J=1 TO 2.
C
C
C
C INPUT:  N          -- THE NUMBER OF EQUATIONS
C         X          -- CURRENT VALUE OF X (IDENTIFIED WITH CONSOL9'S Y VIA
C                       DREAL(X(J))=Y(2*J-1), DIMAG(X(J))=Y(2*J).)
C
C OUTPUT: F          -- F(J) IS THE VALUE OF THE J-TH EQUATION AT X
C         DF         -- DF(J,K) IS THE PARTIAL DERIVATIVE OF F(J) WITH
C                       RESPECT TO X(K)
C
C COMMON:  /COEFS/  IS USED TO PASS COEFFICIENT VALUES FROM SUBROUTINE
C                    INPA  TO SUBROUTINE  FFNN.
C
C
C DECLARATIONS
      INTEGER J,N
      DOUBLE PRECISION A
      COMPLEX*16 DF,F,X,X12,XSQ
C
C DIMENSIONS
      DIMENSION DF(10,1),F(1),X(1),XSQ(2)
C
C COMMON
      COMMON/COEFS/A(2,6)
C
C
      XSQ(1)=X(1)**2
      XSQ(2)=X(2)**2
      X12=X(1)*X(2)
C
      DO 10 J=1,N
         F(J)=A(J,1)*XSQ(1)+A(J,2)*XSQ(2) +A(J,3)*X12
     &      + A(J,4)*X(1) +A(J,5)*X(2) +A(J,6)
         DF(J,1)=2*A(J,1)*X(1)+A(J,3)*X(2)+A(J,4)
         DF(J,2)=2*A(J,2)*X(2)+A(J,3)*X(1)+A(J,5)
  10  CONTINUE
C
      RETURN
      END
```

```fortran
      SUBROUTINE FFNNB3(N,X, F,DF)
C
C PURPOSE:   SAMPLE SYSTEM-EVALUATION ROUTINE FOR THE COMPLEX
C            CONVEX-LINEAR CONTINUATION, FOR USE WITH CONSOL9.
C            SUBROUTINE FFNN RETURNS F AND DF VALUES.
C
C GENERIC NAME: FFNN          FILE NAME: FFNNB3
C
C NOTE: THIS SUBROUTINE IS SET UP FOR FOR BROWN'S ALMOST LINEAR SYSTEM.
C       (SEE BELOW, AND EXPERIMENT 6-17.)
C
C INPUT:  N        -- THE NUMBER OF EQUATIONS
C         X        -- CURRENT VALUE OF X (IDENTIFIED WITH CONSOL9'S Y VIA
C                     DREAL(X(J))=Y(2*J-1), DIMAG(X(J))=Y(2*J).)
C
C OUTPUT: F        -- F(J) IS THE VALUE OF THE J-TH EQUATION AT X
C         DF       -- DF(J,K) IS THE PARTIAL DERIVATIVE OF F(J) WITH
C                     RESPECT TO X(K)
C DECLARATIONS
      INTEGER J,K,N,NM1,NP1
      COMPLEX*16 DF,F,ONE,TWO,X,XX,XP,ZERO
C
C DIMENSIONS
      DIMENSION DF(10,1),F(1),X(1)
C
      NP1=N+1
      NM1=N-1
      ZERO=DCMPLX(0.D0,0.D0)
      ONE =DCMPLX(1.D0,0.D0)
      TWO =DCMPLX(2.D0,0.D0)
C
      XX=ZERO
      XP=ONE
      DO 10 J=1,N
         XX=XX+X(J)
         XP=XP*X(J)
   10 CONTINUE
C
      DO 20 J=1,NM1
         F(J)=X(J)+XX-NP1
   20 CONTINUE
      F(N)=XP-ONE
C
      DO 30 J=1,NM1
      DO 30 K=1,N
         DF(J,K)=ONE
         IF(J.EQ.K)DF(J,K)=TWO
   30 CONTINUE
C
      DO 40 J=1,N
      DF(N,J)=ONE
      DO 40 K=1,N
         IF(J.NE.K)DF(N,J)=DF(N,J)*X(K)
   40 CONTINUE
C
      RETURN
      END
```

Section 6: Utility Programs for Part A

```
      SUBROUTINE LNFN(LDIM,M,EPSO,A, MSG,IROW)
C
C PURPOSE:   GENERATES AN LU FACTORIZATION OF REAL M X M MATRIX A.
C            THUS, LOWER AND UPPER TRIANGULAR MATRICES L AND U ARE
C            GENERATED SUCH THAT A = L*U, WITH U(I,I)=1 FOR I=1 TO M.
C
C REFERENCE: CHAPTER 4 OF "SOLVING POLYNOMIAL SYSTEMS USING
C            CONTINUATION" BY ALEXANDER MORGAN.  BASED ON THE
C            DESCRIPTION OF CROUT'S METHOD IN [DAHLQUIST, ANDERSON]
C            "NUMERICAL METHODS" (PRENTICE HALL, 1974) PP. 157-8,
C            IN [FORSYTHE] "ALGORITHM 16: CROUT WITH PIVOTING,"
C            COMM. ACM, VOL. 3 (1960), PP. 507-8, AND IN
C            [FORSYTHE, MOLER] "COMPUTER SOLUTION OF LINEAR ALGEBRAIC
C            SYSTEMS," (PRENTICE-HALL, 1967) CHAPTERS 10 AND 12.
C
C NOTE: THIS PROGRAM IS DESIGNED TO BE FOLLOWED BY LNSN TO SOLVE
C       THE LINEAR SYSTEM A*Y = B.
C
C INPUT:     LDIM          -- LEADING DIMENSION OF ARRAY A
C            M             -- DIMENSION OF CURRENT PROBLEM
C            EPSO          -- ZERO-EPSILON TO DETERMINE SINGULARITY
C            A             -- MATRIX DIMENSIONED A(LDIM,MM) FOR MM >= M
C                             IN THE CALLING PROGRAM
C
C OUTPUT:    A             -- THE LU FACTORIZATION OF A, STORED IN THE
C                             LOWER AND UPPER TRIANGLES OF A.
C            MSG           -- =1 IF A SINGULAR, =0 OTHERWISE
C            IROW          -- PIVOT ARRAY
C
C SUBROUTINES: DABS
C
C
C DECLARATIONS
      INTEGER   I,IMAX,IR,IROW,ITEMP,J,K,KM1,KP1,KR,
     & LDIM,L,LR,M,MAXN,MSG,P,PR
      DOUBLE PRECISION A,EPSO,PROD,TEMP,TEST
C
C DIMENSIONS
      DIMENSION A(LDIM,1),IROW(1)
C
C
      MSG=0
C
      DO 10 I=1,M
         IROW(I)=I
   10 CONTINUE
C
C K IS THE COLUMN INDEX
      DO 20 K=1,M
         KM1=K-1
         KP1=K+1
```

```
              DO 30 I=K,M
                  IR=IROW(I)
                  PROD=A(IR,K)
                  IF(KM1 .LT. 1) GOTO 36
                  DO 35 P=1,KM1
                      PR=IROW(P)
                      PROD=PROD-A(IR,P)*A(PR,K)
   35             CONTINUE
   36             CONTINUE
                  A(IR,K)=PROD
   30         CONTINUE
C
C FIND THE INDEX, IMAX, OF THE LARGEST OF ABS(A(I,K)) FOR I=K TO N.
          IMAX = K
          KR=IROW(K)
          TEMP  = DABS( A(KR,K) )
          IF(KP1 .GT. M) GOTO 41
          DO 40 I = KP1,M
            IR=IROW(I)
            TEST = DABS( A(IR,K) )
            IF(TEST .GT. TEMP) THEN
                  IMAX = I
                  TEMP = TEST
            ENDIF
   40     CONTINUE
   41     CONTINUE
C
C IF THE LARGEST ENTRY IN THE K-TH COLUMN AT OR BELOW THE (K,K)-TH ENTRY
C IS ZERO, THEN THE MATRIX A IS SINGULAR.
          IF(TEMP .LE. EPSO) THEN
                  MSG=1
                  GOTO 20
          ENDIF
C
C SWITCH INDICES IROW(K) AND IROW(IMAX)
          ITEMP =IROW(K)
          IROW(K)=IROW(IMAX)
          IROW(IMAX)=ITEMP
C
          KR=IROW(K)
          IF(KP1.GT.M) GOTO 51
          DO 50 J= KP1,M
              PROD=A(KR,J)
              IF(KM1. LT. 1) GOTO 61
              DO 60 P=1,KM1
                  PR=IROW(P)
                  PROD=PROD-A(KR,P)*A(PR,J)
   60         CONTINUE
   61         CONTINUE
              A(KR,J)=PROD/A(KR,K)
   50     CONTINUE
   51     CONTINUE
C
   20 CONTINUE
C
      RETURN
      END
```

```
      SUBROUTINE LNSN(LDIM,M,IROW,A,B,Y)
C
C PURPOSE:  SOLVES THE REAL LINEAR SYSTEM L*U * Y = B, WHERE B IS AN
C           M X 1 COLUMN OF REAL NUMBERS AND L AND U ARE M X M LOWER
C           AND UPPER TRIANGULAR REAL MATRICES, RESPECTIVELY, STORED
C           IN ARRAY A BY SUBROUTINE LNFN.
C
C REFERENCE: CHAPTER 4 OF "SOLVING POLYNOMIAL SYSTEMS USING
C            CONTINUATION" BY ALEXANDER MORGAN.
C
C NOTE:  SEE SUBROUTINE LNFN.
C
C INPUT:    LDIM         -- LEADING DIMENSION OF ARRAY A
C           M            -- DIMENSION OF CURRENT PROBLEM
C           IROW         -- PIVOT ARRAY, AS GENERATED BY LNFN
C           A            -- CONTAINS L AND U AS GENERATED BY LNFN
C           B            -- RIGHT HAND SIDE OF THE SYSTEM BEING SOLVED
C
C OUTPUT:   Y            -- SOLUTION TO L*U * Y = B.
C
C
C DECLARATIONS
      INTEGER I,II,IM1,IP1,IR,IROW,J,K,LDIM,M
      DOUBLE PRECISION A,B,PROD,Y
C
C DIMENSIONS
      DIMENSION A(LDIM,1),B(1),IROW(1),Y(1)
C
C FORWARD ELIMINATION TO SOLVE L*Y=B
      DO 10 I=1,M
         IR = IROW(I)
         PROD=B(IR)
         IM1=I-1
         IF(IM1 .LT. 1) GOTO 21
         DO 20 K=1,IM1
            PROD=PROD - A(IR,K)*Y(K)
 20      CONTINUE
 21      CONTINUE
         Y(I)=PROD/A(IR,I)
 10   CONTINUE
C
C BACK SUBSTITUTION TO SOLVE U*Y=Y
      DO 30 II=1,M
         I=M-II+1
         IR=IROW(I)
         PROD=Y(I)
         IP1=I+1
         IF(IP1 .GT. M) GOTO 41
         DO 40 K=IP1,M
            PROD=PROD - A(IR,K)*Y(K)
 40      CONTINUE
 41      CONTINUE
         Y(I)=PROD
 30   CONTINUE
C
      RETURN
      END
```

```
      SUBROUTINE LNFNC(LDIM,M,EPSO,A, MSG,IROW)
C
C PURPOSE:  GENERATES AN LU FACTORIZATION OF NONSINGULAR COMPLEX
C           M X M MATRIX A.  IF A IS NONSINGULAR, LOWER AND UPPER
C           TRIANGULAR MATRICES L AND U ARE GENERATED SUCH THAT A=L*U,
C           WITH U(I,I)=1 FOR I=1 TO M.  IF A IS SINGULAR, THEN THE
C           FACTORIZATION IS ABANDONED AND AN ERROR MESSAGE IS RETURNED.
C
C REFERENCE: CHAPTER 4 OF "SOLVING POLYNOMIAL SYSTEMS USING
C           CONTINUATION" BY ALEXANDER MORGAN.  BASED ON THE
C           DESCRIPTION OF CROUT'S METHOD IN [DAHLQUIST, ANDERSON]
C           "NUMERICAL METHODS" (PRENTICE HALL, 1974) PP. 157-8,
C           IN [FORSYTHE] "ALGORITHM 16: CROUT WITH PIVOTING,"
C           COMM. ACM, VOL. 3 (1960), PP. 507-8, AND IN
C           [FORSYTHE, MOLER] "COMPUTER SOLUTION OF LINEAR ALGEBRAIC
C           SYSTEMS," (PRENTICE-HALL, 1967) CHAPTERS 10 AND 12.
C
C NOTE: THIS PROGRAM IS DESIGNED TO BE FOLLOWED BY LNSNC TO SOLVE
C       THE COMPLEX LINEAR SYSTEM A*Y = B.
C
C INPUT:    LDIM            -- LEADING DIMENSION OF ARRAY A
C           M               -- DIMENSION OF CURRENT PROBLEM
C           EPSO            -- ZERO-EPSILON TO DETERMINE SINGULARITY
C           A               -- MATRIX DIMENSIONED A(LDIM,MM) FOR MM > M
C                              IN THE CALLING PROGRAM
C
C OUTPUT:   A               -- IF MSG=0, CONTAINS THE LU FACTORIZATION OF
C                              THE INPUT A, STORED IN THE LOWER AND UPPER
C                              TRIANGLES OF A.
C           MSG             -- =1 IF A SINGULAR, =0 OTHERWISE
C           IROW            -- PIVOT ARRAY
C
C SUBROUTINES: DABS,XNORM
C
C
C DECLARATIONS
      INTEGER I,IMAX,IR,IROW,ITEMP,J,K,KM1,KP1,KR,
     & LDIM,L,LR,M,MAXN,MSG,P,PR
      DOUBLE PRECISION EPSO,TEMP,TEST,XNORM
      COMPLEX*16 A,PROD
C
C DIMENSIONS
      DIMENSION A(LDIM,1),IROW(1)
C
C
      MSG=0
C
      DO 10 I=1,M
        IROW(I)=I
   10 CONTINUE
C
C K IS THE COLUMN INDEX
      DO 20 K=1,M
        KM1=K-1
        KP1=K+1
```

```
           DO 30 I=K,M
               IR=IROW(I)
               PROD=A(IR,K)
               IF(KM1 .LT. 1) GOTO 36
               DO 35 P=1,KM1
                   PR=IROW(P)
                   PROD=PROD-A(IR,P)*A(PR,K)
   35          CONTINUE
   36          CONTINUE
               A(IR,K)=PROD
   30      CONTINUE
C
C FIND THE INDEX, IMAX, OF THE LARGEST OF ABS(A(I,K)) FOR I=K TO N.
       IMAX = K
       KR=IROW(K)
       TEMP =  XNORM(2, A(KR,K) )
       IF(KP1 .GT. M) GOTO 41
       DO 40 I = KP1,M
         IR=IROW(I)
         TEST = XNORM(2, A(IR,K) )
         IF(TEST .GT. TEMP) THEN
               IMAX = I
               TEMP = TEST
         ENDIF
   40    CONTINUE
   41    CONTINUE
C
C IF THE LARGEST ENTRY IN THE K-TH COLUMN AT OR BELOW THE (K,K)-TH ENTRY
C IS ZERO, THEN THE MATRIX A IS SINGULAR.
           IF(TEMP .LE. EPSO) THEN
               MSG=1
               RETURN
           ENDIF
C
C SWITCH INDICES IROW(K) AND IROW(IMAX)
           ITEMP =IROW(K)
           IROW(K)=IROW(IMAX)
           IROW(IMAX)=ITEMP
C
           KR=IROW(K)
           IF(KP1.GT.M) GOTO 51
           DO 50 J= KP1,M
               PROD=A(KR,J)
               IF(KM1. LT. 1) GOTO 61
               DO 60 P=1,KM1
                   PR=IROW(P)
                   PROD=PROD-A(KR,P)*A(PR,J)
   60          CONTINUE
   61          CONTINUE
               A(KR,J)=PROD/A(KR,K)
   50      CONTINUE
   51      CONTINUE
C
   20  CONTINUE
C
       RETURN
       END
```

```
      SUBROUTINE LNSNC(LDIM,M,IROW,A,B,Y)
C
C PURPOSE:  SOLVES THE COMPLEX LINEAR SYSTEM L*U * Y = B, WHERE B IS AN
C           M X 1 COLUMN OF COMPLEX NUMBERS AND L AND U ARE M X M LOWER
C           AND UPPER TRIANGULAR COMPLEX MATRICES, RESPECTIVELY, STORED
C           IN ARRAY A BY SUBROUTINE LNFNC.
C
C REFERENCE: CHAPTER 4 OF "SOLVING POLYNOMIAL SYSTEMS USING
C            CONTINUATION" BY ALEXANDER MORGAN.
C
C NOTE:  SEE SUBROUTINE LNFNC.
C
C INPUT:    LDIM         -- LEADING DIMENSION OF ARRAY A
C           M            -- DIMENSION OF CURRENT PROBLEM
C           IROW         -- PIVOT ARRAY, AS GENERATED BY LNFNC
C           A            -- CONTAINS L AND U AS GENERATED BY LNFNC
C           B            -- RIGHT HAND SIDE OF THE SYSTEM BEING SOLVED
C
C OUTPUT:   Y            -- SOLUTION TO L*U * Y = B.
C
C
C DECLARATIONS
      INTEGER I,II,IM1,IP1,IR,IROW,J,K,LDIM,M
      COMPLEX*16 A,B,PROD,Y
C
C DIMENSIONS
      DIMENSION A(LDIM,1),B(1),IROW(1),Y(1)
C
C FORWARD ELIMINATION TO SOLVE L*Y=B
      DO 10 I=1,M
         IR = IROW(I)
         PROD=B(IR)
         IM1=I-1
         IF(IM1 .LT. 1) GOTO 21
         DO 20 K=1,IM1
            PROD=PROD - A(IR,K)*Y(K)
   20    CONTINUE
   21    CONTINUE
         Y(I)=PROD/A(IR,I)
   10 CONTINUE
C
C BACK SUBSTITUTION TO SOLVE U*Y=Y
      DO 30 II=1,M
         I=M-II+1
         IR=IROW(I)
         PROD=Y(I)
         IP1=I+1
         IF(IP1 .GT. M) GOTO 41
         DO 40 K=IP1,M
            PROD=PROD - A(IR,K)*Y(K)
   40    CONTINUE
   41    CONTINUE
         Y(I)=PROD
   30 CONTINUE
C
      RETURN
      END
```

```
        SUBROUTINE LNFP(LDIM,M,KSTAR,EPSO,A, MSG,IROW,ICOL)
C
C PURPOSE:    GENERATES AN LU FACTORIZATION OF NONSINGULAR REAL MATRIX
C             A<KSTAR>, WHERE A<KSTAR> IS THE M X M MATRIX FORMED BY
C             OMITTING THE KSTAR-TH COLUMN OF A, AN M X M+1 MATRIX.
C             THUS, IF A<KSTAR> IS NONSINGULAR, THEN LOWER AND UPPER
C             TRIANGULAR MATRICES L AND U ARE GENERATED SUCH THAT
C             A<KSTAR> = L*U, WITH U(I,I)=1 FOR I=1 TO M.  IF A<KSTAR> IS
C             SINGULAR, THEN THE FACTORIZATION IS ABANDONED AND AN ERROR
C             MESSAGE IS GENERATED.
C
C REFERENCE: CHAPTER 4 OF "SOLVING POLYNOMIAL SYSTEMS USING
C             CONTINUATION" BY ALEXANDER MORGAN.  BASED ON THE
C             DESCRIPTION OF CROUT'S METHOD IN [DAHLQUIST, ANDERSON]
C             "NUMERICAL METHODS" (PRENTICE HALL, 1974) PP. 157-8,
C             IN [FORSYTHE] "ALGORITHM 16: CROUT WITH PIVOTING,"
C             COMM. ACM, VOL. 3 (1960), PP. 507-8, AND IN
C             [FORSYTHE, MOLER] "COMPUTER SOLUTION OF LINEAR ALGEBRAIC
C             SYSTEMS," (PRENTICE-HALL, 1967) CHAPTERS 10 AND 12.
C
C
C NOTE: THIS PROGRAM IS DESIGNED TO BE FOLLOWED BY LNSP TO SOLVE
C       THE LINEAR SYSTEM A<KSTAR>*Y = -A(KSTAR), WHERE A(KSTAR) IS
C       THE KSTAR-TH COLUMN OF A.
C
C INPUT:      LDIM           -- LEADING DIMENSION OF ARRAY A
C             M              -- DIMENSION OF CURRENT PROBLEM
C             KSTAR          -- BALANCE COLUMN
C             EPSO           -- ZERO-EPSILON TO DETERMINE SINGULARITY
C             A              -- MATRIX DIMENSIONED A(LDIM,MM) FOR MM >= M+1
C                               IN THE CALLING PROGRAM
C
C OUTPUT:     A              -- IF MSG=0, CONTAINS THE LU DECOMPOSITION OF
C                               THE INPUT A<KSTAR>, STORED IN THE
C                               LOWER AND UPPER TRIANGLES OF A, EXCEPT THAT
C                               THE KSTAR-TH COLUMN IS SKIPPED.  IT IS
C                               UNCHANGED FROM THE INPUT VALUES.
C             MSG            -- =1 IF A SINGULAR, =0 OTHERWISE
C             IROW           -- PIVOT ARRAY
C             ICOL           -- ICOL(I)=I FOR I < KSTAR, =I+1 FOR I >= KSTAR
C
C SUBROUTINES: DABS
C
C
C DECLARATIONS
        INTEGER  I,IC,ICOL,IMAX,IR,IROW,ITEMP,J,JC,K,KM1,KP1,KC,KR,KSTAR,
     &   KSTAR1,LDIM,L,LR,M,MAXN,MSG,P,PC,PR
        DOUBLE PRECISION A,EPSO,PROD,TEMP,TEST
C
C DIMENSIONS
        DIMENSION A(LDIM,1),ICOL(1),IROW(1)
C
C
        MSG=0
C
```

```
          DO 10 I=1,M
            IROW(I)=I
    10    CONTINUE
C
          KSTAR1 = KSTAR-1
          IF(KSTAR1.LT.1) GOTO 13
          DO 12 I=1,KSTAR1
            ICOL(I)=I
    12    CONTINUE
    13    CONTINUE
          IF(KSTAR.GT.M) GOTO 15
          DO 14 I=KSTAR,M
            ICOL(I)=I+1
    14    CONTINUE
    15    CONTINUE
C
C
C K IS THE COLUMN INDEX
          DO 20 K=1,M
            KM1=K-1
            KP1=K+1
            KC=ICOL(K)
            DO 30 I=K,M
                IR=IROW(I)
                IC=ICOL(I)
                PROD=A(IR,KC)
                IF(KM1 .LT. 1) GOTO 36
                DO 35 P=1,KM1
                    PR=IROW(P)
                    PC=ICOL(P)
                    PROD=PROD-A(IR,PC)*A(PR,KC)
    35          CONTINUE
    36          CONTINUE
                A(IR,KC)=PROD
    30      CONTINUE
C
C FIND THE INDEX, IMAX, OF THE LARGEST OF ABS(A(I,K)) FOR I=K TO N.
            IMAX = K
            KR=IROW(K)
            TEMP =  DABS( A(KR,KC) )
            IF(KP1 .GT. M) GOTO 41
            DO 40 I = KP1,M
              IR=IROW(I)
              TEST = DABS( A(IR,KC) )
              IF(TEST .GT. TEMP) THEN
                  IMAX = I
                  TEMP = TEST
              ENDIF
    40      CONTINUE
    41      CONTINUE
C
C IF THE LARGEST ENTRY IN THE K-TH COLUMN AT OR BELOW THE (K,K)-TH ENTRY
C IS ZERO, THEN THE MATRIX A IS SINGULAR.
            IF(TEMP .LE. EPSO) THEN
                MSG=1
                GOTO 20
            ENDIF
```

```
C
C SWITCH INDICES IROW(K) AND IROW(IMAX)
        ITEMP =IROW(K)
        IROW(K)=IROW(IMAX)
        IROW(IMAX)=ITEMP
C
        KR=IROW(K)
        IF(KP1.GT.M) GOTO 51
        DO 50 J= KP1,M
           JC=ICOL(J)
           PROD=A(KR,JC)
           IF(KM1. LT. 1) GOTO 61
           DO 60 P=1,KM1
              PR=IROW(P)
              PC=ICOL(P)
              PROD=PROD-A(KR,PC)*A(PR,JC)
   60         CONTINUE
   61         CONTINUE
           A(KR,JC)=PROD/A(KR,KC)
   50   CONTINUE
   51   CONTINUE
C
   20 CONTINUE
C
        RETURN
        END
```

```
      SUBROUTINE LNSP(LDIM,M,KSTAR,IROW,ICOL,A,Y)
C
C PURPOSE: SOLVES THE LINEAR SYSTEM L*U * Y = -A(KSTAR), WHERE A(KSTAR) IS
C          THE KSTAR-TH COLUMN OF A, AND L AND U ARE M X M LOWER AND UPPER
C          TRIANGULAR MATRICES, STORED IN A<KSTAR> BY LNFP, WHERE A<KSTAR>
C          IS A WITH THE KSTAR-TH COLUMN OMITTED.
C
C REFERENCE: CHAPTER 4 OF "SOLVING POLYNOMIAL SYSTEMS USING
C            CONTINUATION" BY ALEXANDER MORGAN.
C
C NOTE:  SEE SUBROUTINE LNFP.
C
C INPUT:     LDIM            -- LEADING DIMENSION OF ARRAY A
C            M               -- DIMENSION OF CURRENT PROBLEM
C            KSTAR           -- THE BALANCE COLUMN INDEX
C            IROW            -- PIVOT ARRAY, AS GENERATED BY LNFP
C            ICOL            -- ICOL(I)=I FOR I < KSTAR, =I+1 FOR I >= KSTAR
C            A               -- ARRAY CONTAINING ELEMENTS AS DESCRIBED ABOVE.
C
C OUTPUT:    Y               -- SOLUTION TO L*U * Y = -A(KSTAR)
C
C DECLARATIONS
      INTEGER I,IC,ICOL,II,IM1,IP1,IR,IROW,J,K,KC,KSTAR,LDIM,M
      DOUBLE PRECISION A,PROD,Y
C
C DIMENSIONS
      DIMENSION A(LDIM,1),ICOL(1),IROW(1),Y(1)
C
C FORWARD ELIMINATION TO SOLVE L*Y=B
      DO 10 I=1,M
         IR = IROW(I)
         IC = ICOL(I)
         PROD=-A(IR,KSTAR)
         IM1=I-1
         IF(IM1 .LT. 1) GOTO 21
         DO 20 K=1,IM1
            KC=ICOL(K)
            PROD=PROD - A(IR,KC)*Y(K)
   20    CONTINUE
   21    CONTINUE
         Y(I)=PROD/A(IR,IC)
   10 CONTINUE
C
C BACK SUBSTITUTION TO SOLVE U*Y=Y
      DO 30 II=1,M
         I=M-II+1
         IR=IROW(I)
         PROD=Y(I)
         IP1=I+1
         IF(IP1 .GT. M) GOTO 41
         DO 40 K=IP1,M
            KC=ICOL(K)
            PROD=PROD - A(IR,KC)*Y(K)
   40    CONTINUE
   41    CONTINUE
         Y(I)=PROD
   30 CONTINUE
      RETURN
      END
```

```
          SUBROUTINE ICD(LDIM,N,EPSO,A, E,IROW,AINV,COND,DET,MSG)
C
C
C PURPOSE: COMPUTES THE INVERSE, CONDITION, AND DETERMINANT OF THE
C          NONSINGULAR REAL N X N MATRIX A.
C
C
C REFERENCE: CHAPTER 4 OF "SOLVING POLYNOMIAL SYSTEMS USING
C            CONTINUATION" BY ALEXANDER MORGAN.
C
C
C INPUT:  LDIM        -- LEADING DIMENSION OF ARRAY A
C         N           -- DIMENSION OF THE CURRENT PROBLEM
C         EPSO        -- ZERO-EPSILON FOR SINGULARITY
C         A           -- MATRIX DIMENSIONED A(LDIM,NN) WHERE NN >= N
C                        IN THE CALLING PROGRAM.
C
C OUTPUT: E           -- COMPLEX WORKSPACE ARRAY OF LENGTH N.
C         IROW        -- INTEGER WORKSPACE ARRAY OF LENGTH N.
C         AINV        -- IF MSG=0, INVERSE OF A, DIMENSIONED THE SAME AS A.
C         COND        -- IF MSG=0, CONDITION OF A, C(A), WHERE C(A) IS
C                        DEFINED BY C(A) = |A|*|1/A| WHERE 1/A DENOTES
C                        THE INVERSE OF A AND "| |" DENOTES THE MATRIX
C                        NORM DERIVED FROM XNORM (THE "1-NORM").
C         DET         -- IF MSG=0, DETERMINANT OF A
C         MSG         -- =1, IF A IS SINGULAR WITH RESPECT TO EPSO.
C                        =0, OTHERWISE.
C
C
C SUBROUTINES: DABS,LNFN,LNSN,XNORM
C
C
C DECLARATIONS
      INTEGER I,IM1,IPIV,IROW,J,K,KP1,LDIM,MSG,N
      DOUBLE PRECISION A,AINV,BIG,COND,DABS,DET,E,EPSO,TEST,XNORM,XNORMM
C
C DIMENSIONS
      DIMENSION A(LDIM,1),AINV(LDIM,1),E(1),IROW(1)
C
C
C COMPUTE XNORMM(A).  XNORMM IS THE MATRIX NORM ASSOCIATED WITH XNORM.
C (THUS, XNORMM IS THE MAXIMUM OF THE SUMS OF ABSOLUTE VALUES OF THE
C COLUMNS OF A.)
      XNORMM=0.DO
      DO 10 J=1,N
        TEST=XNORM(N,A(1,J))
        IF(TEST .GT. XNORMM) XNORMM=TEST
 10   CONTINUE
C
C STORE XNORMM(A) IN COND.
      COND=XNORMM
C
C
```

```
C GENERATE AINV=1/A
      CALL LNFN(LDIM,N,EPSO,A, MSG,IROW)
      IF (MSG .EQ. 1) RETURN
C
      DO 15 I=1,N
       E(I)=0.DO
  15  CONTINUE
C
      DO 20 I=1,N
       IM1=I-1
       IF(I.GT.1)E(IM1)=0.DO
       E(I)=1.DO
       CALL LNSN(LDIM,N,IROW,A,E, AINV(1,I))
  20  CONTINUE
C
C COMPUTE XNORMM(AINV)
      XNORMM=0.DO
      DO 30 J=1,N
       TEST=XNORM(N,AINV(1,J))
       IF(TEST .GT. XNORMM) XNORMM=TEST
  30  CONTINUE
C
      COND=COND*XNORMM
C
C
C COMPUTE DETERMINANT OF A
      DET=1.DO
      DO 35 I=1,N
       DET=DET*A(IROW(I),I)
  35  CONTINUE
C
      IPIV=1
      DO 40 K=1,N
       IF(IROW(K).EQ.K) GOTO 41
       KP1=K+1
       DO 45 J=KP1,N
       IF(IROW(J).EQ.K) THEN
        IROW(J)=IROW(K)
        IPIV=-IPIV
       END IF
  45    CONTINUE
  41    CONTINUE
  40  CONTINUE
C
      DET=IPIV*DET
C
      RETURN
      END
```

```
      SUBROUTINE ICDC(LDIM,N,EPSO,A, E,IROW,AINV,COND,DET,MSG)
C
C PURPOSE:  COMPUTES THE INVERSE, DETERMINANT, AND CONDITION OF THE
C           NONSINGULAR COMPLEX N X N MATRIX A.
C
C REFERENCE: CHAPTER 4 OF "SOLVING POLYNOMIAL SYSTEMS USING
C            CONTINUATION" BY ALEXANDER MORGAN.
C
C NOTE:  THIS SUBROUTINE USES THE FORTRAN "COMPLEX*16" DECLARATION.
C
C INPUT:  LDIM      -- LEADING DIMENSION OF ARRAY A
C         N         -- DIMENSION OF THE CURRENT PROBLEM
C         EPSO      -- ZERO-EPSILON FOR SINGULARITY
C         A         -- MATRIX DIMENSIONED A(LDIM,NN) WHERE NN >= N
C                      IN THE CALLING PROGRAM.
C
C OUTPUT: E         -- COMPLEX WORKSPACE ARRAY OF LENGTH N.
C         IROW      -- INTEGER WORKSPACE ARRAY OF LENGTH N.
C         AINV      -- IF MSG=0, INVERSE OF A, DIMENSIONED THE SAME AS A.
C         COND      -- IF MSG=0, THE CONDITION OF A, C(A), WHERE C(A) IS
C                      DEFINED BY C(A) = |A|*|1/A| WHERE 1/A DENOTES THE
C                      INVERSE OF A AND "|D|" DENOTES THE MATRIX NORM
C                      DERIVED FROM XNORM (THE "1-NORM").
C         DET       -- IF MSG=0, DETERMINANT OF A
C         MSG       -- =1, IF A IS SINGULAR WITH RESPECT TO EPSO.
C                      =0, OTHERWISE.
C
C SUBROUTINES: DCMPLX,LNFNC,LNSNC,XNORM
C
C
C DECLARATIONS
      INTEGER I,IM1,IPIV,IROW,J,K,KP1,LDIM,MSG,N,N2
      DOUBLE PRECISION BIG,COND,EPSO,TEST,XNORM,XNORMM
      COMPLEX*16 A,AINV,DET,E,ONE,ZERO
C
C DIMENSIONS
      DIMENSION A(LDIM,1),AINV(LDIM,1),E(1),IROW(1)
C
C
      N2=2*N
      ONE =DCMPLX(1.D0,0.D0)
      ZERO=DCMPLX(0.D0,0.D0)
C
C COMPUTE XNORMM(A).  XNORMM IS THE MATRIX NORM ASSOCIATED WITH XNORM.
C (THUS, XNORMM IS THE MAXIMUM OF THE SUMS OF ABSOLUTE VALUES OF THE
C COLUMNS OF A.)
      XNORMM=0.D0
      DO 10 J=1,N
        TEST=XNORM(N2,A(1,J))
        IF(TEST .GT. XNORMM) XNORMM=TEST
   10 CONTINUE
C
C STORE XNORMM(A) IN COND.
      COND=XNORMM
C
```

```
C
C GENERATE AINV=1/A
      CALL LNFNC(LDIM,N,EPSO,A,MSG,IROW)
      IF(MSG .EQ. 1) RETURN
C
      DO 15 I=1,N
        E(I)=ZERO
   15 CONTINUE
C
      DO 20 I=1,N
        IM1=I-1
        IF(I.GT.1)E(IM1)=ZERO
        E(I)=ONE
        CALL LNSNC(LDIM,N,IROW,A,E, AINV(1,I))
   20 CONTINUE
C
C
C COMPUTE XNORMM(AINV)
      XNORMM=0.DO
      DO 30 J=1,N
        TEST=XNORM(N2,AINV(1,J))
        IF(TEST .GT. XNORMM) XNORMM=TEST
   30 CONTINUE
C
      COND=COND*XNORMM
C
C
C COMPUTE DETERMINANT OF A
      DET=1.DO
      DO 35 I=1,N
        DET=DET*A(IROW(I),I)
   35 CONTINUE
C
      IPIV=1
      DO 40 K=1,N
        IF(IROW(K).EQ.K) GOTO 41
        KP1=K+1
        DO 45 J=KP1,N
        IF(IROW(J).EQ.K) THEN
          IROW(J)=IROW(K)
          IPIV=-IPIV
        END IF
   45   CONTINUE
   41   CONTINUE
   40 CONTINUE
C
      DET=IPIV*DET
C
      RETURN
      END
```

```
      FUNCTION XNORM(N,X)
C
C PURPOSE: FINDS THE 1-NORM OF ARRAY X OF LENGTH N.
C
C INPUT: N, X
C
C RETURNS: XNORM = DABS(X(1)) + ... + DABS(X(N))
C
C SUBROUTINES: DABS
C
C DECLARATIONS
      INTEGER J,N
      DOUBLE PRECISION X,XNORM
C
C DIMENSIONS
      DIMENSION X(N)
C
C
      XNORM=0.D0
      DO 10 J=1,N
        XNORM=XNORM + DABS(X(J))
  10  CONTINUE
C
      RETURN
      END
```

```
      FUNCTION XNRM2(N,X)
C
C PURPOSE: FINDS THE EUCLIDEAN NORM OF ARRAY X OF LENGTH N.
C
C INPUT: N, X
C
C RETURNS: XNRM2 = DSQRT((X(1)**2 + ... + X(N)**2)
C
C SUBROUTINES: DSQRT
C
C DECLARATIONS
      INTEGER J,N
      DOUBLE PRECISION XNRM2,X
C
C DIMENSIONS
      DIMENSION X(N)
C
C
      XNRM2=0.D0
      DO 10 J=1,N
        XNRM2=XNRM2 + X(J)**2
   10 CONTINUE
      XNRM2=DSQRT(XNRM2)
C
      RETURN
      END
```

```
      FUNCTION IRMAX(N,X)
C
C PURPOSE: FINDS THE ENTRY OF ARRAY X OF LARGEST ABSOLUTE VALUE
C          AND RETURNS ITS INDEX.  X IS OF LENGTH N.
C
C NOTE: ASSUMES N .GE. 1.
C
C INPUT: N, X
C
C RETURNS: IRMAX = INDEX OF LARGEST OF ABS(X(J)) FOR J=1,N
C
C SUBROUTINES:  DABS
C
C
C DECLARATIONS
      INTEGER J,IRMAX,N
      DOUBLE PRECISION RMAX,TEMP,X
C
C DIMENSIONS
      DIMENSION X(N)
C
C
      RMAX=DABS(X(1))
      IRMAX=1
C
      IF(N.EQ.1) RETURN
C
        DO 10 J=2,N
           TEMP=DABS(X(J))
           IF( TEMP .GT. RMAX ) THEN
             IRMAX=J
             RMAX=TEMP
           ENDIF
   10   CONTINUE
C
        RETURN
        END
```

```
      FUNCTION DREAL(X)
C
C PURPOSE: FINDS THE REAL PART OF A COMPLEX VARIABLE, X
C
C INPUT: X
C
C RETURNS: DREAL = REAL PART OF X
C
C DECLARATIONS
      DOUBLE PRECISION DREAL
      COMPLEX*16 X
C
      DREAL=X
C
      RETURN
      END
```

```
      FUNCTION DIMAG(X)
C
C PURPOSE:  FINDS THE IMAGINARY PART OF A COMPLEX VARIABLE, X
C
C INPUT: X
C
C RETURNS: DIMAG = IMAGINARY PART OF X
C
C SUBROUTINES: DCMPLX
C
C DECLARATIONS
      DOUBLE PRECISION DIMAG
      COMPLEX*16 X
C
      DIMAG=X*DCMPLX(0.D0,-1.D0)
C
      RETURN
      END
```

Section 1: Programs for Chapter 1

```
C PROGRAM NAME:  CONSOL2R
C
C PURPOSE:    SOLVES THE QUADRATIC EQUATION
C
C                  EQ = X**2 + A*X + B
C
C              USING THE CONTINUATION EQUATION
C
C              H   = X**2 + A*T*X + B.
C
C REFERENCE: CHAPTER 1 OF "SOLVING POLYNOMIAL SYSTEMS USING
C            CONTINUATION" BY ALEXANDER MORGAN
C
C NOTE: 1.    THIS PROGRAM DOES NOT USE THE FORTRAN "COMPLEX"
C            DECLARATION.  IT EMULATES COMPLEX OPERATIONS
C            USING REAL ARITHMETIC.  SEE APPENDIX 2.
C
C       2.    THIS PROGRAM GENERATES MANY UNDERFLOWS.  THIS IS
C            NOT AN ERROR CONDITION.  IF YOU NEED TO PREVENT
C            UNDERFLOWS FROM OCCURRING, SEE THE COMMENTS LABELED
C            "UNDERFLOWS" BELOW.
C
C       3.    INPUT IS ON UNIT 5.  OUTPUT IS TO UNIT 6.
C            SAMPLE INPUT FOR DELT=.05, A=10., B=.2:
C                  5.D-02
C                  1.D+01
C                  2.D-01
C
C INPUT VARIABLES:
C
C     DELT         -- THE STEP SIZE FOR T
C     A            -- COEFFICIENT FOR EQ
C     B            -- COEFFICIENT FOR EQ
C
C OUTPUT VARIABLES:
C
C   FOR EACH STEP:
C
C     NUMSTP       -- CURRENT STEP NUMBER
C     T            -- CURRENT VALUE OF T
C     X            -- CURRENT VALUE OF X
C     H            -- CURRENT VALUE OF THE CONTINUATION EQUATION
C     RESID        -- NEWTON METHOD RESIDUAL (SHOULD BE NEAR O)
C     DH           -- CURRENT DERIVATIVE OF CONTINUATION EQUATION
C
C   AT THE END OF THE PATH:
C
C     NUMIT        -- NUMBER OF NEWTON ITERATIONS FOR THE PATH
C
C   AT THE END OF THE RUN:
C
C     ITOTIT       -- TOTAL NUMBER OF NEWTON ITERATIONS OVER ALL PATHS
C
C
C FORTRAN ROUTINES: DABS,DIV,DSQRT,MOD,MUL
```

```
C
C DECLARATIONS
      INTEGER I,ITER,ITOTIT,NS,NUMIT,NUMPAT,NUMSTP
      DOUBLE PRECISION A,B,DELT,EPSOO,T,XRESID
      DOUBLE PRECISION DH,H,ONE,RESID,X,XSQ
C
C DIMENSIONS
      DIMENSION DH(2),H(2),ONE(2),RESID(2),X(2),XSQ(2)
C
C INPUT STATEMENTS FOLLOW.
      READ(5,10) DELT
  10  FORMAT(D22.15)
      READ(5,10) A,B
C
      WRITE(6,20)
  20  FORMAT(/,'  CONSOL2R',//)
      WRITE(6,25) DELT
  25  FORMAT(/,'  DELT =',D22.15,/)
      WRITE(6,30) A,B
  30  FORMAT('  COEFFICIENTS,    A=',D22.15,'  B=',D22.15)
C
C COMPUTE NUMBER OF STEPS PER PATH
      NS = (1./DELT) + 1.
C
      WRITE(6,40)
  40  FORMAT(/)
C
      WRITE(6,50)  NS
  50  FORMAT(' NUMBER OF STEPS PER PATH =',I10//)
C
C INITIALIZE COUNTER FOR TOTAL NUMBER OF CORRECTOR ITERATIONS
      ITOTIT=0
C
      ONE(1)=1.D0
      ONE(2)=0.D0
C
C ****************************************************************
C *****                                                    *****
C *****       BEGINNING OF MAIN PART OF PROGRAM             *****
C *****                                                    *****
C ****************************************************************
C
C PATHS LOOP -- ITERATE THROUGH PATHS
      DO   60 NUMPAT = 1 ,  2
C
         WRITE(6,70) NUMPAT
  70     FORMAT('  PATH NUMBER =',I10)
C
C INITIALIZE COUNTERS AND CONSTANTS FOR THE PATH
         NUMIT=0
         T=0.D0
C
C GENERATE A START POINT, S, FOR THE PATH
         IF(B.LE.0.D0) THEN
            X(1)=DSQRT(-B)
            X(2)=0.D0
         ENDIF
```

450

```
               IF(B.GT.O.DO) THEN
                   X(1)=0.DO
                   X(2)=DSQRT(B)
               ENDIF
               IF(NUMPAT.EQ.2) THEN
                   X(1)=-X(1)
                   X(2)=-X(2)
               ENDIF
C
               WRITE(6,80)X
    80         FORMAT(/,'  START X=',2D22.15,//)
C
C
C STEPS LOOP -- ITERATE STEPS ALONG THE PATH
               DO 90 NUMSTP = 1 , NS
C
                   T=T+DELT
                   IF(T.GE.1.DO)T=1.DO
C
C
C NEWTON'S METHOD LOOP
                   DO 100 ITER = 1,100
C
C EVALUATE THE CONTINUATION EQUATION AND DERIVATIVE
                       CALL MUL(X,X,XSQ)
                       DO 104 I=1,2
                           H(I)  = XSQ(I) + A*T*X(I) + B*ONE(I)
                           DH(I) = 2.*X(I) + A*T*ONE(I)
   104                 CONTINUE
C
C GENERATE THE RESIDUAL AND UPDATE X
                       CALL DIV(H,DH,RESID)
                       DO 106 I=1,2
                           X(I) = X(I) - RESID(I)
   106                 CONTINUE
C
C$$$$$$$$$$$$$$$$$$$$$$$$$$$$$$$$$$$$$$$$$$$$$$$$$$$$$$$$$$$$$$$$$$$$$$$$$$$$$$$C
C UNDERFLOWS:    IF YOU MUST PREVENT UNDERFLOWS, I SUGGEST YOU
C               TAKE THE COMMENTS OFF THE FOLLOWING BLOCK OF
C               CODE.  THE VALUE FOR EPSOO MAY NEED TO BE ADJUSTED.
C
C               EPSOO=1.D-22
C               IF(DABS(X(1)).LT.EPSOO) X(1)=0.DO
C               IF(DABS(X(2)).LT.EPSOO) X(2)=0.DO
C
C$$$$$$$$$$$$$$$$$$$$$$$$$$$$$$$$$$$$$$$$$$$$$$$$$$$$$$$$$$$$$$$$$$$$$$$$$$$$$$$C
C
C UPDATE NEWTON ITERATION COUNTER
                       NUMIT = NUMIT + 1
C
C LEAVE NEWTON LOOP IF RESIDUAL IS SMALL ENOUGH
                       XRESID = DABS(RESID(1))+DABS(RESID(2))
                       IF(XRESID .LT. 1.D-12) GOTO 101
C
   100                 CONTINUE
   101                 CONTINUE
C
```

```
C
              IF(MOD(NUMSTP,NS/10).EQ.0.OR.NUMSTP.GE.(NS-4))THEN
              WRITE(6,110) NUMSTP
110           FORMAT(/' STEP SUMMARY FOR STEP NUMBER',I10,' :')
              WRITE(6,115)
115           FORMAT(' ----------------------------------------------')
              WRITE(6,120) T
120           FORMAT('   T  = ',D22.15)
              WRITE(6,125)
125           FORMAT('                     REAL PART        IMAGINARY PART')
              WRITE(6,130) X
130           FORMAT('  X  =         ',D11.4,'         ',D11.4)
              WRITE(6,135)  H
135           FORMAT('  H  =         ',D11.4,'         ',D11.4)
              WRITE(6,140) RESID
140           FORMAT(' RESIDUAL =    ',D11.4,'         ',D11.4)
              WRITE(6,150) DH
150           FORMAT(' DH  =         ',D11.4,'         ',D11.4)
              WRITE(6,115)
              WRITE(6,40)
              ENDIF
C
 90       CONTINUE
C             BOTTOM OF LOOP -- NUMSTP
C
          ITOTIT=ITOTIT + NUMIT
C
          WRITE(6,160) NUMIT
160       FORMAT(' TOTAL NEWTON ITERATIONS ON PATH =',I5)
          WRITE(6,40)
C
 60    CONTINUE
C          BOTTOM OF LOOP -- NUMPAT
C
       WRITE(6,170) ITOTIT
170    FORMAT(//' TOTAL NEWTON ITERATIONS =',I10//)
C
       STOP
       END
```

452

```
C PROGRAM NAME:   CONSOL3R
C
C PURPOSE:   SOLVES THE QUADRATIC EQUATION
C
C                 EQ = X**2 + A*X + B
C
C            USING THE CONTINUATION EQUATION
C
C                 H = X**2 + A*T*X + (T*B - (1.-T)*Q**2)
C
C REFERENCE: CHAPTER 1 OF "SOLVING POLYNOMIAL SYSTEMS USING
C            CONTINUATION" BY ALEXANDER MORGAN
C
C NOTE: 1.   THIS PROGRAM DOES NOT USE THE FORTRAN "COMPLEX"
C            DECLARATION.  IT EMULATES COMPLEX OPERATIONS
C            USING REAL ARITHMETIC.  SEE APPENDIX 2.
C
C       2.   THIS PROGRAM GENERATES MANY UNDERFLOWS.  THIS IS
C            NOT AN ERROR CONDITION.  IF YOU NEED TO PREVENT
C            UNDERFLOWS FROM OCCURRING, SEE THE COMMENTS LABELED
C            "UNDERFLOWS" BELOW.
C
C       3.   INPUT IS ON UNIT 5.  OUTPUT IS TO UNIT 6.
C            SAMPLE INPUT FOR DELT=.05, A=10., B=.2:
C                 5.D-02
C                 1.D+01
C                 2.D-01
C
C INPUT VARIABLES:
C
C     DELT          -- THE STEP SIZE FOR T
C     A             -- COEFFICIENT FOR EQ
C     B             -- COEFFICIENT FOR EQ
C
C OUTPUT VARIABLES:
C
C     FOR EACH STEP:
C
C     NUMSTP        -- CURRENT STEP NUMBER
C     T             -- CURRENT VALUE OF T
C     X             -- CURRENT VALUE OF X
C     H             -- CURRENT VALUE OF THE CONTINUATION EQUATION
C     RESID         -- NEWTON METHOD RESIDUAL (SHOULD BE NEAR 0)
C     DH            -- CURRENT DERIVATIVE OF CONTINUATION EQUATION
C
C
C     AT THE END OF THE PATH:
C
C     NUMIT         -- NUMBER OF NEWTON ITERATIONS FOR THE PATH
C
C     AT THE END OF THE RUN:
C
C     ITOTIT        -- TOTAL NUMBER OF NEWTON ITERATIONS OVER ALL PATHS
C
C
```

```
C FORTRAN ROUTINES: DABS,DIV,MOD,MUL
C
C
C DECLARATIONS
      INTEGER I,ITER,ITOTIT,NS,NUMIT,NUMPAT,NUMSTP
      DOUBLE PRECISION A,B,DELT,EPSOO,T,XRESID
      DOUBLE PRECISION DH,H,ONE,Q,QSQ,RESID,X,XSQ
C
C DIMENSIONS
      DIMENSION DH(2),H(2),ONE(2),Q(2),QSQ(2),RESID(2),X(2),XSQ(2)
C
C INPUT STATEMENTS FOLLOW.
      READ(5,10) DELT
   10 FORMAT(D22.15)
      READ(5,10) A,B
C
      WRITE(6,20)
   20 FORMAT(/,'   CONSOL3R',//)
      WRITE(6,25) DELT
   25 FORMAT(/,'   DELT =',D22.15,/)
      WRITE(6,30) A,B
   30 FORMAT(' COEFFICIENTS,    A=',D22.15,'  B=',D22.15)
C
C COMPUTE NUMBER OF STEPS PER PATH
      NS = (1./DELT) + 1.
C
      WRITE(6,40)
   40 FORMAT(/)
C
      WRITE(6,50)  NS
   50 FORMAT(' NUMBER OF STEPS PER PATH =',I5//)
C
C INITIALIZE COUNTER AND CONSTANTS FOR THE RUN
      ITOTIT=0
      ONE(1)=1.D0
      ONE(2)=0.D0
      Q(1)=.928746354D0
      Q(2)=.265465386D0
      CALL MUL(Q,Q,QSQ)
C
C ******************************************************************
C *****                                                      *****
C *****        BEGINNING OF MAIN PART OF PROGRAM             *****
C *****                                                      *****
C ******************************************************************
C
C PATHS LOOP -- ITERATE THROUGH PATHS
      DO  60 NUMPAT = 1 ,  2
C
         WRITE(6,70) NUMPAT
   70    FORMAT(' PATH NUMBER =',I5)
C
C INITIALIZE COUNTERS AND CONSTANTS FOR THE PATH
         NUMIT=0
         T=0
C
```

```
C
C GENERATE A START POINT, S, FOR THE PATH
        DO 75 I=1,2
            X(I)=Q(I)
            IF(NUMPAT.EQ.2)X(I)=-X(I)
  75    CONTINUE
C
        WRITE(6,80)X
  80    FORMAT(/,'  START X=',2D22.15,//)
C
C
C STEPS LOOP -- ITERATE STEPS ALONG THE PATH
        DO 90 NUMSTP = 1 , NS
C
        T=T+DELT
        IF(T.GE.1.)T=1.
C
C
C NEWTON'S METHOD LOOP
        DO 100 ITER = 1,10
C
C EVALUATE THE CONTINUATION EQUATION AND DERIVATIVE
            CALL MUL(X,X,XSQ)
            DO 104 I=1,2
                H(I)=XSQ(I)+A*T*X(I)+(T*B*ONE(I)-(1.-T)*QSQ(I))
                DH(I)=2.*X(I)+A*T*ONE(I)
 104        CONTINUE
C
C
C GENERATE THE RESIDUAL AND UPDATE X
            CALL DIV(H,DH,RESID)
C
            DO 106 I=1,2
                X(I) = X(I) - RESID(I)
 106        CONTINUE
C
C$$$$$$$$$$$$$$$$$$$$$$$$$$$$$$$$$$$$$$$$$$$$$$$$$$$$$$$$$$$$$$$$$$$$$$$$$$C
C UNDERFLOWS:  IF YOU MUST PREVENT UNDERFLOWS, I SUGGEST YOU
C              TAKE THE COMMENTS OFF THE FOLLOWING BLOCK OF
C              CODE.  THE VALUE FOR EPSOO MAY NEED TO BE ADJUSTED.
C
C              EPSOO=1.D-22
C              IF(DABS(X(1)).LT.EPSOO) X(1)=0.DO
C              IF(DABS(X(2)).LT.EPSOO) X(2)=0.DO
C$$$$$$$$$$$$$$$$$$$$$$$$$$$$$$$$$$$$$$$$$$$$$$$$$$$$$$$$$$$$$$$$$$$$$$$$$$C
C
C UPDATE NEWTON ITERATION COUNTER
            NUMIT = NUMIT + 1
C
C LEAVE NEWTON LOOP IF RESIDUAL IS SMALL ENOUGH
            XRESID = DABS(RESID(1))+DABS(RESID(2))
            IF(XRESID .LT. 1.D-12) GOTO 101
C
 100        CONTINUE
 101        CONTINUE
C
```

```
C
                IF(MOD(NUMSTP,NS/10).EQ.0.OR.NUMSTP.GE.(NS-4))THEN
                   WRITE(6,110) NUMSTP
  110              FORMAT(/' STEP SUMMARY FOR STEP NUMBER',I5,' :')
                   WRITE(6,115)
  115              FORMAT(' ----------------------------------------------')
                   WRITE(6,120) T
  120              FORMAT(' T  = ',D22.15)
                   WRITE(6,125)
  125              FORMAT('               REAL PART        IMAGINARY PART')
                   WRITE(6,130) X
  130              FORMAT(' X  =          ',D11.4,'          ',D11.4)
                   WRITE(6,135)  H
  135              FORMAT(' H  =          ',D11.4,'          ',D11.4)
                   WRITE(6,140) RESID
  140              FORMAT(' RESIDUAL =    ',D11.4,'          ',D11.4)
                   WRITE(6,150) DH
  150              FORMAT(' DH =          ',D11.4,'          ',D11.4)
                   WRITE(6,115)
                   WRITE(6,40)
                ENDIF
C
   90      CONTINUE
C            BOTTOM OF LOOP -- NUMSTP
C
           ITOTIT=ITOTIT + NUMIT
C
           WRITE(6,160) NUMIT
  160      FORMAT(' TOTAL NEWTON ITERATIONS ON PATH =',I5)
           WRITE(6,40)
C
   60   CONTINUE
C            BOTTOM OF LOOP -- NUMPAT
C
        WRITE(6,170) ITOTIT
  170   FORMAT(//' TOTAL NEWTON ITERATIONS =',I5//)
C
        STOP
        END
```

```
C PROGRAM NAME:  CONSOL4R
C
C PURPOSE:    SOLVES THE FIFTH-DEGREE EQUATION
C
C                   EQ = X**5 + A*X + B
C
C              USING THE CONTINUATION EQUATION
C
C                   H = X**5 + A*T*X + (T*B - (1.-T)*Q**5)
C
C REFERENCE: CHAPTER 1 OF "SOLVING POLYNOMIAL SYSTEMS USING
C               CONTINUATION" BY ALEXANDER MORGAN
C
C NOTE: 1.    THIS PROGRAM DOES NOT USE THE FORTRAN "COMPLEX"
C            DECLARATION.  IT EMULATES COMPLEX OPERATIONS
C            USING REAL ARITHMETIC.  SEE APPENDIX 2.
C
C       2.    THIS PROGRAM GENERATES MANY UNDERFLOWS.  THIS IS
C            NOT AN ERROR CONDITION.  IF YOU NEED TO PREVENT
C            UNDERFLOWS FROM OCCURRING, SEE THE COMMENTS LABELED
C            "UNDERFLOWS" BELOW.
C
C       3.    INPUT IS ON UNIT 5.  OUTPUT IS TO UNIT 6.
C            SAMPLE INPUT FOR DELT=.05, A=10., B=.2:
C                  5.D-02
C                  1.D+01
C                  2.D-01
C
C  INPUT VARIABLES:
C
C     DELT        -- THE STEP SIZE FOR T
C     A           -- COEFFICIENT FOR EQ
C     B           -- COEFFICIENT FOR EQ
C
C OUTPUT VARIABLES:
C
C    FOR EACH STEP:
C
C     NUMSTP      -- CURRENT STEP NUMBER
C     T           -- CURRENT VALUE OF T
C     X           -- CURRENT VALUE OF X
C     H           -- CURRENT VALUE OF THE CONTINUATION EQUATION
C     RESID       -- NEWTON METHOD RESIDUAL (SHOULD BE NEAR 0)
C     DH          -- CURRENT DERIVATIVE OF CONTINUATION EQUATION
C
C
C    AT THE END OF THE PATH:
C
C     NUMIT       -- NUMBER OF NEWTON ITERATIONS FOR THE PATH
C
C    AT THE END OF THE RUN:
C
C     ITOTIT      -- TOTAL NUMBER OF NEWTON ITERATIONS OVER ALL PATHS
C
C
```

```
C  FORTRAN ROUTINES: DABS,DIV,MOD,MUL,POW
C
C
C DECLARATIONS
      INTEGER I,ITER,ITOTIT,NS,NUMIT,NUMPAT,NUMSTP
      DOUBLE PRECISION A,B,DELT,EPSOO,T,XRESID
      DOUBLE PRECISION DH,H,ONE,Q,Q5,RESID,X,X4,X5
C
C DIMENSIONS
      DIMENSION DH(2),H(2),ONE(2),Q(2),Q5(2),RESID(2),
     & X(2),X4(2),X5(2)
C
C INPUT STATEMENTS FOLLOW.
      READ(5,10) DELT
  10  FORMAT(D22.15)
      READ(5,10) A,B
C
      WRITE(6,20)
  20  FORMAT(/,' CONSOL4R',//)
      WRITE(6,25) DELT
  25  FORMAT(/,' DELT =',D22.15,/)
      WRITE(6,30) A,B
  30  FORMAT(' COEFFICIENTS,    A=',D22.15,' B=',D22.15)
C
C COMPUTE NUMBER OF STEPS PER PATH
      NS = (1./DELT) + 1.
C
      WRITE(6,40)
  40  FORMAT(/)
C
      WRITE(6,50)  NS
  50  FORMAT(' NUMBER OF STEPS PER PATH =',I5//)
C
C INITIALIZE COUNTER AND CONSTANTS FOR THE RUN
      ITOTIT=0
      ONE(1)=1.DO
      ONE(2)=0.DO
      Q(1) = .928746354D0
      Q(2) = .265465386D0
      CALL POW(5,Q,Q5)
C
C ****************************************************************
C ****                                                      *****
C ****         BEGINNING OF MAIN PART OF PROGRAM            *****
C ****                                                      *****
C ****************************************************************
C
C PATHS LOOP -- ITERATE THROUGH PATHS
      DO  60 NUMPAT = 1 ,  5
C
         WRITE(6,70) NUMPAT
  70     FORMAT(' PATH NUMBER =',I5)
C
C INITIALIZE COUNTERS AND CONSTANTS FOR THE PATH
         NUMIT=0
         T=0
C
```

```
C GENERATE A START POINT, X, FOR THE PATH
            IF(NUMPAT.EQ.1) X(1)= .034525822D0
            IF(NUMPAT.EQ.1) X(2)= .965323588D0
            IF(NUMPAT.EQ.2) X(1)=-.907408223D0
            IF(NUMPAT.EQ.2) X(2)= .331137401D0
            IF(NUMPAT.EQ.3) X(1)=-.595334945D0
            IF(NUMPAT.EQ.3) X(2)=-.760669419D0
            IF(NUMPAT.EQ.4) X(1)= .539470992D0
            IF(NUMPAT.EQ.4) X(2)=-.801256956D0
            IF(NUMPAT.EQ.5) X(1)= .928746354D0
            IF(NUMPAT.EQ.5) X(2)- .265465386D0
C
            WRITE(6,80)X
   80       FORMAT(/,'  START X=',2D22.15,//)
C
C STEPS LOOP -- ITERATE STEPS ALONG THE PATH
            DO 90 NUMSTP = 1 , NS
C
                T=T+DELT
                IF(T.GE.1.)T=1.
C
C NEWTON'S METHOD LOOP
                DO 100 ITER = 1,10
C
C EVALUATE THE CONTINUATION EQUATION AND DERIVATIVE
                    CALL POW(4,X,X4)
                    CALL MUL(X,X4,X5)
                    DO 104 I=1,2
                        H(I)=X5(I)+A*T*X(I)+(T*B*ONE(I)-(1.-T)*Q5(I))
                        DH(I)=5.*X4(I)+A*T*ONE(I)
  104               CONTINUE
C
C GENERATE THE RESIDUAL AND UPDATE X
                    CALL DIV(H,DH,RESID)
                    DO 106 I=1,2
                        X(I) = X(I) - RESID(I)
  106               CONTINUE
C
C$$$$$$$$$$$$$$$$$$$$$$$$$$$$$$$$$$$$$$$$$$$$$$$$$$$$$$$$$$$$$$$$$$$$$$$$$$C
C UNDERFLOWS:    IF YOU MUST PREVENT UNDERFLOWS, I SUGGEST YOU
C               TAKE THE COMMENTS OFF THE FOLLOWING BLOCK OF
C               CODE.   THE VALUE FOR EPSOO MAY NEED TO BE ADJUSTED.
C
C               EPSOO=1.D-22
C               IF(DABS(X(1)).LT.EPSOO) X(1)=0.D0
C               IF(DABS(X(2)).LT.EPSOO) X(2)=0.D0
C$$$$$$$$$$$$$$$$$$$$$$$$$$$$$$$$$$$$$$$$$$$$$$$$$$$$$$$$$$$$$$$$$$$$$$$$$$C
C
C UPDATE NEWTON ITERATION COUNTER
                    NUMIT = NUMIT + 1
C
C LEAVE NEWTON LOOP IF RESIDUAL IS SMALL ENOUGH
                    XRESID = DABS(RESID(1))+DABS(RESID(2))
                    IF(XRESID .LT. 1.D-12) GOTO 101
C
  100               CONTINUE
  101               CONTINUE
```

```
C
            IF(MOD(NUMSTP,NS/10).EQ.0.OR.NUMSTP.GE.(NS-4))THEN
               WRITE(6,110) NUMSTP
110            FORMAT(/'  STEP SUMMARY FOR STEP NUMBER',I5,' :')
               WRITE(6,115)
115            FORMAT('  -------------------------------------------------')
               WRITE(6,120) T
120            FORMAT('  T  = ',D22.15)
               WRITE(6,125)
125            FORMAT('                    REAL PART         IMAGINARY PART')
               WRITE(6,130) X
130            FORMAT('  X  =           ',D11.4,'          ',D11.4)
               WRITE(6,135)  H
135            FORMAT('  H  =           ',D11.4,'          ',D11.4)
               WRITE(6,140) RESID
140            FORMAT('  RESIDUAL =     ',D11.4,'          ',D11.4)
               WRITE(6,150) DH
150            FORMAT('  DH  =          ',D11.4,'          ',D11.4)
               WRITE(6,115)
               WRITE(6,40)
            ENDIF
C
 90      CONTINUE
C            BOTTOM OF LOOP -- NUMSTP
C
         ITOTIT=ITOTIT + NUMIT
C
         WRITE(6,160) NUMIT
160      FORMAT('  TOTAL NEWTON ITERATIONS ON PATH =',I5)
         WRITE(6,40)
C
 60   CONTINUE
C            BOTTOM OF LOOP -- NUMPAT
C
      WRITE(6,170) ITOTIT
170   FORMAT(//'  TOTAL NEWTON ITERATIONS =',I5//)
C
      STOP
      END
```

```
C PROGRAM NAME:  CONSOL5R
C
C PURPOSE: SOLVES THE GENERAL POLYNOMIAL EQUATION
C
C     EQ = X**IDEG + A(IDEG)*X**(IDEG-1) + ... + A(2)*X + A(1)
C
C USING THE CONTINUATION EQUATION
C
C     H  = X**IDEG + A(IDEG)*T*X**(IDEG-1) + . . .
C
C                           + T*A(2)*X + (T*A(1) - (1.-T)*Q**IDEG)
C
C REFERENCE: CHAPTER 1 OF "SOLVING POLYNOMIAL SYSTEMS USING
C            CONTINUATION" BY ALEXANDER MORGAN
C
C NOTE: 1.   THIS PROGRAM DOES NOT USE THE FORTRAN "COMPLEX"
C            DECLARATION.  IT EMULATES COMPLEX OPERATIONS
C            USING REAL ARITHMETIC.  SEE APPENDIX 2.
C
C       2.   THIS PROGRAM GENERATES MANY UNDERFLOWS.  THIS IS
C            NOT AN ERROR CONDITION.  IF YOU NEED TO PREVENT
C            UNDERFLOWS FROM OCCURRING, SEE THE COMMENTS LABELED
C            "UNDERFLOWS" BELOW.
C
C       3.   INPUT IS ON UNIT 5.  OUTPUT IS TO UNIT 6.
C            SAMPLE INPUT FOR DELT=.08, IDEG=5, A(J)=J/10:
C                8.D-02
C05
C                   1.D-01
C                   2.D-01
C                   3.D-01
C                   4.D-01
C                   5.D-01
C
C INPUT VARIABLES:
C
C     DELT          -- THE STEP SIZE FOR T
C     IDEG          -- DEGREE OF EQ
C     A             -- COEFFICIENTS FOR EQ, AN ARRAY OF LENGTH IDEG
C
C OUTPUT VARIABLES:
C
C   FOR EACH STEP:
C
C     NUMSTP        -- CURRENT STEP NUMBER
C     T             -- CURRENT VALUE OF T
C     X             -- CURRENT VALUE OF X
C     H             -- CURRENT VALUE OF THE CONTINUATION EQUATION
C     RESID         -- NEWTON METHOD RESIDUAL (SHOULD BE NEAR 0)
C     DH            -- CURRENT DERIVATIVE OF CONTINUATION EQUATION
C
C
C   AT THE END OF THE PATH:
C
C     NUMIT         -- NUMBER OF NEWTON ITERATIONS FOR THE PATH
```

461

```
C
C   AT THE END OF THE RUN:
C
C       ITOTIT     -- TOTAL NUMBER OF NEWTON ITERATIONS OVER ALL PATHS
C
C SUBROUTINES: DABS,DATAN,DIV,DCOS,DSIN,MOD,MUL,POW
C
C
C DECLARATIONS
      INTEGER I,IDEG,IDEGM1,IDEGP1,ITER,ITOTIT,J,JP1,NS,NUMIT,NUMPAT,
     &  NUMSTP
      DOUBLE PRECISION A,ANGLE,DELT,EPSOO,T,TWOPI,XRESID
      DOUBLE PRECISION DH,H,Q,QQ,RESID,X,XX,XXX,ONE
C
C DIMENSIONS
      DIMENSION A(100)
      DIMENSION DH(2),H(2),Q(2),QQ(2),RESID(2),X(2),XX(2),XXX(2),ONE(2)
C
C INPUT STATEMENTS FOLLOW.
      READ(5,10) DELT
   10 FORMAT(D22.15)
      READ(5,15) IDEG
   15 FORMAT(I2)
      READ(5,10) (A(I),I=1,IDEG)
C
      WRITE(6,20)
   20 FORMAT(/,'  CONSOL5R',//)
      WRITE(6,25) DELT
   25 FORMAT(/,'  DELT =',D22.15,/)
      WRITE(6,28) IDEG
   28 FORMAT(' DEGREE OF THE EQUATION =',I5)
      WRITE(6,30)
   30 FORMAT(//,'  COEFFICIENTS:',/)
      WRITE(6,35)(I,A(I),I=1,IDEG)
   35 FORMAT('      A(',I2,')=',D22.15)
C
      WRITE(6,40)
   40 FORMAT(/)
C
C COMPUTE NUMBER OF STEPS PER PATH
      NS = (1./DELT) + 1.
C
      WRITE(6,40)
C
      WRITE(6,50)  NS
   50 FORMAT(' NUMBER OF STEPS PER PATH =',I5//)
C
C INITIALIZE COUNTER AND CONSTANTS FOR THE RUN
      ITOTIT=0
      ONE(1)=1.D0
      ONE(2)=0.D0
      IDEGM1=IDEG-1
      IDEGP1=IDEG+1
      Q(1) = .928746354D0
      Q(2) = .265465386D0
      CALL POW(IDEG,Q,QQ)
C
```

```
C ****************************************************************
C *****                                                    *****
C *****        BEGINNING OF MAIN PART OF PROGRAM           *****
C *****                                                    *****
C ****************************************************************
C
C PATHS LOOP -- ITERATE THROUGH PATHS
      DO  60 NUMPAT = 1,IDEG
C
          WRITE(6,70) NUMPAT
   70     FORMAT(' PATH NUMBER -',I5)
C
C INITIALIZE COUNTERS AND CONSTANTS FOR THE PATH
          NUMIT=0
          T=0
C
C GENERATE A START POINT, S, FOR THE PATH
          TWOPI = 8.D0*DATAN(1.D0)
          ANGLE = TWOPI*NUMPAT/IDEG
          XX(1)=DCOS(ANGLE)
          XX(2)=DSIN(ANGLE)
          CALL MUL(Q,XX,X)
C
          WRITE(6,80)X
   80     FORMAT(/,'  START X=',2D22.15,//)
C
C STEPS LOOP -- ITERATE STEPS ALONG THE PATH
          DO 90 NUMSTP = 1 , NS
              T=T+DELT
              IF(T.GE.1.)T=1.
C
C NEWTON'S METHOD LOOP
              DO 100 ITER = 1,10
C EVALUATE THE CONTINUATION EQUATION AND DERIVATIVE
                  CALL POW(IDEGM1,X,XX)
                  CALL MUL(XX,X,XXX)
                  DO 102 I=1,2
                      H(I) = (T*A(1)*ONE(I) - (1.-T)*QQ(I)) + XXX(I)
                      DH(I) = IDEG*(XX(I))
                      XX(I)=ONE(I)
  102             CONTINUE
                  DO 104 J=1,IDEGM1
                      JP1=J+1
                      CALL MUL(X,XX,XXX)
                      DO 106 I=1,2
                          H(I)  =  H(I) +  A(JP1)*T*XXX(I)
                          DH(I) = DH(I) + J*A(JP1)*T*XX(I)
                          XX(I) =XXX(I)
  106                 CONTINUE
  104             CONTINUE
C
C  GENERATE THE RESIDUAL AND UPDATE X
                  CALL DIV(H,DH,RESID)
                  DO 108 I=1,2
                      X(I)=X(I)-RESID(I)
  108             CONTINUE
C
```

```
C$$$$$$$$$$$$$$$$$$$$$$$$$$$$$$$$$$$$$$$$$$$$$$$$$$$$$$$$$$$$$$$$$$$$$$$$$$$$$$C
C UNDERFLOWS :    IF YOU MUST PREVENT UNDERFLOWS, I SUGGEST YOU
C                 TAKE THE COMMENTS OFF THE FOLLOWING BLOCK OF
C                 CODE.  THE VALUE FOR EPSOO MAY NEED TO BE ADJUSTED.
C
C                 EPSOO=1.D-22
C                 IF(DABS(X(1)).LT.EPSOO) X(1)=0.DO
C                 IF(DABS(X(2)).LT.EPSOO) X(2)=0.DO
CS$$$$$$$$$$$$$$$$$$$$$$$$$$$$$$$$$$$$$$$$$$$$$$$$$$$$$$$$$$$$$$$$$$$$$$$$$$$$$C
C
C UPDATE NEWTON ITERATION COUNTER
                  NUMIT = NUMIT + 1
C
C LEAVE NEWTON LOOP IF RESIDUAL IS SMALL ENOUGH
                  XRESID = DABS(RESID(1))+DABS(RESID(2))
                  IF(XRESID .LT. 1.D-12) GOTO 101
  100             CONTINUE
  101             CONTINUE
C
                  IF(MOD(NUMSTP,NS/10).EQ.O.OR.NUMSTP.GE.(NS-4))THEN
                  WRITE(6,110) NUMSTP
  110             FORMAT(/'  STEP SUMMARY FOR STEP NUMBER',I5,' :')
                  WRITE(6,115)
  115             FORMAT(' -----------------------------------------------')
                  WRITE(6,120) T
  120             FORMAT(' T  = ',D22.15)
                  WRITE(6,125)
  125             FORMAT('                 REAL PART        IMAGINARY PART')
                  WRITE(6,130) X
  130             FORMAT(' X =        ',D11.4,'         ',D11.4)
                  WRITE(6,135) H
  135             FORMAT(' H =        ',D11.4,'         ',D11.4)
                  WRITE(6,140) RESID
  140             FORMAT(' RESIDUAL = ',D11.4,'         ',D11.4)
                  WRITE(6,150) DH
  150             FORMAT(' DH =       ',D11.4,'         ',D11.4)
                  WRITE(6,115)
                  WRITE(6,40)
                  ENDIF
C
  90       CONTINUE
C              BOTTOM OF LOOP -- NUMSTP
C
           ITOTIT=ITOTIT + NUMIT
C
           WRITE(6,160) NUMIT
  160      FORMAT(' TOTAL NEWTON ITERATIONS ON PATH =',I5)
           WRITE(6,40)
C
  60    CONTINUE
C           BOTTOM OF LOOP -- NUMPAT
C
        WRITE(6,170) ITOTIT
  170   FORMAT(//' TOTAL NEWTON ITERATIONS =',I5//)
C
        STOP
        END
```

Section 2: Programs for Chapter 2

```
C PROGRAM NAME:  CONSOL6R
C
C PURPOSE:   SOLVES A SYSTEM OF TWO SECOND-DEGREE
C            EQUATIONS IN TWO UNKNOWNS:
C
C EQ(J) = A(J)*X**2 + B(J)*Y**2 + C(J)*X*Y + D(J)*X + E(J)*Y+ F(J)
C
C            FOR J=1,2 USING THE CONTINUATION EQUATION
C
C    H(J)  =  (1-T)*(P(J)**2 * X**2 - Q(J)**2) + T * EQ(J)
C
C REFERENCE: CHAPTER 2 OF "SOLVING POLYNOMIAL SYSTEMS USING
C            CONTINUATION" BY ALEXANDER MORGAN
C
C NOTE: 1.  THIS PROGRAM DOES NOT USE THE FORTRAN "COMPLEX"
C           DECLARATION.  IT EMULATES COMPLEX OPERATIONS
C           USING REAL ARITHMETIC.  SEE APPENDIX 2.
C
C       2.  THIS PROGRAM GENERATES MANY UNDERFLOWS.  THIS IS
C           NOT AN ERROR CONDITION.  IF YOU NEED TO PREVENT
C           UNDERFLOWS FROM OCCURRING, SEE THE COMMENTS LABELED
C           "UNDERFLOWS" BELOW.
C
C       3.  INPUT IS ON UNIT 5.  OUTPUT IS TO UNITS 6 AND 8.
C           (UNIT 6 IS THE FULL OUTPUT.  UNIT 8 CONTAINS THE SOLUTION LIST.)
C           SAMPLE INPUT FOLLOWS:
CTITLE:    SYSTEM OF TWO QUADRICS,   NO SOLUTIONS AT INFINITY
C              1.D-04     DELT
C           -.00098D 00   X(1)**2     *
C           978000.D+00   X(2)**2     *
C             -9.8D+00    X(1)*X(2)   *
C           -235.0D+00    X(1)        *   EQUATION 1
C           88900.0D+00   X(2)        *
C           -1.000D+00    CONSTANT    *
C           -.0100D+00    X(1)**2     $
C           -.9840D+00    X(2)**2     $
C           -29.70D+00    X(1)*X(2)   $
C            .00987D+00   X(1)        $   EQUATION 2
C           -.1240D+00    X(2)        $
C           -.2500D+00    CONSTANT    $
C
C INPUT VARIABLES:
C
C    TITLE                     -- ANY 60 CHARACTERS
C    DELT                      -- THE STEP SIZE FOR T
C    A(J),B(J),C(J),D(J),E(J),F(J)  -- COEFFICIENTS FOR EQ(J)
C
C OUTPUT VARIABLES:
C
C   FOR EACH STEP:
C
C    NUMSTP     -- CURRENT STEP NUMBER
C    T          -- CURRENT VALUE OF T
C    X,Y        -- CURRENT VALUES OF X AND Y
C    H(J)       -- CURRENT VALUES OF THE CONTINUATION EQUATIONS
```

```
C      XRESID      -- NORM OF NEWTON'S METHOD RESIDUAL (SHOULD BE NEAR 0)
C      XNRMDH      -- MATRIX NORM OF JACOBIAN, DH
C      DET         -- DETERMINANT OF JACOBIAN, DH
C
C
C   AT THE END OF THE PATH:
C
C      NUMIT       -- NUMBER OF NEWTON ITERATIONS FOR THE PATH
C
C   AT THE END OF THE RUN:
C
C      ITOTIT      -- TOTAL NUMBER OF NEWTON ITERATIONS OVER ALL PATHS
C
C
C SUBROUTINES: DABS,DIV,MOD,MUL
C
C
C DECLARATIONS
       INTEGER I,ITER,ITOTIT,J,NS,NUMIT,NUMPAT,NUMSTP,TITLE
       DOUBLE PRECISION A,B,C,D,DELT,E,EPSOO,F,T,TEST,TT,TWOPI,XNRMDH,
      & XRESID
       DOUBLE PRECISION DEQ,DET,DH,EQ,H,ONE,P,PP,PPX,PPXX,PPY,
      & PPYY,Q,QQ,RESID,TEMP1,TEMP2,TEMP3,X,XX,XY,Y,YY
C
C DIMENSIONS
       DIMENSION A(2),B(2),C(2),D(2),DEQ(2,2,2),DH(2,2,2),E(2),EQ(2,2),
      & F(2),H(2,2),P(2,2),PP(2,2),Q(2,2),QQ(2,2),RESID(2,2),TITLE(60)
       DIMENSION DET(2),ONE(2),PPX(2),PPXX(2),PPY(2),PPYY(2),
      & TEMP1(2),TEMP2(2),TEMP3(2),X(2),XX(2),XY(2),Y(2),YY(2)
C
C
C INPUT STATEMENTS FOLLOW.
       READ(5,5) TITLE
    5  FORMAT(60A1)
       READ(5,10) DELT
   10  FORMAT(D22.15)
       READ(5,10) (  A(J),B(J),C(J),D(J),E(J),F(J),J=1,2)
C
       WRITE(6,20)
   20  FORMAT(/,'   CONSOL6R',//)
       WRITE(6,5) TITLE
       WRITE(6,25) DELT
   25  FORMAT(/,'  DELT =',D22.15,/)
       WRITE(6,30)(J,A(J),B(J),C(J),D(J),E(J),F(J),J=1,2)
   30  FORMAT('  COEFFICIENTS FOR EQUATION ',I1,//,'  A=',D22.15,/,
      &'  B=',D22.15,/,'  C=',D22.15,/,'  D=',D22.15,/,'  E=',D22.15,/,
      &'  F=',D22.15,///)
C
C COMPUTE NUMBER OF STEPS PER PATH
       NS = (1.D0/DELT) + 1.D0
C
       WRITE(6,40)
   40  FORMAT(/)
C
       WRITE(6,50)  NS
   50  FORMAT(' NUMBER OF STEPS PER PATH =',I8//)
C
```

466

```
C INITIALIZE COUNTER AND CONSTANTS FOR THE RUN
      ITOTIT=0
      ONE(1)=1.DO
      ONE(2)=0.DO
C
      P(1,1)= .12324754231DO
      P(2,1)= .76253746298DO
      P(1,2)= .93857838950DO
      P(2,2)=-.99375892810DO
C
      Q(1,1)= .58720452864DO
      Q(2,1)= .01321964722DO
      Q(1,2)= .97884134700DO
      Q(2,2)=-.14433009712DO
C
      CALL MUL(P(1,1),P(1,1),PP(1,1))
      CALL MUL(P(1,2),P(1,2),PP(1,2))
      CALL MUL(Q(1,1),Q(1,1),QQ(1,1))
      CALL MUL(Q(1,2),Q(1,2),QQ(1,2))
C
C ****************************************************************
C *****                                                      *****
C *****       BEGINNING OF MAIN PART OF PROGRAM              *****
C *****                                                      *****
C ****************************************************************
C
C PATHS LOOP -- ITERATE THROUGH PATHS
      DO  60 NUMPAT = 1,4
C
          WRITE(6,70) NUMPAT
   70     FORMAT(' PATH NUMBER =',I8)
C
C INITIALIZE COUNTERS AND CONSTANTS FOR THE PATH
          NUMIT=0
          T=0.DO
C
C   GENERATE A START POINT, (X,Y), FOR THE PATH
          CALL DIV(Q(1,1),P(1,1),X)
          CALL DIV(Q(1,2),P(1,2),Y)
          IF(NUMPAT.EQ.2 .OR. NUMPAT.EQ.4) THEN
              X(1)=-X(1)
              X(2)=-X(2)
          ENDIF
          IF(NUMPAT.EQ.3 .OR. NUMPAT.EQ.4) THEN
              Y(1)=-Y(1)
              Y(2)=-Y(2)
          ENDIF
C
          WRITE(6,80)X,Y
   80     FORMAT(/,' START X,Y=',2D11.4,'   ',2D11.4,//)
C
C STEPS LOOP -- ITERATE STEPS ALONG THE PATH
          DO 90 NUMSTP = 1 , NS
C
              T=T+DELT
              IF(T.GE.1.DO)T=1.DO
C
```

```
C NEWTON'S METHOD LOOP
              DO 100 ITER = 1,10
C
C   EVALUATE THE ORIGINAL SYSTEM EQUATIONS AND DERIVATIVES.
C   THE SECOND INDEX, 1 OR 2, INDICATES PARTIAL DERIVATIVE
C   WITH RESPECT TO X OR Y RESPECTIVELY.
              CALL MUL(X,X,XX)
              CALL MUL(Y,Y,YY)
              CALL MUL(X,Y,XY)
C
              DO 102 I=1,2
              DO 102 J=1,2
              EQ(I,J) = A(J)*XX(I) + B(J)*YY(I) + C(J)*XY(I)
        &            + D(J)*X(I) + E(J)*Y(I)+ F(J)*ONE(I)
              DEQ(I,J,1)= 2.*A(J)*X(I)+ C(J)*Y(I)+ D(J)*ONE(I)
              DEQ(I,J,2)= 2.*B(J)*Y(I)+ C(J)*X(I)+ E(J)*ONE(I)
  102         CONTINUE
C
C   EVALUATE THE CONTINUATION EQUATIONS AND DERIVATIVES.
              CALL MUL(PP(1,1),XX,PPXX)
              CALL MUL(PP(1,1), X,PPX )
              CALL MUL(PP(1,2),YY,PPYY)
              CALL MUL(PP(1,2), Y,PPY )
C
              TT=1.-T
              DO 104 I=1,2
              H(I,1)    = TT*(PPXX(I) - QQ(I,1)) + T*EQ(I,1)
              DH(I,1,1)= TT*2.*PPX(I)            + T*DEQ(I,1,1)
              DH(I,1,2) =                          T*DEQ(I,1,2)
C
              H(I,2)    =   TT*(PPYY(I) - QQ(I,2)) + T*EQ(I,2)
              DH(I,2,1) =                            T*DEQ(I,2,1)
              DH(I,2,2) = TT*2.*PPY(I)            + T*DEQ(I,2,2)
  104         CONTINUE
C
C   GENERATE THE RESIDUAL AND UPDATE X,Y.
C   SOLVE THE RESIDUAL EQUATION DH*RESID=H USING CRAMER'S RULE.
              CALL MUL(DH(1,1,1),DH(1,2,2),TEMP1)
              CALL MUL(DH(1,1,2),DH(1,2,1),TEMP2)
              DET(1)=TEMP1(1)-TEMP2(1)
              DET(2)=TEMP1(2)-TEMP2(2)
C
              CALL MUL( H(1,1)   ,DH(1,2,2),TEMP1)
              CALL MUL( H(1,2)   ,DH(1,1,2),TEMP2)
              TEMP3(1)=TEMP1(1)-TEMP2(1)
              TEMP3(2)=TEMP1(2)-TEMP2(2)
              CALL DIV(TEMP3,DET,RESID(1,1))
C
              CALL MUL( H(1,2)   ,DH(1,1,1),TEMP1)
              CALL MUL( H(1,1)   ,DH(1,2,1),TEMP2)
              TEMP3(1)=TEMP1(1)-TEMP2(1)
              TEMP3(2)=TEMP1(2)-TEMP2(2)
              CALL DIV(TEMP3,DET,RESID(1,2))
C
```

```
              DO 106 I=1,2
                  X(I) = X(I) - RESID(I,1)
                  Y(I) = Y(I) - RESID(I,2)
 106          CONTINUE
C$$$$$$$$$$$$$$$$$$$$$$$$$$$$$$$$$$$$$$$$$$$$$$$$$$$$$$$$$$$$$$$$$$$$$$$$$$$$$C
C UNDERFLOWS:   IF YOU MUST PREVENT UNDERFLOWS, I SUGGEST YOU
C              TAKE THE COMMENTS OFF THE FOLLOWING BLOCK OF
C              CODE.  THE VALUE FOR EPSOO MAY NEED TO BE ADJUSTED.
C
C              EPSOO=1.D-22
C              DO 108 I=1,2
C                  IF(DABS(X(I)).LT.EPSOO) X(I)=0.DO
C                  IF(DABS(Y(I)).LT.EPSOO) Y(I)=0.DO
C108          CONTINUE
C$$$$$$$$$$$$$$$$$$$$$$$$$$$$$$$$$$$$$$$$$$$$$$$$$$$$$$$$$$$$$$$$$$$$$$$$$$$$$C
C UPDATE NEWTON ITERATION COUNTER
              NUMIT = NUMIT + 1
C
C LEAVE NEWTON LOOP IF RESIDUAL IS SMALL ENOUGH
              XRESID = DABS(RESID(1,1))+DABS(RESID(2,1))
     &               + DABS(RESID(1,2))+DABS(RESID(2,2))
              IF(XRESID .LT. 1.D-12) GOTO 101
 100          CONTINUE
 101          CONTINUE
C
              IF(MOD(NUMSTP,NS/10).EQ.O.OR.NUMSTP.GE.(NS-4))THEN
              WRITE(6,110) NUMSTP
 110          FORMAT(/' STEP SUMMARY FOR STEP NUMBER',I8,' :')
              WRITE(6,115)
 115          FORMAT(' ---------------------------------------------')
              WRITE(6,120) T
 120          FORMAT(' T   = ',D22.15)
              WRITE(6,125)
 125          FORMAT('           REAL PART        IMAGINARY PART')
              WRITE(6,130) X
 130          FORMAT(' X   =        ',D11.4,'        ',D11.4)
              WRITE(6,131) Y
 131          FORMAT(' Y   =        ',D11.4,'        ',D11.4)
              WRITE(6,135)  (J,(H(I,J),I=1,2),J=1,2)
 135          FORMAT(' H(',I1,') =       ',D11.4,'        ',D11.4)
              WRITE(6,140) XRESID
 140          FORMAT(' NORM OF RESIDUAL = ',D11.4)
C
C COMPUTE THE MATRIX NORM OF DH, XNRMDH.
              XNRMDH = DABS(DH(1,1,1))+DABS(DH(2,1,1))
     &               + DABS(DH(1,2,1))+DABS(DH(2,2,1))
              TEST   = DABS(DH(1,1,2))+DABS(DH(2,1,2))
     &               + DABS(DH(1,2,2))+DABS(DH(2,2,2))
              IF(TEST.GT.XNRMDH) XNRMDH = TEST
              WRITE(6,145) XNRMDH
 145          FORMAT(' NORM OF DH =',D11.4)
C
              WRITE(6,150) DET
 150          FORMAT(' DET(DH) =       ',2D11.4)
              WRITE(6,115)
              WRITE(6,40)
              ENDIF
```

```
C
                  IF(NUMSTP.EQ.NS) THEN
                      WRITE(8,70) NUMPAT
                      WRITE(8,130) X
                      WRITE(8,131) Y
                      WRITE(8,140) XRESID
                      WRITE(8,40)
                  ENDIF
C
  90        CONTINUE
C              BOTTOM OF LOOP -- NUMSTP
C
            ITOTIT=ITOTIT + NUMIT
C
            WRITE(6,160) NUMIT
 160        FORMAT(' TOTAL NEWTON ITERATIONS ON PATH =',I8)
            WRITE(6,40)
C
  60    CONTINUE
C              BOTTOM OF LOOP -- NUMPAT
C
        WRITE(6,170) ITOTIT
 170  FORMAT(//' TOTAL NEWTON ITERATIONS =',I8//)
C
        STOP
        END
C
```

```
C PROGRAM NAME:   CONSOL7R
C
C
C PURPOSE:    SOLVES A SYSTEM OF TWO POLYNOMIAL
C             EQUATIONS IN TWO UNKNOWNS:
C
C                   EQ(J)=0
C
C             FOR J=1,2, USING THE CONTINUATION EQUATIONS
C
C H(1) = (1-T)*(P(1)**IDEG(1) * X**IDEG(1)- Q(1)**IDEG(1))+ T*EQ(1)
C H(2) = (1-T)*(P(2)**IDEG(2) * Y**IDEG(2)- Q(2)**IDEG(2))+ T*EQ(2)
C
C             WHERE DEGREE(EQ(J))=IDEG(J).
C
C REFERENCE: CHAPTER 2 OF "SOLVING POLYNOMIAL SYSTEMS USING
C             CONTINUATION" BY ALEXANDER MORGAN
C
C NOTE: 1. THIS CODE REQUIRES THE USER TO PROVIDE SUBROUTINE "FFUN7R"
C          TO RETURN EQ AND DEQ VALUES.  THREE SAMPLE SUBROUTINE
C          FFUN7R'S ARE INCLUDED FOR PROBLEMS (2-19), (2-26), AND
C          (2-27), NAMED FFUN7RA, FFUN7RB, FFUN7RC RESPECTIVELY.
C          TO USE ONE OF THESE SUBROUTINES, CHANGE THE NAME TO FFUN7R.
C
C       2. THIS PROGRAM DOES NOT USE THE FORTRAN "COMPLEX"
C          DECLARATION.  IT EMULATES COMPLEX OPERATIONS
C          USING REAL ARITHMETIC.  SEE APPENDIX 2.
C
C       3. THIS PROGRAM GENERATES MANY UNDERFLOWS.  THIS IS
C          NOT AN ERROR CONDITION.  IF YOU NEED TO PREVENT
C          UNDERFLOWS FROM OCCURRING, SEE THE COMMENTS LABELED
C          "UNDERFLOWS" BELOW.
C
C       4. INPUT IS ON UNIT 5.  OUTPUT IS TO UNITS 6 AND 8.
C          (UNIT 6 IS THE FULL OUTPUT.  UNIT 8 CONTAINS THE SOUTION LIST.)
C          SAMPLE INPUT FOLLOWS:
CTHIS IS THE SAMPLE TITLE.  SIXTY COLUMNS ARE ALLOWED FOR IT.
C                   1.1D-02      DELT
C05                              IDEG(1)
C07                              IDEG(2)
C                   1.D-01       A(1)
C                   2.D-01       B(1)
C                   3.D-01       C(1)
C                   4.D-01       D(1)
C                   5.D-01       E(1)
C                   6.D-01       F(1)
C                   7.D-01       G(1)
C                   1.D+01       A(2)
C                   2.D-01       B(2)
C                   3.D+01       C(2)
C                   4.D-01       D(2)
C                   5.D+01       E(2)
C                   6.D-01       F(2)
C                   7.D+01       G(2)
C
```

```
C
C INPUT VARIABLES:
C
C     DELT                                -- STEP SIZE FOR T
C     IDEG(J)                             -- DEGREE OF EQ(J)
C     A(J),B(J),C(J),D(J),E(J),F(J),G(J)  -- COEFFICIENTS FOR EQ(J)
C
C
C OUTPUT VARIABLES:
C
C   FOR EACH STEP:
C
C     NUMSTP      -- CURRENT STEP NUMBER
C     T           -- CURRENT VALUE OF T
C     X,Y         -- CURRENT VALUES OF X AND Y
C     H(J)        -- CURRENT VALUES OF THE CONTINUATION EQUATIONS
C     XRESID      -- NORM OF NEWTON'S METHOD RESIDUAL  (SHOULD BE NEAR 0)
C     XNRMDH      -- MATRIX NORM OF JACOBIAN, DH
C     DET         -- DETERMINANT OF JACOBIAN, DH
C
C   AT THE END OF THE PATH:
C
C     NUMIT       -- NUMBER OF NEWTON ITERATIONS FOR THE PATH
C
C   AT THE END OF THE RUN:
C
C     ITOTIT      -- TOTAL NUMBER OF NEWTON ITERATIONS OVER ALL PATHS
C
C
C SUBROUTINES: DABS,DATAN,DCOS,DSIN,FFUN7R,MOD
C
C
C DECLARATIONS
      INTEGER I,ICOUNT,IDEG,IDEGM1,ITER,ITOTDG,ITEST,ITOTIT,J,NS,NUMIT,
     & NUMPAT,NUMSTP,TITLE
      DOUBLE PRECISION A,ANGLE,B,C,D,DELT,E,EPSOO,F,G,T,TEST,TT,TWOPI,
     & XNRMDH,XRESID
      DOUBLE PRECISION DEQ,DET,DH,EQ,H,P,PP,PPXDG,PPXDG1,PPYDG,PPYDG1,
     & Q,QQ,RESID,TEMP1,TEMP2,TEMP3,X,XDG,XDGM1,Y,YDG,YDGM1
C
C DIMENSIONS
      DIMENSION DEQ(2,2,2),DH(2,2,2),EQ(2,2),H(2,2),ICOUNT(2),IDEG(2),
     & P(2,2),PP(2,2),Q(2,2),QQ(2,2),RESID(2,2),TITLE(60)
      DIMENSION DET(2),IDEGM1(2),PPXDG(2),PPXDG1(2),PPYDG(2),PPYDG1(2),
     & TEMP1(2),TEMP2(2),TEMP3(2),X(2),XDG(2),XDGM1(2),Y(2),YDG(2),
     & YDGM1(2)
C
C COMMON
      COMMON/COEFS/ A(2),B(2),C(2),D(2),E(2),F(2),G(2)
C
C
C INPUT STATEMENTS FOLLOW.
      READ(5,5) TITLE
    5 FORMAT(60A1)
      READ(5,10) DELT
   10 FORMAT(D22.15)
```

```
          READ(5,15) IDEG
    15    FORMAT(I2)
          READ(5,10) ( A(J),B(J),C(J),D(J),E(J),F(J),G(J),J=1,2 )
C
          WRITE(6,20)
    20    FORMAT(/,'  CONSOL7R',//)
          WRITE(6,5) TITLE
          WRITE(6,25) DELT
    25    FORMAT(/,'  DELT =',D22.15,/)
          WRITE(6,30) (J,IDEG(J),J=1,2)
    30    FORMAT(/,'  IDEG(',I1,')=',I2,'   IDEG(',I1,')=',I2,//)
          WRITE(6,35)(J,A(J),B(J),C(J),D(J),E(J),F(J),G(J),J=1,2)
    35    FORMAT('  COEFFICIENTS FOR EQUATION ',I1,//,'  A=',D22.15,/,
         &'  B=',D22.15,/,'  C=',D22.15,/,'  D=',D22.15,/,'  E=',D22.15,/,
         &'  F=',D22.15,/,'  G=',D22.15,///)
C
C
C COMPUTE NUMBER OF STEPS PER PATH
          NS = (1.D0/DELT) + 1.D0
C
          WRITE(6,40)
    40    FORMAT(/)
C
          WRITE(6,50)  NS
    50    FORMAT(' NUMBER OF STEPS PER PATH =',I8//)
C
C INITIALIZE COUNTER AND CONSTANTS FOR THE RUN
C
C    COUNTER FOR TOTAL NUMBER OF CORRECTOR ITERATIONS
          ITOTIT=0
C
C    TOTAL DEGREE FOR THE SYSTEM
          ITOTDG=IDEG(1)*IDEG(2)
C
C    ICOUNT IS A COUNTER USED FOR GENERATING START POINTS.
          ICOUNT(1)=0
          ICOUNT(2)=1
C
          TWOPI=8.D0*DATAN(1.D0)
C
          P(1,1)= .12324754231D0
          P(2,1)= .76253746298D0
          P(1,2)= .93857838950D0
          P(2,2)=-.99375892810D0
C
          Q(1,1)= .58720452864D0
          Q(2,1)= .01321964722D0
          Q(1,2)= .97884134700D0
          Q(2,2)=-.14433009712D0
C
          CALL POW(IDEG(1),P(1,1),PP(1,1))
          CALL POW(IDEG(2),P(1,2),PP(1,2))
          CALL POW(IDEG(1),Q(1,1),QQ(1,1))
          CALL POW(IDEG(2),Q(1,2),QQ(1,2))
C
```

```
C **************************************************************
C *****                                                    *****
C *****        BEGINNING OF MAIN PART OF PROGRAM           *****
C *****                                                    *****
C **************************************************************
C
C PATHS LOOP -- ITERATE THROUGH PATHS
      DO  60 NUMPAT = 1,ITOTDG
C
          WRITE(6,70) NUMPAT
   70     FORMAT(' PATH NUMBER =',I8)
C
C INITIALIZE COUNTERS AND CONSTANTS FOR THE PATH
          NUMIT=0
          T=0.D0
C
C GENERATE A START POINT, (X,Y), FOR THE PATH
C
C ICOUNT IS A COUNTER USED TO INCREMENT EACH VARIABLE AROUND
C THE UNIT CIRCLE SO THAT EVERY COMBINATION OF START VALUES
C IS CHOSEN BY THE END OF THE RUN.  ICOUNT IS INITIALIZED ABOVE.
C
          ITEST = ICOUNT(1)
          IF( ITEST  .GE. IDEG(1) ) ICOUNT(1)=1
          IF( ITEST  .GE. IDEG(1) ) ICOUNT(2)=ICOUNT(2)+1
          IF( ITEST  .LT. IDEG(1) ) ICOUNT(1)=ICOUNT(1)+1
C
          ANGLE = (TWOPI*ICOUNT(1))/IDEG(1)
          CALL DIV(Q(1,1),P(1,1),TEMP1)
          TEMP2(1)=DCOS(ANGLE)
          TEMP2(2)=DSIN(ANGLE)
          CALL MUL(TEMP1,TEMP2,X)
C
          ANGLE = (TWOPI*ICOUNT(2))/IDEG(2)
          CALL DIV(Q(1,2),P(1,2),TEMP1)
          TEMP2(1)=DCOS(ANGLE)
          TEMP2(2)=DSIN(ANGLE)
          CALL MUL(TEMP1,TEMP2,Y)
C
C STEPS LOOP -- ITERATE STEPS ALONG THE PATH
          DO 90 NUMSTP = 1 , NS
              T=T+DELT
              IF(NUMSTP.EQ.NS)T=1.D0
C
C NEWTON'S METHOD LOOP
              DO 100 ITER = 1,10
C
C  COMPUTE POWERS OF X AND Y TO BE USED IN H AND DH AND (OPTIONALLY)
C  BY FFUN7R
                  IDEGM1(1)=IDEG(1)-1
                  IDEGM1(2)=IDEG(2)-1
                  CALL POW(IDEGM1(1),X,XDGM1)
                  CALL POW(IDEGM1(2),Y,YDGM1)
                  CALL MUL(X,XDGM1,XDG)
                  CALL MUL(Y,YDGM1,YDG)
C
                  CALL FFUN7R(X,Y,XDGM1,YDGM1,XDG,YDG, EQ,DEQ)
```

```
C
C   EVALUATE THE CONTINUATION EQUATIONS AND DERIVATIVES.
C
                  CALL MUL(PP(1,1),XDGM1,PPXDG1)
                  CALL MUL(X,PPXDG1,PPXDG)
                  CALL MUL(PP(1,2),YDGM1,PPYDG1)
                  CALL MUL(Y,PPYDG1,PPYDG)
C
                  TT=1.-T
                  DO 104 I=1,2
                     H(I,1)    = TT*(PPXDG(I) - QQ(I,1)) + T*EQ(I,1)
                     DH(I,1,1) = TT*IDEG(1)*PPXDG1(I) + T*DEQ(I,1,1)
                     DH(I,1,2) =                         T*DEQ(I,1,2)
C
                     H(I,2)    = TT*(PPYDG(I) - QQ(I,2)) + T*EQ(I,2)
                     DH(I,2,1) =                         T*DEQ(I,2,1)
                     DH(I,2,2) = TT*IDEG(2)*PPYDG1(I) + T*DEQ(I,2,2)
  104             CONTINUE
C
C   GENERATE THE RESIDUAL AND UPDATE X,Y.
C   SOLVE THE RESIDUAL EQUATION DH*RESID=H USING CRAMER'S RULE.
                  CALL MUL(DH(1,1,1),DH(1,2,2),TEMP1)
                  CALL MUL(DH(1,1,2),DH(1,2,1),TEMP2)
                  DET(1)=TEMP1(1)-TEMP2(1)
                  DET(2)=TEMP1(2)-TEMP2(2)
C
                  CALL MUL( H(1,1)   ,DH(1,2,2),TEMP1)
                  CALL MUL( H(1,2)   ,DH(1,1,2),TEMP2)
                  TEMP3(1)=TEMP1(1)-TEMP2(1)
                  TEMP3(2)=TEMP1(2)-TEMP2(2)
                  CALL DIV(TEMP3,DET,RESID(1,1))
C
                  CALL MUL( H(1,2)   ,DH(1,1,1),TEMP1)
                  CALL MUL( H(1,1)   ,DH(1,2,1),TEMP2)
                  TEMP3(1)=TEMP1(1)-TEMP2(1)
                  TEMP3(2)=TEMP1(2)-TEMP2(2)
                  CALL DIV(TEMP3,DET,RESID(1,2))
C
                  DO 106 I=1,2
                     X(I) = X(I) - RESID(I,1)
                     Y(I) = Y(I) - RESID(I,2)
  106             CONTINUE
C
C$$$$$$$$$$$$$$$$$$$$$$$$$$$$$$$$$$$$$$$$$$$$$$$$$$$$$$$$$$$$$$$$$$$$$$$$$$$$
C UNDERFLOWS:    IF YOU MUST PREVENT UNDERFLOWS, I SUGGEST YOU
C               TAKE THE COMMENTS OFF THE FOLLOWING BLOCK OF
C               CODE.  THE VALUE FOR EPSOO MAY NEED TO BE ADJUSTED.
C
C               EPSOO=1.D-22
C               DO 108 I=1,2
C                  IF(DABS(X(I)).LT.EPSOO) X(I)=0.D0
C                  IF(DABS(Y(I)).LT.EPSOO) Y(I)=0.D0
C108             CONTINUE
C$$$$$$$$$$$$$$$$$$$$$$$$$$$$$$$$$$$$$$$$$$$$$$$$$$$$$$$$$$$$$$$$$$$$$$$$$$$$
C
C UPDATE NEWTON ITERATION COUNTER
                  NUMIT = NUMIT + 1
```

```
C
C LEAVE NEWTON LOOP IF RESIDUAL IS SMALL ENOUGH
                XRESID = DABS(RESID(1,1))+DABS(RESID(2,1))
     &                 + DABS(RESID(1,2))+DABS(RESID(2,2))
                IF(XRESID .LT. 1.D-12) GOTO 101
C
 100           CONTINUE
 101           CONTINUE
C
C
                IF(MOD(NUMSTP,NS/10).EQ.0.OR.NUMSTP.GE.(NS-4))THEN
                 WRITE(6,110) NUMSTP
 110             FORMAT(/'  STEP SUMMARY FOR STEP NUMBER',I8,' :')
                 WRITE(6,115)
 115             FORMAT(' --------------------------------------------')
                 WRITE(6,120) T
 120             FORMAT(' T    = ',D22.15)
                 WRITE(6,125)
 125             FORMAT('            REAL PART          IMAGINARY PART')
                 WRITE(6,130) X
 130             FORMAT(' X   =        ',D11.4,'           ',D11.4)
                 WRITE(6,131) Y
 131             FORMAT(' Y   =        ',D11.4,'           ',D11.4)
                 WRITE(6,135)  (J,(H(I,J),I=1,2),J=1,2)
 135             FORMAT(' H(',I1,') =        ',D11.4,'           ',D11.4)
                 WRITE(6,140) XRESID
 140             FORMAT(' NORM OF RESIDUAL = ',D11.4)
C
C COMPUTE THE MATRIX NORM OF DH, XNRMDH.
                XNRMDH = DABS(DH(1,1,1))+DABS(DH(2,1,1))
     &                 + DABS(DH(1,2,1))+DABS(DH(2,2,1))
                TEST   = DABS(DH(1,1,2))+DABS(DH(2,1,2))
     &                 + DABS(DH(1,2,2))+DABS(DH(2,2,2))
                IF(TEST.GT.XNRMDH) XNRMDH = TEST
                 WRITE(6,145) XNRMDH
 145             FORMAT(' NORM OF DH =',D11.4)
C
                 WRITE(6,150) DET
 150             FORMAT(' DET(DH) =      ',2D11.4)
                 WRITE(6,115)
                 WRITE(6,40)
                ENDIF
C
                IF(NUMSTP.EQ.NS) THEN
                    WRITE(8,70) NUMPAT
                    WRITE(8,130) X
                    WRITE(8,131) Y
                    WRITE(8,140) XRESID
                    WRITE(8,40)
                ENDIF
C
 90            CONTINUE
C                 BOTTOM OF LOOP -- NUMSTP
C
                ITOTIT=ITOTIT + NUMIT
C
C
```

```
            WRITE(6,160) NUMIT
  160       FORMAT(' TOTAL NEWTON ITERATIONS ON PATH =',I8)
            WRITE(6,40)
C
   60   CONTINUE
C               BOTTOM OF LOOP -- NUMPAT
C
        WRITE(6,170) ITOTIT
  170   FORMAT(//' TOTAL NEWTON ITERATIONS =',I8//)
C
        STOP
        END
```

```
      SUBROUTINE FFUN7RA(X,Y,XDGM1,YDGM1,XDG,YDG, EQ,DEQ)
C
C PURPOSE:  EVALUATE THE SYSTEM EQUATIONS AND DERIVATIVES
C           FOR CONSOL7R
C
C REFERENCE: CHAPTER 2 OF "SOLVING POLYNOMIAL SYSTEMS USING
C            CONTINUATION" BY ALEXANDER MORGAN.
C
C GENERIC NAME: FFUN7R        FILE NAME: FFUN7RA
C
C NOTE:   1. THE SECOND INDEX, 1 OR 2, INDICATES THE PARTIAL DERIVATIVE
C            WITH RESPECT TO X OR Y, RESPECTIVELY.
C
C         2. THIS SUBROUTINE IS SET UP FOR PROBLEM (2-19).
C
C INPUT:  X        -- INDEPENDENT VARIABLE
C         Y        -- INDEPENDENT VARIABLE
C         XDGM1    -- X**(IDEG(1)-1), WHERE IDEG(1) IS THE DEGREE OF EQ(1)
C         YDGM1    -- Y**(IDEG(2)-1), WHERE IDEG(2) IS THE DEGREE OF EQ(2)
C         XDG      -- X**IDEG(1)
C         YDG      -- Y**IDEG(2)
C
C OUTPUT: EQ       -- EQUATION VALUES
C         DEQ      -- DERIVATIVE VALUES
C
C DECLARATIONS
      INTEGER I
      DOUBLE PRECISION A,B,C,D,E,F,G
      DOUBLE PRECISION EQ,DEQ,X,X2,X3,X4,X5,XDG,XDGM1,
     & Y,Y3,Y4,Y5,Y6,YDG,YDGM1
      DOUBLE PRECISION ONE,XY,XY5,XY6,X2Y,X2Y4,X3Y3,X3Y4
C
C DIMENSIONS
      DIMENSION EQ(2,2),DEQ(2,2,2)
      DIMENSION X(2),X2(2),X3(2),X4(2),X5(2),XDG(2),XDGM1(2),
     & Y(2),Y3(2),Y4(2),Y5(2),Y6(2),YDG(2),YDGM1(2)
      DIMENSION ONE(2),XY(2),XY5(2),XY6(2),X2Y(2),
     & X2Y4(2),X3Y3(2),X3Y4(2)
C
C COMMON
      COMMON/COEFS/ A(2),B(2),C(2),D(2),E(2),F(2),G(2)
C
C
      ONE(1)=1.D0
      ONE(2)=0.D0
C
      CALL MUL(X,X,X2)
      CALL MUL(X,X2,X3)
      X4(1)=XDGM1(1)
      X4(2)=XDGM1(2)
      X5(1)=XDG(1)
      X5(2)=XDG(2)
C
      CALL POW(3,Y,Y3)
      CALL MUL(Y,Y3,Y4)
```

```fortran
      CALL MUL(Y,Y4,Y5)
      Y6(1)=YDGM1(1)
      Y6(2)=YDGM1(2)
C
      CALL MUL(X2, Y, X2Y)
      CALL MUL( X, Y,  XY)
      CALL MUL(X3,Y4,X3Y4)
      CALL MUL( X,Y6, XY6)
      CALL MUL(X2,Y4,X2Y4)
      CALL MUL(X3,Y3,X3Y3)
      CALL MUL( X,Y5, XY5)
C
      DO 10 I=1,2
          EQ(I,1)= A(1)*X5(I)+ B(1)*X2Y(I)+ C(1)*Y4(I)
     &        + D(1)*X (I)+ E(1)*Y (I)+ F(1)*ONE(I)
          DEQ(I,1,1)= 5.*A(1)*X4(I)+ 2.*B(1)*XY(I)+ D(1)*ONE(I)
          DEQ(I,1,2)= B(1)*X2(I)+ 4.*C(1)*Y3(I)+ E(1)*ONE(I)
C
          EQ(I,2)= A(2)*X3(I)+ B(2)*X3Y4(I)+ C(2)*XY6(I)
     &        + D(2)*X(I)+ E(2)*Y(I)+ F(2)*ONE(I)
          DEQ(I,2,1)= 3.*A(2)*X2(I)+ 3.*B(2)*X2Y4(I)
     &        + C(2)*Y6(I)+ D(2)*ONE(I)
          DEQ(I,2,2)= 4.*B(2)*X3Y3(I)+ 6.*C(2)*XY5(I)+ E(2)*ONE(I)
   10 CONTINUE
C
      RETURN
      END
```

```
      SUBROUTINE FFUN7RB(X,Y,XDGM1,YDGM1,XDG,YDG, EQ,DEQ)
C
C PURPOSE:  EVALUATE THE SYSTEM EQUATIONS AND DERIVATIVES
C           FOR CONSOL7R
C
C REFERENCE: CHAPTER 2 OF "SOLVING POLYNOMIAL SYSTEMS USING
C            CONTINUATION" BY ALEXANDER MORGAN.
C
C GENERIC NAME: FFUN7R          FILE NAME: FFUN7RB
C
C NOTE:  1. THE SECOND INDEX, 1 OR 2, INDICATES THE PARTIAL DERIVATIVE
C           WITH RESPECT TO X OR Y, RESPECTIVELY.
C
C        2. THIS SUBROUTINE IS SET UP FOR PROBLEM (2-26).
C
C
C INPUT:  X       -- INDEPENDENT VARIABLE
C         Y       -- INDEPENDENT VARIABLE
C         XDGM1   -- X**(IDEG(1)-1), WHERE IDEG(1) IS THE DEGREE OF EQ(1)
C         YDGM1   -- Y**(IDEG(2)-1), WHERE IDEG(2) IS THE DEGREE OF EQ(2)
C         XDG     -- X**IDEG(1)
C         YDG     -- Y**IDEG(2)
C
C OUTPUT: EQ      -- EQUATION VALUES
C         DEQ     -- DERIVATIVE VALUES
C
C
C DECLARATIONS
      INTEGER I
      DOUBLE PRECISION A,B,C,D,E,F,G
      DOUBLE PRECISION EQ,DEQ,X,X2,XDG,XDGM1,
     & Y,Y2,Y3,YDG,YDGM1
      DOUBLE PRECISION ONE,XY,XY2,X2Y,X2Y2
C
C DIMENSIONS
      DIMENSION EQ(2,2),DEQ(2,2,2)
      DIMENSION X(2),X2(2),XDG(2),XDGM1(2),
     & Y(2),Y2(2),Y3(2),YDG(2),YDGM1(2)
      DIMENSION ONE(2),XY(2),XY2(2),X2Y(2),X2Y2(2)
C
C COMMON
      COMMON/COEFS/ A(2),B(2),C(2),D(2),E(2),F(2),G(2)
C
C
      ONE(1)=1.D0
      ONE(2)=0.D0
C
      X2(1)=XDGM1(1)
      X2(2)=XDGM1(2)
C
      Y2(1)=YDGM1(1)
      Y2(2)=YDGM1(2)
      Y3(1)=YDG(1)
      Y3(2)=YDG(2)
C
```

```
          CALL MUL( X, Y,  XY)
          CALL MUL( X,Y2, XY2)
          CALL MUL(X2, Y ,X2Y)
          CALL MUL(X2,Y2,X2Y2)
C
      DO 10 I=1,2
          EQ(I,1)= A(1)*X2Y(I)+ B(1)*XY2(I)+ C(1)*XY(I)
     &          + D(1)*Y3(I)+E(1)*Y2(I)+F(1)*Y(I)+G(1)*ONE(I)
          DEQ(I,1,1)= 2.*A(1)*XY(I)+ B(1)*Y2(I)+ C(1)*Y(I)
          DEQ(I,1,2)= A(1)*X2(I)+ 2.*B(1)*XY(I)+ C(1)*X(I)
     &          + 3.*D(1)*Y2(I)+ 2.*E(1)*Y(I)+ F(1)*ONE(I)
C
          EQ(I,2)= A(2)*X2Y(I)+ B(2)*X2(I)+ C(2)*XY(I)
     &          + D(2)*X(I)+ E(2)*Y(I)
          DEQ(I,2,1)= 2.*A(2)*XY(I)+ 2.*B(2)*X(I)+ C(2)*Y(I)
     &          + D(2)*ONE(I)
          DEQ(I,2,2)= A(2)*X2(I)+ C(2)*X(I)+ E(2)*ONE(I)
   10 CONTINUE
C
      RETURN
      END
```

```
      SUBROUTINE FFUN7RC(X,Y,XDGM1,YDGM1,XDG,YDG, EQ,DEQ)
C
C PURPOSE:  EVALUATE THE SYSTEM EQUATIONS AND DERIVATIVES
C           FOR CONSOL7R
C
C REFERENCE: CHAPTER 2 OF "SOLVING POLYNOMIAL SYSTEMS USING
C            CONTINUATION" BY ALEXANDER MORGAN.
C
C GENERIC NAME: FFUN7R           FILE NAME: FFUN7RC
C
C NOTE:  1. THE SECOND INDEX, 1 OR 2, INDICATES THE PARTIAL DERIVATIVE
C           WITH RESPECT TO X OR Y, RESPECTIVELY.
C
C        2. THIS SUBROUTINE IS SET UP FOR PROBLEM (2-27).
C
C INPUT:  X        -- INDEPENDENT VARIABLE
C         Y        -- INDEPENDENT VARIABLE
C         XDGM1    -- X**(IDEG(1)-1), WHERE IDEG(1) IS THE DEGREE OF EQ(1)
C         YDGM1    -- Y**(IDEG(2)-1), WHERE IDEG(2) IS THE DEGREE OF EQ(2)
C         XDG      -- X**IDEG(1)
C         YDG      -- Y**IDEG(2)
C
C OUTPUT: EQ       -- EQUATION VALUES
C         DEQ      -- DERIVATIVE VALUES
C
C DECLARATIONS
      INTEGER I
      DOUBLE PRECISION A,B,C,D,E,F,G
      DOUBLE PRECISION EQ,DEQ,X,X2,XDG,XDGM1,
     & Y,Y2,Y3,YDG,YDGM1
      DOUBLE PRECISION ONE,XY,XY2,X2Y,X2Y2
C
C DIMENSIONS
      DIMENSION EQ(2,2),DEQ(2,2,2)
      DIMENSION X(2),X2(2),XDG(2),XDGM1(2),
     & Y(2),Y2(2),Y3(2),YDG(2),YDGM1(2)
      DIMENSION ONE(2),XY(2),XY2(2),X2Y(2),X2Y2(2)
C
C COMMON
      COMMON/COEFS/ A(2),B(2),C(2),D(2),E(2),F(2),G(2)
C
C
      ONE(1)=1.DO
      ONE(2)=0.DO
C
      Y2(1)=YDGM1(1)
      Y2(2)=YDGM1(2)
      Y3(1)=YDG(1)
      Y3(2)=YDG(2)
C
      CALL MUL( X, X,   X2)
      CALL MUL( X, Y,   XY)
      CALL MUL( X,Y2,  XY2)
      CALL MUL(X2, Y,  X2Y)
      CALL MUL(X2,Y2, X2Y2)
```

```
C
      DO 10 I=1,2
      EQ(I,1)= A(1)*X2Y2(I)+ B(1)*XY2(I)+ C(1)*XY(I)
     &        + D(1)*Y3(I)+E(1)*Y2(I)+F(1)*Y(I)+G(1)*ONE(I)
      DEQ(I,1,1)= 2.*A(1)*XY2(I)+ B(1)*Y2(I)+C(1)*Y(I)
      DEQ(I,1,2)= 2.*A(1)*X2Y(I)+ 2.*B(1)*XY(I)+ C(1)*X(I)
     &        + 3.*D(1)*Y2(I)+ 2.*E(1)*Y(I)+ F(1)*ONE(I)
C
      EQ(I,2)= A(2)*X2Y(I)+ B(2)*X2(I)+ C(2)*XY(I)
     &        + D(2)*X(I)+ E(2)*Y(I)
      DEQ(I,2,1)= 2.*A(2)*XY(I)+ 2.*B(2)*X(I)+ C(2)*Y(I)
     &        + D(2)*ONE(I)
      DEQ(I,2,2)= A(2)*X2(I)+ C(2)*X(I)+ E(2)*ONE(I)
  10  CONTINUE
C
      RETURN
      END
```

```
C PROGRAM NAME: CONSOL8R
C
C
C PURPOSE:  SOLVES A SYSTEM OF N POLYNOMIAL EQUATIONS IN N UNKNOWNS
C           USING THE CONSOL CONTINUATION FOR POLYNOMIAL SYSTEMS.
C
C REFERENCE: CHAPTER 4 OF "SOLVING POLYNOMIAL SYSTEMS USING
C            CONTINUATION" BY ALEXANDER MORGAN
C
C NOTE:  1.   THIS PROGRAM DOES NOT USE THE FORTRAN "COMPLEX"
C            DECLARATION.  IT EMULATES COMPLEX OPERATIONS
C            USING REAL ARITHMETIC.  SEE APPENDIX 2.
C
C        2.   THIS PROGRAM GENERATES MANY UNDERFLOWS.  THIS IS
C            NOT AN ERROR CONDITION.  IF YOU NEED TO PREVENT
C            UNDERFLOWS FROM OCCURRING, SEE THE COMMENTS LABELED
C            "UNDERFLOWS" IN SUBROUTINE CORCTR.
C
C        3.   THIS PROGRAM COMES WITH SUBROUTINES INPTA, INPTBR, FFUNR,
C            AND OTPUTR SET UP TO SOLVE TWO SECOND-DEGREE EQUATIONS
C            IN TWO UNKNOWNS.  SEE SUBROUTINE INPTA.
C
C        4.   INPUT IS ON UNIT 5.  OUTPUT IS TO UNITS 6 AND 8.
C            SEE SUBROUTINE INPTA FOR SAMPLE INPUT.
C
C        5.   THE DIMENSIONS OF VARIABLES ARE SET TO ALLOW FOR A
C            MAXIMUM OF TEN EQUATIONS.  IF YOU WISH TO ALTER THIS
C            MAXIMUM, THEN ALL DIMENSIONS THAT ARE CURRENTLY SET
C            TO 10, 11, 20, AND 21 CAN BE CHANGED TO M, M+1,
C            2*M, AND 2*M+1, RESPECTIVELY, WHERE M IS AN UPPER BOUND
C            ON THE NUMBER OF EQUATIONS.
C
C
C USER SUPPLIED SUBROUTINES: INPTA, INPTBR, FFUNR, OTPUTR
C
C     "INPTA" SUPPLIES FREQUENTLY CHANGED CONSTANTS, INCLUDING
C     THE COEFFICIENTS A(I,J) FOR THE SYSTEM.
C
C     "INPTBR" SUPPLIES LESS FREQUENTLY CHANGED CONSTANTS, AND
C     ORDINARILY NEED NOT BE CHANGED BY THE USER.
C
C     "FFUNR" SUPPLIES THE EQUATION AND PARTIAL DERIVATIVE VALUES
C     FOR THE SYSTEM TO BE SOLVED.
C
C     "OTPUTR" IS CALLED AT THE END OF EACH PATH.  IT CAN BE USED
C     TO REPROCESS THE X VALUES, AT THE USER'S DISCRETION.
C
C
C INPUT VARIABLES:  SEE SUBROUTINE INPTA FOR FORMAT.
C
C     TITLE      -- FIRST 60 CHARACTERS OF FIRST LINE OF INPUT
C     IFLGCR     -- IF =1, THEN CORRECTOR ITERATIONS ARE WRITTEN TO
C                   UNIT 8.  (SEE SUBROUTINE CORCT.)
C     IFLGST     -- IF =1, THEN STEP SUMMARY OUTPUT IS PRINTED AFTER
C                   EVERY STEP.
```

```
C      MAXNS      -- MAXIMUM NUMBER OF STEPS FOR A PATH
C      MAXIT      -- MAXIMUM NUMBER OF ITERATIONS TO CORRECT A STEP
C      EPSBIG     -- THE CORRECTOR RESIDUAL MUST BE LESS THAN EPS FOR
C                    A STEP TO "SUCCEED."  EPS=EPSBIG UNTIL T > .95
C      SSZBEG     -- BEGINNING STEP LENGTH
C      N          -- NUMBER OF EQUATIONS
C      IDEG       -- IDEG(I) IS THE DEGREE OF EQUATION I.
C      A          -- COEFFICIENTS FOR F (PASSED FROM SUBROUTINE INPTA
C                    TO SUBROUTINE FFUNR VIA "COMMON/COEFS/")
C
C OUTPUT VARIABLES:
C
C    FOR EACH STEP (OPTIONAL OUTPUT: PRINTED IF IFLGST=1)
C
C      NUMSTP     -- CURRENT STEP NUMBER
C      STPSZE     -- CURRENT STEP SIZE
C      T          -- CURRENT VALUE OF T
C      X          -- X(J)  IS THE J-TH INDEPENDENT VARIABLE
C      XRESID     -- NORM OF THE RESIDUAL OF THE LAST CORRECTOR ITERATION
C
C    FOR EACH PATH:
C
C      NUMPAT     -- PATH NUMBER
C      NUMSTP     -- TOTAL NUMBER OF STEPS
C      NSSUCC     -- NUMBER OF STEPS THAT SUCCEEDED
C      NSFAIL     -- NUMBER OF STEPS THAT FAILED
C      NUMIT      -- TOTAL CORRECTOR ITERATIONS
C      NUMLN      -- TOTAL NUMBER OF LINEAR SYSTEMS SOLVED
C      ARCLN2     -- ARC LENGTH OF PATH
C      T          -- VALUE AT END OF PATH (T=1, IF PATH CONVERGED)
C      X          -- VALUE AT END OF PATH (SEE ABOVE)
C      XRESID     -- VALUE AT END OF PATH (SEE ABOVE)
C      COND       -- THE CONDITION OF DF(X) (PRINTED IF DF(X) IS
C                    NONSINGULAR)
C      DET        -- DETERMINANT OF DF(X) (PRINTED IF DF(X) IS
C                    NONSINGULAR)
C
C    TOTALS OVER ALL PATHS:
C
C      ITOTIT     -- NUMBER OF CORRECTOR ITERATIONS
C      ITOTLN     -- NUMBER OF LINEAR SYSTEMS
C      TOTAR      -- ARC LENGTH
C
C
C SUBROUTINES (CALLED DIRECTLY OR INDIRECTLY):
C
C    USER SUPPLIED (OR MODIFIED)
C      FFUNR
C      INPTA
C      INPTBR
C      OTPUTR
C
C    INITIALIZATION
C      INITR
C      STRPTR
C
```

```
C      MAIN SUBROUTINES
C        CORCTR
C        GFUNR
C        HFUNR
C        LINNR
C        PREDCR
C
C      UTILITY SUBROUTINES
C        DABS
C        DATAN
C        DCOS
C        DIV
C        DSIN
C        ICDCR
C        LNFNCR
C        LNSNCR
C        MAXNCR
C        MUL
C        POW
C        PRNTSR
C        PRNTVR
C        SQR
C        XNORM
C        XNRM2
C
C DECLARATIONS
       INTEGER I,ICOUNT,IDEG,IFLGCR,IFLGST,IJ,ITOTDG,ITOTIT,ITOTLN,
      & J,J2,J2M1,MAXIT,MAXNS,MSG,N,N2,N2P1,NADV,NP1,NSFAIL,NSSUCC,
      & NUMI,NUMIT,NUMPAT,NUMSTP
       DOUBLE PRECISION ARCLN2,COND,EPS,EPSO,EPSBIG,EPSSML,FACTOR,OLDT,
      & SSZBEG,SSZMIN,STPSZE,T,TOTAR,XNRM2,XRESID,Z
       DOUBLE PRECISION DET,OLDX,P,PDG,Q,QDG,R,X
C
C DIMENSIONS
       DIMENSION DET(2),ICOUNT(10),IDEG(10),OLDX(2,10),P(2,10),PDG(2,10),
      & Q(2,10),QDG(2,10),R(2,10),X(2,11),Z(21)
C
C CALL INPUT AND INITIALIZATION ROUTINES.
       CALL INPTA(IFLGCR,IFLGST,MAXNS,MAXIT,EPSBIG,SSZBEG,N,IDEG)
       CALL INPTBR(N,SSZBEG,EPSBIG,SSZMIN,EPSSML,EPSO,P,Q)
       CALL INITR(N,IDEG,P,Q, ITOTDG,PDG,QDG,R)
C
C INITIALIZE COUNTERS AND CONSTANTS FOR THE RUN.
       ITOTIT=0
       ITOTLN=0
       TOTAR=0.DO
C
C ICOUNT IS A COUNTER USED BY SUBROUTINE STRPTR.
       ICOUNT(1)=0
       DO 10 J=2,N
           ICOUNT(J)=1
   10  CONTINUE
C
       N2=2*N
       N2P1=N2+1
C
C
```

```
C
C     ****************************************************************
C     *****                                                      *****
C     *****          BEGINNING OF MAIN PART OF PROGRAM           *****
C     *****                                                      *****
C     ****************************************************************
C
C     PATHS LOOP -- ITERATE THROUGH PATHS
C
      DO 20 NUMPAT = 1,ITOTDG
C
          WRITE(6,30) NUMPAT
   30     FORMAT(' PATH NUMBER =',I5)
C
C INITIALIZE COUNTERS AND CONSTANTS FOR THE PATH.
          NSSUCC=0
          NSFAIL=0
          NUMIT=0
          STPSZE=SSZBEG
          T=0.DO
C
C THE COUNTER NADV CONTROLS THE "STEP INCREASE" LOGIC
          NADV=0
C
          ARCLN2=0.DO
C
          CALL STRPTR(N,ICOUNT,IDEG,R ,X)
C                     RETURNS A START POINT, X, FOR THE PATH
C
C THE FOLLOWING STATEMENT CAN BE USED TO SKIP SELECTED PATHS.
C THIS IS ESPECIALLY USEFUL IN SETTING UP RERUNS.
          IF(NUMPAT .EQ. 0 ) GOTO 51
C                                   JUST AFTER BOTTOM OF NUMSTP LOOP
C
C SAVE VALUES OF T AND X IN CASE INITIAL STEP MUST BE REDONE
          OLDT=T
          DO 40 I=1,2
          DO 40 J=1,N
              OLDX(I,J)=X(I,J)
   40     CONTINUE
C
C
C STEPS LOOP -- ITERATE STEPS ALONG THE PATH.
C
          DO 50 NUMSTP = 1,MAXNS
C
              CALL PREDCR(N,IDEG,PDG,QDG,EPSO,STPSZE, T,X,MSG)
C
              EPS=EPSBIG
              IF( T .GT. .95DO ) EPS=EPSSML
C
              CALL CORCTR(IFLGCR,N,IDEG,MAXIT,EPS,EPSO,PDG,QDG,
     &                                    T,X,NUMI,MSG,XRESID)
C
              NUMIT = NUMIT + NUMI
C
```

487

```fortran
              IF( MSG .EQ. 0 ) THEN
C STEP SUCCEEDED
C
              NSSUCC=NSSUCC+1
C
              DO 60 I=1,2
              DO 60 J=1,N
                IJ=2*J+I-2
                Z(IJ)=X(I,J)-OLDX(I,J)
   60         CONTINUE
              Z(N2P1)=T-OLDT
              ARCLN2=ARCLN2+XNRM2(N2P1,Z)
C
              IF(IFLGST.EQ.1)  CALL PRNTSR(N,NUMSTP,MSG,STPSZE,
     &                                 T,X,XRESID)
C
              IF( T .GE. 1.DO ) THEN
C MAKE A FINAL CORRECTION AND LEAVE LOOP.
              WRITE(6,70)
   70         FORMAT(/' PATH CONVERGED')
C
              FACTOR=(1.DO-OLDT)/(T-OLDT)
              DO 80 I=1,2
              DO 80 J=1,N
                X(I,J)=OLDX(I,J)+FACTOR*(X(I,J)-OLDX(I,J))
   80         CONTINUE
              T=1.DO
C
              CALL CORCTR(IFLGCR,N,IDEG,10,EPSO,EPSO,PDG,QDG,
     &                                 T,X,NUMI,MSG,XRESID)
C
              GOTO 51
C                 LEAVE LOOP
              ENDIF
C
              IF( NUMSTP .GT. MAXNS ) THEN
              WRITE(6,90)
   90         FORMAT(' PATH FAILED, MAX NUM STEPS EXCEEDED')
              GOTO 51
C                 LEAVE LOOP
              ENDIF
C
              NADV=NADV+1
C                 NADV CONTROLS THE "STEP INCREASE" LOGIC.
C                 AFTER 5 SUCCESSFUL STEPS, STPSZE IS DOUBLED.
C                 WHEN A STEP FAILS, NADV IS SET TO 0.
              IF( NADV .EQ. 5 ) STPSZE=2.DO*STPSZE
              IF( NADV .EQ. 5 ) NADV=0
C
C SAVE VALUES OF T AND X IN CASE STEP MUST BE REDONE
C
              OLDT=T
              DO 100 J=1,N
              DO 100 I=1,2
                OLDX(I,J)=X(I,J)
  100         CONTINUE
C
```

```
                ELSE
C STEP FAILED
C
                NSFAIL=NSFAIL+1
C
                IF(IFLGST.EQ.1)   CALL PRNTSR(N,NUMSTP,MSG,STPSZE,
     &                                         T,X,XRESID)
C
C RESET "STEP INCREASE COUNTER"
                NADV=0
                STPSZE=STPSZE/2.D0
C
                IF( NUMSTP  .GT.  MAXNS ) THEN
                    WRITE(6,90)
                    GOTO 51
C                        LEAVE LOOP
                ENDIF
C
                IF( STPSZE  .LT.  SSZMIN ) THEN
                    WRITE(6,110)
 110                FORMAT(/' PATH FAILED, SSZMIN EXCEEDED'/)
                    GOTO 51
C                        LEAVE LOOP
                ENDIF
C
                T=OLDT
                DO 120 I=1,2
                DO 120 J=1,N
                    X(I,J)=OLDX(I,J)
 120            CONTINUE
C
            ENDIF
C
 50     CONTINUE
C                 BOTTOM OF LOOP -- NUMSTP
C
 51     CONTINUE
C            LEAVE LOOP -- NUMSTP
C
C
        ITOTIT=ITOTIT + NUMIT
        ITOTLN=ITOTLN + NUMIT + NUMSTP
        TOTAR= TOTAR + ARCLN2
C
        WRITE(6,190)
 190    FORMAT(/' FINAL VALUES FOR PATH'/)
        WRITE(6,195) NUMPAT
 195    FORMAT(' PATH NUMBER = ',I5)
        WRITE(6,200) NUMSTP
 200    FORMAT(' TOTAL NUMBER OF STEPS = ',I5)
        WRITE(6,205) NSSUCC
 205    FORMAT('       NUMBER OF STEPS THAT SUCCEEDED = ',I5)
        WRITE(6,210) NSFAIL
 210    FORMAT('       NUMBER OF STEPS THAT FAILED    = ',I5)
C
        WRITE(6,215) NUMIT
 215    FORMAT(/'  NUMBER OF CORRECTOR ITERATIONS ON PATH =',I5)
```

```
C
        NUMIT=NUMIT+NUMSTP
C
        WRITE(6,216) NUMIT
 216    FORMAT(/'   NUMBER OF LINEAR SYSTEMS SOLVED ON PATH =',I5)
C
        WRITE(6,217) ARCLN2
 217    FORMAT(/'  PATH ARC LENGTH =',D11.4)
C
        CALL OTPUTR(N,NUMPAT,IDEG,EPSO,X, COND,DET)
C
        CALL PRNTVR(N,T,X,XRESID)
C
        IF (COND .GE. 0.D0) THEN
           WRITE(6,218) COND,DET
 218       FORMAT(
     &          /,'  CONDITION AND DETERMINANT OF DF(X) AT END OF PATH',
     &          /,'      COND =', D11.4,
     &          /,'      DET  =',2D11.4)
        ELSE
           WRITE(6,219) EPSO
 219       FORMAT(' DF(X) IS SINGULAR WITH RESPECT TO EPSO =',D11.4)
        ENDIF
C
        WRITE(6,999)
 999    FORMAT(/)
C
 20   CONTINUE
C           BOTTOM OF LOOP -- NUMPAT
C
        WRITE(6,220) ITOTIT
 220    FORMAT(//'  TOTAL CORRECTOR ITERATIONS  =',I10)
C
        WRITE(6,225) ITOTLN
 225    FORMAT(/'  TOTAL LINEAR SYSTEMS SOLVED =',I10)
C
        WRITE(6,230) TOTAR
 230    FORMAT(/'  TOTAL ARC LENGTH =',D11.4//)
C
      STOP
      END
```

```
      SUBROUTINE PREDCR(N,IDEG,PDG,QDG,EPSO,STPSZE,T,X, MSG)
C
C
C PURPOSE:  ON ENTRY TO PREDCR, Z=(X,T) IS A POINT ON A
C           CONTINUATION CURVE.  PREDCR FINDS A POINT,
C           DZ, ON THE TANGENT TO THE CURVE AT Z.  THEN
C           Z + DZ WILL BE THE "PREDICTED POINT."
C
C
C INPUT:  N        -- NUMBER OF EQUATIONS
C         IDEG     -- IDEG(J) IS THE DEGREE OF THE J-TH EQUATION
C         PDG      -- CONSTANTS TO DEFINE G, THE START SYSTEM
C         QDG      -- CONSTANTS TO DEFINE G
C         EPSO     -- ZERO-EPSILON FOR SINGULARITY
C         STPSZE   -- CURRENT STEPSIZE
C         T        -- CURRENT VALUE OF T
C         X        -- CURRENT VALUE OF X
C
C OUTPUT: T        -- IF MSG=0, PREDICTED VALUE OF T
C         X        -- IF MSG=0, PREDICTED VALUE OF X
C         MSG      -- =1, IF SYSTEM JACOBIAN IS SINGULAR.
C                     =0, OTHERWISE.
C
C
C SUBROUTINES:  HFUNR,LINNR,XNORM
C
C
C DECLARATIONS
      INTEGER I,IDEG,J,MSG,N,NP1,N2P1
      DOUBLE PRECISION EPSO,FACTOR,STPSZE,T,XLNGTH,XNORM
      DOUBLE PRECISION DHT,DHX,DZ,H,PDG,QDG,RHS,X
C
C DIMENSIONS
      DIMENSION DHT(2,10),DHX(2,10,10),DZ(2,11),H(2,10),IDEG(1),
     & PDG(2,1),QDG(2,1),RHS(2,10),X(2,1)
C
C
      NP1=N+1
      N2P1=2*N+1
      MSG = 0
C
      CALL HFUNR(N,IDEG,PDG,QDG,T,X, H,DHX,DHT)
C
      DO 10 I=1,2
      DO 10 J=1,N
         RHS(I,J)=-DHT(I,J)
   10 CONTINUE
C
C SOLVE DHX*DZ=RHS.  MSG=1 MEANS DHX IS SINGULAR WITH RESPECT TO EPS
      CALL LINNR(N,EPSO,DHX,RHS, MSG,DZ)
      IF( MSG .EQ. 1 ) WRITE(6,15)
   15 FORMAT(//' FROM PREDCR, MSG=1, ERROR '//)
      IF( MSG .EQ. 1 ) RETURN
C
      DZ(1,NP1)=1.D0
C
```

```
C DZ IS NOW THE TANGENT POINT TO BE ADDED TO Z=(X,T), EXCEPT
C THAT ITS LENGTH MAY NEED TO BE CHANGED.
C
C ADJUST THE LENGTH OF DZ BY THE STEP LENGTH, STPSZE.
C
      XLNGTH = XNORM(N2P1,DZ(1,1))
C
      FACTOR=STPSZE/XLNGTH
C
C IF THIS FACTOR WILL MAKE T > 1, THEN SHORTEN FACTOR SO THAT T
C WILL EQUAL 1.
      IF(T+FACTOR .GT. 1.DO) FACTOR = 1.DO-T
C
C
      DO 20 I=1,2
      DO 20 J=1,N
         DZ(I,J)=DZ(I,J)*FACTOR
  20  CONTINUE
      DZ(1,NP1)=FACTOR
C
C THE LENGTH OF DZ IS NOW CORRECT.
C
C UPDATE Z=(X,T)
      DO 30 I=1,2
      DO 30 J=1,N
         X(I,J)=X(I,J)+DZ(I,J)
  30  CONTINUE
      T=T+DZ(1,NP1)
C
      RETURN
      END
```

```
          SUBROUTINE CORCTR(IFLGCR,N,IDEG,MAXIT,EPS,EPSO,PDG,QDG,
      &                                        T,X,NUMI,MSG,XRESID)
C
C PURPOSE: MOVE THE POINT Z=(X,T) BACK TO THE CONTINUATION CURVE.
C
C INPUT:   IFLGCR   -- =1, IF CORRECTOR ITERATIONS ARE TO BE WRITTEN
C                          TO UNIT 8.
C                       =0, OTHERWISE.
C          N        -- NUMBER OF EQUATIONS
C          IDEG     -- IDEG(J) IS THE DEGREE OF THE J-TH EQUATION
C          MAXIT    -- MAXIMUM NUMBER OF ITERATIONS ALLOWED
C          EPS      -- ZERO-EPSILON FOR CORRECTOR RESIDUAL
C          EPSO     -- ZERO-EPSILON FOR SINGULARITY
C          PDG      -- CONSTANTS TO DEFINE G, THE START SYSTEM
C          QDG      -- CONSTANTS TO DEFINE G
C          T        -- PREDICTED VALUE OF T
C          X        -- PREDICTED VALUE OF X
C
C OUTPUT:  T        -- IF MSG=0, CORRECTED VALUE OF T
C          X        -- IF MSG=0, CORRECTED VALUE OF X
C          NUMI     -- NUMBER OF ITERATIONS USED FOR CORRECTION
C          MSG      -- =1, IF SYSTEM JACOBIAN IS SINGULAR.
C                       =2, IF CORRECTOR FAILED TO REDUCE RESIDUAL
C                          TO BE LESS THAN EPS WITHIN MAXIT ITERATIONS.
C                       =0, OTHERWISE.
C          XRESID   -- NORM OF CORRECTOR RESIDUAL
C
C SUBROUTINES: DABS,HFUNR,LINNR,XNORM
C
C
C DECLARATIONS
      INTEGER I,IDEG,IFLGCR,J,MAXIT,MSG,N,N2,NUMI
      DOUBLE PRECISION DABS,EPS,EPSO,EPSOO,T,XL,XNORM,XRESID
      DOUBLE PRECISION DHX,DHT,H,PDG,QDG,RESID,X
C
C DIMENSIONS
      DIMENSION  DHT(2,10),DHX(2,10,10),H(2,10),IDEG(1),
     & PDG(2,1),QDG(2,1),RESID(2,10),X(2,1)
C
      N2=2*N
      MSG = 0
C
      DO 10 NUMI = 1,MAXIT
C
          CALL HFUNR(N,IDEG,PDG,QDG,T,X, H,DHX,DHT)
C
C SOLVE DHX*RESID=H. MSG=1 MEANS DHX IS SINGULAR WITH RESPECT TO EPSO.
          CALL LINNR(N,EPSO,DHX,H, MSG,RESID)
C
          IF( MSG .EQ. 1 ) WRITE(6,20)
 20       FORMAT(//' FROM CORCTR, MSG=1, ERROR '//)
          IF( MSG .EQ. 1 ) RETURN
C
          DO 30 I=1,2
          DO 30 J=1,N
              X(I,J)= X(I,J)-RESID(I,J)
 30       CONTINUE
```

```
C
C$$$$$$$$$$$$$$$$$$$$$$$$$$$$$$$$$$$$$$$$$$$$$$$$$$$$$$$$$$$$$$$$$$$$$$$$$$
C UNDERFLOWS :   IF YOU MUST PREVENT UNDERFLOWS, I SUGGEST YOU
C                TAKE THE COMMENTS OFF THE FOLLOWING BLOCK OF
C                CODE.  THE VALUE FOR EPSOO MAY NEED TO BE ADJUSTED.
C
C          EPSOO=1.D-22
C
C          DO 35 J=1,N
C              IF(DABS(X(1,J)).LE.EPSOO) X(1,J)=0.DO
C              IF(DABS(X(2,J)).LE.EPSOO) X(2,J)=0.DO
C 35       CONTINUE
C$$$$$$$$$$$$$$$$$$$$$$$$$$$$$$$$$$$$$$$$$$$$$$$$$$$$$$$$$$$$$$$$$$$$$$$$$$
C
C COMPUTE THE LENGTH OF RESID
          XRESID=XNORM(N2,RESID(1,1))
C
          IF(IFLGCR.EQ.1) THEN
             XL=XNORM(N2,X(1,1))
             WRITE(8,40) T,XL,XRESID
   40        FORMAT(' T=',D11.4,' XL=',D11.4,' XRESID =',D11.4)
             IF( XRESID .LT. EPS ) WRITE(8,999)
  999        FORMAT(/)
          END IF
C
          IF( XRESID .LT. EPS ) RETURN
C
   10     CONTINUE
C
      MSG=2
C
      RETURN
      END
```

```
      SUBROUTINE HFUNR(N,IDEG,PDG,QDG,T,X, H,DHX,DHT)
C
C PURPOSE: EVALUATES H, THE CONTINUATION EQUATION
C
C INPUT:    N        -- NUMBER OF EQUATIONS
C           IDEG     -- IDEG(J) IS THE DEGREE OF THE J-TH EQUATION
C           PDG      -- CONSTANTS TO DEFINE G, THE START SYSTEM
C           QDG      -- CONSTANTS TO DEFINE G
C           T        -- CURRENT VALUE OF T
C           X        -- CURRENT VALUE OF X
C
C OUTPUT:   H        -- H IS THE CONTINUATION SYSTEM, A SYSTEM OF N
C                       COMPLEX POLYNOMIALS IN THE N+1 UNKNOWNS X AND T.
C           DHX      -- DHX IS THE N X N (COMPLEX) MATRIX OF PARTIAL
C                       DERIVATIVES OF H WITH RESPECT TO X.  DHX(J,K) IS
C                       THE PARTIAL OF H(J) WITH RESPECT TO X(K).
C           DHT      -- DHT IS THE N X 1 MATRIX OF PARTIAL DERIVATIVES
C                       OF H WITH RESPECT TO T.
C
C SUBROUTINES:  FFUNR,GFUNR,MUL,POW
C
C DECLARATIONS
      INTEGER I,IDEG,J,K,N,NNNN
      DOUBLE PRECISION ONEMT,T
      DOUBLE PRECISION DF,DG,DHT,DHX,F,G,H,PDG,QDG,X,XDG,XDGM1
C
C DIMENSIONS
      DIMENSION DF(2,10,11),DG(2,10),DHT(2,10),DHX(2,10,10),F(2,10),
     & G(2,10),H(2,10),IDEG(1),PDG(2,1),QDG(2,1),X(2,1),XDG(2,10),
     & XDGM1(2,10)
C
C COMPUTE THE (IDEG-1)-TH AND IDEG-TH POWER OF X
C
      DO 10 J=1,N
        NNNN=IDEG(J)-1
        CALL POW(NNNN,X(1,J), XDGM1(1,J))
        CALL MUL(X(1,J),XDGM1(1,J), XDG(1,J))
   10 CONTINUE
C
      CALL GFUNR(N,IDEG,PDG,QDG,  XDGM1,XDG, G,DG)
C
      CALL FFUNR(N,              X,XDGM1,XDG, F,DF)
C
      ONEMT=1.D0 - T
      DO 20 I=1,2
      DO 20 J=1,N
        DO 30 K=1,N
          DHX(I,J,K)= T*DF(I,J,K)
   30   CONTINUE
        DHX(I,J,J)= DHX(I,J,J) + ONEMT*DG(I,J)
C
        DHT(I,J)=  F(I,J)      -        G(I,J)
        H(I,J)  = T*F(I,J)    + ONEMT* G(I,J)
   20 CONTINUE
C
      RETURN
      END
```

```
      SUBROUTINE GFUNR(N,IDEG,PDG,QDG,XDGM1,XDG, G,DG)
C
C PURPOSE:  EVALUATES G, THE START EQUATION
C
C INPUT:  N        -- NUMBER OF EQUATIONS
C         IDEG     -- IDEG(J) IS THE DEGREE OF THE J-TH EQUATION
C         PDG      -- CONSTANTS TO DEFINE G
C         QDG      -- CONSTANTS TO DEFINE G
C         XDGM1    -- XDGM1(I) = X(I)**(IDEG(I)-1)
C         XDG      -- XDG(I) = X(I)**IDEG(I)
C
C OUTPUT: G        -- G IS THE START SYSTEM.
C         DG       -- DG(J) IS THE PARTIAL DERIVATIVE OF G(J) WITH
C                     RESPECT TO X(J).
C
C SUBROUTINES:  MUL
C
C DECLARATIONS
      INTEGER I,IDEG,J,N
      DOUBLE PRECISION DG,G,PDG,PXDGM1,PXDG,QDG,XDG,XDGM1
C
C DIMENSIONS
      DIMENSION  DG(2,1),G(2,1),IDEG(1),PDG(2,1),PXDG(2,10),
     & PXDGM1(2,10),QDG(2,1),XDG(2,1),XDGM1(2,1)
C
C
C COMPUTE THE PRODUCT OF PDG AND XDGM1
C
      DO 10 J=1,N
          CALL MUL( PDG(1,J), XDGM1(1,J), PXDGM1(1,J) )
   10 CONTINUE
C
C COMPUTE THE PRODUCT OF PDG AND XDG
C
      DO 20 J=1,N
          CALL MUL( PDG(1,J), XDG(1,J), PXDG(1,J) )
   20 CONTINUE
C
      DO 30 I=1,2
      DO 30 J=1,N
              G(I,J)=PXDG(I,J)-QDG(I,J)
              DG(I,J)= IDEG(J)*PXDGM1(I,J)
   30 CONTINUE
C
      RETURN
      END
```

```
      SUBROUTINE LINNR(N,EPSO,DHX,RHS, MSG,RESID)
C
C PURPOSE: SOLVES THE COMPLEX LINEAR SYSTEM DHX*RESID=RHS
C
C INPUT:    N       -- NUMBER OF EQUATIONS
C           EPSO    -- ZERO-EPSILON FOR SINGULARITY
C           DHX     -- N X N COMPLEX JACOBIAN OF H WITH RESPECT TO X.
C           RHS     -- COMPLEX VECTOR OF LENGTH N
C
C OUTPUT: MSG       -- =1, IF DHX IS SINGUALAR WITH RESPECT TO EPSO.
C                      =0, OTHERWISE
C         RESID     -- IF MSG=0, SOLUTION TO THE LINEAR SYSTEM
C                      DHX*RESID=RHS
C
C SUBROUTINES: LNFNCR,LINSNC
C
C
C DECLARATIONS
      INTEGER IR,IROW,J,LDIM,MSG,N
      DOUBLE PRECISION EPSO
      DOUBLE PRECISION DHX,RESID,RHS
C
C DIMENSIONS
      DIMENSION DHX(2,10,1),IROW(10),RESID(2,1),RHS(2,1)
C
      LDIM=10
C
      CALL LNFNCR(LDIM,N,EPSO,DHX, MSG,IROW)
      IF( MSG .EQ. 1) RETURN
C
      CALL LNSNCR(LDIM,N,IROW,DHX,RHS, RESID)
C
      RETURN
      END
```

```
      SUBROUTINE INITR(N,IDEG,P,Q, ITOTDG,PDG,QDG,R)
C
C PURPOSE: COMPUTES POWERS OF P AND Q,
C          COMPUTES R=Q/P, USED IN SUBROUTINE STRPT, AND
C          COMPUTES THE TOTAL DEGREE, ITOTDG.
C
C INPUT:   N        -- NUMBER OF EQUATIONS
C          IDEG     -- IDEG(J) IS THE DEGREE OF THE J-TH EQUATION
C          P        -- CONSTANTS TO DEFINE G, THE START SYSTEM
C          Q        -- CONSTANTS TO DEFINE G
C
C OUTPUT:  ITOTDG   -- THE TOTAL DEGREE OF THE SYSTEM
C          PDG      -- CONSTANTS TO DEFINE G
C          QDG      -- CONSTANTS TO DEFINE G
C          R        -- R(J)=Q(J)/P(J), USED IN SUBROUTINE STRPT
C
C
C SUBROUTINES:  POW,DIV
C
C
C DECLARATIONS
      INTEGER IDEG,ITOTDG,J,N
      DOUBLE PRECISION P,PDG,Q,QDG,R
C
C DIMENSIONS
      DIMENSION IDEG(1),P(2,1),PDG(2,1),Q(2,1),QDG(2,1),R(2,1)
C
C COMPUTE THE IDEG-TH POWER OF P
      DO 10 J=1,N
         CALL POW(IDEG(J),P(1,J),PDG(1,J))
  10  CONTINUE
C
C COMPUTE THE IDEG-TH POWER OF Q
      DO 20 J=1,N
         CALL POW(IDEG(J),Q(1,J),QDG(1,J))
  20  CONTINUE
C
C COMPUTE R=Q/P
      DO 30 J=1,N
         CALL DIV(Q(1,J),P(1,J),R(1,J))
  30  CONTINUE
C
C COMPUTE ITOTDG
      ITOTDG=1
      DO 40 J=1,N
         ITOTDG=ITOTDG*IDEG(J)
  40  CONTINUE
C
      RETURN
      END
```

```
      SUBROUTINE STRPTR(N,ICOUNT,IDEG,R ,X)
C
C PURPOSE: COMPUTES START POINT FOR EACH PATH.
C
C
C INPUT:   N          -- NUMBER OF EQUATIONS
C          ICOUNT     -- A COUNTER USED TO INCREMENT EACH (NORMALIZED)
C                        VARIABLE AROUND THE UNIT CIRCLE, SO THAT EVERY
C                        COMBINATION OF START VALUES IS CHOSEN.  ICOUNT
C                        IS INITIALIZED IN THE MAIN ROUTINE.
C          IDEG       -- IDEG(J) IS THE DEGREE OF THE J-TH EQUATION.
C          R          -- CONSTANTS TO DEFINE START POINTS
C
C OUTPUT: X           -- START POINT FOR A PATH
C
C
C SUBROUTINES:  DATAN,DCOS,DSIN,MUL
C
C
C DECLARATIONS
      INTEGER ICOUNT,IDEG,ITEST,J,N
      DOUBLE PRECISION ANGLE,TWOPI,XCOUNT,XDEG,XXXX
      DOUBLE PRECISION R,X
C
C DIMENSIONS
      DIMENSION ICOUNT(1),IDEG(1),R(2,1),X(2,1),XXXX(2)
C
C
      DO 10 J=1,N
         ITEST = ICOUNT(J)
         IF(  ITEST  .GE.   IDEG(J) ) ICOUNT(J)=1
         IF(  ITEST  .LT.   IDEG(J) ) ICOUNT(J)=ICOUNT(J)+1
         IF(  ITEST  .LT.   IDEG(J) ) GOTO 20
C                                           LEAVE LOOP
  10  CONTINUE
  20  CONTINUE
C
      TWOPI=8.DO*DATAN(1.DO)
C
      DO 30 J=1,N
         ANGLE = (TWOPI/IDEG(J))*ICOUNT(J)
         XXXX(1) = DCOS(ANGLE)
         XXXX(2) = DSIN(ANGLE)
         CALL MUL(XXXX,R(1,J),X(1,J))
  30  CONTINUE
C
      RETURN
      END
```

```
      SUBROUTINE PRNTSR(N,NUMSTP,MSG,STPSZE,T,X,XRESID)
C
C PURPOSE: PRINTS STEP SUMMARY.
C
C INPUT:   N         -- NUMBER OF EQUATIONS
C          NUMSTP    -- STEP NUMBER
C          MSG       -- ERROR MESSAGE
C          STPSZE    -- STEPSIZE
C          T         -- CURRENT VALUE OF T
C          X         -- CURRENT VALUE OF X
C          XRESID    -- NORM OF CORRECTOR RESIDUAL
C
C OUTPUT: TO UNIT 6
C
C SUBROUTINES: PRNTVR
C
C
C DECLARATIONS
      INTEGER MSG,N,NUMSTP
      DOUBLE PRECISION STPSZE,T,XRESID
      DOUBLE PRECISION X
C
C DIMENSIONS
      DIMENSION X(2,1)
C
C
      WRITE(6,100) NUMSTP
 100  FORMAT(/' STEP SUMMARY FOR STEP NUMBER',I5,' :')
C
      IF(MSG.EQ.0) WRITE(6,110) NUMSTP
 110  FORMAT('  STEP ',I5,' SUCCEEDED')
      IF(MSG.EQ.1) WRITE(6,120) NUMSTP
 120  FORMAT('  STEP ',I5,' FAILED DUE TO SINGULARITY')
      IF(MSG.EQ.2) WRITE(6,130) NUMSTP
 130  FORMAT('  STEP ',I5,' FAILED DUE TO NON-CONVERGENCE')
C
      WRITE(6,140) STPSZE
 140  FORMAT('  STPSZE =',D11.4)
      CALL PRNTVR(N,T,X,XRESID)
C
      RETURN
      END
```

```
      SUBROUTINE PRNTVR(N,T,X,XRESID)
C
C PURPOSE: PRINTS SELECTED VARIABLE VALUES
C
C INPUT:   N         -- NUMBER OF EQUATIONS
C          T         -- CURRENT VALUE OF T
C          X         -- CURRENT VALUE OF X
C          XRESID    -- NORM OF CORRECTOR RESIDUAL
C
C OUTPUT:  TO UNIT 6
C
C DECLARATIONS
      INTEGER J,N
      DOUBLE PRECISION T,XRESID
      DOUBLE PRECISION X
C
C DIMENSIONS
      DIMENSION  X(2,1)
C
C
      WRITE(6,100)
 100  FORMAT(' ----------------------------------------------------')
C
      WRITE(6,110) T
 110  FORMAT('  T  = ',D22.15)
C
      WRITE(6,120)
 120  FORMAT('                  REAL PART          IMAGINARY PART' )
C
      DO 125 J=1,N
         WRITE(6,130) J,X(1,J),X(2,J)
 125  CONTINUE
 130  FORMAT('  X(',I5,')  = ', D11.4,'          ',D11.4)
C
      WRITE(6,150) XRESID
 150  FORMAT(' NORM OF RESIDUAL = ',D11.4)
C
      WRITE(6,100)
C
      RETURN
      END
```

```
      SUBROUTINE INPTBR(N,SSZBEG,EPSBIG,SSZMIN,EPSSML,EPSO,P,Q)
C
C PURPOSE: INITIALIZES LESS FREQUENTLY CHANGED CONSTANTS
C
C NOTE:   IF YOU CHANGE THE DIMENSIONS OF VARIABLES IN CONSOL8
C         IN ORDER TO CHANGE THE MAXIMUM NUMBER OF EQUATIONS,
C         DO NOT CHANGE ANY OF THE DIMENSIONS IN THIS SUBROUTINE.
C
C INPUT:  N         -- NUMBER OF EQUATIONS
C         SSZBEG    -- BEGINNING STEPSIZE
C         EPSBIG    -- ZERO-EPSILON FOR PATH TRACKING RESIDUAL,
C                      UNTIL T > .95
C
C OUTPUT: SSZMIN    -- MINIMUM STEPSIZE ALLOWED
C         EPSSML    -- ZERO-EPSILON FOR PATH TRACKING RESIDUAL,
C                      AFTER T > .95
C         EPSO      -- ZERO-EPSILON FOR SINGULARITY
C         P         -- "RANDOM" CONSTANTS TO DEFINE G, THE START SYSTEM
C         Q         -- "RANDOM" CONSTANTS TO DEFINE G
C
C
C SUBROUTINES: MOD
C
C
C DECLARATIONS
      INTEGER I,J,JJ,N
      DOUBLE PRECISION EPSO,EPSBIG,EPSSML,PP,QQ,SSZBEG,SSZMIN
      DOUBLE PRECISION P,Q
C
C DIMENSIONS
      DIMENSION  P(2,1),PP(2,10),Q(2,1),QQ(2,10)
C
C
      SSZMIN = 1.D-5*SSZBEG
      EPSSML = 1.D-3*EPSBIG
      EPSO   = 1.D-22
C
      PP(1, 1)= .12324754231D0
         PP(2, 1)= .76253746298D0
      PP(1, 2)= .93857838950D0
         PP(2, 2)=-.99375892810D0
      PP(1, 3)=-.23467908356D0
         PP(2, 3)= .39383930009D0
      PP(1, 4)= .83542556622D0
         PP(2, 4)=-.10192888288D0
      PP(1, 5)=-.55763522521D0
         PP(2, 5)=-.83729899911D0
      PP(1, 6)=-.78348738738D0
         PP(2, 6)=-.10578234903D0
      PP(1, 7)= .03938347346D0
         PP(2, 7)= .04825184716D0
      PP(1, 8)=-.43428734331D0
         PP(2, 8)= .93836289418D0
      PP(1, 9)=-.99383729993D0
         PP(2, 9)=-.40947822291D0
      PP(1,10)= .09383736736D0
         PP(2,10)= .26459172298D0
```

```fortran
C
      QQ(1, 1)= .58720452864D0
         QQ(2, 1)= .01321964722D0
      QQ(1, 2)= .97884134700D0
         QQ(2, 2)=-.14433009712D0
      QQ(1, 3)= .39383737289D0
         QQ(2, 3)= .41543223411D0
      QQ(1, 4)=-.03938376373D0
         QQ(2, 4)=-.61253112318D0
      QQ(1, 5)= .39383737388D0
         QQ(2, 5)=-.26454678861D0
      QQ(1, 6)=-.00938376766D0
         QQ(2, 6)= .34447867861D0
      QQ(1, 7)=-.04837366632D0
         QQ(2, 7)= .48252736790D0
      QQ(1, 8)= .93725237347D0
         QQ(2, 8)=-.54356527623D0
      QQ(1, 9)= .39373957747D0
         QQ(2, 9)= .65573434564D0
      QQ(1,10)=-.39380038371D0
         QQ(2,10)= .98903450052D0
C
      DO 10 I=1,2
      DO 10 J=1,N
         JJ=MOD(J-1,10)+1
         P(I,J)=PP(I,JJ)
         Q(I,J)=QQ(I,JJ)
   10 CONTINUE
C
      WRITE(6,100) SSZMIN
  100 FORMAT(/' MINIMUM STPSZE = ',D11.4)
      WRITE(6,105) EPSSML
  105 FORMAT(/' EPSSML =',D11.4)
      WRITE(6,110) EPSO
  110 FORMAT(/' EPSO   =',D11.4)
C
      WRITE(6,115) ((P(I,J),I=1,2),J=1,N)
  115 FORMAT(/' P   =',2D11.4)
C
      WRITE(6,120) ((Q(I,J),I=1,2),J=1,N)
  120 FORMAT(/' Q   =',2D11.4)
C
      WRITE(6,999)
  999 FORMAT(/)
C
      RETURN
      END
```

```
      SUBROUTINE FFUNR(N,X,XDGM1,XDG, F,DF)
C
C PURPOSE: EVALUATES F AND DF.
C
C INPUT:   N          -- NUMBER OF EQUATIONS
C          X          -- CURRENT VALUE OF X
C          XDGM1      -- XDGM1(J) = X(J)**(IDEG(J)-), WHERE IDEG(J) IS
C                        THE DEGREE OF THE J-TH EQUATIONS, F(J).
C          XDG        -- XDG(J) = X(J)**IDEG(J)
C
C OUTPUT:  F          -- F(J) IS THE J-TH EQUATION IN THE SYSTEM WE WANT
C                        TO SOLVE.
C          DF         -- DF(J,K) IS THE PARTIAL DERIVATIVE OF F(J) WITH
C                        RESPECT TO X(K).
C
C COMMON:  /COEFS/  IS USED TO PASS COEFFICIENT VALUES FROM
C                     SUBROUTINE INPTA TO SUBROUTINE FFUNR.
C
C *************************************************************
C THIS SUBROUTINE IS FOR TWO QUADRIC EQUATIONS
C
C     A(J,1)*XDG(1) +A(J,2)*XDG(2) +A(J,3)*X(1)*X(2)
C
C                    +A(J,4)*X(1) +A(J,5)*X(2) +A(J,6) = 0
C
C FOR J=1 TO 2.
C *************************************************************
C
C DECLARATIONS
      INTEGER I,J,N
      DOUBLE PRECISION A
      DOUBLE PRECISION DF,F,ONE,X,X12,XDG,XDGM1
C
C DIMENSIONS
      DIMENSION DF(2,10,1),F(2,1),ONE(2),X(2,1),X12(2),
     & XDG(2,1),XDGM1(2,1)
C
C COMMON
      COMMON/COEFS/A(2,6)
C
C
      ONE(1)=1.D0
      ONE(2)=0.D0
C
      CALL MUL(X(1,1),X(1,2),X12)
C
      DO 10 I=1,2
      DO 10 J=1,N
      F(I,J)= A(J,1)*XDG(I,1) +A(J,2)*XDG(I,2) +A(J,3)*X12(I)
     &      + A(J,4)*X(I,1)   +A(J,5)*X(I,2)   +A(J,6)*ONE(I)
C
      DF(I,J,1)=2*A(J,1)*X(I,1)+A(J,3)*X(I,2)+A(J,4)*ONE(I)
      DF(I,J,2)=2*A(J,2)*X(I,2)+A(J,3)*X(I,1)+A(J,5)*ONE(I)
   10 CONTINUE
C
      RETURN
      END
```

```
      SUBROUTINE OTPUTR(N,NUMPAT,IDEG,EPSO,X, COND,DET)
C
C PURPOSE: OTPUT IS CALLED BY THE MAIN ROUTINE AFTER EACH PATH IS COMPLETED.
C          OUTPUT CAN BE USED TO TRANSFORM THE VARIABLES OR FOR ANY OTHER
C          PURPOSE.  THIS VERSION OF OTPUT COMPUTES THE CONDITION AND
C          DETERMINANT OF THE JACOBIAN OF F AT X.
C
C
C INPUT:   N        -- NUMBER OF EQUATIONS
C          NUMPAT   -- CURRENT PATH NUMBER
C          IDEG     -- IDEG(J) IS THE DEGREE OF THE J-TH EQUATION.
C          EPSO     -- ZERO-EPSILON FOR SINGULARITY.
C          X        -- CURRENT X VALUE
C
C OUTPUT:  COND     -- CONDITION OF THE JACOBIAN, DF, AT X.  COND=-1
C                      MEANS THAT DF IS SINGULAR WITH RESPECT TO EPSO.
C          DET      -- IF COND >= 0, DETERMINANT OF DF.
C
C
C SUBROUTINES: DCMPLX,FFUN,ICDC
C
C DECLARATIONS
      INTEGER IDEG,IWORK,J,MSG,N,NNNN,NUMPAT
      DOUBLE PRECISION COND,EPSO
      DOUBLE PRECISION DF,F,DET,WORK,WORK2,X,XDG,XDGM1
C
C DIMENSIONS
      DIMENSION DET(2),DF(2,10,10),F(2,10),IDEG(1),IWORK(10),WORK(2,10),
     & WORK2(2,10,10),X(2,1),XDG(2,10),XDGM1(2,10)
C
C
      DO 10 J=1,N
        NNNN=IDEG(J)-1
        CALL POW(NNNN,X(1,J), XDGM1(1,J))
        CALL MUL(X(1,J),XDGM1(1,J), XDG(1,J))
   10 CONTINUE
C
      CALL FFUNR(N,X,XDGM1,XDG, F,DF)
C
      CALL ICDCR(10,N,EPSO,DF, WORK,IWORK,WORK2,COND,DET,MSG)
C
      IF(MSG.EQ.1) COND=-1
C
      RETURN
      END
```

```
      SUBROUTINE INPTBRT(N,SSZBEG,EPSBIG, SSZMIN,EPSSML,EPSO,P,Q)
C
C PURPOSE: INITIALIZES LESS FREQUENTLY CHANGED CONSTANTS
C
C GENERIC NAME: INPTBR           FILE NAME: INPTBRT
C
C REFERENCE:  CHAPTER 5 OF "SOLVING POLYNOMIAL SYSTEMS USING
C             CONTINUATION" BY ALEXANDER MORGAN.
C
C NOTE:  1. FOR SOLVING A GENERAL POLYNOMIAL SYSTEM, USING TABLEAU
C           INPUT.  CHANGE TO GENERIC NAME AND USE WITH CONSOL8R.
C
C        2. IF YOU CHANGE THE DIMENSIONS OF VARIABLES IN CONSOL8R
C           IN ORDER TO CHANGE THE MAXIMUM NUMBER OF EQUATIONS,
C           CHANGE THE DIMENSION OF  CL  IN COMMON BLOCK "/PROJT/"
C           FROM CL(2,11) TO CL(2,M+1) WHERE M IS THE NEW UPPER BOUND
C           ON THE NUMBER OF EQUATIONS.  DO NOT CHANGE ANY OF
C           THE OTHER DIMENSIONS IN THIS SUBROUTINE.
C
C
C INPUT:  N        -- NUMBER OF EQUATIONS
C         SSZBEG   -- BEGINNING STEPSIZE
C         EPSBIG   -- ZERO-EPSILON FOR PATH TRACKING RESIDUAL,
C                     UNTIL T > .95
C
C OUTPUT: SSZMIN   -- MINIMUM STEPSIZE ALLOWED
C         EPSSML   -- ZERO-EPSILON FOR PATH TRACKING RESIDUAL,
C                     AFTER T > .95
C         EPSO     -- ZERO-EPSILON FOR SINGULARITY
C         P        -- "RANDOM" CONSTANTS TO DEFINE G, THE START SYSTEM
C         Q        -- "RANDOM" CONSTANTS TO DEFINE G
C
C
C COMMON:  /FLGPT/  IS USED TO PASS THE PROJECTIVE TRANSFORMATION
C                   FLAG FROM SUBROUTINE INPTART TO SUBROUTINE INPTBRT.
C          /PROJT/  IS USED TO PASS PROJECTIVE TRANSFORMATION
C                   CONSTANTS FROM SUBROUTINE INPTBRT TO SUBROUTINES
C                   FFUNRT AND OTPUTRT.
C
C SUBROUTINES: MOD
C
C
C  DECLARATIONS
      INTEGER I,IFLGPT,J,JJ,N,NP1
      DOUBLE PRECISION CCL,EPSO,EPSBIG,EPSSML,PP,QQ,SSZBEG,SSZMIN
      DOUBLE PRECISION CL,P,Q
C
C  DIMENSIONS
      DIMENSION CCL(2,11),P(2,1),PP(2,10),Q(2,1),QQ(2,10)
C
C  COMMON
      COMMON/PROJT/CL(2,11)
      COMMON/FLGPT/IFLGPT
C
C
      NP1=N+1
      SSZMIN = 1.D-5*SSZBEG
```

```
      EPSSML = 1.D-3*EPSBIG
      EPSO  = 1.D-22
C
      PP(1, 1)= .12324754231D0
         PP(2, 1)= .76253746298D0
      PP(1, 2)= .93857838950D0
         PP(2, 2)=-.99375892810D0
      PP(1, 3)=-.23467908356D0
         PP(2, 3)= .39383930009D0
      PP(1, 4)= .83542556622D0
         PP(2, 4)=-.10192888288D0
      PP(1, 5)=-.55763522521D0
         PP(2, 5)=-.83729899911D0
      PP(1, 6)=-.78348738738D0
         PP(2, 6)=-.10578234903D0
      PP(1, 7)= .03938347346D0
         PP(2, 7)= .04825184716D0
      PP(1, 8)=-.43428734331D0
         PP(2, 8)= .93836289418D0
      PP(1, 9)=-.99383729993D0
         PP(2, 9)=-.40947822291D0
      PP(1,10)= .09383736736D0
         PP(2,10)= .26459172298D0
C
      QQ(1, 1)= .58720452864D0
         QQ(2, 1)= .01321964722D0
      QQ(1, 2)= .97884134700D0
         QQ(2, 2)=-.14433009712D0
      QQ(1, 3)= .39383737289D0
         QQ(2, 3)= .41543223411D0
      QQ(1, 4)=-.03938376373D0
         QQ(2, 4)=-.61253112318D0
      QQ(1, 5)= .39383737388D0
         QQ(2, 5)=-.26454678861D0
      QQ(1, 6)=-.00938376766D0
         QQ(2, 6)= .34447867861D0
      QQ(1, 7)=-.04837366632D0
         QQ(2, 7)= .48252736790D0
      QQ(1, 8)= .93725237347D0
         QQ(2, 8)=-.54356527623D0
      QQ(1, 9)= .39373957747D0
         QQ(2, 9)= .65573434564D0
      QQ(1,10)=-.39380038371D0
         QQ(2,10)= .98903450052D0
C
      CCL(1, 1)=-.03485644332D0
         CCL(2, 1)= .28554634336D0
      CCL(1, 2)= .91453454766D0
         CCL(2, 2)= .35354566613D0
      CCL(1, 3)=-.36568737635D0
         CCL(2, 3)= .45634642477D0
      CCL(1, 4)=-.89089767544D0
         CCL(2, 4)= .34524523544D0
      CCL(1, 5)= .13523462465D0
         CCL(2, 5)= .43534535555D0
      CCL(1, 6)=-.34523544445D0
         CCL(2, 6)= .00734522256D0
```

```
          CCL(1, 7)=-.80004678763D0
             CCL(2, 7)=-.009387123644D0
          CCL(1, 8)=-.875432124245D0
             CCL(2, 8)= .00045687651D0
          CCL(1, 9)= .65256352333D0
             CCL(2, 9)=-.12356777452D0
          CCL(1,10)= .09986798321548D0
             CCL(2,10)=-.56753456577D0
          CCL(1,11)= .29674947394739D0
             CCL(2,11)= .93274302173D0
C
          DO 10 I=1,2
          DO 10 J=1,N
            JJ=MOD(J-1,10)+1
            P(I,J)=PP(I,JJ)
            Q(I,J)=QQ(I,JJ)
   10     CONTINUE
C
          DO 20 I=1,2
          DO 20 J=1,NP1
            JJ=MOD(J-1,11)+1
            CL(I,J)=CCL(I,JJ)
   20     CONTINUE
C
C IF WE WISH NOT TO USE THE PROJECTIVE TRANSFORMATION, THEN WE
C RESET CL TO THE NOMINAL VALUES BELOW.
          IF(IFLGPT .NE. 1) THEN
              DO 30 I=1,2
              DO 30 J=1,N
                CL(I,J)=0.D0
   30         CONTINUE
              CL(1,NP1)=1.D0
              CL(2,NP1)=0.D0
          END IF
C
          WRITE(6,100) SSZMIN
  100     FORMAT(/' MINIMUM STPSZE = ',D11.4)
          WRITE(6,105) EPSSML
  105     FORMAT(/' EPSSML =',D11.4)
          WRITE(6,110) EPSO
  110     FORMAT(/' EPSO   =',D11.4)
C
          WRITE(6,115) ((P(I,J),I=1,2),J=1,N)
  115     FORMAT(/' P   =',2D11.4)
C
          WRITE(6,120) ((Q(I,J),I=1,2),J=1,N)
  120     FORMAT(/' Q   =',2D11.4)
C
          WRITE(6,125) ((CL(I,J),I=1,2),J=1,NP1)
  125     FORMAT(/' CL  =',2D22.15)
C
C
          WRITE(6,999)
  999     FORMAT(/)
C
          RETURN
          END
```

```
      SUBROUTINE FFUNRT(N,X,XDGM1,XDG, F,DF)
C
C PURPOSE: EVALUATES F AND DF.
C
C GENERIC NAME: FFUNR           FILE NAME: FFUNRT
C
C REFERENCE:  CHAPTER 5 OF "SOLVING POLYNOMIAL SYSTEMS USING
C             CONTINUATION" BY ALEXANDER MORGAN.
C
C NOTE:  1. FOR SOLVING A GENERAL POLYNOMIAL SYSTEM, USING TABLEAU
C           INPUT.  CHANGE TO GENERIC NAME AND USE WITH CONSOL8R.
C
C        2. THIS SUBROUTINE DOESN'T USE XDGM1 OR XDG.
C
C        3. EQUATION DEFINITION PARAMETERS, EQUATION COEFFICIENTS,
C           AND THE PROJECTIVE TRANSFORMATION CONSTANTS ARE
C           PASSED TO FFUNRT VIA COMMON, AS NOTED BELOW.
C
C
C INPUT:  N       -- NUMBER OF EQUATIONS
C         X       -- CURRENT VALUE OF X
C         XDGM1   -- XDGM1(J) = X(J)**(IDEG(J)-), WHERE IDEG(J) IS
C                    THE DEGREE OF THE J-TH EQUATIONS, F(J).
C         XDG     -- XDG(J) = X(J)**IDEG(J)
C
C OUTPUT: F       -- F(J) IS THE J-TH EQUATION IN THE SYSTEM WE WANT
C                    TO SOLVE.
C         DF      -- DF(J,K) IS THE PARTIAL DERIVATIVE OF F(J) WITH
C                    RESPECT TO X(K).
C
C COMMON:  /EQUAT/  IS USED TO PASS EQUATION DEFINITION PARAMETERS
C                   FROM SUBROUTINE INPTART TO SUBROUTINE FFUNRT.
C          /COEFS/  IS USED TO PASS COEFFICIENT VALUES FROM
C                   SUBROUTINE INPTART TO SUBROUTINE FFUNRT.
C          /PROJT/  IS USED TO PASS PROJECTIVE TRANSFORMATION
C                   CONSTANTS FROM SUBROUTINE INPTBRT TO SUBROUTINES
C                   FFUNRT AND OTPUTRT.
C
C SUBROUTINES: DIV,MUL,POW,XNORM
C
C DECLARATIONS
      INTEGER I,IFLGPT,J,K,KDEG,L,M,N,NNNN,NP1,NT,NUMT
      DOUBLE PRECISION ASCL,EPSDIV,XNORM
      DOUBLE PRECISION CL,CLX,DF,DSTRM,DTRM,DXNP1,F,FACV,STRM,
     & TEMP,TEMPP,TRM,XDG,XDGM1,X,XX,ZERO
C
C IN THE DIMENSIONED VARIABLES BELOW, "10" DENOTES THE NUMBER
C OF EQUATIONS, AND "30" THE MAXIMUM NUMBER OF TERMS
C OVER ALL EQUATIONS.  ("11" IS THE NUMBER OF VARIABLES, INCLUDING
C XNP1.)
C
C DIMENSIONS
      DIMENSION CLX(2),DF(2,10,1),DSTRM(2,10,11,30),
     & DTRM(2,10,11,30),DXNP1(2,10),F(2,1),STRM(2,10,30),
     & TEMP(2),TEMPP(2),TRM(2,10,30),X(2,1),XDG(2,1),XDGM1(2,1),
     & XX(2,10,11,30)
C
```

```
C COMMON
      COMMON/EQUAT/NUMT(10),KDEG(10,11,30)
      COMMON/COEFS/ASCL(10,30)
      COMMON/PROJT/CL(2,11)
C
C EPSDIV IS USED TO DETERMINE IF PARTIAL DERIVATIVES OF F CAN
C BE SAFELY DERIVED FROM THE TERMS OF F BY DIVISION BY X.
C THE FOLLOWING AD HOC CRITERION IS USED:  "A/B" IS DEFINED
C IF ABS(B) .GE. EPSDIV.  A MORE EXACT CRITERION CAN BE DEVELOPED
C AS FOLLOWS.  LET MACHMAX BE THE LARGEST FLOATING POINT NUMBER.
C THEN "A/B" IS DEFINED IF ABS(A)/MACHMAX .LT. ABS(B).
      EPSDIV=1.D-8
C
      NP1=N+1
C
      TEMP(1)=CL(1,NP1)
      TEMP(2)=CL(2,NP1)
      DO 3  J=1,N
          CALL MUL(CL(1,J),X(1,J),CLX)
          DO 4 I=1,2
              TEMP(I)=TEMP(I)+CLX(I)
              DXNP1(I,J)=CL(I,J)
    4     CONTINUE
    3 CONTINUE
      X(1,NP1)=TEMP(1)
      X(2,NP1)=TEMP(2)
C
      DO 10 J=1,N
      DO 10 L=1,NP1
          NT=NUMT(J)
          DO 10 K=1,NT
              CALL POW(KDEG(J,L,K),X(1,L),XX(1,J,L,K))
   10 CONTINUE
C
      DO 20 J=1,N
          NT=NUMT(J)
          DO 20 K=1,NT
              STRM(1,J,K)=1.D0
              STRM(2,J,K)=0.D0
              DO 22 L=1,NP1
                  CALL MUL(XX(1,J,L,K),STRM(1,J,K),TEMPP)
                  STRM(1,J,K)=TEMPP(1)
                  STRM(2,J,K)=TEMPP(2)
   22         CONTINUE
          TRM(1,J,K)=ASCL(J,K)*STRM(1,J,K)
          TRM(2,J,K)=ASCL(J,K)*STRM(2,J,K)
   20 CONTINUE
C
      DO 30 J=1,N
          F(1,J)=0.D0
          F(2,J)=0.D0
          NT=NUMT(J)
          DO 30 K=1,NT
          DO 30 I=1,2
              F(I,J)= F(I,J) + TRM(I,J,K)
   30 CONTINUE
C
```

```
          DO 40 J=1,N
          DO 40 M=1,NP1
             NT=NUMT(J)
             DO 40 K=1,NT
                IF(KDEG(J,M,K).EQ.0) THEN
                    DSTRM(1,J,M,K)=0.DO
                    DSTRM(2,J,M,K)=0.DO
                    GOTO 40
                ENDIF
                IF( XNORM(2,X(1,M)) .GE .EPSDIV) THEN
                    CALL DIV(STRM(1,J,K),X(1,M),DSTRM(1,J,M,K))
                    DSTRM(1,J,M,K)=KDEG(J,M,K)*DSTRM(1,J,M,K)
                    DSTRM(2,J,M,K)=KDEG(J,M,K)*DSTRM(2,J,M,K)
                    GOTO 40
                ENDIF
                DSTRM(1,J,M,K)=1.DO
                DSTRM(2,J,M,K)=0.DO
                DO 44 L=1,NP1
                    IF(L.NE.M)THEN
                        CALL MUL(XX(1,J,L,K),DSTRM(1,J,M,K),TEMPP)
                        DSTRM(1,J,M,K)=TEMPP(1)
                        DSTRM(2,J,M,K)=TEMPP(2)
                    ENDIF
   44           CONTINUE
                NNNN=KDEG(J,M,K)-1
                CALL POW(NNNN,X(1,M),TEMP)
                CALL MUL(TEMP,TEMPP,DSTRM(1,J,M,K))
                DSTRM(1,J,M,K)=KDEG(J,M,K)*DSTRM(1,J,M,K)
                DSTRM(2,J,M,K)=KDEG(J,M,K)*DSTRM(2,J,M,K)
   40 CONTINUE
C
      DO 50 J=1,N
      DO 50 M=1,NP1
         NT=NUMT(J)
         DO 50 K=1,NT
         DTRM(1,J,M,K)=ASCL(J,K)*DSTRM(1,J,M,K)
         DTRM(2,J,M,K)=ASCL(J,K)*DSTRM(2,J,M,K)
   50 CONTINUE
      DO 60 J=1,N
      DO 60 M=1,NP1
         DF(1,J,M)=0.DO
         DF(2,J,M)=0.DO
         NT=NUMT(J)
         DO 60 I=1,2
         DO 60 K=1,NT
            DF(I,J,M)= DF(I,J,M) + DTRM(I,J,M,K)
   60 CONTINUE
C
      DO 70 J=1,N
      DO 70 K=1,N
         CALL MUL(DF(1,J,NP1),DXNP1(1,K),TEMP)
         DO 70 I=1,2
            DF(I,J,K)=DF(I,J,K)+TEMP(I)
   70 CONTINUE
C
      RETURN
      END
```

```
      SUBROUTINE OTPUTRT(N,NUMPAT,IDEG,EPSO,X, COND,DET)
C
C PURPOSE: OTPUTRT IS CALLED BY CONSOL8R AFTER EACH PATH IS COMPLETED.
C          OTPUTRT COMPUTES THE CONDITION AND DETERMINANT OF THE
C          JACOBIAN OF F AT THE SOLUTION X, AND THEN UNTRANSFORMS AND
C          UNSCALES X. FINALLY, IT IDENTIFIES SOLUTIONS AS REAL OR
C          COMPLEX, AND WRITES A BRIEF SOLUTION SUMMARY TO UNIT 8.
C
C GENERIC NAME: OTPUTR          FILE NAME: OTPUTRT
C
C REFERENCE:   CHAPTER 5 OF "SOLVING POLYNOMIAL SYSTEMS USING
C              CONTINUATION" BY ALEXANDER MORGAN.
C
C NOTE:   1. FOR SOLVING A GENERAL POLYNOMIAL SYSTEM, USING TABLEAU
C            INPUT.  CHANGE TO GENERIC NAME AND USE WITH CONSOL8R.
C         2. TWO CONSTANTS MAY NEED TO BE CHANGED: THE
C            ZERO-EPSILON TO DETERMINE IF A SOLUTION IS AT
C            INFINITY (EPSFIN) AND THE ZERO-EPSILON TO DETERMINE
C            IF A SOLUTION IS REAL (EPSREL).  SEE BELOW.
C
C INPUT:  N        -- NUMBER OF EQUATIONS
C         NUMPAT   -- CURRENT PATH NUMBER
C         IDEG     -- IDEG(J) IS THE DEGREE OF THE J-TH EQUATION.
C         EPSO     -- ZERO-EPSILON FOR SINGULARITY.
C         X        -- COMPUTED PATH ENDPOINT
C
C OUTPUT: X        -- IF THE PROJECTIVE COORDINATE, XNP1, HAS NORM
C                     GREATER THAN EPSFIN, THEN X IS THE PATH ENDPOINT
C                     UNTRANSFORMED AND UNSCALED.  OTHERWISE, X IS
C                     UNCHANGED.
C         COND     -- CONDITION OF THE JACOBIAN, DF, AT THE COMPUTED
C                     PATH ENDPOINT (THAT IS, TRANSFORMED AND SCALED).
C                     COND=-1 MEANS THAT DF IS SINGULAR WITH RESPECT
C                     TO EPSO.
C         DET      -- IF COND >= 0, DETERMINANT OF DF.
C
C COMMON:  /OTPVT/  IS USED TO PASS VARIABLE SCALE FACTORS FROM
C                   SUBROUTINE INPTART TO SUBROUTINE OTPUTRT.
C          /PROJT/  IS USED TO PASS PROJECTIVE TRANSFORMATION CONSTANTS
C                   FROM SUBROUTINE INPTBRT TO SUBROUTINES FFUNRT AND
C                   OTPUTRT.
C
C SUBROUTINES: DABS,DIV,FFUNRT(CALLED AS FFUNR),ICDCR,MUL,XNORM
C
C DECLARATIONS
      INTEGER I,IDEG,IFLGPT,ITEST,IWORK,J,MSG,N,NP1,NUMPAT
      DOUBLE PRECISION COND,DABS,EPSO,EPSFIN,EPSREL,FAC,FACV,T,XNORM
      DOUBLE PRECISION CL,CLX,DET,DF,DUM1,DUM2,F,ONE,TEMP,WORK,WORK2,X
C
C DIMENSIONS
      DIMENSION DF(2,10,11),DUM1(2,10),DUM2(2,10),F(2,10),IDEG(1),
     & IWORK(10),WORK(2,10),WORK2(2,10,10),X(2,1)
      DIMENSION CLX(2),DET(2),ONE(2),TEMP(2)
C
C COMMON
      COMMON/OTPVT/FACV(10)
      COMMON/PROJT/CL(2,11)
```

```
         NP1=N+1
         EPSREL=1.D-4
         EPSFIN=1.D-22
         ONE(1)=1.D0
         ONE(2)=0.D0
C
         CALL FFUNR(N,X,DUM1,DUM2, F,DF)
         CALL ICDCR(10,N,EPSO,DF, WORK,IWORK,WORK2,COND,DET,MSG)
         IF(MSG.EQ.1) COND=-1
C
         TEMP(1)=CL(1,NP1)
         TEMP(2)=CL(2,NP1)
         DO 20 J=1,N
             CALL MUL(CL(1,J),X(1,J),CLX)
             TEMP(1)=TEMP(1)+CLX(1)
             TEMP(2)=TEMP(2)+CLX(2)
  20     CONTINUE
         X(1,NP1)=TEMP(1)
         X(2,NP1)=TEMP(2)
         WRITE(6,30) X(1,NP1),X(2,NP1)
  30     FORMAT(//,' XNP1=',2D20.10,/)
         IF(XNORM(2,X(1,NP1)) .LE. EPSFIN) RETURN
C
C UNTRANSFORM VARIABLES
         DO 40 J=1,N
             CALL DIV(X(1,J),X(1,NP1),TEMP)
             X(1,J)=TEMP(1)
             X(2,J)=TEMP(2)
  40     CONTINUE
C UNSCALE VARIABLES
         DO 50 J=1,N
             FAC=10.D0**FACV(J)
             DO 50 I=1,2
             X(I,J)=FAC*X(I,J)
  50     CONTINUE
C
         WRITE(8,60) NUMPAT
  60     FORMAT(' NUMPAT=',I5)
C DESIGNATE SOLUTIONS "REAL" OR "COMPLEX"
         ITEST=0
         DO 70 J=1,N
             IF(DABS(X(2,J)) .GE. EPSREL) ITEST=1
  70     CONTINUE
         IF(ITEST.EQ.1) THEN
                 WRITE(6,80)
                 WRITE(8,80)
  80             FORMAT(' COMPLEX SOLUTION   ')
         ELSE
                 WRITE(6,90)
                 WRITE(8,90)
  90             FORMAT(' REAL SOLUTION   ')
         ENDIF
         WRITE(8,100) ((X(I,J),I=1,2),J=1,N)
 100     FORMAT(2D22.15)
C
         RETURN
         END
```

Section 5: Programs for Chapter 6

```
      SUBROUTINE INPB3R(M,SSZBEG,EPSBIG,SSZMIN,EPSSML,EPSO)
C
C PURPOSE:   SAMPLE INPUT ROUTINE FOR THE COMPLEX CONVEX-LINEAR
C            CONTINUATION, FOR USE WITH CONSOL9.  INITIALIZES
C            LESS FREQUENTLY CHANGED CONSTANTS.
C
C GENERIC NAME: INPB          FILE NAME: INPB3R
C
C
C NOTE: 1. THIS PROGRAM DOES NOT USE THE FORTRAN "COMPLEX" DECLARATION.
C          IT EMULATES COMPLEX OPERATIONS USING REAL ARITHMETIC.  (SEE
C          APPENDIX 2.)  THIS ROUTINE SHOULD BE USED WITH INPA3, STPT3R,
C          HFNN3R, AND FFNNQ3R OR FFNNB3R.
C
C       2. WHILE INPB NEED NOT BE CUSTOMIZED FOR EACH EXPERIMENT,
C          THE CHOICE OF THE PARAMETERS P AND Q CAN EFFECT THE
C          CONVERGENCE OF THE METHOD.  NOTE THAT IN THIS VERSION OF
C          THE SUBROUTINE, I HAVE CHOSEN P TO HAVE OFF-DIAGONAL ELEMENTS
C          EQUAL TO ZERO.  CODE IS IN PLACE, BUT COMMENTED OUT, TO READ
C          Q AND THE DIAGONAL OF P FROM UNIT 5.
C
C INPUT:  M       -- TWICE THE NUMBER OF EQUATIONS
C         SSZBEG  -- BEGINNING STEPSIZE
C         EPSBIG  -- ZERO-EPSILON FOR PATH TRACKING RESIDUAL, UNTIL T > .95
C
C OUTPUT: SSZMIN  -- MINIMUM STEPSIZE ALLOWED
C         EPSSML  -- ZERO-EPSILON FOR PATH TRACKING RESIDUAL, AFTER T > .95
C         EPSO    -- ZERO-EPSILON FOR SINGULARITY
C
C COMMON:  /RANPAR/  IS USED TO PASS RANDOM PARAMETERS FROM SUBROUTINE
C                    INPB TO SUBROUTINES STPT AND HFNN.
C
C DECLARATIONS
      INTEGER I,J,K,M,MD2
      DOUBLE PRECISION EPSO,EPSBIG,EPSSML,P,Q,SSZBEG,SSZMIN
C
C COMMON
      COMMON/RANPAR/ P(2,10,10),Q(2,10)
C
      MD2=M/2
      SSZMIN = 1.D-5*SSZBEG
      EPSSML = 1.D-3*EPSBIG
      EPSO   = 1.D-22
C
      P( 1, 1, 1)= .12324754231D0
      P( 2, 1, 1)= .76253746298D0
      P( 1, 2, 2)= .93857838950D0
      P( 2, 2, 2)= .99375892810D0
      P( 1, 3, 3)= .23467908356D0
      P( 2, 3, 3)= .39383930009D0
      P( 1, 4, 4)= .83542556622D0
      P( 2, 4, 4)= .10192888288D0
      P( 1, 5, 5)= .55763522521D0
      P( 2, 5, 5)= .83729899911D0
      P( 1, 6, 6)= .78348738738D0
      P( 2, 6, 6)= .10578234903D0
      P( 1, 7, 7)= .03938347346D0
```

514

```
            P( 2, 7, 7)= .04825184716D0
            P( 1, 8, 8)= .43428734331D0
            P( 2, 8, 8)= .93836289418D0
            P( 1, 9, 9)= .99383729993D0
            P( 2, 9, 9)= .40947822291D0
            P( 1,10,10)= .09383736736D0
            P( 2,10,10)= .26459172298D0
      C
            Q( 1, 1)= .58720452864D0
            Q( 2, 1)= .01321964722D0
            Q( 1, 2)= .97884134700D0
            Q( 2, 2)=-.14433009712D0
            Q( 1, 3)= .39383737289D0
            Q( 2, 3)= .41543223411D0
            Q( 1, 4)=-.03938376373D0
            Q( 2, 4)=-.61253112318D0
            Q( 1, 5)= .39383737388D0
            Q( 2, 5)=-.26454678861D0
            Q( 1, 6)=-.00938376766D0
            Q( 2, 6)= .34447867861D0
            Q( 1, 7)=-.04837366632D0
            Q( 2, 7)= .48252736790D0
            Q( 1, 8)= .93725237347D0
            Q( 2, 8)=-.54356527623D0
            Q( 1, 9)= .39373957747D0
            Q( 2, 9)= .65573434564D0
            Q( 1,10)=-.39380038371D0
            Q( 2,10)= .98903450052D0
      C
            DO 4 I=1,2
            DO 4 J=1,MD2
            DO 4 K=1,MD2
            IF(J.NE.K)P(I,J,K)=0.D0
        4   CONTINUE
      C
      CC    READ(5,8)((P(I,J,J),I=1,2),J=1,MD2)
      CC  8 FORMAT(D22.15)
      CC    READ(5,8)((Q(I,J),I=1,2),J=1,MD2)
      C
            WRITE(6,10) SSZMIN
       10   FORMAT(/' MINIMUM STPSZE = ',D22.15)
            WRITE(6,20) EPSSML
       20   FORMAT(/' EPSSML =',D22.15)
            WRITE(6,30) EPSO
       30   FORMAT(/' EPSO   =',D22.15)
            WRITE(6,40) ((J,K,P(1,J,K),P(2,J,K),K=1,MD2),J=1,MD2)
       40   FORMAT(/' P(',I2,',',I2,')  =',2D22.15)
            WRITE(6,50) (J,Q(1,J),Q(2,J),J=1,MD2)
       50   FORMAT(/' Q(',I2,')  =',2D22.15)
            WRITE(6,999)
      999   FORMAT(/)
            RETURN
            END
```

```
      SUBROUTINE STPT3R(M,Y)
C
C PURPOSE:   GENERATE A START POINT FOR THE PATH FOR THE COMPLEX
C            CONVEX-LINEAR CONTINUATION,FOR USE WITH CONSOL9.
C
C GENERIC NAME: STPT              FILE NAME: STPT3R
C
C NOTE: THIS PROGRAM DOES NOT USE THE FORTRAN "COMPLEX" DECLARATION.
C       IT EMULATES COMPLEX OPERATIONS USING REAL ARITHMETIC.  (SEE
C       APPENDIX 2.)  THIS ROUTINE SHOULD BE USED WITH INPA3, INPB3R,
C       HFNN3R, AND FFNNQ3R OR FFNNB3R.
C
C
C INPUT:  M         -- TWICE THE NUMBER OF EQUATIONS
C
C OUTPUT: Y         -- START POINT FOR PATH
C
C COMMON:  /RANPAR/  IS USED TO PASS RANDOM PARAMETERS FROM SUBROUTINE
C                    INPB TO SUBROUTINES STPT AND HFNN.
C
C SUBROUTINES: LNFNCR,LNSNCR
C
C
C DECLARATIONS
      INTEGER I,IROW,J,K,M,MD2,MSG
      DOUBLE PRECISION Y
      DOUBLE PRECISION P,PCOPY,Q
C
C DIMENSIONS
      DIMENSION IROW(10),PCOPY(2,10,10),Y(1)
C
C COMMON
      COMMON/RANPAR/P(2,10,10),Q(2,10)
C
C
      MD2=M/2
C
      DO 10 I=1,2
      DO 10 J=1,MD2
      DO 10 K=1,MD2
        PCOPY(I,J,K)=P(I,J,K)
   10 CONTINUE
      CALL LNFNCR(10,MD2,0.D0,PCOPY, MSG,IROW)
      CALL LNSNCR(10,MD2,IROW,PCOPY,Q, Y)
C
C
      WRITE(6,20)
   20 FORMAT(/,'  START POINT:')
      WRITE(6,30) ( J,Y(J),J=1,M )
   30 FORMAT('    Y(',I2,') = ',D22.15)
C
      RETURN
      END
```

```
      SUBROUTINE HFNN3R(M,T,Y, H,DH)
C
C PURPOSE:  EVALUATES THE CONTINUATION EQUATION,H, AND ITS PARTIAL
C           DERIVATIVES, DH, FOR THE COMPLEX CONVEX-LINEAR CONTINUATION,
C           FOR USE WITH CONSOL9.
C
C
C GENERIC NAME: HFNN           FILE NAME: HFNN3R
C
C
C NOTE: 1. THIS PROGRAM DOES NOT USE THE FORTRAN "COMPLEX" DECLARATION.
C          IT EMULATES COMPLEX OPERATIONS USING REAL ARITHMETIC.  (SEE
C          APPENDIX 2.)  THIS ROUTINE SHOULD BE USED WITH INPA3, INPB3R,
C          STPT3R, AND FFNNQ3R OR FFNNB3R.
C
C       2. THIS SUBROUTINE EVALUATES THE COMPLEX CONTINUATION AND
C          THEN "REALIFIES" IT FOR USE BY CONSOL9.  THUS, FIRST
C          IT EVALUATES THE COMPLEX CONVEX-LINEAR CONTINUATION
C          DEFINED BY CH(X,T) =  (1-T)*(P*X - Q) + T*F(X), WHERE
C          X IS A COMPLEX VECTOR OF LENGTH M/2 IDENTIFIED WITH Y.
C          THEN IT "REALIFIES" CH AND DCH, TO OBTAIN H AND DH,
C          RESPECTIVELY.
C
C
C INPUT:  M        -- TWICE THE NUMBER OF EQUATIONS
C         T        -- CURRENT VALUE OF T
C         Y        -- CURRENT VALUE OF Y
C
C OUTPUT: H        -- H IS THE REALIFICATION OF THE COMPLEX CONTINUATION, CH.
C                     THUS, H IS A SYSTEM OF M REAL POLYNOMIALS IN THE M+1
C                     UNKNOWNS Y AND T.
C         DH       -- DH IS AN M X (M+1) MATRIX, THE REALIFICATION OF THE
C                     JACOBIAN, DCH, OF CH.  DCH(J,K) IS THE PARTIAL OF
C                     CH(J) WITH RESPECT TO X(K), FOR J=1 TO M/2 AND K=1
C                     TO M/2.  DCH(J,M+1) IS THE PARTIAL DERIVATIVE
C                     OF CH(J) WITH RESPECT TO T, FOR J=1 TO M/2.
C
C
C COMMON:  /RANPAR/  IS USED TO PASS RANDOM PARAMETERS FROM SUBROUTINE
C                    INPB TO SUBROUTINES STPT AND HFNN.
C
C
C SUBROUTINES: FFNN,MUL
C
C
C DECLARATIONS
      INTEGER J,J2,J2M1,K,K2,K2M1,M,MD2,MP1
      DOUBLE PRECISION DH,H,ONEMT,T,Y
      DOUBLE PRECISION CH,DCH,DF,F,ONE,P,PL,Q,TEMP,X,ZERO
C
C DIMENSIONS
      DIMENSION CH(2,10),DCH(2,10,11),DF(2,10,10),DH(20,1),F(2,10),H(1),
     & ONE(2),PL(2,10),TEMP(2),X(2,10),Y(1),ZERO(2)
C
C COMMON
      COMMON/RANPAR/P(2,10,10),Q(2,10)
C
```

```
      MD2 =M/2
      ONE(1) = 1.D0
      ONE(2) = 0.D0
      ZERO(1) = 0.D0
      ZERO(2) = 0.D0
C
      MP1=M+1
      ONEMT=1.D0-T
C
      DO 5 J=1,MD2
          J2=2*J
          J2M1=J2-1
          X(1,J)=Y(J2M1)
          X(2,J)=Y(J2)
   5  CONTINUE
C
      CALL FFNN(MD2,X, F,DF)
C
      DO 10 J=1,MD2
          PL(1,J)=ZERO(1)
          PL(2,J)=ZERO(2)
          DO 20 K=1,MD2
              DCH(1,J,K)=ONEMT*P(1,J,K) + T*DF(1,J,K)
              DCH(2,J,K)=ONEMT*P(2,J,K) + T*DF(2,J,K)
              CALL MUL(P(1,J,K),X(1,K),TEMP)
              PL(1,J)=PL(1,J) + TEMP(1)
              PL(2,J)=PL(2,J) + TEMP(2)
  20      CONTINUE
          DCH(1,J,MP1) =    - ( PL(1,J) - Q(1,J) ) +   F(1,J)
          DCH(2,J,MP1) =    - ( PL(2,J) - Q(2,J) ) +   F(2,J)
          CH(1,J)    = ONEMT*( PL(1,J) - Q(1,J) ) + T*F(1,J)
          CH(2,J)    = ONEMT*( PL(2,J) - Q(2,J) ) + T*F(2,J)
  10  CONTINUE
C
C
      DO 30 J=1,MD2
          J2=2*J
          J2M1=J2-1
          DO 40 K=1,MD2
              K2=2*K
              K2M1=K2-1
              DH(J2M1,K2M1)= DCH(1,J,K)
              DH(J2  ,K2 )= DH(J2M1,K2M1)
              DH(J2  ,K2M1)= DCH(2,J,K)
              DH(J2M1,K2 )=-DH(J2  ,K2M1)
  40      CONTINUE
C
          DH(J2M1,MP1)= DCH(1,J,MP1)
          DH(J2  ,MP1)= DCH(2,J,MP1)
          H(J2M1)    =   CH(1,J)
          H(J2  )    =   CH(2,J)
  30  CONTINUE
C
      RETURN
      END
```

```
      SUBROUTINE FFNNQ3R(N,X, F,DF)
C
C
C PURPOSE:   SAMPLE SYSTEM-EVALUATION ROUTINE FOR THE COMPLEX
C            CONVEX-LINEAR CONTINUATION, FOR USE WITH CONSOL9.
C            SUBROUTINE FFNN RETURNS F AND DF VALUES.
C
C GENERIC NAME: FFNN              FILE NAME: FFNNQ3R
C
C NOTE: 1. THIS PROGRAM DOES NOT USE THE FORTRAN "COMPLEX" DECLARATION.
C          IT EMULATES COMPLEX OPERATIONS USING REAL ARITHMETIC.  (SEE
C          APPENDIX 2.)  THIS ROUTINE SHOULD BE USED WITH INPA3, INPB3,
C          STPT3R, AND HFNN3R.
C
C       2. THIS SUBROUTINE IS SET UP FOR THE CASE THAT F IS TWO SECOND
C          DEGREE POLYNOMIALS IN TWO UNKNOWNS WITH REAL COEFFICIENTS
C          AND COMPLEX X:
C
C          F(J) = A(J,1)*X(1)**2 +A(J,2)*X(2)**2 +A(J,3)*X(1)*X(2)
C
C                                 +A(J,4)*X(1)  +A(J,5)*X(2)  +A(J,6)
C
C          FOR  J=1 TO 2.
C
C
C INPUT:  N        -- THE NUMBER OF EQUATIONS
C         X        -- CURRENT VALUE OF X (IDENTIFIED WITH CONSOL9'S Y VIA
C                     X(1,J)=Y(2*J-1), X(2,J))=Y(2*J).)
C
C OUTPUT: F        -- F(J) IS THE VALUE OF THE J-TH EQUATION AT X
C         DF       -- DF(J,K) IS THE PARTIAL DERIVATIVE OF F(J) WITH
C                     RESPECT TO X(K)
C
C COMMON: /COEFS/  IS USED TO PASS COEFFICIENT VALUES FROM SUBROUTINE
C                  INPA  TO SUBROUTINE  FFNN.
C
C SUBROUTINE:  MUL
C
C
C DECLARATIONS
      INTEGER I,J,N
      DOUBLE PRECISION A
      DOUBLE PRECISION DF,F,X,X12,XSQ,ONE
C
C DIMENSIONS
      DIMENSION DF(2,10,1),F(2,1),X(2,1),X12(2),XSQ(2,2),ONE(2)
C
C COMMON
      COMMON/COEFS/A(2,6)
C
      ONE(1)=1.D0
      ONE(2)=0.D0
C
      CALL MUL(X(1,1),X(1,1),XSQ(1,1))
      CALL MUL(X(1,2),X(1,2),XSQ(1,2))
      CALL MUL(X(1,1),X(1,2),X12)
C
```

```fortran
      DO 10 I=1,2
      DO 10 J=1,N
         F(I,J)=A(J,1)*XSQ(I,1)+A(J,2)*XSQ(I,2) +A(J,3)*X12(I)
     &        + A(J,4)*  X(I,1)+A(J,5)*  X(I,2) +A(J,6)*ONE(I)
         DF(I,J,1)=2*A(J,1)*X(I,1)+A(J,3)*X(I,2)+A(J,4)*ONE(I)
         DF(I,J,2)=2*A(J,2)*X(I,2)+A(J,3)*X(I,1)+A(J,5)*ONE(I)
   10 CONTINUE
C
      RETURN
      END
```

```
      SUBROUTINE FFNNB3R(N,X, F,DF)
C
C PURPOSE:  SAMPLE SYSTEM-EVALUATION ROUTINE FOR THE COMPLEX
C           CONVEX-LINEAR CONTINUATION, FOR USE WITH CONSOL9.
C           SUBROUTINE FFNN RETURNS F AND DF VALUES.
C
C
C GENERIC NAME: FFNN          FILE NAME: FFNNB3R
C
C
C NOTE: 1. THIS PROGRAM DOES NOT USE THE FORTRAN "COMPLEX" DECLARATION.
C          IT EMULATES COMPLEX OPERATIONS USING REAL ARITHMETIC.  (SEE
C          APPENDIX 2.)  THIS ROUTINE SHOULD BE USED WITH INPA3, STPT3R,
C          HFNN3R, AND FFNNQ3R OR FFNNB3R.
C
C       2. THIS SUBROUTINE IS SET UP FOR FOR BROWN'S ALMOST LINEAR SYSTEM.
C          (SEE BELOW, AND EXPERIMENT 6-17.)
C INPUT:  N        -- THE NUMBER OF EQUATIONS
C         X        -- CURRENT VALUE OF X (IDENTIFIED WITH CONSOL9'S Y VIA
C                     X(1,J))=Y(2*J-1), X(2,J))=Y(2*J).)
C
C OUTPUT: F        -- F(J) IS THE VALUE OF THE J-TH EQUATION AT X
C         DF       -- DF(J,K) IS THE PARTIAL DERIVATIVE OF F(J) WITH
C                     RESPECT TO X(K)
C
C DECLARATIONS
      INTEGER I,J,K,N,NM1,NP1
      DOUBLE PRECISION DF,F,ONE,TEMP,TWO,X,XX,XP,ZERO,ZNP1
C
C DIMENSIONS
      DIMENSION DF(2,10,1),F(2,1),ONE(2),TEMP(2),TWO(2),X(2,1),XX(2),
     & XP(2),ZERO(2),ZNP1(2)
C
      NP1=N+1
      NM1=N-1
      ZERO(1)=0.D0
      ZERO(2)=0.D0
      ONE(1) =1.D0
      ONE(2) =0.D0
      TWO(1) =2.D0
      TWO(2) =0.D0
      ZNP1(1)=NP1
      ZNP1(2)=0.D0
C
      XX(1)=ZERO(1)
      XX(2)=ZERO(2)
      XP(1)=ONE(1)
      XP(2)=ONE(2)
      DO 10 J=1,N
         XX(1)=XX(1)+X(1,J)
         XX(2)=XX(2)+X(2,J)
         CALL MUL(XP(1),X(1,J),TEMP)
         XP(1)=TEMP(1)
         XP(2)=TEMP(2)
   10 CONTINUE
C
```

```
      DO 22 I=1,2
      DO 20 J=1,NM1
        F(I,J)=X(I,J)+XX(I)-ZNP1(I)
   20 CONTINUE
      F(I,N)=XP(I)-ONE(I)
   22 CONTINUE
C
      DO 30 I=1,2
      DO 30 J=1,NM1
      DO 30 K=1,N
       DF(I,J,K)=ONE(I)
       IF(J.EQ.K)DF(I,J,K)=TWO(I)
   30 CONTINUE
C
      DO 40 J=1,N
      DF(1,N,J)=ONE(1)
      DF(2,N,J)=ONE(2)
      DO 40 K=1,N
       IF(J.NE.K) THEN
         CALL MUL(DF(1,N,J),X(1,K),TEMP)
         DF(1,N,J)=TEMP(1)
         DF(2,N,J)=TEMP(2)
       ENDIF
   40 CONTINUE
C
      RETURN
      END
```

```
      SUBROUTINE LNFNCR(LDIM,M,EPSO,A, MSG,IROW)
C
C PURPOSE:  GENERATES AN LU FACTORIZATION OF NONSINGULAR COMPLEX
C           M X M MATRIX A.  IF A IS NONSINGULAR, LOWER AND UPPER
C           TRIANGULAR MATRICES L AND U ARE GENERATED SUCH THAT A=L*U,
C           WITH U(I,I)=1 FOR I=1 TO M.  IF A IS SINGULAR, THEN THE
C           FACTORIZATION IS ABANDONED AND AN ERROR MESSAGE IS RETURNED.
C
C REFERENCE: CHAPTER 4 OF "SOLVING POLYNOMIAL SYSTEMS USING
C           CONTINUATION" BY ALEXANDER MORGAN.  BASED ON THE
C           DESCRIPTION OF CROUT'S METHOD IN [DAHLQUIST, ANDERSON]
C           "NUMERICAL METHODS" (PRENTICE HALL, 1974) PP. 157-8,
C           IN [FORSYTHE] "ALGORITHM 16: CROUT WITH PIVOTING,"
C           COMM. ACM, VOL. 3 (1960), PP. 507-8, AND IN
C           [FORSYTHE, MOLER] "COMPUTER SOLUTION OF LINEAR ALGEBRAIC
C           SYSTEMS," (PRENTICE-HALL, 1967) CHAPTERS 10 AND 12.
C
C
C NOTES: 1. REAL AND IMAGINARY PARTS OF ALL COMPLEX VARIABLES
C           AND ARRAYS ARE GIVEN BY THE LEADING INDEX.  FOR
C           EXAMPLE, A(1,I,J) IS THE REAL PART OF THE
C           (I,J)-TH ENTRY OF MATRIX A, AND A(2,I,J) IS THE
C           IMAGINARY PART.  COMPLEX OPERATIONS ARE IMPLEMENTED
C           IN REAL ARITHMETIC USING SUBROUTINES MUL AND DIV.
C        2. THIS PROGRAM IS DESIGNED TO BE FOLLOWED BY LNSNCR TO
C           SOLVE THE LINEAR SYSTEM A*Y = B.
C
C
C INPUT:    LDIM         -- LEADING DIMENSION OF ARRAY A
C           M            -- DIMENSION OF CURRENT PROBLEM
C           EPSO         -- ZERO-EPSILON TO DETERMINE SINGULARITY
C           A            -- MATRIX DIMENSIONED A(2,LDIM,MM) FOR MM > M
C                           IN THE CALLING PROGRAM
C
C OUTPUT:   A            -- IF MSG=0, CONTAINS THE LU FACTORIZATION OF
C                           THE INPUT A, STORED IN THE LOWER AND UPPER
C                           TRIANGLES OF A.
C           MSG          -- =1 IF A SINGULAR, =0 OTHERWISE
C           IROW         -- PIVOT ARRAY
C
C
C SUBROUTINES: DABS,DIV,MUL,XNORM
C
C
C DECLARATIONS
      INTEGER I,IMAX,IR,IROW,ITEMP,J,K,KM1,KP1,KR,
     & LDIM,L,LR,M,MAXN,MSG,P,PR
      DOUBLE PRECISION EPSO,TEMP,TEST,XNORM
      DOUBLE PRECISION A,CTEMP,PROD
C
C DIMENSIONS
      DIMENSION A(2,LDIM,1),CTEMP(2),IROW(1),PROD(2)
C
C
      MSG=0
C
```

```
      DO 10 I=1,M
         IROW(I)=I
  10  CONTINUE
C
C K IS THE COLUMN INDEX
      DO 20 K=1,M
         KM1=K-1
         KP1=K+1
         DO 30 I=K,M
            IR=IROW(I)
            PROD(1)=A(1,IR,K)
            PROD(2)=A(2,IR,K)
            IF(KM1 .LT. 1) GOTO 36
            DO 35 P=1,KM1
               PR=IROW(P)
               CALL MUL(A(1,IR,P),A(1,PR,K),CTEMP)
               PROD(1)=PROD(1)-CTEMP(1)
               PROD(2)=PROD(2)-CTEMP(2)
  35        CONTINUE
  36        CONTINUE
            A(1,IR,K)=PROD(1)
            A(2,IR,K)=PROD(2)
  30     CONTINUE
C
C FIND THE INDEX, IMAX, OF THE LARGEST OF ABS(A(I,K)) FOR I=K TO N.
         IMAX = K
         KR=IROW(K)
         TEMP  = XNORM(2, A(1,KR,K) )
         IF(KP1 .GT. M) GOTO 41
         DO 40 I = KP1,M
            IR=IROW(I)
            TEST = XNORM(2, A(1,IR,K) )
            IF(TEST .GT. TEMP) THEN
               IMAX = I
               TEMP = TEST
            ENDIF
  40     CONTINUE
  41     CONTINUE
C
C IF THE LARGEST ENTRY IN THE K-TH COLUMN AT OR BELOW THE (K,K)-TH ENTRY
C IS ZERO, THEN THE MATRIX A IS SINGULAR.
         IF(TEMP .LE. EPSO) THEN
            MSG=1
            RETURN
         ENDIF
C
C SWITCH INDICES IROW(K) AND IROW(IMAX)
         ITEMP =IROW(K)
         IROW(K)=IROW(IMAX)
         IROW(IMAX)=ITEMP
C
```

```
            KR=IROW(K)
            IF(KP1.GT.M) GOTO 51
            DO 50 J= KP1,M
                PROD(1)=A(1,KR,J)
                PROD(2)=A(2,KR,J)
                IF(KM1. LT. 1) GOTO 61
                DO 60 P=1,KM1
                    PR=IROW(P)
                    CALL MUL(A(1,KR,P),A(1,PR,J),CTEMP)
                    PROD(1)=PROD(1)-CTEMP(1)
                    PROD(2)=PROD(2)-CTEMP(2)
   60           CONTINUE
   61           CONTINUE
                CALL DIV(PROD,A(1,KR,K),A(1,KR,J))      .
   50       CONTINUE
   51       CONTINUE
C
   20   CONTINUE
C
        RETURN
        END
```

```
      SUBROUTINE LNSNCR(LDIM,M,IROW,A,B,Y)
C
C
C PURPOSE:   SOLVES THE COMPLEX LINEAR SYSTEM L*U * Y = B, WHERE B IS AN
C            M X 1 COLUMN OF COMPLEX NUMBERS AND L AND U ARE M X M LOWER
C            AND UPPER TRIANGULAR COMPLEX MATRICES, RESPECTIVELY, STORED
C            IN ARRAY A BY SUBROUTINE LNFNCR.
C
C
C REFERENCE: CHAPTER 4 OF "SOLVING POLYNOMIAL SYSTEMS USING
C            CONTINUATION" BY ALEXANDER MORGAN.
C
C
C NOTE:   COMPLEX NUMBERS AND ARRAYS ARE REPRESENTED BY REAL ARRAYS WITH
C         A LEADING DIMENSION OF "2." SEE SUBROUTINE LNFNCR.
C
C
C INPUT:     LDIM          -- LEADING DIMENSION OF ARRAY A
C            M             -- DIMENSION OF CURRENT PROBLEM
C            IROW          -- PIVOT ARRAY, AS GENERATED BY LNFNCR
C            A             -- CONTAINS L AND U AS GENERATED BY LNFNCR
C            B             -- RIGHT HAND SIDE OF THE SYSTEM BEING SOLVED
C
C OUTPUT:    Y             -- SOLUTION TO L*U * Y = B.
C
C
C SUBROUTINES: DIV,MUL
C
C
C DECLARATIONS
      INTEGER I,II,IP1,IR,IM1,IROW,J,K,LDIM,M
      DOUBLE PRECISION A,B,CTEMP,PROD,Y
C
C
C DIMENSIONS
      DIMENSION A(2,LDIM,1),B(2,1),CTEMP(2),IROW(1),PROD(2),Y(2,1)
C
C
C FORWARD ELIMINATION TO SOLVE L*Y=B
      DO 10 I=1,M
        IR = IROW(I)
        PROD(1)=B(1,IR)
        PROD(2)=B(2,IR)
        IM1=I-1
        IF(IM1 .LT. 1) GOTO 21
        DO 20 K=1,IM1
           CALL MUL(A(1,IR,K),Y(1,K),CTEMP)
           PROD(1)=PROD(1)-CTEMP(1)
           PROD(2)=PROD(2)-CTEMP(2)
 20     CONTINUE
 21     CONTINUE
        CALL DIV(PROD,A(1,IR,I),Y(1,I))
 10   CONTINUE
C
```

```
C
C BACK SUBSTITUTION TO SOLVE U*Y=Y
      DO 30 II=1,M
          I=M-II+1
          IR=IROW(I)
          PROD(1)=Y(1,I)
          PROD(2)=Y(2,I)
          IP1=I+1
          IF(IP1 .GT. M) GOTO 41
          DO 40 K=IP1,M
              CALL MUL(A(1,IR,K),Y(1,K),CTEMP)
              PROD(1)=PROD(1)-CTEMP(1)
              PROD(2)=PROD(2)-CTEMP(2)
   40     CONTINUE
   41     CONTINUE
          Y(1,I)=PROD(1)
          Y(2,I)=PROD(2)
   30 CONTINUE
C
      RETURN
      END
```

```
      SUBROUTINE ICDCR(LDIM,N,EPSO,A, WORK,IWORK,AINV,COND,DET,MSG)
C
C PURPOSE:  COMPUTES THE INVERSE, DETERMINANT, AND CONDITION OF THE
C           NONSINGULAR COMPLEX N X N MATRIX A.
C
C REFERENCE: CHAPTER 4 OF "SOLVING POLYNOMIAL SYSTEMS USING
C            CONTINUATION" BY ALEXANDER MORGAN
C
C NOTE: THIS PROGRAM DOES NOT USE THE FORTRAN "COMPLEX" DECLARATION.  IT
C       EMULATES COMPLEX OPERATIONS USING REAL ARITHMETIC.  SEE APPENDIX 2.
C
C INPUT:  LDIM        -- LEADING DIMENSION OF ARRAY A
C         N           -- DIMENSION OF THE CURRENT PROBLEM
C         EPSO        -- ZERO-EPSILON FOR SINGULARITY
C         A           -- MATRIX DIMENSIONED A(2,LDIM,NN) WHERE NN >= N
C                        IN THE CALLING PROGRAM.
C
C OUTPUT: WORK        -- REAL WORKSPACE ARRAY OF LENGTH 2 X N.
C         IWORK       -- INTEGER WORKSPACE ARRAY OF LENGTH N.
C         AINV        -- IF MSG=0, INVERSE OF A, DIMENSIONED THE SAME AS A.
C         COND        -- IF MSG=0, CONDITION OF A, C(A), WHERE C(A) IS
C                        DEFINED BY C(A) = |A|*|1/A| WHERE 1/A DENOTES
C                        THE INVERSE OF A AND "| |" DENOTES THE MATRIX
C                        NORM DERIVED FROM XNORM (THE "1-NORM").
C         DET         -- IF MSG=0, DETERMINANT OF A
C         MSG         -- =1, IF A IS SINGULAR WITH RESPECT TO EPSO.
C                        =0, OTHERWISE.
C
C SUBROUTINES: LNFNCR,LNSNCR,MUL,XNORM
C
C
C DECLARATIONS
      INTEGER I,IM1,IPIV,IWORK,J,K,KP1,LDIM,MSG,N,N2
      DOUBLE PRECISION BIG,COND,EPSO,TEST,XNORM,XNORMM
      DOUBLE PRECISION A,AINV,DET,WORK,TEMP
C
C DIMENSIONS
      DIMENSION A(2,LDIM,1),AINV(2,LDIM,1),DET(2),WORK(2,1),IWORK(1),
     & TEMP(2)
C
C
      N2=2*N
C
C COMPUTE XNORMM(A).  XNORMM IS THE MATRIX NORM ASSOCIATED WITH XNORM.
C (THUS,XNORMM IS THE MAXIMUM OF THE SUMS OF ABSOLUTE VALUES OF THE
C COLUMNS OF A.)
      XNORMM=0.DO
      DO 10 J=1,N
        TEST=XNORM(N2,A(1,1,J))
        IF(TEST .GT. XNORMM) XNORMM=TEST
   10 CONTINUE
C
C STORE XNORMM(A) IN COND.
      COND=XNORMM
C
```

```
C
C GENERATE AINV=1/A
      CALL LNFNCR(LDIM,N,EPSO,A,MSG,IWORK)
      IF(MSG .EQ. 1) RETURN
C
      DO 15 I=1,N
        WORK(1,I)=0.D0
        WORK(2,I)=0.D0
   15 CONTINUE
C
      DO 20 I=1,N
        IM1=I-1
        IF(I.GT.1)WORK(1,IM1)=0.D0
        WORK(1,I)=1.D0
        CALL LNSNCR(LDIM,N,IWORK,A,WORK, AINV(1,1,I))
   20 CONTINUE
C
C
C COMPUTE XNORMM(AINV)
      XNORMM=0.D0
      DO 30 J=1,N
        TEST=XNORM(N2,AINV(1,1,J))
        IF(TEST .GT. XNORMM) XNORMM=TEST
   30 CONTINUE
C
      COND=COND*XNORMM
C
C
C COMPUTE DETERMINANT OF A
      DET(1)=1.D0
      DET(2)=0.D0
      DO 35 I=1,N
        CALL MUL(DET,A(1,IWORK(I),I),TEMP)
        DET(1)=TEMP(1)
        DET(2)=TEMP(2)
   35 CONTINUE
C
      IPIV=1
      DO 40 K=1,N
        IF(IWORK(K).EQ.K) GOTO 41
        KP1=K+1
        DO 45 J=KP1,N
        IF(IWORK(J).EQ.K) THEN
          IWORK(J)=IWORK(K)
          IPIV=-IPIV
        END IF
   45   CONTINUE
   41   CONTINUE
   40 CONTINUE
      DET(1)=IPIV*DET(1)
      DET(2)=IPIV*DET(2)
C
      RETURN
      END
```

```fortran
      SUBROUTINE MUL(X,Y,Z)
C
C PURPOSE: EMULATES COMPLEX MULTIPLICATION IN REAL COMPUTER ARITHMETIC
C
C INPUT: X, Y
C
C OUTPUT: Z = X * Y
C
C NOTE: 1.  X(1) AND X(2) REPRESENT THE REAL AND IMAGINARY PARTS OF X,
C           RESPECTIVELY.  SIMILARLY, FOR Y AND Z.
C       2.  Z SHOULD NOT BE EQUIVALENCED TO X OR Y.  "CALL MUL(X,Y,X)"
C           WILL NOT GENERATE X = X * Y CORRECTLY.
C
C DECLARATIONS
      DOUBLE PRECISION X,Y,Z
C
C DIMENSIONS
      DIMENSION X(2),Y(2),Z(2)
C
C
      Z(1) = X(1)*Y(1) - X(2)*Y(2)
      Z(2) = X(1)*Y(2) + X(2)*Y(1)
C
      RETURN
      END
```

```
      SUBROUTINE POW(N,X,Y)
C
C PURPOSE: EMULATES COMPLEX RAISING TO INTEGRAL POWERS IN REAL COMPUTER
C          ARITHMETIC
C
C INPUT: N, X
C
C OUTPUT: Y = X**N
C
C NOTE:   X(1) AND X(2) REPRESENT THE REAL AND IMAGINARY PARTS OF X,
C         RESPECTIVELY.  SIMILARLY, FOR Y.  N IS A NON-NEGATIVE INTEGER.
C
C SUBROUTINES: MUL
C
C
C DECLARATIONS
      INTEGER N,NM1
      DOUBLE PRECISION X,Y,Z
C
C DIMENSIONS
      DIMENSION X(2),Y(2),Z(2)
C
C
      IF(N.EQ.0) THEN
         Y(1)=1.D0
         Y(2)=0.D0
         RETURN
      ENDIF
C
      IF(N.EQ.1) THEN
         Y(1)=X(1)
         Y(2)=X(2)
         RETURN
      ENDIF
C
      NM1=N-1
      Y(1)=X(1)
      Y(2)=X(2)
C
      DO 10 J=1 , NM1
         CALL MUL(X,Y,Z)
         Y(1)=Z(1)
         Y(2)=Z(2)
   10 CONTINUE
C
      RETURN
      END
```

```
      SUBROUTINE DIV(X,Y,Z)
C
C PURPOSE: EMULATES COMPLEX DIVISION IN REAL COMPUTER ARITHMETIC
C
C INPUT: X, Y, Z
C
C OUTPUT: Z = X / Y
C
C NOTE: 1.  X(1) AND X(2) REPRESENT THE REAL AND IMAGINARY PARTS OF X,
C            RESPECTIVELY.  SIMILARLY, FOR Y AND Z.
C       2.  Z SHOULD NOT BE EQUIVALENCED TO X OR Y.  "CALL DIV(X,Y,X)"
C            WILL NOT GENERATE X = X / Y CORRECTLY.
C
C
C DECLARATIONS
      DOUBLE PRECISION X,Y,Z,DENOM
C
C DIMENSIONS
      DIMENSION X(2),Y(2),Z(2)
C
C
      DENOM = Y(1)*Y(1) + Y(2)*Y(2)
      Z(1) = ( X(1)*Y(1) + X(2)*Y(2) ) / DENOM
      Z(2) = ( X(2)*Y(1) - X(1)*Y(2) ) / DENOM
C
      RETURN
      END
```

ORDER FORM

Name ——

Address ——

City-State-Zip ——

Enclosed is $35 for two IBM-PC compatible double-sided DOS 2.0 diskettes containing FORTRAN source code for the programs in *Solving Polynomial Systems Using Continuation for Engineering and Scientific Problems*. I have read and acknowledge the "limitations of liability" disclaimer printed on the copyright page of this book.

Signature ——

Mail to: Alexander Morgan
        687 Davis Street
        Birmingham, MI 48009

BRIEF BIBLIOGRAPHY

The following provide basic additional information on the main topics covered in the text. Each is introductory, readable, and recommended. No algebraic geometry book is included, because I know of none that is simple and oriented to computational issues.

ALLGOWER, E. L., A survey of homotopy methods for smooth mappings, in *Numerical Solution of Nonlinear Equations*, ed. E. L. Allgower, K. Glashoff, and H.-O. Peitgen, Lecture Notes in Mathematics No. 878, Springer-Verlag, New York, 1981.

FORSYTHE, G. E., and C. B. MOLER, *Computer Solution of Linear Algebraic Systems*, Prentice-Hall, Englewood Cliffs, N.J., 1967.

FORSYTHE, G. E., M. A. MALCOMB, and C. B. MOLER, *Computer Methods for Mathematical Computation*, Prentice-Hall, Englewood Cliffs, N.J., 1977.

GARCIA, C. B., and W. I. ZANGWILL, *Pathways to Solutions, Fixed Points, and Equilibria*, Prentice-Hall, Englewood Cliffs, N. J., 1981.

GUILLEMIN, V., and A. POLLACK, *Differential Topology*, Prentice-Hall, Englewood Cliffs, N. J., 1974.

RHEINBOLDT, W. C., *Methods for Solving Systems of Nonlinear Equations*, Society for Industrial and Applied Mathematics, Philadelphia, 1974.

ADDITIONAL BIBLIOGRAPHY

The following provide additional information on various topics related to issues discussed in the text. The literature on continuation and on solving nonlinear systems is vast, and I have in no way done justice to it. See [Allgower, 1981], [Allgower and Georg, 1980], [Watson, 1981], and [Watson et al., 1986] for further references on continuation, and see [Grewank, 1985] and [Ortega and Rheinboldt, 1970] for local methods.

ALLGOWER, E. L., and K. GEORG, Simplicial and continuation methods for approximating fixed points and solutions to systems of equations, *SIAM Rev.*, 22:28–85 (1980).

GREWANK, A., On solving nonlinear equations with simple singularities or nearly singular solutions, *SIAM Rev.*, 27:537–563 (1985).

ORTEGA, J. M., and W. C. RHEINBOLDT, *Iterative Solution of Nonlinear Equations in Several Variables*, Academic Press, New York, 1970.

WATSON, L. T., Engineering applications of the Chow–Yorke algorithm, *Appl. Math. Comput.*, 9:111–133 (1981).

WATSON, L. T., S. C. BILLUPS, and A. P. MORGAN, HOMPACK: a suite of codes for globally convergent homotopy algorithms, Research Publication GMR-5344, G. M. Research Laboratories, Warren, Mich., 1986.

REFERENCES

ABBOTT, J. P., Algorithm 110: Computing solution arcs of nonlinear equations with a parameter, *Comput. J.* 23:85–89 (1980).

ABBOTT, J. P., and R. P. BRENT, Fast local convergence with single and multistep methods for nonlinear equations, *J. Aust. Math. Soc.* 19B:173–199 (1975).

BENSON, S. W., *Chemical Calculations*, Wiley, New York, 1971.

BLANCHARD, P., Complex analytic dynamics on the Riemann sphere, *Bull. Amer. Math. Soc.*, 11:85–141 (1984).

BOGGS, P., The solution of nonlinear systems of equations by A-stable integration techniques, *SIAM J. Numer. Anal.*, 8:767–785 (1971).

BROYDEN, C. G., A new method of solving nonlinear simultaneous equations, *Comput. J.*, 12:94–99 (1969).

BOYCE, W. E., and R. C. DiPRIMA, *Elementary Differential Equations*, Wiley, New York, 1977.

BOYSE, J. W., and J. E. GILCHRIST, GMSOLID—Interactive modeling for design and analysis of solids, *IEEE Comput. Graphics Appl.*, 2:27–40 (1982).

BRANIN, F. H., Widely convergent methods for finding multiple solutions of simultaneous nonlinear equations, *IBM J. Research Develop.*, 16:434–436 (1972).

BRINKLEY, S. R., Note on the conditions of equilibrium for systems of many constituents, *J. Chem. Phys.*, 14 (1946).

BROWN, K. M., and J. E. DENNIS, Jr., On Newton-like iteration functions: general convergence theorems and a specific algorithm, *Numer. Math.*, 12:186–191 (1968).

BROWN, K. M., and W. B. GEARHART, Deflation techniques for the calculation of further solutions of a nonlinear system, *Numer. Math.*, 16:334–342 (1971).

BRUNOVSKÝ, P., and R. MERAVÝ, Solving systems of polynomial equations by bounded and real homotopy, *Numer. Math.*, 43:397–418 (1984).

CHAO, K. S., D. K. LIU, and C. T. PAN, A systematic search method for obtaining multiple solutions of simultaneous nonlinear equations, *IEEE Trans. Circuits Sys.*, CAS-22:748–753 (1975).

CHIEN, M. J., Searching for multiple solutions of nonlinear systems, *IEEE Trans. Circuits Sys.*, CAS-26:817–827 (1979).

CHOW, N. S., J. MALLET-PARET, and J. A. YORKE, Finding zeros of maps: homotopy methods that are constructive with probability one, *Math. Comp.*, 32:887–899 (1978).

CHOW, N. S., J. MALLET-PARET, and J. A. YORKE, A homotopy method for locating all zeros of a system of polynomials, in *Functional Differential Equations and Approximation of Fixed Points*, ed. H.-O. Peitgen and H. O. Walther, Lecture Notes in Mathematics No. 730, Springer-Verlag, New York, 1979, 228–237.

CULLEN, C., *Matrices and Linear Transformations*, Addison-Wesley, Reading, Mass., 1966.

CURRY, J. A., L. GARNETT, and D. SULLIVAN, On the iteration of a rational function:

computer experiments with Newton's method, *Commun. Math. Phys.*, 91:267–277 (1983).

DECKER, D. W., and C. T. KELLEY, Newton's method at singular points: I, *SIAM J. Numer. Anal.*, 17:66–70 (February 1980).

DECKER, D. W., and C. T. KELLEY, Newton's method at singular points: II, *SIAM J. Numer. Anal.*, 17:465–471 (June 1980).

DECKER, D. W., and C. T. KELLEY, Broyden's method for a class of problems having singular Jacobian at the root, *SIAM J. Numer. Anal.*, 22:566–574 (1985).

DEVANEY, R. L., *Introduction to Chaotic Dynamical Systems*, Benjamin/Cummings, 1986.

DONGARRA, J. J., C. B. MOLER, J. R. BUNCH, and G. W. STEWART, *LINPACK User's Guide*, Society for Industrial and Applied Mathematics, Philadelphia, 1979.

DREXLER, F. J., Eine Methode zur Berechung sämtlicher Lösunger von Polynom-gleichungessystemen, *Numer. Math.*, 29:45–58 (1977).

DREXLER, F. J., A homotopy method for the calculation of zeros of zero-dimensional polynomial ideals, in *Continuation Methods*, ed. H. G. Wacker, Academic Press, New York, 1978, 69–93.

DUFFY, J., *Analysis of Mechanisms and Robot Manipulators*, Wiley, New York, 1980.

DUFFY, J., and C. CRANE, A displacement analysis of the general spatial 7R mechanism, *Mech. Mach. Theory*, 15:153–169 (1980).

DUFFY, J., and J. ROONEY, A foundation for a unified theory of analysis of spacial mechanisms, *J. Eng. Ind., Trans. ASME*, 97B:1159–1164 (1975).

DUGUNDJI, J., *Topology*, Allyn and Bacon, Boston, 1967.

FISCHER, G., *Complex Analytic Geometry*, Lecture Notes in Mathematics No. 538, Springer-Verlag, New York, 1977.

FLETCHER, R., *Practical Methods of Optimization*, 2 vols., Wiley, New York, 1980, 1981.

FLETCHER, R., and M. POWELL, A rapidly convergent descent method for minimization, *Comput. J.*, 6:163–168 (1963).

FULTON, W., *Intersection Theory*, Springer-Verlag, New York, 1984.

GARCIA, C. B., and T. Y. LI, On the number of solutions to polynomial systems of equations, *SIAM J. Numer. Anal.*, 17:540–546 (1980).

GARCIA, C. B., and W. I. ZANGWILL, Finding all solutions to polynomial systems and other systems of equations, *Math. Programming*, 16:159–176 (1979).

GARCIA, C. B., and W. I. ZANGWILL, An approach to homotopy and degree theory, *Math. Oper. Res.*, 4:390–405 (1979).

GARCIA, C. B., and W. I. ZANGWILL, Determining all solutions to certain systems of nonlinear equations, *Math. Oper. Res.*, 4:1–14 (1979).

GRENANDER, U., and O. SHISHA, Experimental mathematics, *J. Approx. Theory*, 43:99–104 (1985).

GRIEWANK, A., and M. R. OSBORNE, Analysis of Newton's method at irregular singularities, *SIAM J. Numer. Anal.*, 20:747–773 (1983).

GOLDMAN, R. N., Two approaches to a computer model for quadric surfaces, *IEEE Trans. Comput. Graphics Appl.*, 3:21–24 (1983).

HARTENBERG, R. S., and J. DENAVIT, *Kinematic Synthesis of Linkages*, McGraw-Hill, New York, 1964.

HAZEWINKEL, M., Experimental mathematics, *Math. Modelling*, 6:175–211 (1985).

HEARN, A. C., *REDUCE User's Manual*, Rand Publication CP78(4/83), The RAND Corporation, Santa Monica, Calif., 1983.

DEN HEIJER, C., and W. C. RHEINBOLDT, On steplength algorithms for a class of continuation methods, *SIAM J. Numer. Anal.*, 18:925–948 (1981).

HINDMARSH, A. C., "GEAR" ordinary differential equation system solver, Report UCID-30001, Rev. 3, Lawrence Livermore Laboratory, Livermore, Calif., 1974.

HOFFMAN, K., and R. KUNZE, *Linear Algebra*, Prentice-Hall, Englewood Cliffs, N.J., 1961.

INCERTI, S., V. PARISI, and F. ZIRILLI, Algorithm 111: A FORTRAN subroutine for solving systems of nonlinear simultaneous equations, *Comput. J.*, 24:87–91 (1981).

JENKINS, M. A., and J. F. TRAUB, A three-stage variable-shift iteration for polynomial zeros and its relation to generalized Rayleigh iteration, *Numer. Math.*, 14:252–263 (1970).

KANTOROVICH, L. V., and G. P. AKILOV, *Functional Analysis in Normed Spaces*, Pergamon Press, Elmsford, N.Y., 1964.

KAPLAN, W., *Advanced Calculus*, 2nd ed., Addison-Wesley, Reading, Mass., 1973.

KEARFOTT, R. B., A proof of convergence and an error bound for the method of bisection in R^n, *Math. Comp.*, 32:1147–1153 (1978).

KEARFOTT, R. B., An efficient degree-computation method for a generalized method of bisection, *Numer. Math.*, 32:109–127 (1979).

KUBÍČEK, M., Algorithm 502, Dependence of solution of nonlinear systems on a parameter, *ACM Trans. Math. Software*, 2:98–107 (1976).

KUL'CHITSKII, O. Y., and L. I., SHIMELEVICH, Determination of the initial approximation for Newton's method, *Zh. Vychisl. Mat. Mat. Fiz.*, 14:1016–1018 (1974).

LI, T. Y., On Chow, Mallet-Paret and Yorke homotopy for solving system of polynomials, *Bull. Inst. Math. Academia Sinica*, 11:433–437 (1983).

MEINTJES, K., and A. P. MORGAN, Performance of algorithms for calculating the equilibrium composition of a mixture of gases, *J. Comput. Phys.*, 60:219–234 (1985).

MEINTJES, K., and A. P. MORGAN, A methodology for solving chemical equilibrium systems, *Appl. Math. Comput.*, 22 (1987).

MORÉ, J. J., and M. Y. COSNARD, Numerical solution of nonlinear equations, *ACM Trans. Math. Software*, 5:64–85 (1979).

MOORE, R. E., and S. T. JONES, Safe starting regions for iterative methods, *SIAM J. Numer. Anal.*, 14:1051–1065 (1977).

MORGAN, A. P., A method for computing all solutions to systems of polynomial equations, *ACM Trans. Math. Software*, 9:1–17 (1983).

MORGAN, A. P., A transformation to avoid solutions at infinity for polynomial systems, *Appl. Math. Comput.*, 18:77–86 (1986).

MORGAN, A. P., A homotopy for solving polynomial systems, *Appl. Math. Comput.*, 18:87–92 (1986).

MORGAN, A. P., and A. SOMMESE, A homotopy for solving general polynomial systems that respects *m*-homogenous structures, Research Publication GMR-5437, GM Research Laboratories, Warren, Mich., 1986. Accepted for publication in *Appl. Math. Comput.*

MORGAN, A. P. and A. SOMMESE, Computing all solutions to polynominal systems using homotopy continuation, Research Publication GMR-5692, GM Research Laboratories, Warren, Mich., 1987. Accepted for publication in *Appl. Math. Comput.*

PETEK, P., A nonconverging Newton sequence, *Math. Mag.*, 56:43–45 (1983).

PIEPER, D. L., and B. ROTH, The kinematics of manipulators under computer control, *Proc. Int. Cong. Theory Mach. Mech.*, 2:159–168 (1969).

RALL, L. B., Convergence of the Newton process to multiple solutions, *Numer. Math.*, 9:23–37 (1966).

RALL, L. B., A note on the convergence of Newton's method, *SIAM J. Numer. Anal.*, 11:34–36 (1974).

REDDIEN, G. W., On Newton's method for singular problems, *SIAM J. Numer. Anal.*, 15:993–996 (1978).

REQUICHA, A. A. G. and H. B. VOELCKER, Solid modeling: current status and research directions, *IEEE Comput. Graphics Appl.*, 3:25–37 (1983).

RHEINBOLDT, W. C., An adaptive continuation process for solving systems of nonlinear equations, Polish Academy of Science, *Banach Ctr. Publ.*, 3:129–142 (1977).

RHEINBOLDT, W. C., Solution fields of nonlinear equations and continuation methods, *SIAM J. Numer. Anal.*, 17:221–237 (1980).

ROTH, S., Ray casting for modeling solids, *Comput. Graphics Image Process.*, 18:109–144 (1982).

ROTH, B., J. RASTEGAR, and V. SCHEINMAN, On the design of computer controlled manipulators, in *On the Theory and Practice of Robots and Manipulators*, Vol. 1, First CISM-IFToMM Symposium, 1973, 93–113.

SCHAFAREVICH, I. R., *Basic Algebraic Geometry*, Springer-Verlag, New York, 1977.

SCHUNCK, B. G., Gaussian filters and edge detection, Research Publication GMR-5586, G. M. Research Laboratories, Warren, Mich., 1986.

SHAMPINE, L. F., and M. K. GORDON, *Computer Solution of Ordinary Differential Equations*, W. H. Freeman, San Francisco, 1975.

SHIELDS, P. C., *Elementary Linear Algebra*, Worth, New York, 1967.

SPIVAK, M., *Calculus on Manifolds*, W. A. Benjamin, New York, 1965.

SPIVAK, M., *Differential Geometry*, Vol. 1, Publish or Perish, Inc., Berkeley, Cal., 1970.

TSAI, L.-W., and A. P. MORGAN, Solving the kinematics of the most general six- and five-degree-of-freedom manipulators by continuation methods, *ASME J. Mech. Transmissions and Automation in Design*, 107:189–200 (1985).

VOELCKER, H. B., and A. A. G. REQUICHA, Geometric modeling of mechanical parts and processes, *Computer*, 10:48–57 (1977).

VAN DER WAERDEN, B. L., *Modern Algebra*, 2 vols., trans. F. Blum, Ungar, New York, 1953.

WATSON, L. T., A globally convergent algorithm for computing fixed points of C^2 maps, *Appl. Math. Comput.*, 5:297–311 (1979).

WATSON, L. T., Computational experience with the Chow-Yorke algorithm, *Math. Programming*, 19:92–101 (1980).

WATSON, L. T., Numerical linear algebra aspects of globally convergent homotopy methods, *SIAM Rev.*, 28:529–545 (1986).

WILKINSON, J. H., *Rounding Errors in Algebraic Processes*, Prentice-Hall, Englewood Cliffs, N.J., 1963.

WRIGHT, A. H., Finding all solutions to a system of polynomial equations, *Math. Comp.*, 44:125–133 (1985).

Index